What Makes the Life of Stressed Plants a Little Easier? Defense Mechanisms against Adverse Conditions

What Makes the Life of Stressed Plants a Little Easier? Defense Mechanisms against Adverse Conditions

Editors

Ewa Muszyńska
Kinga Dziurka
Mateusz Labudda

MDPI • Basel • Beijing • Wuhan • Barcelona • Belgrade • Manchester • Tokyo • Cluj • Tianjin

Editors
Ewa Muszyńska
Warsaw University of Life Sciences-SGGW
Poland

Kinga Dziurka
Polish Academy of Sciences
Poland

Mateusz Labudda
Warsaw University of Life Sciences-SGGW
Poland

Editorial Office
MDPI
St. Alban-Anlage 66
4052 Basel, Switzerland

This is a reprint of articles from the Special Issue published online in the open access journal *Plants* (ISSN 2223-7747) (available at: https://www.mdpi.com/journal/plants/special_issues/Defense_Against_Adverse).

For citation purposes, cite each article independently as indicated on the article page online and as indicated below:

LastName, A.A.; LastName, B.B.; LastName, C.C. Article Title. *Journal Name* **Year**, *Volume Number*, Page Range.

ISBN 978-3-0365-7830-9 (Hbk)
ISBN 978-3-0365-7831-6 (PDF)

Cover image courtesy of Ewa Muszyńska

Contents

About the Editors

Ewa Muszyńska

Ewa Muszyńska received her post-doctoral habilitation degree in the discipline of biological sciences in 2021. She is currently a researcher in the Department of Botany, Institute of Biology, Warsaw University of Life Sciences (WULS), Poland. Her scientific interests concern the adaptation strategies of metallophytes to elevated levels of metallic elements and other abiotic stresses. In her interdisciplinary research, she focuses primarily on reactive oxygen species metabolism and their deactivation pathways under stress, as well as on visualization of different compounds within plant cells and tissues by various microscopic imaging methods. She has developed in vitro protocols for many native metal-tolerant species, which were successfully micropropagated under fully-controlled conditions. She is a laureate of several Team and Individual Awards of the Rector of WULS for scientific achievements, and a member of scientific organizations including The Polish Botanical Society.

Kinga Dziurka

Kinga Dziurka received her PhD in biology from The W. Szafer Institute of Botany, Polish Academy of Sciences, in Cracow in 2009. Her doctoral thesis was devoted to the reproduction of endangered species: *Astragalus penduliflorus*, *Stipa joannis*, and *Stipa pulcherrima*. She is currently a researcher at The F. Górski Institute of Plant Physiology, Polish Academy of Sciences in Cracow, and works in the Group of Distant Hybridization. She became a laureate of the Team Award, granted by the Minister of Agriculture and Rural Development for outstanding national achievement of importance for the implementation of progress in agricultural practice and dissemination of the results of scientific work, for her work: "Obtaining the lines of doubled haploid of oat (*Avena sativa* L.) with the method of distant crossing". Her research interests are focused on the stress physiology of plants, micropropagation, and haploidization. She is particularly interested in phenylpropanoids, phytohormones, polyamines, and sugars, and their occurrence and activity in plants.

Mateusz Labudda

Mateusz Labudda received his post-doctoral habilitation degree in the discipline of biological sciences in 2021. He is currently a researcher in the Department of Biochemistry and Microbiology, Institute of Biology, Warsaw University of Life Sciences (WULS), Poland. His scientific interests include the molecular, biochemical, physiological, and structural changes in cereals as host plants inhabited by parasitic cyst nematodes. He is also interested in the holistic defense responses of plants subjected to heavy metals, drought, and biotic stress factors. He pays special attention to the regulation of proteolysis and proteome modifications, as well as the biochemistry of oxidative stress induced by various stress conditions. He has been appreciated by the Rector of WULS for scientific achievements and received many team and individual awards. He is also involved in the activities of national and foreign scientific societies, such as the Polish Botanical Society and the Polish Biochemical Society.

Preface to "What Makes the Life of Stressed Plants a Little Easier? Defense Mechanisms against Adverse Conditions"

This Special Issue, entitled "What Makes the Life of Stressed Plants a Little Easier? Defense Mechanisms against Adverse Conditions", is devoted to various aspects of the effects of stress factors on plants. Thanks to an interdisciplinary approach to the subject, our Special Issue disseminates new research in the fields of anatomy, genetics, biochemistry, and physiology of both abiotic and biotic stresses. It covers all plant species, from model plants, such as *Arabidopsis thaliana* and *Brachypodium distachyon*, through the medicinal (*Bryophyllum calycinum, Pelargonium graveolens, Rhodiola semenowii*) and ornamental (*Armeria maritima*), and ending with cultivated plants, including *Beta vulgaris, Olea europaea*, and *Triticum aestivum*. It provides information on stresses related to drought, waterlogging, low and high temperatures, alkalinity, salinity, heavy metals, oxidizing species, UV radiation, wounds, and fungal diseases. We hope that this Special Issue will find a wide audience.

We would like to sincerely thank Ms. Snow Liu for her support in editing this Special Issue, as well as the Authors and Reviewers of the articles for their cooperation. Special words of thanks go to our families and relatives for their understanding and invaluable support.

Ewa Muszyńska, Kinga Dziurka, and Mateusz Labudda
Editors

plants

MDPI

Editorial

What Makes the Life of Stressed Plants a Little Easier? Defense Mechanisms against Adverse Conditions

Ewa Muszyńska [1,*], Kinga Dziurka [2] and Mateusz Labudda [3]

1 Department of Botany, Institute of Biology, Warsaw University of Life Sciences-SGGW, Nowoursynowska 159, 02-776 Warsaw, Poland
2 Department of Biotechnology, The Franciszek Górski Institute of Plant Physiology, Polish Academy of Sciences, Niezapominajek 21, 30-239 Kraków, Poland
3 Department of Biochemistry and Microbiology, Institute of Biology, Warsaw University of Life Sciences-SGGW, Nowoursynowska 159, 02-776 Warsaw, Poland
* Correspondence: ewa_muszynska@sggw.edu.pl; Tel.: +48-22-59326-60

Citation: Muszyńska, E.; Dziurka, K.; Labudda, M. What Makes the Life of Stressed Plants a Little Easier? Defense Mechanisms against Adverse Conditions. *Plants* **2023**, *12*, 1040. https://doi.org/10.3390/plants12051040

Received: 21 February 2023
Accepted: 23 February 2023
Published: 24 February 2023

1. Introduction

Plants experience a wide array of external factors, some of which negatively affect their metabolism, growth, and development. Such unfavorable circumstances, collectively called environmental stress, require plants' mechanisms to overcome these pressures and reach homeostasis if they want to survive and reproduce in the stressful environments in which they find themselves. The current Special Issue, entitled 'What makes the life of stressed plants a little easier? Defense mechanisms against adverse conditions' is a compilation of fifteen papers (original and review) related to structural, metabolomics, proteomics, and genomics changes occurring under various stress conditions. In this way, it covers the latest advances in our multifaceted understanding of responses and adaptation strategies of plants, mainly to water scarcity and elements' imbalance, but also to other stressors such as non-optimal temperature or fungal disease.

2. Drought Stress

The problem of drought and maintaining efficient crop production in a changing climate is a challenge for people around the world. Research is therefore focused on understanding the mechanisms of plant defense reactions against water deficit, as well as on the search for markers of drought resistance and substances modulating plant response to a stress factor. Expanding the knowledge in this field by integrating different approaches not only prevents crop losses, but also accelerates the process of breeding drought-resistant varieties. Wahab et al. [1] reviewed the effects of drought stress on earlier germination and flowering, the morphology of leaves, stems, and roots, as well as crop yields. The authors discussed changes in physiological and biochemical parameters, such as leaf relative water content (RWC), photosynthesis efficiency, stomatal conductivity, respiration, membrane stability, accumulation of osmolytes (proline, soluble sugars, glycine betaine) and antioxidant enzymes, and non-enzymatic reactive oxygen species (ROS) scavengers (carotenoids, phenolics). Finally, they showed the possibility of inducing tolerance to drought using exogenous phytohormones and changes in the content of endogenous phytohormones under drought stress (auxins, cytokinins, gibberellins, abscisic acid (ABA), ethylene, jasmonates).

Going deeper into the details, Hassan et al. [2] studied the application of nano-silicon (nSi) as a regulator improving the biochemical and physiological parameters of drought-sensitive 'Kalamata' olives grown in semi-arid climates with insufficient rainfall. The 'Kalamata' olive reacted to drought stress with oxidative stress manifested by an increase in H_2O_2, malondialdehyde (MDA), electrolyte leakage, and fruit drop. The use of nSi in the form of foliar spray at a concentration of 150–200mg/L^{-3} under moderate drought (90%

of irrigation water requirements) resulted in improved yield, total chlorophyll content, and RWC, reduced fruit drop, and simultaneously lowered the content of proline, soluble sugars, H_2O_2, MDA, ABA, and electrolyte leakage. The authors postulated that nSi increases plant tolerance to drought by reducing the production of ROS and thus alleviating oxidative stress.

Drought stress usually does not occur alone but is accompanied by temperature stress: cold or heat. Hence, the influence of several stressors on the condition and physiology of plants is being studied more and more often. Terletskaya et al. [3] studied the effect of water deficit and cold on the morphology, physiology, and biochemistry of the medicinal plant *Rhodiola semenowii* Boriss. The authors noticed non-specific reactions, such as a decrease in photosynthesis efficiency or the accumulation of tocopherol in shoots, beta-sitosterol in roots, and squalene in both shoots and roots, regardless of the type of stress. At the same time, they observed specific changes in the anatomy of *R. semenowii* under studied stress factors. The root, shoot, and leaf cells were flattened, with reduced turgor under drought stress, and more hydrated, rounded, and densely packed in cold than control.

Drought is often accompanied by osmotic stress. Paunescu et al. [4] applied a polyethylene glycol (PEG) solution in hydroponic cultivation to mimic drought stress to screen winter wheat varieties of various origins for the activity of antioxidant enzymes. Increased peroxidase activity under the 25% PEG treatment might suggest their resistance to drought in most Romanian varieties. On the other hand, varieties considered to be drought-tolerant showed low ascorbate peroxidase activity compared to the control. The authors concluded that none of the tested biochemical markers can be a clear indicator of drought tolerance due to the lack of correlation between the yield index and the ratio of antioxidant enzyme content between PEG and control.

3. Chemical Elements' Stress

The effect of elevated levels of metallic elements resulting from strong environmental contamination of anthropogenic origin is one of the most frequently studied issues of both basic and applied sciences. In the experiment by Niedziela et al. [5], triticale (×*Triticosecale* Wittm. ex A. Camus) lines differing in aluminum (Al) tolerance were compared regarding their ability to root regrowth and proteomic changes that may help in recovery after Al stress. The authors demonstrated that the roots of the Al-tolerant triticale line could recover without modification of proteome profiles. On the contrary, Al-sensitive genotypes maintained the proteome alteration caused by unfavorable environments. The highest upregulation was detected for proteins involved in protein folding (i.e., protein disulfide-isomerase), stress-related response (such as glutathione *S*-transferase, oxalate oxidase, and 1-Cys peroxiredoxin), and flavonoid metabolism (flavone O-methyltransferase 1), while proteins involved in cell division (tubulin) and metabolic pathways associated with amino acid metabolism and methylation control (*S*-adenosyl-L-homocysteine hydrolase), as well as ascorbic acid biosynthesis (phosphomannomutase), were downregulated. In turn, Chrysargyris et al. [6] noted that the physiological markers of plant response to metal-induced stress are often beneficial bioactive secondary metabolites, mainly antioxidants employed in the food, pharmaceutical, and cosmetic industries. Thus, they hypothesized that the treatment of the medicinal plant *Pelargonium graveolens* with different levels of copper (Cu) ions could stimulate the synthesis of phenolic compounds with high ability to scavenge of ROS. *P. graveolens* plants; these proved to be surprisingly tolerant to Cu stress, and in general, the applied concentrations of Cu up to 100 μM did not have a negative influence on biomass production and the tested physiological parameters, such as stomatal resistance, chlorophyll *a* fluorescence, and photosynthetic pigment contents. The observed tolerance of *P. graveolens* was attributed to both the detoxification of most of the Cu ions in roots, the restriction of their translocation to shoots, and an efficient antioxidant system based on flavonoids and other, undetected classes of phenolics with strong ROS neutralization properties. Taken together, these results showed that the leaves of *P. graveolens* can be safely exploited as an herbal raw material which can be used for the extraction of bioactive compounds, the biosynthesis of which might be stimulated by plant exposure to Cu. On the other hand, the review paper of Labudda et al. [7] showed the significance of basic sciences in the development of applied sciences. According to the authors, the understanding of plant adaptation and acclimation mechanisms in response to harsh conditions is necessary to relieve the pressure of environmental changes and ensure global food security for an increasing population, as well as to restore areas degraded by human activity. Therefore, special attention was paid to the natural strategies of metallophytes and hyperaccumulators that exhibit microevolutionary adaptation to high concentrations of metals in the soil. The authors highlighted the possibility of the practical application of metal-tolerant species in various phytoremediation techniques, as well as indicating similarities between metal tolerance mechanisms and those activated during salinity and drought. Finally, they discussed in detail the latest scientific achievements in priming methods regarding the use of phytohormones, nanoparticles, reactive chemical species, and (non-)ionizing radiations, which may effectively induce a defense response to subsequently occurring stress, and provide plant acclimation and improved resistance to heavy metals, salinity, and water deficit.

The chemical ion imbalance in soil can also result from the excessive accumulation of Na^+, K^+, Cl^-, NO_3^-, and SO_4^{2-} in the soil solution, and this disturbance is responsible for salinity stress. The study conducted by Purmale et al. [8] aimed to compare salinity tolerance and ion accumulation ability between the *Armeria maritima* subsp. *elongata* accession from geographically isolated, salt-affected habitats. Interestingly, the authors used both in vitro shoot culture and soil-cultivated plants for examination. It was found that the increasing concentration of NaCl under in vitro treatment negatively influenced shoot multiplication and biomass production, probably due to the lack of a natural mechanical barrier (i.e., roots with Casparian strips) to ion translocation to shoots. On the contrary, the growth of *A. maritima* under greenhouse cultivation was not significantly affected by increasing salinity, and Na^+ or K^+ treatment had a similar effect on plants. However, there were differences in osmotic adjustment (inorganic ions vs. organic osmolytes) between these two cations. Tolerance to salinity in *A. maritima* was achieved by the deposition of salt crystals on the leaf surface and flower stalks, as well as by the storage of toxic ions in the oldest leaves, and therefore they were discarded by the plants.

4. Varia

Delving into the molecular basis of plant responses to biotic and abiotic stress factors, Kesawat et al. [9] studied the proline-rich extensin-like receptor kinases (PERKs) gene family in *Triticum aestivum* L. PERKs genes are known to be involved in plant development and stress responses. The authors identified thirty seven genes in wheat from this family (TaPERKs) and confirmed their increased expression under stress conditions. Many of the TaPERK genes were up-regulated during the Septoria tritici blotch, powdery mildew, or stripe rust infections. A few of them were also expressed under hot and cold stress. The TaPERK gene family does not appear to directly participate in drought stress. In addition, different genes of the TaPERK family have been shown to respond to various stress factors. Their molecular role in plant responses was also assigned to the SKP1-CUL1-F-box (SCF) type E3 ubiquitin ligases, which participate in multiple specific protein degradation, including those that permit survival under stress conditions. SCF complexes use F-box (FBX) proteins as renewable substrate adaptors to levy proteins for ubiquitylation. The F-BOX STRESS INDUCED (FBS) subfamily of plant FBX proteins has an abnormal structure, nevertheless, with an F-box domain placed in the center and extra conserved regions at the N- and C-termini. Sepulveda-Garcia et al. [10] showed two WD40 repeat-like proteins in Arabidopsis that are highly conserved in plants and interact with FBS proteins, which were named FBS INTERACTING PROTEINs (FBIPs). FBIPs react only with the N-terminus of FBS proteins, and this interaction takes place in the nucleus. The authors concluded that FBS proteins may act in stress-responsive nuclear events, and described two

WD40 repeat-like proteins as new tools to study how the abnormal SCF complex, SCFFBS, functions through the interaction events of N-terminal FBX proteins.

Continuing with stress responses at the cellular level, the review paper of Li et al. [11] indicated that calcium ions are one of the important regulators of plant reactions to various abiotic stresses, such as water deficit and excess, salt stress, light stress, heavy metals, non-optimal temperatures, and mechanical stimuli. In addition, the authors presented the crosstalk between Ca^{2+} and other signaling molecules (ROS, ABA, nitric oxide, inositol 1,4,5-trisphosphate, cyclic ADP-ribose, cyclic guanosine 30,50-monophosphate) in plants under environmental challenges. In turn, from an organismal point of view, plants can react to stressors through the formation of the secondary abscission zone (SAZ), which allows them to eliminate a damaged organ or its fragment from parent individuals. Dziurka et al. [12] elucidated the mode of action of methyl jasmonate (JA-Me) in inducing the formation of SAZ, as well as analyzing changes in the content of endogenous phytohormones above and below the SAZ in the stems of *Bryophyllum calycinum*. It was found that changes in the metabolism of auxin and jasmonic acid-related compounds accompanied the formation of SAZ under the influence of JA-Me. At the same time, the authors did not observe any modification of the indole-3-acetic acid (IAA) content (which can also induce SAZ), suggesting that the mode of action of JA-Me in inducing SAZ is different than in the case of IAA.

A promising way to relieve plant stress is the involvement of microbiota, which—working with the host plant—improves the condition of the plant and often enables it to survive. A perfect example of this *Brachypodium distachyon*, in which the composition of the leaf microbiome changed under the influence of acclimatization to cold temperatures [13]. The relative abundance of *Streptomyces* sp. M2, which releases antibiotics and siderophores known to inhibit the growth of pathogens, increased by over 200-fold. At the same time, the amount of *Pseudomonas syringe*, an ice nucleation active pathogen that can incite frost injury to crops, drastically decreased. *Streptomyces* spp. produced heat shock proteins and cryoprotectants which could protect the host plant during acclimatization to cold temperatures. On the other hand, many microorganisms can be pathogenic and bring about a serious threat to crop productivity. Regarding this issue, Rabby et al. [14] screened a new bioactive secondary metabolites against the most damaging fungal disease of wheat. The authors noted that two marine secondary metabolites—bonactin and feigrisolide C, isolated from the marine bacteria *Streptomyces* spp.—significantly inhibited the growth of *Magnaporthe oryzae* hyphae in vitro. In further analyses, the authors found that bonactin and feigrisolide C decreased the mycelial growth of this pathogen in a dose-dependent manner. Bonactin reduced mycelium growth more effectively than feigrisolide C. The authors noted that, to be able to recommend these molecules as fungicides for wheat blight control, the further study of additional *M. oryzae* isolates is needed to determine their exact mode of action and disease control effectiveness under various field conditions.

Next, a literature review, which is more species-specific in its scope, concerns the response to abiotic stress of cultivated beet (*Beta vulgaris* L.) [15]; this review shows selected aspects of molecular, biochemical, physiological, and structural responses to heavy metals, alkalines, non-optimal temperatures, and UV stresses. Here, attention was also paid to changes in the expression of selected genes that seem to play important roles in response to the abiotic stresses of beet plants. The use of this kind of approach and description regarding this topic, as the authors also noted, may be useful in future works on the topic of breeding.

5. Conclusions and Future Perspectives

The excellent articles published in this Special Issue summarize and broaden the latest achievements in the field of stress biology. The shared knowledge gathered here, showing what can make the lives of stressed plants a little easier, can provide clues for the development of new strategies (or for the improvement of those already in existence) to combat the global problems faced by the modern world, including drought, metal pollution, salinity, and climate changes. Such a multi-perspective view of plant complexity may bring about a promising solution for the restoration of degraded lands, as well as the conduction of sustainable agriculture and improved quality and productivity of plants growing under diverse, even stressful conditions. On the other hand, more research is needed on at least two different stresses acting simultaneously to fill the gaps in our knowledge concerning plant responses to external stimuli that usually occur collectively in nature. Another scientific challenge is also to discover both crosstalk connections between particular metabolic pathways and signaling cascades responsible for efficient and rapid plant adaptation and acclimation to changing environments.

As Guest Editors, we are grateful to all the authors for choosing to publish their vulnerable research in our Special Issue. We also greatly appreciate the work of the reviewers who took the time to review all of the submitted manuscripts. We believe that, together, we have contributed to a better cognition and understanding of compound plant reactions to multiple stresses and the determination of future perspectives and trends.

Author Contributions: Conceptualization, E.M.; formal analysis, E.M., K.D. and M.L.; writing—original draft preparation, E.M., K.D. and M.L; supervision, E.M. All authors have read and agreed to the published version of the manuscript.

Funding: This research received no external funding.

Institutional Review Board Statement: Not applicable.

Informed Consent Statement: Not applicable.

Data Availability Statement: Not applicable.

Conflicts of Interest: The authors declare no conflict of interest.

References

1. Wahab, A.; Abdi, G.; Saleem, M.H.; Ali, B.; Ullah, S.; Shah, W.; Mumtaz, S.; Yasin, G.; Muresan, C.C.; Marc, R.A. Plants' Physio-Biochemical and Phyto-Hormonal Responses to Alleviate the Adverse Effects of Drought Stress: A Comprehensive Review. *Plants* **2022**, *11*, 1620. [CrossRef] [PubMed]
2. Hassan, I.F.; Ajaj, R.; Gaballah, M.S.; Ogbaga, C.C.; Kalaji, H.M.; Hatterman-Valenti, H.M.; Alam-Eldein, S.M. Foliar Application of Nano-Silicon Improves the Physiological and Biochemical Characteristics of 'Kalamata' Olive Subjected to Deficit Irrigation in a Semi-Arid Climate. *Plants* **2022**, *11*, 1561. [CrossRef] [PubMed]
3. Terletskaya, N.V.; Korbozova, N.K.; Kudrina, N.O.; Kobylina, T.N.; Kurmanbayeva, M.S.; Meduntseva, N.D.; Tolstikova, T.G. The Influence of Abiotic Stress Factors on the Morphophysiological and Phytochemical Aspects of the Acclimation of the Plant *Rhodiola semenowii* Boriss. *Plants* **2021**, *10*, 1196. [CrossRef] [PubMed]
4. Paunescu, R.A.; Bonciu, E.; Rosculete, E.; Paunescu, G.; Rosculete, C.A.; Babeanu, C. The Variability for the Biochemical Indicators at the Winter Wheat Assortment and Identifying the Sources with a High Antioxidant Activity. *Plants* **2021**, *10*, 2443. [CrossRef] [PubMed]
5. Niedziela, A.; Domżalska, L.; Dynkowska, W.M.; Pernisová, M.; Rybka, K. Aluminum stress induces irreversible proteomic changes in the roots of the sensitive but not the tolerant genotype of triticale seedlings. *Plants* **2022**, *11*, 165. [CrossRef] [PubMed]
6. Chrysargyris, A.; Maggini, R.; Incrocci, L.; Pardossi, A.; Tzortzakis, N. Copper Tolerance and Accumulation on *Pelargonium graveolens* L'Hér. Grown in Hydroponic Culture. *Plants* **2021**, *10*, 1663. [CrossRef] [PubMed]
7. Labudda, M.; Dziurka, K.; Fidler, J.; Gietler, M.; Rybarczyk-Płońska, A.; Nykiel, M.; Prabucka, B.; Morkunas, I.; Muszyńska, E. The Alleviation of Metal Stress Nuisance for Plants—A Review of Promising Solutions in the Face of Environmental Challenges. *Plants* **2022**, *11*, 2544. [CrossRef] [PubMed]
8. Purmale, L.; Jēkabsone, A.; Andersone-Ozola, U.; Ievinsh, G. Salinity Tolerance, Ion Accumulation Potential and Osmotic Adjustment In Vitro and In Planta of Different *Armeria maritima* Accessions from a Dry Coastal Meadow. *Plants* **2022**, *11*, 2570. [CrossRef] [PubMed]
9. Kesawat, M.S.; Kherawat, B.S.; Singh, A.; Dey, P.; Routray, S.; Mohapatra, C.; Saha, D.; Ram, C.; Siddique, K.H.M.; Kumar, A.; et al. Genome-Wide Analysis and Characterization of the Proline-Rich Extensin-like Receptor Kinases (PERKs) Gene Family Reveals Their Role in Different Developmental Stages and Stress Conditions in Wheat (*Triticum aestivum* L.). *Plants* **2022**, *11*, 496. [CrossRef]
10. Sepulveda-Garcia, E.; Fulton, E.C.; Parlan, E.V.; O'Connor, L.E.; Fleming, A.A.; Replogle, A.J.; Rocha-Sosa, M.; Gendron, J.M.; Thines, B. Unique N-Terminal Interactions Connect F-BOX STRESS INDUCED (FBS) Proteins to a WD40 Repeat-like Protein Pathway in Arabidopsis. *Plants* **2021**, *10*, 2228. [CrossRef] [PubMed]

11. Li, Y.; Liu, Y.; Jin, L.; Peng, R. Crosstalk between Ca^{2+} and Other Regulators Assists Plants in Responding to Abiotic Stress. *Plants* **2022**, *11*, 1351. [CrossRef] [PubMed]
12. Dziurka, M.; Góraj-Koniarska, J.; Marasek-Ciolakowska, A.; Kowalska, U.; Saniewski, M.; Ueda, J.; Miyamoto, K. A Possible Mode of Action of Methyl Jasmonate to Induce the Secondary Abscission Zone in Stems of Bryophyllum Calycinum: Relevance to Plant Hormone Dynamics. *Plants* **2022**, *11*, 360. [CrossRef] [PubMed]
13. Juurakko, C.L.; diCenzo, G.C.; Walker, V.K. Cold Acclimation in Brachypodium Is Accompanied by Changes in Above-Ground Bacterial and Fungal Communities. *Plants* **2021**, *10*, 2824. [CrossRef] [PubMed]
14. Rabby, S.M.F.; Chakraborty, M.; Gupta, D.R.; Rahman, M.; Paul, S.K.; Mahmud, N.U.; Rahat, A.A.M.; Jankuloski, L.; Islam, T. Bonactin and Feigrisolide C Inhibit *Magnaporthe oryzae Triticum* Fungus and Control Wheat Blast Disease. *Plants* **2022**, *11*, 2108. [CrossRef] [PubMed]
15. Yolcu, S.; Alavilli, H.; Ganesh, P.; Asif, M.; Kumar, M.; Song, K. An Insight into the Abiotic Stress Responses of Cultivated Beets (*Beta vulgaris* L.). *Plants* **2021**, *11*, 12. [CrossRef] [PubMed]

plants

MDPI

Review

Plants' Physio-Biochemical and Phyto-Hormonal Responses to Alleviate the Adverse Effects of Drought Stress: A Comprehensive Review

Abdul Wahab [1], Gholamreza Abdi [2], Muhammad Hamzah Saleem [3,*], Baber Ali [4], Saqib Ullah [5], Wadood Shah [6], Sahar Mumtaz [7], Ghulam Yasin [8], Crina Carmen Muresan [9] and Romina Alina Marc [9,*]

1 Shanghai Center for Plant Stress Biology, CAS Center for Excellence in Molecular Plant Sciences, Chinese Academy of Sciences, Shanghai 200032, China; wahabcrop_science@mails.ucas.ac.cn
2 Department of Biotechnology, Persian Gulf Research Institute, Persian Gulf University, Bushehr 75169, Iran; abdi@pgu.ac.ir
3 College of Plant Science and Technology, Huazhong Agricultural University, Wuhan 430070, China
4 Department of Plant Sciences, Quaid-i-Azam University, Islamabad 45320, Pakistan; baberali@bs.qau.edu.pk
5 Department of Botany, Islamia College, Peshawar 25120, Pakistan; saqibullahstd@icp.edu.pk
6 Department of Botany, University of Peshawar, Peshawar 25120, Pakistan; wadood0301@gmail.com
7 Department of Botany, Division of Science and Technology, University of Education, Lahore 54770, Pakistan; sahar_botany@yahoo.com
8 Department of Botany, Bahauddin Zakariya University, Multan 60800, Pakistan; yasingmn_bzu@yahoo.com
9 Food Engineering Department, Faculty of Food Science and Technology, University of Agricultural Science and Veterinary Medicine Cluj-Napoca, 3-5 Calea Mănăştur Street, 400372 Cluj-Napoca, Romania; crina.muresan@usamvcluj.ro
* Correspondence: saleemhamza312@webmail.hzau.edu.cn (M.H.S.); romina.vlaic@usamvcluj.ro (R.A.M.)

Citation: Wahab, A.; Abdi, G.; Saleem, M.H.; Ali, B.; Ullah, S.; Shah, W.; Mumtaz, S.; Yasin, G.; Muresan, C.C.; Marc, R.A. Plants' Physio-Biochemical and Phyto-Hormonal Responses to Alleviate the Adverse Effects of Drought Stress: A Comprehensive Review. *Plants* **2022**, *11*, 1620. https://doi.org/10.3390/plants11131620

Academic Editors: Kinga Dziurka, Mateusz Labudda and Ewa Muszyńska

Received: 23 May 2022
Accepted: 14 June 2022
Published: 21 June 2022

Abstract: Water, a necessary component of cell protoplasm, plays an essential role in supporting life on Earth; nevertheless, extreme changes in climatic conditions limit water availability, causing numerous issues, such as the current water-scarce regimes in many regions of the biome. This review aims to collect data from various published studies in the literature to understand and critically analyze plants' morphological, growth, yield, and physio-biochemical responses to drought stress and their potential to modulate and nullify the damaging effects of drought stress via activating natural physiological and biochemical mechanisms. In addition, the review described current breakthroughs in understanding how plant hormones influence drought stress responses and phytohormonal interaction through signaling under water stress regimes. The information for this review was systematically gathered from different global search engines and the scientific literature databases Science Direct, including Google Scholar, Web of Science, related studies, published books, and articles. Drought stress is a significant obstacle to meeting food demand for the world's constantly growing population. Plants cope with stress regimes through changes to cellular osmotic potential, water potential, and activation of natural defense systems in the form of antioxidant enzymes and accumulation of osmolytes including proteins, proline, glycine betaine, phenolic compounds, and soluble sugars. Phytohormones modulate developmental processes and signaling networks, which aid in acclimating plants to biotic and abiotic challenges and, consequently, their survival. Significant progress has been made for jasmonates, salicylic acid, and ethylene in identifying important components and understanding their roles in plant responses to abiotic stress. Other plant hormones, such as abscisic acid, auxin, gibberellic acid, brassinosteroids, and peptide hormones, have been linked to plant defense signaling pathways in various ways.

Keywords: drought stress; abiotic stress; osmolytes; antioxidant enzymes; phytohormones; photosynthesis

1. Introduction

Drought Stress

Changing climatic regimes are posing a threat to life on Earth because meeting the rising food demand and achieving sustainable agriculture for a growing population is becoming an uphill task in the present scenario of changing climatic conditions [1], which include droughts, heavy floods, earthquakes, and temperature variations [2,3]. Drought stress interrupts many physio-biochemical processes, hindering plant growth and development [4,5]. Plants can frequently withstand limited water conditions but at the cost of substantial loss in total biomass and productivity. Drought affects around half of the world's semi-arid and arid areas. Photosynthesis, growth, and other critical physiological and biochemical activities are interrupted under drought stress conditions [6,7]. Previous studies [8–10] found that drought stress causes oxidative stress, damaging biological membranes and macromolecules (DNA, proteins, lipids, and photosynthetic pigments). Plants engage their natural defense systems in response to oxidative stress and create osmolytes [11], such as soluble proteins, proline, soluble sugars, and glycine betaine [12].

Osmolytes, also known as osmoprotectants, are found mainly in the cytoplasm and prevent cellular deterioration by maintaining the cell's osmoregulation. Because osmolytes are non-toxic and highly soluble, they do not interfere with other physiological and biochemical processes [13,14]. Plants generate antioxidant substances such as flavonoids, carotenoids, vitamins, and antioxidative enzymes such as glutathione reductase (GTX), superoxide dismutase (SOD), catalase (CAT), peroxidase (POD), and ascorbate peroxidase (APX) in response to abiotic stress [15,16]. Water deprivation causes reduced turgor pressure and oxidative damage from reactive oxygen species (ROS), including superoxide and hydroxyl radicals, nitric oxide and singlet oxygen, causing alterations in leaf gas exchange rates [17] (Figure 1). Natural drought-resistance mechanisms in plants have been well developed, including morphological, physiological, and biochemical adaptations, such as drought-resistant epigenetic plasticity and gene activation [18]. Drought resistance and transformation in food legumes and crop plants are maintained through morphological, physiological, and biochemical changes. These characteristics may assist crops in adapting to harsh environmental conditions. Imbalances in nutrition are caused by drought stress, causing significant ecological constraints on agricultural output worldwide [19].

Drought stress is the most challenging issue to agricultural productivity and has a pronounced negative effect on plant growth, development, and productivity. Making it difficult to maintain a sustainable agricultural system worldwide [19,20]. Drought-induced changes in wheat characteristics were investigated, and their impact on agronomic attributes and yield were studied. Spikelet fertility and grain filling were affected negatively by drought stress [21]. Maize (*Zea mays* L.) and wheat (*Triticum aestivum* L.) crops water limitations lead to reduced crop yields and quality]. Water stressed conditions reduce agricultural output and put food production at risk [22,23]. Reduction in agricultural productivity leads to shrinking revenue for local farmers. The loss of production substantially impacts farmers' livelihoods and economies [24]. Wheat (*Triticum aestivum* L.) crop is a major cereal crop and a common food source worldwide. Wheat (*Triticum aestivum* L.) crop with improved drought tolerance is essential for long-term food production and global food security [25]. Many critical genes and transcription regulators controlling morpho-physiological and biochemical features have been discovered due to recent developments in drought tolerance research [26] (Figure 2).

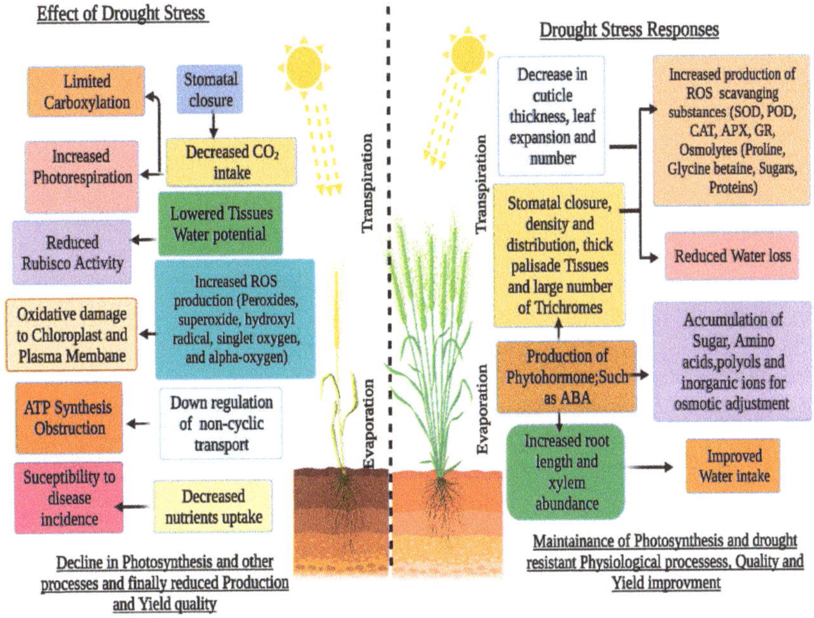

Figure 1. Effects of drought stress on sensitive and tolerant wheat (*Triticum aestivum* L.) crops.

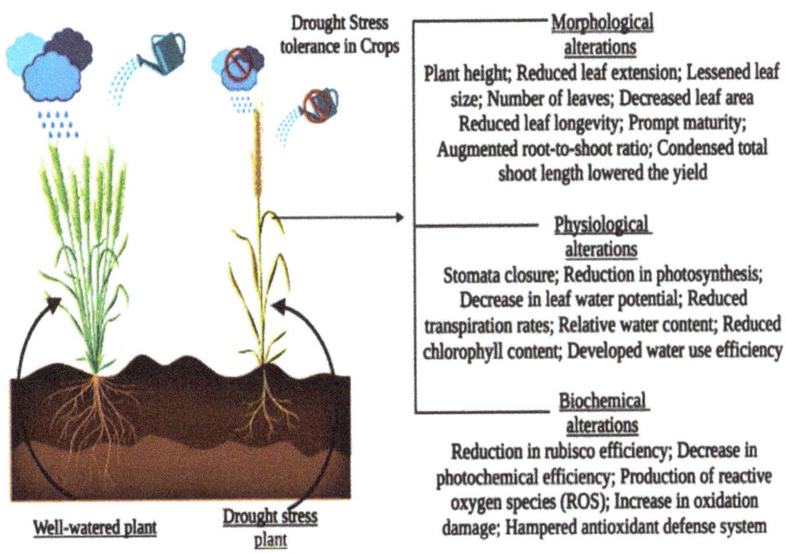

Figure 2. Drought stress impacts plants' morphological, physiological, and biochemical processes.

2. Drought-Induced Changes in Plant Morphology

Drought stress adversely affects morphological aspects of plants, such as early germination, plant height, relative root length, root diameter, the total biomass of leaves and roots, number of leaves/plants, and branch number/plant [27,28].

2.1. Early Seed Germination and Flowering

Water is essential for seed germination; however, while other conditions may be ideal, drought stress inhibits the imbibition of seeds and, consequently, hinders germination [29]. Similarly, it reduces seedling vigour and impacts germination by lowering water intake [30]. In the early stages of crop development, drought stress manifests through reduced seed germination resulting in poor stand establishment [31]. Poor seedling germination was observed under exposure to drought stress in two crops: rice (*Oryza sativa* L.) and pea (*Pisum sativum* L.) [30,32]. Low water content in the soil combined with other environmental factors can alter germination success. Drought stress considerably influences *Zea mays* L. seedling germination [33,34]. Some field crops are particularly vulnerable to cold and dryness, especially during germination and seedling development (early phases). Every seed has optimal soil moisture levels and temperature for germination [35].

2.2. Plant Morphological Characteristics of Leaves under Drought Stress

Drought stress substantially influences the internal plant components that increase plant height [35]. Plant height loss might be related to decreased cell growth, a high rate of leaf abscission under dryness, and poor mitosis [36,37]. Water stressed conditions considerably reduced the number of leaves in *Zea mays* L. [34,38]. The study by [33] showed that sweet basil (*Ocimum basilicum*) leaves are significantly more critical than shoot and roots because leaves are responsible for photosynthesis and contain photosynthetic pigments. Drought regimes reduce leaf area and plant total biomass [39]; by limiting leaf growth and affecting the photosynthetic process. Previous research studies reported that leaf area was significantly decreased under drought stress conditions in many crops, including *Triticum aestivum* L. and *Oryza sativa* L. [39,40]. Loss of water from the upper epidermis of the leaf results in diminishing leaf pressure potential, which causes the leaf to roll. Reduced leaf temperature, increased interception of the incident light, and increased transpiration rate benefit this phenomenon. Under drought stress regimes, leaf area and leaf rolling were dramatically enhanced in maize (*Zea mays* L.) crop leaves [41].

2.3. Plants Shoot Morphology and Architecture under Drought Stress

Drought stress has a negative impact on shoot length and fresh weight. In *Phaseolus vulgaris* L., however, there was a considerable drop in the dry weight of the shoot [42]. Conversely, the shoot length in maize (*Zea mays* L.) crop was discovered, which needs to recover by supplying adequate water and nutrients for survival and defeating drought stress conditions [43]. Similarly, it was observed that the seedling length of maize (*Zea mays* L.) was dramatically reduced under drought stress. The water shortage tremendously affects maize crops' dry weight after drying in shades [44]. The results showed drought stress considerably affects maize (*Zea mays* L.) crops' fresh weight compared to control. In such situations, the plant needs a well-developed root system attaching themselves and collecting water and nutrients from their environment [45].

2.4. Plant Root Morphology and Architecture under Drought Stress

Drought stress alters agricultural plants' root architecture and morphology. During abiotic stress conditions, many plants' root biomass increases as the roots' length become more prolonged, and more water and minerals are absorbed from the soil [18,46]. Furthermore, polyethylene glycol-induced drought stress decreased hypocotyl length and fresh and dry weight roots in maize (*Zea mays* L.) while increasing root length [47]. Sometimes, moderate drought has no pronounced negative impacts on root development [48]. Root development in maize, for example, was unaffected by water stress [49]. Previous research found that drought stress boosted root development in two plants, *Catharanthus roseus* L. and *Helianthus anuus* L. [49]. Drought stress affects crops, but the most relevant characteristic is increased legumes, shoot, and root-shoot ratios in different plant species [16]. Root architecture plays a crucial role in plant growth and development. When plants are subjected to water-stressed conditions, their roots elongate into the ground, reaching deeper and absorbing enough water and minerals to survive [43].

2.5. Yield

In some plant species, yields may be reduced depending on the period and intensity of the limited water condition; nevertheless, the lack after anthesis is deleterious to crop yields regardless of the severity and time of the deficit. Drought stress lowers yields in several ways [50]. In barley (*Hordeum vulgare* L.) and wheat (*Triticum aestivum* L.) crops, drought stress reduced the number of spikes, tillers, and grains per plant, as well as the grain weight [51,52]. Reduced millet (*Pennisetum glaucum* L.) production by drought stress caused silking to be delayed and the anthesis-to-silking gap to be lengthened [53]. Drought stress harmed soybean seed production and influenced the physiology and yield of crop germplasms [50]. This trait was substantially related to grain production, namely the number of ears and kernels per plant [54]. The association was investigated between grain yield, grasslands, and harvest index [55].

Here we are discussing that drought stress dramatically impacts plants' morphological characteristics of wheat corps (*Triticum aestivum* L.). Drought stress significantly affects the early stages of seed germinations [56]. In leaf morphology, drought stress also plays a critical role, such as; a reduction in leaf expansions and leaf rolling [57]. Productions of Yield quality and yield losses in wheat crop (*Triticum aestivum* L.) species have been linked to a limited water conditions, with the severity and duration being the focus factors in this association, as shown in Figure 3 [58].

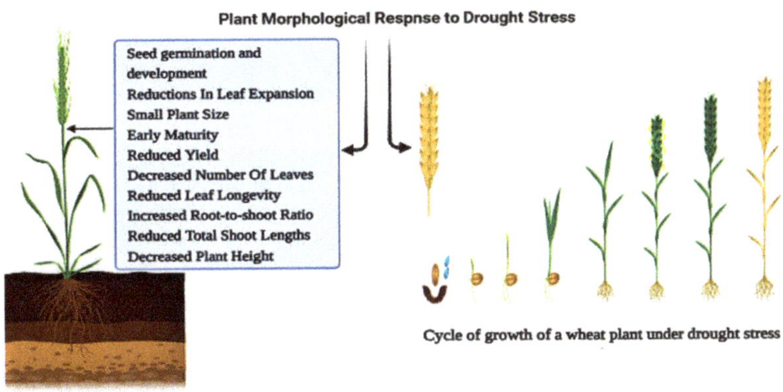

Figure 3. Impact of drought stress on morphological aspects: Cycle of growth of a *Triticum aestivum* L. plant.

3. Physiological and Biochemical Responses under Drought Stress

Drought causes water shortage since there is not enough water in the soil. A water shortage in the soil is not always the cause of the physiological drought [59]. A physiological drought occurs when a plant cannot get enough water; plants react to water stress in various ways [50]. Physiological, biochemical, anatomical, morphological, and long- and short-term developmental and growth-related adaptable techniques might be involved (Figure 4) [60–62]. Reducing leaf relative water content, turgor loss, and stomatal closure are the frequent consequences of drought stress in Barley (*Hordeum vulgare* L.) [51]. During drought, leaf wilting and abscission reduce water loss via transpiration [17,63]. When there is a significant water shortage, cell enlargement in higher plants is hampered by the interruption of xylem water flow. When drought stress is minimal to nonexistent, stomatal closure, cell membrane structural damage, and plant metabolic disturbances occur [64,65]. The results suggested by [43] concluded that numerous internal and external conditions govern internal plant water interactions in *Zea mays* L., such as the stomatal resistance,

RWC, rate of transpiration, leaf temperature of wheat crop (*Triticum aestivum* L.), leaf water potential, and the canopy temperature just above the plant [21,66,67].

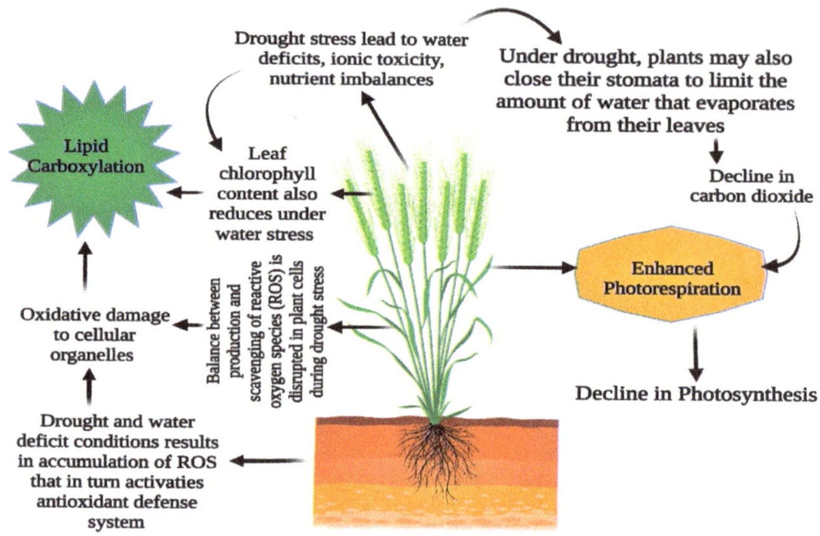

Figure 4. Drought stress's effects on *Triticum aestivum* L. plant morpho-physiological and metabolic processes.

3.1. *Leaf Relative Water Content (RWC)*

Leaf RWC is a crucial controller of physiological processes in plants. RWC reduction is the first symptom of the drought stress response [68]. The relative water content of leaves strongly correlates simultaneously with leaf tissue growth rate and rate of transpiration [69]. Lower RWC reduces leaf water potential, causing stomata to contract. Transpiration is the primary mechanism governing leaf temperature; increasing stomatal resistance minimizes the transpiration rate in rice leaves due to ABA content and increases leaf temperature [63]. In a *Triticum aestivum* L., the leaf's relative water content increases throughout development and decreases as dry matter accumulates as the leaf ages [70]. Water-stressed wheat and rice plants contained less water content as compared to those wheat and rice plants that were grown under controlled conditions [71]. A decline in relative water content induced a drop in water content and osmotic potential under stress regimes. In wheat (*Triticum aestivum*), the state of reduced leaf turgor pressure disrupts plant metabolic functions. Under drought conditions, crop development is impeded by a lower soil water potential, and the resulting lowered plant osmotic potential leads to low nutrient absorption [72].

3.2. *Effect of Drought Stress Conditions on Photosynthesis and Stomatal Aperture*

In photosynthesis, CO_2 and H_2O within the chloroplast of plant cells produce sugars and O_2 as a by-product in the presence of light. Chlorophyll is an essential component of chloroplasts required for photosynthesis [73,74]. Chlorophyll pigments are essential for photosynthesis, affected by water-stressed conditions during stomatal closure and openings in *Nicotiana tabacum* L. [75]. Plants must capture light and use it during the photosynthesis process. Under drought stress, the chlorophyll concentration is dramatically reduced due to increased oxidative stress, degeneration, or photo-oxidation of chlorophyll pigments [76]. Drought sensitivity in (*Triticum aestivum* L.) was predominantly connected with reductions in stomatal conductance, which decreased the delivery of carbon dioxides to chloroplasts and, consequently, reduced net photosynthesis [77,78]. The results determined that drought stress affected plant growth and development by lowering the rate of photosynthesis [79]. The major factors responsible for slowing photosynthesis might be stomatal closure (reduced stomatal CO_2 fixation), non-stomatal (decreased photosynthesis activity in mesophyll tissues), or both [80,81]. Water stressed condition is one of the numerous environmental variables that impede photosynthesis. The high sensitivity of connecting photosystems II (PS-II) following limiting tensions induced by external variables motivates drought stress in harming these systems, which are reaction locations. Methods of chlorophyll fluorescence revealed a hazard and suggested that manufacturing operations were not balanced [82,83]. Drought stress causes plants to be adapted accordingly by regulating their stomata movement, adjusting their osmotic balance, and mounting an antioxidatant defense [8,84]. However, a protracted period of high-intensity limited water conditions might slow plant development, alter the morphological structure and biomass distribution pattern in tomato crops (*Solanum lycopersicum* L.), or cause mortality [15,85]. Drought stress significantly influences the photosynthetic system and its pigments, such as chlorophyll a, b, and carotenoids [72,86,87]. Drought stress also impacts complex systems such as photosystems I and II. Drought stress significantly influences plant starch production by affecting the Calvin cycle and enzyme activity (*Ribulose phosphate*) [86]. The first sign of a plant's drought stress response is closing its stomata. When drought stress becomes more severe during the day, stomata progressively close in sugar beet (*Beta vulgaris* L.) [88]. Stomata are entirely closed in extreme drought stress conditions. Still, full closure varies among plant species depending on their specific tolerances to drought conditions, as shown in pea crops (*Pisum sativum* L.) [30,89] (Figure 5). As a result, plant species tolerance influences the stomatal mechanism, which regulates carbon fixation rates, photosynthesis, and water usage efficiency. When stomata restrict CO_2 uptake into the leaves, more electrons are available to produce active oxygen species [90]. When physiological processes at the stomata are reduced by environmental conditions that increase transpiration rates, then the pH of the leaf sap is elevated; [91] observed reductions of photosynthesis, ROS production regulations decreased, and stomatal conductance under drought stress could be recovered by following re-watering [92].

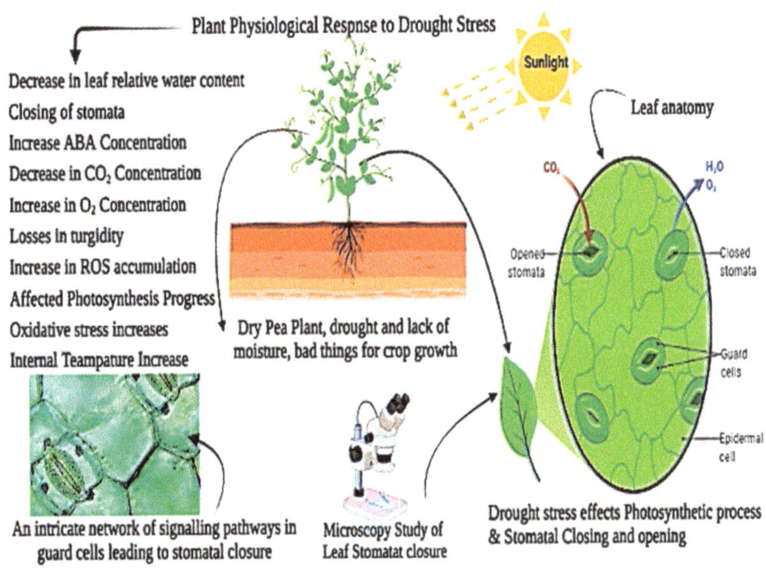

Figure 5. Drought stress and morpho-physiological responses in pea plants; drought stress affects photosynthetic pigments and leaf stomatal openings and closings in pea crops (*Pisum sativum* L.).

3.3. Carotenoids

According to [34,93] studies, drought stress has reduced the concentration of carotenoids in higher plants. Carotenes are classified into two types: hydro-carbon carotenes [83], which include lycopene and xanthophylls, and carotene, which differs from the former due to the inclusion of lutein. The enzymatic antioxidant system contains carotenes, tocopherol, ascorbate, and enzymes such as APX, POD, SOD, polyphenol oxidase, glutathione reductase (GR), and CAT was, protecting carotenoids from the damaging of ROS [8,88,94–97]. The enzymatic antioxidant system, which contains carotenoids, also protects carotenoids from ROS. Beta-carotene, which is involved in the breakdown of triple chlorophyll, prevents singlet oxygen formation, which helps in protecting the plant cells from oxidative stress. In addition, carotene is required to avoid and maintain photochemical reactions [14].

3.4. *Cell Size, Cell Membrane Stability, and Respiration*

Many developmental processes and all aspects of the growth have been adversely affected by droughts, such as cell division, cell expansion, cell differentiation, and genetic, ecological, and physio-morphological approaches [57]. These events, influenced by limited water regimes, govern the amount and quality of plant growth. As a result of the drought, one of the most drought-sensitive physiological processes is cell development as turgor pressure drops [62]. Drought stress is characterized by the limitations of a water path from the xylem to the neighboring elongating cells, which ultimately results in the plant's death; it may impair cell elongation in higher plants [95]. According to [22], drought stress reduces cell size in winter wheat crops (*Triticum aestivum* L.) verities; and enhances interactions between Protein-protein aggregation and denaturation [96]. It is possible that increasing solute concentrations, particularly in the presence of photosynthetic equipment, will be harmful to enzyme activity, as evidenced by an increase in cytoplasmic viscosity [57]. Drought stress reduces the respiration rate in various plant components, including leaves, shoots, and the whole plant [15,97]. According to research, plants' respiration rates remain unaltered or even increase [98]. Drought seems part of a systemic metabolic response when dryness significantly restricts CO_2 availability inside leaf cells, raising the danger of secondary oxidative stress [17]. Root respiration and biomass may decrease during excessive soil drying, resulting in more significant drought-resistant wheat growth, physiological activity, and grain yield [21]. The drought-resistant wheat crop (*Triticum aestivum* L.) spring varieties should be favored over drought-sensitive wheat (*Triticum aestivum* L.) in dry settings [99,100]. The cell membrane stability (CMS) test can identify genotypes susceptible to drought stress. CMS and cell membrane integrity are indicators of resistance to limited water availability under water-stress situations. Lower CMS genotypes were more sensitive to water deficit stress and vice versa. Similarly, the CMS index is essential in breeding programs since it predicts drought tolerance or sensitivity requirements. Drought sensitivity is higher in genotypes with a low CMS value, but drought sensitivity is higher in genotypes with high CMs in wheat crops [23,101–103].

Even with limited water availability, CMS indicated a positive relationship between wheat crops (*Triticum aestivum* L.) tillering ability and grain output but a negative relationship between grain weight measured in kilograms (1000-grain weight) and grain yield [86]. As a side note, drought has been demonstrated to increase the oxidative process among plant species. This results in reduced membrane stability due to lipid peroxidation and, as a result, cell membrane damage [9].

4. Biochemical Responses under Drought Stress Conditions

Accumulating biochemicals such as proline, protein, sugar and glycine betaine (GB) improve crop production by scavenging ROS-generated oxidative stress [10]. Moreover, physiological processes including cellular respiration, rate of photosynthesis, mineral nutrition, enzymatic activities, and, Redox (oxidation/reduction) homeostasis are influenced by drought stress regimes. Likewise, biochemicals, including membrane lipo-proteins and DNA and cellular protein content, deteriorate under water-limited conditions [98]. Plants withstand drought stress regimes by developing various biochemical, structural, and molecular strategies, including the accumulation of certain osmolytes such as proline, proteins, sugars and glycine betaine. Applying salicylic acid improved drought-stress tolerance by upholding redox potential and activating proline biosynthesis [104,105]. Compatible solutes such as proteins, proline, glycine betaine, phenolic compounds, soluble sugars and organic acids accumulated chiefly in the cytoplasm in response to limited water availability by scavenging ROS, improving the water potential, and protecting biological molecules from lipid peroxidation [106]. Plant cells collect soluble chemicals during drought stress and increase cytoplasm viscosity. Under some situations, the content of these unique chemicals may become toxic, causing issues with enzyme development and the entire photosynthetic process [107]. The rate of regeneration of ribulose-1,5-bisphosphate, the maximum rate of ribulose-1,5-carboxylate, NADP-malic enzyme, phosphoenolpyruvate carboxylase, Rubisco, fructose-1,6-bisphosphatase, and orthophosphate-Di kinase pyruvate are all reduced as a result of the rapid decrease in "dry" photosynthesis [108]. Noncyclic electron transport is similarly lowered to satisfy the needs of decreased NADPH synthesis, ATP production, and ROS production. Different cultivars may respond and adapt differently to drought stress [109]. According to transcriptome studies, drought-tolerant and sensitive wheat genotypes may use distinct molecular processes to deal with drought stress. Differential expression of numerous drought-inducible genes involved in regulation, cell defense, and cellular component remodeling is one of the most noticeable changes [92]. According to transcriptome research, drought-tolerant and sensitive wheat (*Triticum aestivum* L.) genotypes may use molecular methods to cope with drought stress [69,84]. One of the most noticeable changes is the differential expression of several drought-inducible genes involved in cell defense regulation and cellular component remodeling [110]. While many of these genes are activated in drought-sensitive wheat (*Triticum aestivum* L.) genotypes and contribute to limiting drought impacts and perception, many of these genes are expressed constitutively in tolerant genotypes [111].

Furthermore, signal transduction and hormone-dependent regulation mechanisms change amongst *Triticum aestivum* L. genotypes [112]. Drought stress-tolerant genotypes perceive drought quickly and activate signal transduction pathways that trigger downstream components, helping them withstand drought stress [113]. When there is a lack of water, chemicals and metabolites including proline, glycine betaine (GB), and soluble sugar accumulate in the cytoplasm, assisting in osmotic adjustment and preparing the plant to cope with the adverse effects of oxidative stress in *Triticum aestivum* L. [62,114]. These metabolites are significant because their distinct biochemical processes promote plant tolerance—drought signaling results in crosstalk between various biological molecules and metabolites. Proline is an essential metabolite that accumulates in higher amounts in water-stressed environments [83].

4.1. Reactive Oxygen Species (ROS)

Water scarcity is the primary constraint on agricultural growth and development in irrigated and non-irrigated zones. This is because climatic conditions in irrigated and non-irrigated agricultural regions have changed [115]. ROS production is combined with a normal metabolic function in a drought-stressed climate, such as aerobic metabolism [116]. The reaction of plants to drought stress, whether through photosynthesis or other means, results in oxidative damage in proteins, lipids, and nucleic acids. Because plants are sessile creatures, they have devised techniques to assist them in surviving, adapting, or tolerating drought stress [58]. Under drought stress environments, increased ROS formation is unavoidable; phytotoxic levels of ROS are hazardous [117], resulting in cellular damage and even death [94,118]. However, they function as an essential signaling molecule at low concentrations, stimulating multiple stress-responsive pathways and initiating crosstalk between them. ROS-producing and scavenging enzymes and the antioxidant system fine-tune these for maintaining the cell's redox state by removing or changing the intracellular ROS concentration [119].

4.2. Total Soluble Phenolic, Antioxidatant Enzymatic, and Osmolyte Regulation under Drought Stress Conditions

According to previous findings, there was a 100% increase in phenolic content under drought stress conditions [117,120,121]. Drought-stressed tomatoes had more total phenolic (46.4 mg GAE/100 g DM) than well-watered tomatoes [122]. Total phenolic rather than individual polyphenol concentrations were used in this study because of the wide range of phenolic compounds and the structural diversity of phenolic compounds [123]. Food polyphenol content cannot be determined using a single method, and the Folin-Ciocalteu reagent can be affected by other reducing agents, such as ascorbic acid [124]. High phenolic compounds in tomato fruits protect cells from oxidative damage. Peppers are a popular vegetable worldwide [14,125]. Drought stress reduces pepper fruit pithiness and reproductive development parameters; however, antioxidant activity was boosted after 45 days of blooming [79]. The coordination and management of multiple antioxidant enzymes in tea plants during drought stress is not well understood; despite all the stressful situations, foliar antioxidant content was noticed. Chemically reactive oxygen species are scavenged by enzymes that maintain membrane integrity and modify the osmotic pressure via signaling pathways that regulate gene expression and transcription [126].

Maize (*Zea mays* L.) crops under drought stress had the highest levels of antioxidant enzymes (POD), hydrogen peroxide (H_2O_2), glutathione (GSH), proline, and malondialdehyde of any crop tested (MDA) [86,127]. The finishing purpose of this study, according to the authors, was to assess the number of antioxidant chemicals discovered in the flesh of tomato fruits that had either been well-watered or had been subjected to a 10-day drought cycle throughout their development [128]. GPX produces lignin, guaiacol, and pyrogallol, which function as electron donors to scavenge hydrogen peroxide inside and outside the cell. Many studies have shown that GPX levels increase in drought-stricken plants like wheat crops (*Triticum aestivum* L.) [58]. The report concluded that drought stress increases GPX activity in rice and has been extensively researched and confirmed as a helpful screening approach for tolerance characteristics [129]. Proline is known for its vital role in osmoprotectants [130]. It is suggested that proline regulates cellular redox status and directly acts as a ROS scavenger under oxidative stress conditions. High proline concentration is associated with drought tolerance and a powerful defensive antioxidant system. The rainfed genotypes exhibited a greater proline concentration than irrigated or humid genotypes. Agricultural plants undergo various internal physiological processes [8,68].

Similarly, wheat (*Triticum aestivum* L.) cultivars with a high proline content in the leaves efficiently utilized water. Proline accumulates more significantly in response to various abiotic environmental challenges, including abiotic stress such as drought stress [131]. It is widely recognized that higher proline concentrations in agricultural plants cultivated under water-stress conditions relate to drought tolerance. Those drought-tolerant varieties have higher proline concentrations than drought-sensitive cultivars [132]. Many investigators identified a buildup of soil proline in the leaves of saline-stressed higher halophytic plants. However, plants subjected to drought stress showed significantly higher proline concentrations in the plants' leaves, shoots, desiccating pollen, and root apical regions. Increasing the quantity of proline in the plant saves less water potential, resulting in the buildup of osmolytes in the osmoregulation process, allowing the plant to take up water for growth and metabolic activities [103,110,124].

The previous study explained that several antioxidant defence system enzymes' activity changes when the wheat crop (*Triticum aestivum* L.) is exposed to oxidative stress caused by environmental stresses [133]. Guaiacol peroxidase, peroxiredoxins, SOD, CAT, GPX, ascorbate-glutathione cycle enzymes, including dehydro-ascorbate reductase, monodehydroascorbate reductase, APX, and glutathione reductase are among the enzymatic activities [15,134]. Tocopherols, carotenoids, and phenolic chemicals are non-enzymatic components, as are the primary cellular redox buffers ascorbate and glutathione. The wheat crop (*Triticum aestivum* L.), which is grown in the field and the lab, the activity of peroxidase, superoxide dismutase, ascorbate glutathione reductase, catalase, and guaiacol peroxidase, as well as the amount of ROS, were discovered [86,135,136]. Furthermore, multiple investigations show that abiotic stress has a genotype-specific effect on *Triticum aestivum* L., with different genotypes reacting differentially to the limited water supply. Drought-tolerant genotypes have a better antioxidant capability, which results in less oxidative damage [78,91]. Wheat crop (*Triticum aestivum* L.) responses vary by tissue type, duration, the severity of stress, and developmental stage, demonstrating the intricacy of ROS generation and detoxifying pathways and the impact of ROS on antioxidant systems [137].

5. Improvement of Drought Tolerance Using Molecular Tools

Rather than a qualitative feature, drought tolerance combines quantitative plant features regulated by several genes and other plant variables with minor individual impacts [138]. Understanding drought stress responses has necessitated the development of molecular regulatory understanding in recent years [139,140]. Transcriptome research, for example, has improved performance and aided the discovery of potential genes that might be used in plant breeding [141,142]. However, it was evident that the translational and post-translational machinery, particularly for immediate molecular activity during abiotic stress adaptation, is essential [112]. Understand stress-induced signal receipt and transduction, translational movement, and induced protein levels. In addition to transcriptome investigations, proteomics has emerged as the most direct and consequential approach for acquiring protein expression information on plants' responses to drought stress [93]. Comparing proteomics of drought-tolerant and sensitive wheat (*Triticum aestivum* L.) genotypes is one technique for assessing the complexity of molecular pathways in wheat (*Triticum aestivum* L.) crop in response to drought stress [143]. In irrigation water shortage and climate change, efforts to enhance crop drought tolerance and related soil salinity are critical [144]. Specific chromosomal sites (quantitative trait loci (QTL) were connected to express traits using a combination of DNA fingerprints from various genotypes and phenotypic evaluations. Using marker-assisted selection (MAS) technology, some DNA markers have been linked to favorable QTLs [145]. Because of advancements in next-generation sequencing, the synthesis of many genetic markers, such as single nucleotide polymorphisms (SNPs) [146]; and insertion-deletions (InDels), provides a realistic option for increasing drought tolerance in cereal crops [147]. Drought-responsive genes and QTLs have recently been discovered in wheat (*Triticum aestivum* L.) crop, revealing that QTLs have been the focus of research over the last decade to identify the gene loci governing crops' adaptive response to drought stress [148]. In addition to traditional and molecular plant breeding methods, the transfer of genes and gene regulatory sites vital for plant water management has emerged as an essential strategy [149]. Candidate genes have been thoroughly investigated in transgenic approaches [150].

In the previous research, many drought stress response genes were discovered and introduced into cultivated plants [132]; drought-resistant like *Triticum aestivum* L., *Oryza sativa* L., and *Zea mays* L. transgenic crops. Only a few drought-resistant grain cultivars developed through genetic transformation have been approved commercially [106,138,151]. The cspB gene, which encodes the cold shock protein B, was introduced into maize to give drought tolerance [129]. The cspB transgenic plant retains RNA stability and translation during drought stress, maintaining normal cellular function [152,153]. More profound knowledge of interactions between growth-promoting microbes and plants is another promising approach to the abiotic stress problem in many plants (PGPM) [132,154]. Plants can be protected against abiotic stress's adverse effects, mainly drought and salinity stress. The biotechnology approach may be used to improve plant-microbe interactions. Plants inhabited by genetically changed soil bacteria that overproduce trehalose benefit from genetically modified PGPM [20,54,155].

6. Phytohormonal Modulation under Drought Stress

Phytohormones play an essential role in the development and growth of plants and their responses to environmental stress [156]. While not all plant cells respond to hormones simultaneously, those genetically programmed to do so at certain moments throughout the plant's growth cycle. Plants need hormones at certain times and sites throughout their development and reproduction [157]. Hormones must disengage their effects when no longer required. Plants may also chemically break down hormones, leading to death [158]. Plant hormones oversee regulating the levels of other plant hormones [159]. Plant hormones are among the most significant biochemical influencing plant development and yield production in various environments, including drought stress [160,161]. Plant hormones are essential in developing and growing a plant when under water deficit stress [162]. Water stressed-induced responses in plant growth regulators such as salicylic acid, gibberellins, Cytokinin, and abscisic acid have been observed [149]. Besides stress responses, phytohormones also control internal and external stimuli and signal transduction pathways. Difficulty growing plants and low output are caused by different abiotic stresses, with drought stress most prevalent worldwide [163,164].

For this reason, the drought tolerance mechanism understanding in plants is essential for enhancing drought resistance in plants. According to [109] the growing body of research, phytohormones appear to be critical signaling molecules that modulate various wheat plant (*Triticum aestivum* L.) development processes and growth stages when plants are subjected to drought stress. The production of phytohormones regulates wheat plant (*Triticum aestivum* L.) growth in response to drought stress [106,165].

6.1. Salicylic Acid

Johann Buchner, a German scientist, first isolated SA from the bark of a Salix species (willow tree) in 1928 and named the glucoside of salicylic alcohol "silicon" [126]. SA is a phenolic molecule generated by secondary metabolism [166]; that plays a role in many biological processes, including CO_2 assimilation, antioxidation, stomatal regulation, and photosynthesis [167,168]. Though SA's role in abiotic and biotic stress has been thoroughly studied, evidence of its impact on drought stress is limited. Several studies, however, suggest that it may have a role in drought stress by modifying regulating drought-related genes through transcriptional regulation and stomatal aperture; depending on the amount of SA utilized, drought tolerance and sensitivity are affected [156,169,170].

Similarly, a higher SA treatment concentration reduces maize plants' capacity to withstand drought. Water shortage increased endogenous SA levels significantly in *Phillyrea angustifolia* L. plants [171,172]. SA (500 M) applied externally to drought-stressed barley enhanced stomatal conductance and CO2 assimilation, leading to a dry matter increase [173]. According to [174], SA controls proline production and maintains the cellular redox state in the *Brassica rapa* L. plant. According to Castro et al., the light-induced stomatal opening was reduced in plants with high SA levels and the siz1 mutant (impaired function in SUMO E3 ligase, SIZ1), minimizing water loss and giving drought resistance [175].

Similarly, drought stress tolerance, increased SA buildup, and lower stomatal conductance was observed in cpr5 and acd6 mutants. Furthermore, many essential proteins were revealed for drought stress physiology and metabolism by priming the wheat seedlings with SA (0.5 mM) [133]. Proteins such as carbohydrate metabolism, photosynthesis, anti-stress proteins, and the signaling cascade are differentially expressed in primed seedlings, resulting in drought tolerance and improved growth [176–178]. SA applied exogenously has also been found to boost plant drought resilience. Plants overexpress CBP60g (a transcription regulator of SA biosynthesis) are more sensitive to ABA, accumulate more SA, and have a robust drought resistance phenotype [179].

Applying Salicylic acid (SA) to the leaves has induced plant stress tolerance. Several studies have found that Salicylic acid (SA) has beneficial effects on plants in terms of resistance to salinity, drought, and high temperatures [105,180,181]. The previous results suggested that Salicylic acid (SA) helps plants adapt to abiotic stresses [182]. Salicylic acid (SA) and exogenously applied substances develop dry period resilience and upgrade the submerged plants' development and harvest [126,183]. Under drought-stressed conditions, salicylic acid (SA) application increased wheat crops (*Triticum aestivum* L.) catalase activity [184]. Salicylic acid (SA) and its derivatives in foliar and seed treatments improved drought tolerance in drought-stressed wheat crops (*Triticum aestivum* L.) [179]. Purslane (*Portulaca oleracea* L.) was utilized as a model plant in this study to see how foliar salicylic acid (SA) affected plant drought tolerance. According to the findings, Salicylic acid (SA) promoted purslane (*Portulaca oleracea* L.) growth by improving the pigments of photosynthetic apparatus and secondary metabolites production; suitable solutes and gas exchanges [185,186].

6.2. Cytokinin and Auxin

Another prominent phytohormone is cytokinin, which functions critically in the plant's life cycle [81,187]. This low molecular weight plant hormone was initially found in maize (*Zea mays* L.) and is now recognized to serve many essential roles in plant growth and development [188]. Isoprenoid cytokinin contains an isoprenoid-derived side chain and aromatic cytokinin, which has an aromatic side chain at the N6 terminus [189]. The investigation of [190,191] revealed the existence of meta-tooling, a very active growth component that belongs to aromatic CKs, suggesting that aromatic cytokinin is far more significant than PGRs. The adenine moiety and the side chain are modified during CK metabolism. The central location of CK synthesis is the root tips, from which it is delivered to xylem sap by transpiration pull in an acropetal manner [192]. Cell division control, photosynthetic sink strength, unit stability, cell differentiation, delayed senescence, nutrient absorption, flower and seed germination and development, and prevention of lateral root initiation are just a few of the many functions of cytokinin in plant physiological processes [14,193,194]. The first phytohormone identified, Auxin, impacts some plant processes, including cell dedifferentiation and differentiation, root morphology or architecture, geotropism, root growth, floral organ development, and seed dormancy [195]. Recently, a tangible link between auxin content and plant drought stress response has been discovered. It has also been shown that auxin homeostasis regulates ABA production and drought stress responses [196]. TAA transforms tryptophan to IPA, which is then converted to IAA by YUCCA (YUC) flavin monooxygenase-like proteins in the auxin biosynthesis pathway (Arabidopsis) [197]. Drought-stressed rice (*Oryza sativa* L.) showed considerably decreased transcript abundance of IAA biosynthesis genes (YUCCAs) but dramatically increased transcription of IAA conjugating genes [198].

6.3. Gibberellins

Gibberellic acids (GAs) (tetracyclic diterpenoid carboxylic acid) can enhance plant growth and development in a different stage of the life cycle by boosting the cell division and elongation [199]. The most bioactive versions of the other GAs generated by plants are GA1 and Gas [200]. The GAs hormone is related to drought stress tolerance and is associated with seed germination, stem elongation, and reproductive development in the rice (*Oryza sativa* L.) plant [201]. Growth inhibitors imparted drought resistance to plants by lowering endogenous GA production, providing the first evidence of GA's role in abiotic stress tolerance. In growth-retarded plants GA- and deficient mutants, GA treatment corrected dwarf growth and stress tolerance responses [202]. Plastid, endoplasmic reticulum (ER), and cytoplasm are involved in gas generation, with trans-geranylgeranyl diphosphate being the starting point in the chloroplast [203]. The overwhelming evidence implies that dioxygenases control GAs synthesis and that GA2ox genes in plants are primarily vulnerable to abiotic stress. Inhibition of plant growth and development by gibberellins (GAs), which are carboxylic acids that can regulate plant growth and development, has been observed. Gibberellins (GAs) affect leaf growth, seed germination, stem lengthening, flower development, and trichome formation [204,205]. Genetically altered (GA) hormones may interact with other hormones and impact several developmental processes [206]. These interactions may entail both negative and positive regulatory activities. Gibberellins (GAs) are a type of endogenous hormone found in plants that regulate the development of the plant's vegetative and reproductive systems [207]. When controlling stem elongation, the effects of gibberellins (GAs) processes on cell growth and division are critical [208]. Compared to the shoot, Gibberellin insufficiency promotes the partitioning of reserves to the root [209]. Impaired GAs biosynthesis causes significant changes in primary metabolism, mainly due to drought stress [158,205]. Gibberellin deficiency enhances water content maintenance, improving drought stress tolerance [210]. Gibberellins (GAs) deficient symptoms look phenotypically like drought stress symptoms [211]. Under prolonged drought stress, plants show reduced height, leaf development, and flowering/fruit development [212].

Dwarfed plants with diminished stem elongation, leaf development, aberrant flowering, and fruit set occur from a decrease in endogenous GA concentration [213]. Water deficit stress lowers the rate of gene expression involved in GA biosynthesis, lowering the amount of bioactive GAs produced [214,215]. Under drought stress conditions, gibberellin content can be reduced, resulting in decreased internode elongation based on the degree of Gas reduction. Plants with less elongation may be more suited to situations where drought stress is standard. It inhibits stem cell elongation and growth [214]. Because GAs are critical regulators of cell elongation, the goal of the previous research was to see if the loss in development caused by drought is linked to changes in GA metabolism or signaling [106]. Drought stress, we postulated, influences plant development and stem elongation through its interaction with GAs metabolism, based on earlier research [216]. As a result, the main aim of this study was to find out how water-deficient stress affected stem elongation and Gas metabolism-related gene expression in tomato plants [217].

6.4. Abscise Acid

The natural plant stress hormone abscisic acid (ABA) regulates various physiological processes (Figure 6). Plants' osmotic stress is linked to low water availability, triggering ABA production and adaptation mechanisms [156,218,219]. Abscisic acid production begins in the plastids once the plasma membrane receives stress signals, with the xanthorin transition to ABA being excluded; and happens in the cytoplasm. Most ABA is created in the roots and then transported to the plant's upper portions via vascular tissues [220,221]. The former is a crucial player in expressing stress-responsive genes with the help of ABA under many situations, including osmotic stress [134,222,223]. Several receptors have been identified in the cytosol, plasma membrane, chloroplast envelope, and nucleus. Protein phosphatase 2C (PP2C) inhibits the action of non-fermenting sucrose 1-linked protein kinase 2 (SnRK2) proteins in plants with low ABA levels, resulting in dephosphorylation [224]. Antibiotics increase tolerance to drought in cotton (*Gossypium hirsutism* L.) plants by ABA, which regulates a stress-related gene [150]. In the Arabidopsis (*Arabidopsis thaliana* L.) plant, overexpression of the ABA-induced cotton gene (GhCBF3) leads to the high drought tolerance in transgenic lines by maintaining Ch, RW, and proline levels more significant than in the wild-type plant [225,226]. The stress hormone abscisic acid (ABA) is implicated in plants' leaf abscission and abiotic stress [227]. ABA has the primary and critical role in plants' developmental and physiological activities, including seed dormancy [228], tumor cell maintenance, stomatal opening, embryo morphogenesis, and fat and stored protein production. Abscisic acid affects the expression of protein-coding genes [229]. ABA is also required for root development and structural changes in nitrogen-deficient plants. Dehydrins, osmoprotectants, and protective proteins are all made by this enzyme. ABA plays two roles in drought stress: water balance and cell dehydration tolerance. Water balance is achieved in virtually all cells by controlling guard cells and the expression of genes that produce dehydration tolerance proteins [14,139,230].

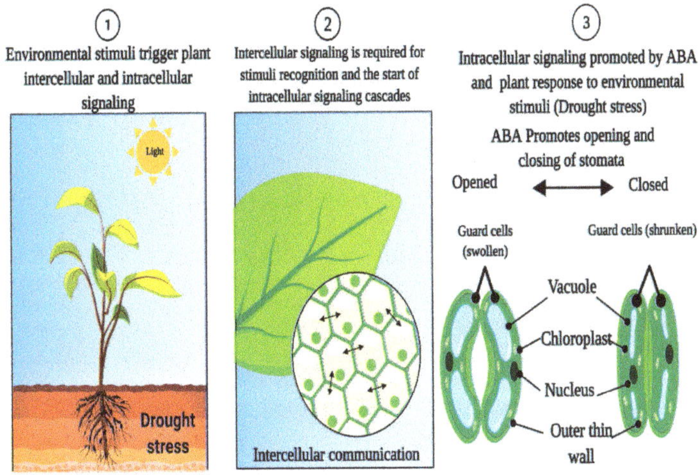

Figure 6. Abscisic acid (ABA) is vital for plant development and stress response. In response to biotic and abiotic stimuli, ABA transfer to guard cells triggers stomatal closure in leaves.

Wheat crops (*Triticum aestivum* L.) with lower amounts of ABA in their leaves are more drought tolerant than those with higher proline levels [231]. When plants are drying out, soil moisture levels are more critical than leaf water levels, controlled mainly by ABA production in the roots [17,232]. Under drought stress, the phytohormone abscisic acid regulates crop morpho-physiology and biochemistry. Stomata closure is the most effective and essential response to ABA in drought-stressed crops [233]. Plants employ ABA as a signal molecule to help them cope with environmental stresses such as cold, salt, drought, heat, and phosphate deficiency in the olive tree (*Olea europaea* L.) [234]. Exogenous ABA treatment on leaves has been shown to elicit many adaptive changes in response to water scarcity, including the enhanced GR, SOD, APX, and CAT activity in tomato plants (*Solanum lycopersicum* L.) [81].

The exogenous ABA can also minimize ROS and increase cell membrane stability (CMS) to aid plants in their recovery after being subjected to stress [117,235]. Exogenous ABA spraying has been shown in some studies to improve plant stress tolerance in various crop species. However, research evaluating the responses of different *Zea mays* L. and *Glycine max* L. to drought stress using exogenous ABA and fluoridone is severely limited (ABA synthesis inhibitor) [236,237]. ABA substantially enhanced the activities of SOD and POD during drought stress, with a considerable drop after re-watering [235]. Under drought stress, ABA priming substantially raised the relative water content in both wheat cultivars [195]. Plant drought pathways use ABA as a primary stress sensor to improve the plant's response to desiccation. The rise in ABA concentration coincided with the accumulation of lycopene and carotene in the fruits [238,239].

6.5. Ethylene

Gaseous phytohormone ethylene regulates the floral senescence, fruit ripening, petal and leaf abscission, and plant stress responses [240]. ET plays a vital role in biotic and abiotic stressors [28,241]. However, in these newly found activities of ethylene, there has been significantly less investigation on the drought stress response. According to a recent study, the dry shoot weight of six wheat genotypes ranging from tolerant to sensitive was more significant in the tolerant group under mild drought stress, related to an increase in ethylene [242,243]. Interestingly, several investigations on the influence of ethylene on stomata closure have shown contradictory results. For example, Arabidopsis eto1 mutants with higher ethylene accumulation have slower stomatal closure under drought stress conditions than control plants, even though ethylene has been considered to improve stomatal closure in guard cells [244,245]. More ethylene accumulates in the rice etol1 mutant, resulting in more drought-tolerant plants than OsETOL1 plants susceptible to drought stress treatment. Drought-tolerant transgenic plants were generated by modifying genes in the ethylene signaling pathway. Our findings underscore the need to understand and eventually use stress tolerance-related features in crops by interpreting ethylene signaling under abiotic stressors [246,247].

6.6. *Jasmonates Acid (JAs)*

Jasmonic acid (JA) is a phytohormone found in plants, and its active derivatives are known as jasmonates. It is essential in the fight against a variety of biotic and abiotic stressors [248]. Furthermore, JA is linked to improved root structure, pollen production, tendril coiling, and fruit ripening in many species [249]. Exogenously applied JA has improved plant performance and modulated stomatal dynamics in dry surroundings. JA signaling route and production have been extensively researched [250,251].

Nonetheless, in the absence of water, JAZ proteins are destroyed, resulting in active transcription factors such as MYC_2, which up-regulate genes associated with stress tolerance [252]. Plant hormones, in most cases, do not function in a single route but rather interact with one another at different stages to control environmental and developmental pathways [253]. Signal transduction emerges in plants and may coordinate a complex set of events to adapt to a hostile environment. Jasmonates (JAs) are complex phytohormones created by the breakdown of lipids in the cell membrane in various plant species [162,254]. Plant growth regulators known as JAs may be found in almost every country. Jasmonates have also been shown to interact with other phytohormones to regulate plant growth and development and adapt to biotic and abiotic stimuli [250]. Seed dormancy and germination are affected by JAs in different ways. Jasmonates (JAs) treatment has negatively impaired seed germination in several species, including Solanum Lycopersicum, under water-stressed conditions. However, we know little about how JAs impact germination water deficit stress and salinity stress regimes [255,256].

7. Conclusions

Drought is a severe environmental stressor that threatens crop productivity worldwide. However, drought is more damaging during the reproductive and grain-filling stages (terminal drought). Terminal drought impacts grain set, pace, duration, yield production, and quality. Drought influences grain yield depending on the crop stage, length, and intensity. Drought-resistant genotypes and accompanying crop management practices can help reduce drought stress's adverse effects. Improving drought resilience requires a thorough grasp of the impact of terminal drought. Although, research focusing on the physiological and molecular components of the drought response has helped improve wheat resistance to terminal drought. New advances in sequencing, marker creation, and genomic analysis have opened the door to tackling drought-resistant components. Drought stress has a long-term effect on CO_2 absorption rates because it causes stomatal conductance to decrease. Deteriorated photosynthetic pigments, and restricted gaseous exchange, resulting in decreased plant growth and productivity. Plant growth, development, dry matter, and harvestable yield are all affected by drought stress, even though each species responds differently. Ramified root systems have been linked to drought resistance and high biomass production due to their capacity to collect more water from the soil and transport it to above-ground areas for photosynthesis. Many factors, including changes in photosynthetic pigments, influence the amount of water available to plants during drought stress regimes. A variety of roles in drought tolerance are played by carotenoids, one of two families of photosynthetic pigments. These roles include light-harvesting and oxidative damage avoidance, among other things. The phytohormone ABA influences drought stress responses and resistance in plants, which acts at the cellular and intercellular levels.

However, it is unclear how plants detect drought stress and communicate that information into the cell to regulate ABA accumulation to withstand drought stress conditions.

8. Concluding Remarks and Outlook

Climate change and anthropogenic activity create a global danger to crop yield, exacerbated by shrinking agricultural areas, posing severe food security and safety challenge. Drought severely affects plant productivity and lowers the overall economic viability of agriculture. Many methodologies have been developed to challenge drought; each has its advantages and limitations. Though plants have an inherent defense system to deal with adverse environmental conditions, the genetic composition of the plant, the stage at which stress is identified, and the duration and degree of the stress all impact the plant's reaction. The drought stress response is more than just a defense mechanism; it is also a means of achieving long-term development and ensuring a healthy ecological succession for future generations. Several studies have discovered several molecular markers associated with drought stress, with the phytohormonal syndicate having an important role. Because of their inter-crosstalk response, phytohormone signaling modules promote a complex cascade. The complicated reaction is enabled to improve the cellular potential to withstand adversities when multiple phytohormones are juxtaposed in a single frame of the event. As a result, proper drought stress response necessitates the interaction of these phytohormones and their communication and coordination. The discovery of PGRs crosstalk adds a new dimension to their previously well-understood functions and control. However, a thorough knowledge of these phytohormones' molecular interactions remains completely unexplored. Although ABA helped relieve drought stress, the current work demonstrates the significance of hormone crosstalk throughout the drought stress response. Although most drought stress tolerance gene function research has been undertaken in the model plant Arabidopsis thaliana, the target gene(s) must be tweaked in economically relevant crops to benefit the end consumer directly.

Author Contributions: Conceptualization, methodology, investigation, A.W., M.H.S.; Validation, B.A., R.A.M. and C.C.M.; Formal analysis, G.A., S.U., A.W. and M.H.S.; Resources, B.A., M.H.S. and S.M.; Data curation, G.Y.; writing—original draft preparation, A.W.; writing—review and editing, M.H.S., G.Y., S.M. and G.A.; Supervision, M.H.S.; Project administration, W.S. and S.U.; Funding acquisition, B.A., C.C.M., R.A.M. All authors have read and agreed to the published version of the manuscript.

Funding: The study was supported by the National Research Development Projects to finance excellence (PFE)-14/2022-2024 granted by the Romanian Ministry of Research and Innovation.

Institutional Review Board Statement: Not applicable.

Informed Consent Statement: Not applicable.

Data Availability Statement: Not applicable.

Conflicts of Interest: The authors declare no conflict of interest.

References

1. Amna; Ali, B.; Azeem, M.A.; Qayyum, A.; Mustafa, G.; Ahmad, M.A.; Javed, M.T.; Chaudhary, H.J. Bio-Fabricated Silver Nanoparticles: A Sustainable Approach for Augmentation of Plant Growth and Pathogen Control. In *Sustainable Agriculture Reviews 53*; Springer: Berlin/Heidelberg, Germany, 2021; pp. 345–371.
2. Zainab, N.; Amna; Khan, A.A.; Azeem, M.A.; Ali, B.; Wang, T.; Shi, F.; Alghanem, S.M.; Hussain Munis, M.F.; Hashem, M.; et al. PGPR-Mediated Plant Growth Attributes and Metal Extraction Ability of *Sesbania sesban* L. in Industrially Contaminated Soils. *Agronomy* **2021**, *11*, 1820. [CrossRef]
3. Hussain, S.Q.; Rasheed, M.; Saleem, M.H.; Ahmed, Z.I.; Hafeez, A.; Jilani, G.; Alamri, S.; Hashem, M.; Ali, S. Salt tolerance in maize with melatonin priming to achieve sustainability in yield on salt affected soils. *Pak. J. Bot.* **2022**, *55*, 1. [CrossRef]
4. Yahaya, M.A.; Shimelis, H. Drought Stress in Sorghum: Mitigation Strategies, Breeding Methods and Technologies—A Review. *J. Agron. Crop. Sci.* **2022**, *208*, 127–142. [CrossRef]
5. Muhammad Aslam, M.; Waseem, M.; Jakada, B.H.; Okal, E.J.; Lei, Z.; Saqib, H.S.A.; Yuan, W.; Xu, W.; Zhang, Q. Mechanisms of Abscisic Acid-Mediated Drought Stress Responses in Plants. *Int. J. Mol. Sci.* **2022**, *23*, 1084. [CrossRef]

6. McDowell, N.G.; Sapes, G.; Pivovaroff, A.; Adams, H.D.; Allen, C.D.; Anderegg, W.R.L.; Arend, M.; Breshears, D.D.; Brodribb, T.; Choat, B.; et al. Mechanisms of Woody-Plant Mortality under Rising Drought, CO_2 and Vapour Pressure Deficit. *Nat. Rev. Earth Environ.* **2022**, *3*, 294–308. [CrossRef]
7. Pepe, M.; Crescente, M.F.; Varone, L. Effect of Water Stress on Physiological and Morphological Leaf Traits: A Comparison among the Three Widely-Spread Invasive Alien Species Ailanthus Altissima, Phytolacca Americana, and Robinia Pseudoacacia. *Plants* **2022**, *11*, 899. [CrossRef]
8. Zandi, P.; Schnug, E. Reactive Oxygen Species, Antioxidant Responses and Implications from a Microbial Modulation Perspective. *Biology (Basel)* **2022**, *11*, 155. [CrossRef]
9. Sofy, M.R.; Aboseidah, A.A.; Heneidak, S.A.; Ahmed, H.R. ACC Deaminase Containing Endophytic Bacteria Ameliorate Salt Stress in Pisum Sativum through Reduced Oxidative Damage and Induction of Antioxidative Defense Systems. *Environ. Sci. Pollut. Res.* **2021**, *28*, 40971–40991. [CrossRef]
10. Perveen, S.; Hussain, S.A. Methionine-Induced Changes in Growth, Glycinebetaine, Ascorbic Acid, Total Soluble Proteins and Anthocyanin Contents of Two *Zea mays* L. Varieties under Salt Stress. *J. Anim. Plant Sci.* **2020**, *31*, 131–142. [CrossRef]
11. Ali, B.; Wang, X.; Saleem, M.H.; Sumaira; Hafeez, A.; Afridi, M.S.; Khan, S.; Zaib-Un-Nisa; Ullah, I.; Amaral Júnior, A.T.; et al. PGPR-Mediated Salt Tolerance in Maize by Modulating Plant Physiology, Antioxidant Defense, Compatible Solutes Accumulation and Bio-Surfactant Producing Genes. *Plants* **2022**, *11*, 345. [CrossRef]
12. Ozturk, M.; Turkyilmaz Unal, B.; García-Caparrós, P.; Khursheed, A.; Gul, A.; Hasanuzzaman, M. Osmoregulation and Its Actions during the Drought Stress in Plants. *Physiol. Plant.* **2021**, *172*, 1321–1335. [CrossRef] [PubMed]
13. Ali, B.; Wang, X.; Saleem, M.H.; Azeem, M.A.; Afridi, M.S.; Nadeem, M.; Ghazal, M.; Batool, T.; Qayyum, A.; Alatawi, A.; et al. Bacillus mycoides PM35 Reinforces Photosynthetic Efficiency, Antioxidant Defense, Expression of Stress-Responsive Genes, and Ameliorates the Effects of Salinity Stress in Maize. *Life* **2022**, *12*, 219. [CrossRef] [PubMed]
14. Fahad, S.; Nie, L.; Chen, Y.Y.; Wu, C.; Xiong, D.; Saud, S.; Hongyan, L.; Cui, K.; Huang, J.; Bogati, K.; et al. The Solanum Melongena COP1LIKE Manipulates Fruit Ripening and Flowering Time in Tomato (*Solanum lycopersicum*). *Front. Plant Sci.* **2021**, *10*, 369–382. [CrossRef]
15. Chai, Q.; Gan, Y.; Zhao, C.; Xu, H.L.; Waskom, R.M.; Niu, Y.; Siddique, K.H.M. Regulated Deficit Irrigation for Crop Production under Drought Stress. A Review. *Agron. Sustain. Dev.* **2016**, *36*, 3. [CrossRef]
16. Khatun, M.; Sarkar, S.; Era, F.M.; Islam, A.K.M.M.; Anwar, M.P.; Fahad, S.; Datta, R.; Islam, A.K.M.A. Drought Stress in Grain Legumes: Effects, Tolerance Mechanisms and Management. *Agronomy* **2021**, *11*, 2374. [CrossRef]
17. Alam, H.; Khattak, J.Z.; Ksiksi, T.S.; Saleem, M.H.; Fahad, S.; Sohail, H.; Ali, Q.; Zamin, M.; El-Esawi, M.A.; Saud, S. Negative impact of long-term exposure of salinity and drought stress on native *Tetraena mandavillei* L. *Physiol. Plant.* **2021**, *172*, 1336–1351. [CrossRef]
18. Ghafar, M.A.; Akram, N.A.; Saleem, M.H.; Wang, J.; Wijaya, L.; Alyemeni, M.N. Ecotypic Morphological and Physio-Biochemical Responses of Two Differentially Adapted Forage Grasses, *Cenchrus ciliaris* L. and *Cyperus arenarius* Retz. to Drought Stress. *Sustainability* **2021**, *13*, 8069. [CrossRef]
19. Azeem, M.; Haider, M.Z.; Javed, S.; Saleem, M.H.; Alatawi, A. Drought Stress Amelioration in Maize (*Zea mays* L.) by Inoculation of Bacillus spp. Strains under Sterile Soil Conditions. *Agriculture* **2022**, *12*, 50. [CrossRef]
20. Ma, Y.; Vosátka, M.; Freitas, H. Editorial: Beneficial Microbes Alleviate Climatic Stresses in Plants. *Front. Plant Sci.* **2019**, *10*, 595. [CrossRef]
21. Grzesiak, M.T.; Hordyńska, N.; Maksymowicz, A.; Grzesiak, S.; Szechyńska-Hebda, M. Variation among Spring Wheat (*Triticum aestivum* L.) Genotypes in Response to the Drought Stress. Ii—Root System Structure. *Plants* **2019**, *8*, 584. [CrossRef]
22. Kamal, N.M.; Gorafi, Y.S.A.; Abdelrahman, M.; Abdellatef, E.; Tsujimoto, H. Stay-Green Trait: A Prospective Approach for Yield Potential, and Drought and Heat Stress Adaptation in Globally Important Cereals. *Int. J. Mol. Sci.* **2019**, *20*, 5837. [CrossRef] [PubMed]
23. Abid, M.; Ali, S.; Qi, L.K.; Zahoor, R.; Tian, Z.; Jiang, D.; Snider, J.L.; Dai, T. Physiological and Biochemical Changes during Drought and Recovery Periods at Tillering and Jointing Stages in Wheat (*Triticum aestivum* L.). *Sci. Rep.* **2018**, *8*, 4615. [CrossRef] [PubMed]
24. Boudiar, R.; Casas, A.M.; Gioia, T.; Fiorani, F.; Nagel, K.A.; Igartua, E. Effects of Low Water Availability on Root Placement and Shoot Development in Landraces and Modern Barley Cultivars. *Agronomy* **2020**, *10*, 134. [CrossRef]
25. Mehmood, S.; Khatoon, Z.; Amna; Ahmad, I.; Muneer, M.A.; Kamran, M.A.; Ali, J.; Ali, B.; Chaudhary, H.J.; Munis, M.F.H. Bacillus sp. PM31 harboring various plant growth-promoting activities regulates Fusarium dry rot and wilt tolerance in potato. *Arch. Agron. Soil Sci.* **2021**, *2021*, 1–15. [CrossRef]
26. Singh, B.; Norvell, E.; Wijewardana, C.; Wallace, T.; Chastain, D.; Reddy, K.R. Assessing Morphological Characteristics of Elite Cotton Lines from Different Breeding Programmes for Low Temperature and Drought Tolerance. *J. Agron. Crop. Sci.* **2018**, *204*, 467–476. [CrossRef]
27. Queiroz, M.S.; Oliveira, C.E.S.; Steiner, F.; Zuffo, A.M.; Zoz, T.; Vendruscolo, E.P.; Silva, M.V.; Mello, B.F.F.R.; Cabral, R.C.; Menis, F.T. Drought Stresses on Seed Germination and Early Growth of Maize and Sorghum. *J. Agric. Sci.* **2019**, *11*, 310. [CrossRef]
28. Wagaw, K. Isolation and Screening of ACC Deaminase-Producing Microbes for Drought Stress Management in Crops. *Acad. Res. J. Agri. Sci. Res.* **2019**, *7*, 87–99. [CrossRef]

29. Islam, M.M.; Kayesh, E.; Zaman, E.; Urmi, T.A.; Haque, M.M. Evaluation of Rice (*Oryza sativa* L.) Genotypes for Drought Tolerance at Germination and Early Seedling Stage. *Agriculturists* **2018**, *16*, 44–54. [CrossRef]
30. Al-Quraan, N.A.; Al-Ajlouni, Z.I.; Qawasma, N.F. Physiological and Biochemical Characterization of the Gaba Shunt Pathway in Pea (*Pisum sativum* L.) Seedlings under Drought Stress. *Horticulturae* **2021**, *7*, 125. [CrossRef]
31. Sabagh, A.E.L.; Hossain, A.; Barutçular, C.; Iqbal, M.A.; Islam, M.S.; Fahad, S.; Sytar, O.; Çiğ, F.; Meena, R.S.; Erman, M. Consequences of Salinity Stress on the Quality of Crops and Its Mitigation Strategies for Sustainable Crop Production: An Outlook of Arid and Semi-Arid Regions. In *Environment, Climate, Plant and Vegetation Growth*; Springer: Berlin/Heidelberg, Germany, 2020; pp. 503–533.
32. Liang, Y.; Tabien, R.E.; Tarpley, L.; Mohammed, A.R.; Septiningsih, E.M. Transcriptome Profiling of Two Rice Genotypes under Mild Field Drought Stress during Grain-Filling Stage. *AoB Plants* **2021**, *13*, plab043. [CrossRef]
33. Damalas, C.A. Improving Drought Tolerance in Sweet Basil (*Ocimum basilicum*) with Salicylic Acid. *Sci. Hortic. (Amsterdam)* **2019**, *246*, 360–365. [CrossRef]
34. Ahmad, S.; Kamran, M.; Ding, R.; Meng, X.; Wang, H.; Ahmad, I.; Fahad, S.; Han, Q. Exogenous Melatonin Confers Drought Stress by Promoting Plant Growth, Photosynthetic Capacity and Antioxidant Defense System of Maize Seedlings. *PeerJ* **2019**, *2019*, e7793. [CrossRef] [PubMed]
35. Lei, C.; Bagavathiannan, M.; Wang, H.; Sharpe, S.M.; Meng, W.; Yu, J. Osmopriming with Polyethylene Glycol (Peg) for Abiotic Stress Tolerance in Germinating Crop Seeds: A Review. *Agronomy* **2021**, *11*, 2194. [CrossRef]
36. Liang, B.; Gao, T.; Zhao, Q.; Ma, C.; Chen, Q.; Wei, Z.; Li, C.; Li, C.; Ma, F. Effects of Exogenous Dopamine on the Uptake, Transport, and Resorption of Apple Ionome under Moderate Drought. *Front. Plant Sci.* **2018**, *9*, 755. [CrossRef]
37. Elnaggar, A.; El-Keblawy, A.; Mosa, K.A.; Soliman, S. Drought Tolerance during Germination Depends on Light and Temperature of Incubation in Salsola Imbricata, a Desert Shrub of Arabian Deserts. *Flora Morphol. Distrib. Funct. Ecol. Plants* **2018**, *249*, 156–163. [CrossRef]
38. Paponov, M.; Kechasov, D.; Lacek, J.; Verheul, M.J.; Paponov, I.A. Supplemental Light-Emitting Diode Inter-Lighting Increases Tomato Fruit Growth Through Enhanced Photosynthetic Light Use Efficiency and Modulated Root Activity. *Front. Plant Sci.* **2020**, *10*, 1656. [CrossRef]
39. Kumar, S.; Islam, A.R.M.T.; Islam, H.M.T.; Hasanuzzaman, M.; Ongoma, V.; Khan, R.; Mallick, J. Water Resources Pollution Associated with Risks of Heavy Metals from Vatukoula Goldmine Region, Fiji. *J. Environ. Manag.* **2021**, *293*, 112868. [CrossRef]
40. Naz, S.; Perveen, S. Response of Wheat (*Triticum aestivum* L. Var. Galaxy-2013) to Pre-Sowing Seed Treatment with Thiourea under Drought Stress. *Pakistan J. Bot.* **2021**, *53*, 1209–1217. [CrossRef]
41. Cai, F.; Zhang, Y.; Mi, N.; Ming, H.; Zhang, S.; Zhang, H.; Zhao, X. Maize (*Zea mays* L.) Physiological Responses to Drought and Rewatering, and the Associations with Water Stress Degree. *Agric. Water Manag.* **2020**, *241*, 106379. [CrossRef]
42. Widuri, L.I.; Lakitan, B.; Sodikin, E.; Hasmeda, M.; Meihana, M.; Kartika, K.; Siaga, E. Shoot and Root Growth in Common Bean (*Phaseolus vulgaris* L.) Exposed to Gradual Drought Stress. *Agrivita* **2018**, *40*, 442–452. [CrossRef]
43. Tůmová, L.; Tarkowská, D.; Řřová, K.; Marková, H.; Kočová, M.; Rothová, O.; čečetka, P.; Holá, D. Drought-Tolerant and Drought-Sensitive Genotypes of Maize (*Zea mays* L.) Differ in Contents of Endogenous Brassinosteroids and Their Drought-Induced Changes. *PLoS ONE* **2018**, *13*, e0197870. [CrossRef] [PubMed]
44. Bocchini, M.; D'Amato, R.; Ciancaleoni, S.; Fontanella, M.C.; Palmerini, C.A.; Beone, G.M.; Onofri, A.; Negri, V.; Marconi, G.; Albertini, E.; et al. Soil Selenium (Se) Biofortification Changes the Physiological, Biochemical and Epigenetic Responses to Water Stress in *Zea mays* L. by Inducing a Higher Drought Tolerance. *Front. Plant Sci.* **2018**, *9*, 389. [CrossRef] [PubMed]
45. Begum, N.; Ahanger, M.A.; Su, Y.; Lei, Y.; Mustafa, N.S.A.; Ahmad, P.; Zhang, L. Improved Drought Tolerance by AMF Inoculation in Maize (*Zea mays*) Involves Physiological and Biochemical Implications. *Plants* **2019**, *8*, 579. [CrossRef] [PubMed]
46. Bhattacharya, A. Effect of Soil Water Deficit on Growth and Development of Plants: A Review. *Soil Water Deficit Physiol. Issues Plants* **2021**, *2021*, 393–488. [CrossRef]
47. Hu, Y.; Chen, B. Arbuscular Mycorrhiza Induced Putrescine Degradation into γ-Aminobutyric Acid, Malic Acid Accumulation, and Improvement of Nitrogen Assimilation in Roots of Water-Stressed Maize Plants. *Mycorrhiza* **2020**, *30*, 329–339. [CrossRef]
48. Valliere, J.M.; Zhang, J.; Sharifi, M.R.; Rundel, P.W. Can We Condition Native Plants to Increase Drought Tolerance and Improve Restoration Success? *Ecol. Appl.* **2019**, *29*, e01863. [CrossRef]
49. Sharma, M.; Delta, A.K.; Kaushik, P. Glomus Mosseae and Pseudomonas Fluorescens Application Sustains Yield and Promote Tolerance to Water Stress in *Helianthus annuus* L. *Stresses* **2021**, *1*, 305–316. [CrossRef]
50. Malinowska, M.; Donnison, I.; Robson, P. Morphological and Physiological Traits That Explain Yield Response to Drought Stress in Miscanthus. *Agronomy* **2020**, *10*, 1194. [CrossRef]
51. Istanbuli, T.; Baum, M.; Touchan, H.; Hamwieh, A. Evaluation of Morpho-Physiological Traits under Drought Stress Conditions in Barley (*Hordeum vulgare* L.). *Photosynthetica* **2020**, *58*, 1059–1067. [CrossRef]
52. Nofouzi, F. Evaluation of Seed Yield of Durum Wheat (*Triticum durum*) under Drought Stress and Determining Correlation among Some Yield Components Using Path Coefficient Analysis. *UNED Res. J.* **2018**, *10*, 179–183. [CrossRef]
53. Kalagare, V.S.; Ganesan, N.M.; Iyanar, K.; Chitdeshwari, T.; Chandrasekhar, C.N. Strategy of Multiple Selection Indices for Discrimination of Potential Genotypes and Associated Traits for Yield Improvement in Pearl Millet [*Pennisetum glaucum* (L.) R.Br.]. *Electron. J. Plant Breed.* **2021**, *12*, 895–906. [CrossRef]

54. Ullah, A.; Farooq, M. The Challenge of Drought Stress for Grain Legumes and Options for Improvement. *Arch. Agron. Soil Sci.* **2021**, *2021*, 1–18. [CrossRef]
55. Wellstein, C.; Poschlod, P.; Gohlke, A.; Chelli, S.; Campetella, G.; Rosbakh, S.; Canullo, R.; Kreyling, J.; Jentsch, A.; Beierkuhnlein, C. Effects of Extreme Drought on Specific Leaf Area of Grassland Species: A Meta-Analysis of Experimental Studies in Temperate and Sub-Mediterranean Systems. *Glob. Chang. Biol.* **2017**, *23*, 2473–2481. [CrossRef] [PubMed]
56. Seleiman, M.F.; Al-Suhaibani, N.; Ali, N.; Akmal, M.; Alotaibi, M.; Refay, Y.; Dindaroglu, T.; Abdul-Wajid, H.H.; Battaglia, M.L. Drought Stress Impacts on Plants and Different Approaches to Alleviate Its Adverse Effects. *Plants* **2021**, *10*, 259. [CrossRef] [PubMed]
57. Ahmad, Z.; Waraich, E.A.; Akhtar, S.; Anjum, S.; Ahmad, T.; Mahboob, W.; Hafeez, O.B.A.; Tapera, T.; Labuschagne, M.; Rizwan, M. Physiological Responses of Wheat to Drought Stress and Its Mitigation Approaches. *Acta Physiol. Plant.* **2018**, *40*, 80. [CrossRef]
58. Caverzan, A.; Casassola, A.; Brammer, S.P. Antioxidant Responses of Wheat Plants under Stress. *Genet. Mol. Biol.* **2016**, *39*, 1–6. [CrossRef]
59. Gupta, A.; Rico-Medina, A.; Caño-Delgado, A.I. The Physiology of Plant Responses to Drought. *Science* **2020**, *368*, 266–269. [CrossRef]
60. Chaudhry, S.; Sidhu, G.P.S. Climate Change Regulated Abiotic Stress Mechanisms in Plants: A Comprehensive Review. *Plant Cell Rep.* **2022**, *41*, 1–31. [CrossRef]
61. Kijowska-Oberc, J.; Staszak, A.M.; Kamiński, J.; Ratajczak, E. Adaptation of Forest Trees to Rapidly Changing Climate. *Forests* **2020**, *11*, 123. [CrossRef]
62. Elansary, H.O.; Abdel-Hamid, A.M.E.; Yessoufou, K.; Al-Mana, F.A.; El-Ansary, D.O.; Mahmoud, E.A.; Al-Yafrasi, M.A. Physiological and Molecular Characterization of Water-Stressed Chrysanthemum under Robinin and Chitosan Treatment. *Acta Physiol. Plant.* **2020**, *42*, 31. [CrossRef]
63. Wu, J.; Wang, J.; Hui, W.; Zhao, F.; Wang, P.; Su, C.; Gong, W. Physiology of Plant Responses to Water Stress and Related Genes: A Review. *Forests* **2022**, *13*, 324. [CrossRef]
64. Correia, M.J.; Rodrigues, M.L.; Ferreira, M.I.; Pereira, J.S. Diurnal Change in the Relationship between Stomatal Conductance and Abscisic Acid in the Xylem Sap of Field-Grown Peach Trees. *J. Exp. Bot.* **1997**, *48*, 1727–1736. [CrossRef]
65. Kim, K.H.; Kabir, E.; Jahan, S.A. Exposure to Pesticides and the Associated Human Health Effects. *Sci. Total Environ.* **2017**, *575*, 525–535. [CrossRef] [PubMed]
66. Badr, A.; Brüggemann, W. Comparative Analysis of Drought Stress Response of Maize Genotypes Using Chlorophyll Fluorescence Measurements and Leaf Relative Water Content. *Photosynthetica* **2020**, *58*, 638–645. [CrossRef]
67. Fang, Y.; Du, Y.; Wang, J.; Wu, A.; Qiao, S.; Xu, B.; Zhang, S.; Siddique, K.H.M.; Chen, Y. Moderate Drought Stress Affected Root Growth and Grain Yield in Old, Modern and Newly Released Cultivars of Winter Wheat. *Front. Plant Sci.* **2017**, *8*, 672. [CrossRef]
68. Hussain, S.; Khalid, M.F.; Saqib, M.; Ahmad, S.; Zafar, W.; Rao, M.J.; Morillon, R.; Anjum, M.A. Drought Tolerance in Citrus Rootstocks Is Associated with Better Antioxidant Defense Mechanism. *Acta Physiol. Plant.* **2018**, *40*, 135. [CrossRef]
69. Kapoor, D.; Bhardwaj, S.; Landi, M.; Sharma, A.; Ramakrishnan, M.; Sharma, A. The Impact of Drought in Plant Metabolism: How to Exploit Tolerance Mechanisms to Increase Crop Production. *Appl. Sci.* **2020**, *10*, 5692. [CrossRef]
70. Alghory, A.; Yazar, A. Evaluation of Crop Water Stress Index and Leaf Water Potential for Deficit Irrigation Management of Sprinkler-Irrigated Wheat. *Irrig. Sci.* **2019**, *37*, 61–77. [CrossRef]
71. Yan, F.; Sun, Y.; Xu, H.; Yin, Y.; Wang, H.; Wang, C.; Guo, C.; Yang, Z.; Sun, Y.; Ma, J. Effects of Wheat Straw Mulch Application and Nitrogen Management on Rice Root Growth, Dry Matter Accumulation and Rice Quality in Soils of Different Fertility. *Paddy Water Environ.* **2018**, *16*, 507–518. [CrossRef]
72. Ashrafi, M.; Azimi-Moqadam, M.R.; Mohsenifard, E.; Shekari, F.; Jafary, H.; Moradi, P.; Pucci, M.; Abate, G.; Mastinu, A. Physiological and Molecular Aspects of Two Thymus Species Differently Sensitive to Drought Stress. *BioTech* **2022**, *11*, 8. [CrossRef]
73. Duursma, R.A.; Blackman, C.J.; Lopéz, R.; Martin-StPaul, N.K.; Cochard, H.; Medlyn, B.E. On the Minimum Leaf Conductance: Its Role in Models of Plant Water Use, and Ecological and Environmental Controls. *New Phytol.* **2019**, *221*, 693–705. [CrossRef] [PubMed]
74. Misson, L.; Limousin, J.M.; Rodriguez, R.; Letts, M.G. Leaf Physiological Responses to Extreme Droughts in Mediterranean Quercus Ilex Forest. *Plant Cell Environ.* **2010**, *33*, 1898–1910. [CrossRef] [PubMed]
75. Hu, W.; Tian, S.B.; Di, Q.; Duan, S.H.; Dai, K. Effects of Exogenous Calcium on Mesophyll Cell Ultrastructure, Gas Exchange, and Photosystem II in Tobacco (*Nicotiana tabacum* Linn.) under Drought Stress. *Photosynthetica* **2018**, *56*, 1204–1211. [CrossRef]
76. Allakhverdiev, S.I. Optimising Photosynthesis for Environmental Fitness. *Funct. Plant Biol.* **2020**, *47*. [CrossRef]
77. Soares, J.C.; Santos, C.S.; Carvalho, S.M.P.; Pintado, M.M.; Vasconcelos, M.W. Preserving the Nutritional Quality of Crop Plants under a Changing Climate: Importance and Strategies. *Plant Soil* **2019**, *443*, 1–26. [CrossRef]
78. Devi, M.J.; Bhatnagar-Mathur, P.; Sharma, K.K.; Serraj, R.; Anwar, S.Y.; Vadez, V. Relationships Between Transpiration Efficiency and Its Surrogate Traits in the Rd29A:DREB1A Transgenic Lines of Groundnut. *J. Agron. Crop Sci.* **2011**, *197*, 272–283. [CrossRef]
79. Ferrara, A.; Lovelli, S.; Di Tommaso, T.; Perniola, M. Flowering, Growth and Fruit Setting in Greenhouse Bell Pepper under Water Stress. *J. Agron.* **2011**, *10*, 12–19. [CrossRef]
80. Tátrai, Z.A.; Sanoubar, R.; Pluhár, Z.; Mancarella, S.; Orsini, F.; Gianquinto, G. Morphological and Physiological Plant Responses to Drought Stress in Thymus Citriodorus. *Int. J. Agron.* **2016**, *2016*, 4165750. [CrossRef]

81. Li, S.; Liu, J.; Liu, H.; Qiu, R.; Gao, Y.; Duan, A. Corrigendum: Role of Hydraulic Signal and ABA in Decrease of Leaf Stomatal and Mesophyll Conductance in Soil Drought-Stressed Tomato (Frontiers in Plant Science, (2021), 12, (653186), 10.3389/Fpls.2021.653186). *Front. Plant Sci.* **2021**, *12*, 711. [CrossRef]
82. Ings, J.; Mur, L.A.J.; Robson, P.R.H.; Bosch, M. Physiological and Growth Responses to Water Deficit in the Bioenergy Crop Miscanthus x Giganteus. *Front. Plant Sci.* **2013**, *4*, 468. [CrossRef]
83. Pourghasemian, N.; Moradi, R.; Naghizadeh, M.; Landberg, T. Mitigating Drought Stress in Sesame by Foliar Application of Salicylic Acid, Beeswax Waste and Licorice Extract. *Agric. Water Manag.* **2020**, *231*, 105997. [CrossRef]
84. Kamran, M.; Parveen, A.; Ahmar, S.; Malik, Z.; Hussain, S.; Chattha, M.S.; Saleem, M.H.; Adil, M.; Heidari, P.; Chen, J.T. An Overview of Hazardous Impacts of Soil Salinity in Crops, Tolerance Mechanisms, and Amelioration through Selenium Supplementation. *Int. J. Mol. Sci.* **2020**, *21*, 148. [CrossRef] [PubMed]
85. Ahanger, M.A.; Alyemeni, M.N.; Wijaya, L.; Alamri, S.A.; Alam, P.; Ashraf, M.; Ahmad, P. Potential of Exogenously Sourced Kinetin in Protecting *Solanum lycopersicum* from NaCl-Induced Oxidative Stress through up-Regulation of the Antioxidant System, Ascorbate-Glutathione Cycle and Glyoxalase System. *PLoS ONE* **2018**, *13*, e0202175. [CrossRef]
86. Uarrota, V.G.; Stefen, D.L.V.; Leolato, L.S.; Gindri, D.M.; Nerling, D. Revisiting Carotenoids and Their Role in Plant Stress Responses: From Biosynthesis to Plant Signaling Mechanisms during Stress. In *Antioxidants and Antioxidant Enzymes in Higher Plants*; Springer: Berlin/Heidelberg, Germany, 2018; pp. 207–232. ISBN 9783319750880.
87. Riaz, M.; Zia-Ul-Haq, M.; Dou, D. Chemistry of Carotenoids. In *Carotenoids: Structure and Function in the Human Body*; Springer: Berlin/Heidelberg, Germany, 2021; pp. 43–76.
88. Islam, M.J.; Kim, J.W.; Begum, M.K.; Sohel, M.A.T.; Lim, Y.S. Physiological and Biochemical Changes in Sugar Beet Seedlings to Confer Stress Adaptability under Drought Condition. *Plants* **2020**, *9*, 1511. [CrossRef] [PubMed]
89. Hashmat, S.; Shahid, M.; Tanwir, K.; Abbas, S.; Ali, Q.; Niazi, N.K.; Akram, M.S.; Saleem, M.H.; Javed, M.T. Elucidating Distinct Oxidative Stress Management, Nutrient Acquisition and Yield Responses of *Pisum sativum* L. Fertigated with Diluted and Treated Wastewater. *Agric. Water Manag.* **2021**, *247*, 106720. [CrossRef]
90. Yang, Y.J.; Bi, M.H.; Nie, Z.F.; Jiang, H.; Liu, X.D.; Fang, X.W.; Brodribb, T.J. Evolution of Stomatal Closure to Optimize Water-Use Efficiency in Response to Dehydration in Ferns and Seed Plants. *New Phytol.* **2021**, *230*, 2001–2010. [CrossRef]
91. Foyer, C.H. Reactive Oxygen Species, Oxidative Signaling and the Regulation of Photosynthesis. *Environ. Exp. Bot.* **2018**, *154*, 134–142. [CrossRef]
92. Yang, X.; Lu, M.; Wang, Y.; Wang, Y.; Liu, Z.; Chen, S. Response Mechanism of Plants to Drought Stress. *Horticulturae* **2021**, *7*, 50. [CrossRef]
93. Naeem, M.; Shahzad, K.; Saqib, S.; Shahzad, A.; Nasrullah; Younas, M.; Afridi, M.I. The Solanum Melongena COP1LIKE Manipulates Fruit Ripening and Flowering Time in Tomato (*Solanum lycopersicum*). *Plant Growth Regul.* **2022**, *96*, 369–382. [CrossRef]
94. Wang, Z.; Li, G.; Sun, H.; Ma, L.; Guo, Y.; Zhao, Z.; Gao, H.; Mei, L. Effects of Drought Stress on Photosynthesis and Photosynthetic Electron Transport Chain in Young Apple Tree Leaves. *Biol. Open* **2018**, *7*, bio035279. [CrossRef]
95. Ali, Q.; Shahid, S.; Nazar, N.; Hussain, A.I.; Ali, S.; Chatha, S.A.S.; Perveen, R.; Naseem, J.; Haider, M.Z.; Hussain, B.; et al. Use of Phytohormones in Conferring Tolerance to Environmental Stress. In *Plant Ecophysiology and Adaptation under Climate Change: Mechanisms and Perspectives II: Mechanisms of Adaptation and Stress Amelioration*; Springer: Berlin/Heidelberg, Germany, 2020; pp. 245–355, ISBN 9789811521720.
96. Simova-Stoilova, L.; Pecheva, D.; Kirova, E. Drought Stress Response in Winter Wheat Varieties—Changes in Leaf Proteins and Proteolytic Activities. *Acta Bot. Croat.* **2020**, *7*, 121–130. [CrossRef]
97. Wright, I.J.; Reich, P.B.; Westoby, M. Strategy Shifts in Leaf Physiology, Structure and Nutrient Content between Species of High- and Low-Rainfall and High- and Low-Nutrient Habitats. *Funct. Ecol.* **2001**, *15*, 423–434. [CrossRef]
98. Khan, S.; Basit, A.; Hafeez, M.B.; Irshad, S.; Bashir, S.; Bashir, S.; Maqbool, M.M.; Saddiq, M.S.; Hasnain, Z.; Aljuaid, B.S.; et al. Moringa Leaf Extract Improves Biochemical Attributes, Yield and Grain Quality of Rice (*Oryza sativa* L.) under Drought Stress. *PLoS ONE* **2021**, *16*, e0254452. [CrossRef] [PubMed]
99. Djanaguiraman, M.; Boyle, D.L.; Welti, R.; Jagadish, S.V.K.; Prasad, P.V.V. Decreased Photosynthetic Rate under High Temperature in Wheat Is Due to Lipid Desaturation, Oxidation, Acylation, and Damage of Organelles. *BMC Plant Biol.* **2018**, *18*, 55. [CrossRef]
100. Ommen, O.E.; Donnelly, A.; Vanhoutvin, S.; Van Oijen, M.; Manderscheid, R. Chlorophyll Content of Spring Wheat Flag Leaves Grown under Elevated CO2 Concentrations and Other Environmental Stresses within the 'ESPACE-Wheat'Project. *Eur. J. Agron.* **1999**, *10*, 197–203. [CrossRef]
101. Mahajan, S.; Tuteja, N. Cold, Salinity and Drought Stresses: An Overview. *Arch. Biochem. Biophys.* **2005**, *444*, 139–158. [CrossRef]
102. Talaat, N.B. Role of Reactive Oxygen Species Signaling in Plant Growth and Development. *React. Oxyg. Nitrogen Sulfur Species Plants Prod. Metab. Signal. Def. Mech.* **2019**, *2019*, 225–266. [CrossRef]
103. Shahid, M.A.; Sarkhosh, A.; Khan, N.; Balal, R.M.; Ali, S.; Rossi, L.; Gómez, C.; Mattson, N.; Nasim, W.; Garcia-Sanchez, F. Insights into the Physiological and Biochemical Impacts of Salt Stress on Plant Growth and Development. *Agronomy* **2020**, *10*, 938. [CrossRef]
104. Sytar, O.; Kumari, P.; Yadav, S.; Brestic, M.; Rastogi, A. Phytohormone Priming: Regulator for Heavy Metal Stress in Plants. *J. Plant Growth Regul.* **2019**, *38*, 739–752. [CrossRef]

105. Abdelaal, K.A.A.; Attia, K.A.; Alamery, S.F.; El-Afry, M.M.; Ghazy, A.I.; Tantawy, D.S.; Al-Doss, A.A.; El-Shawy, E.S.E.; Abu-Elsaoud, A.M.; Hafez, Y.M. Exogenous Application of Proline and Salicylic Acid Can Mitigate the Injurious Impacts of Drought Stress on Barley Plants Associated with Physiological and Histological Characters. *Sustainability* **2020**, *12*, 1736. [CrossRef]
106. Raza, A.; Mehmood, S.S.; Tabassum, J.; Batool, R. Targeting Plant Hormones to Develop Abiotic Stress Resistance in Wheat. In *Wheat Production in Changing Environments*; Springer: Berlin/Heidelberg, Germany, 2019; pp. 557–577.
107. Mahmood, T.; Rana, R.M.; Ahmar, S.; Saeed, S.; Gulzar, A.; Khan, M.A.; Wattoo, F.M.; Wang, X.; Branca, F.; Mora-Poblete, F.; et al. Effect of Drought Stress on Capsaicin and Antioxidant Contents in Pepper Genotypes at Reproductive Stage. *Plants* **2021**, *10*, 1286. [CrossRef] [PubMed]
108. Wu, P.; Xiao, C.; Cui, J.; Hao, B.; Zhang, W.; Yang, Z.; Ahammed, G.J.; Liu, H.; Cui, H. Nitric Oxide and Its Interaction with Hydrogen Peroxide Enhance Plant Tolerance to Low Temperatures by Improving the Efficiency of the Calvin Cycle and the Ascorbate–Glutathione Cycle in Cucumber Seedlings. *J. Plant Growth Regul.* **2021**, *40*, 2390–2408. [CrossRef]
109. Abhinandan, K.; Skori, L.; Stanic, M.; Hickerson, N.M.N.; Jamshed, M.; Samuel, M.A. Abiotic Stress Signaling in Wheat—An Inclusive Overview of Hormonal Interactions during Abiotic Stress Responses in Wheat. *Front. Plant Sci.* **2018**, *9*, 734. [CrossRef] [PubMed]
110. Iqbal, M.J. Role of Osmolytes and Antioxidant Enzymes for Drought Tolerance in Wheat. *Glob. Wheat Prod.* **2018**, *51*. [CrossRef]
111. Rauf, M.; Munir, M.; Ul Hassan, M.; Ahmad, M.; Afzal, M. Performance of Wheat Genotypes under Osmotic Stress at Germination and Early Seedling Growth Stage. *African J. Biotechnol.* **2007**, *6*, 971–975. [CrossRef]
112. Ghatak, A.; Chaturvedi, P.; Weckwerth, W. Cereal Crop Proteomics: Systemic Analysis of Crop Drought Stress Responses towards Marker-Assisted Selection Breeding. *Front. Plant Sci.* **2017**, *8*, 757. [CrossRef]
113. Kaur, H.; Chowrasia, S.; Gaur, V.S.; Mondal, T.K. Allantoin: Emerging Role in Plant Abiotic Stress Tolerance. *Plant Mol. Biol. Rep.* **2016**, *39*, 648–661. [CrossRef]
114. Dhanda, S.S.; Sethi, G.S.; Behl, R.K. Indices of Drought Tolerance in Wheat Genotypes at Early Stages of Plant Growth. *J. Agron. Crop. Sci.* **2004**, *190*, 6–12. [CrossRef]
115. Talbi, S.; Rojas, J.A.; Sahrawy, M.; Rodríguez-Serrano, M.; Cárdenas, K.E.; Debouba, M.; Sandalio, L.M. Effect of Drought on Growth, Photosynthesis and Total Antioxidant Capacity of the Saharan Plant Oudeneya Africana. *Environ. Exp. Bot.* **2020**, *176*, 104099. [CrossRef]
116. Miller, G.; Suzuki, N.; Ciftci-Yilmaz, S.; Mittler, R. Reactive Oxygen Species Homeostasis and Signalling during Drought and Salinity Stresses. *Plant Cell Environ.* **2010**, *33*, 453–467. [CrossRef]
117. Hasanuzzaman, M.; Parvin, K.; Bardhan, K.; Nahar, K.; Anee, T.I.; Masud, A.A.C.; Fotopoulos, V. Biostimulants for the Regulation of Reactive Oxygen Species Metabolism in Plants under Abiotic Stress. *Cells* **2021**, *10*, 2537. [CrossRef] [PubMed]
118. Franchina, D.G.; Dostert, C.; Brenner, D. Reactive Oxygen Species: Involvement in T Cell Signaling and Metabolism. *Trends Immunol.* **2018**, *39*, 489–502. [CrossRef] [PubMed]
119. Verma, G.; Srivastava, D.; Tiwari, P.; Chakrabarty, D. ROS Modulation in Crop Plants under Drought Stress. *React. Oxyg. Nitrogen Sulfur Species Plants Prod. Metab. Signal. Def. Mech.* **2019**, *2019*, 311–336.
120. Madzikane-Mlungwana, O.; Moyo, M.; Aremu, A.O.; Plíhalová, L.; Doležal, K.; Van Staden, J.; Finnie, J.F. Differential Responses to Isoprenoid, N 6-Substituted Aromatic Cytokinins and Indole-3-Butyric Acid in Direct Plant Regeneration of Eriocephalus Africanus. *Plant Growth Regul.* **2017**, *82*, 103–110. [CrossRef]
121. Blum, A. Plant Water Relations, Plant Stress and Plant Production. In *Plant Breeding for Water-Limited Environments*; Springer: Berlin/Heidelberg, Germany, 2011; pp. 11–52.
122. Hura, T.; Grzesiak, S.; Hura, K.; Thiemt, E.; Tokarz, K.; Wędzony, M. Physiological and Biochemical Tools Useful in Drought-Tolerance Detection in Genotypes of Winter Triticale: Accumulation of Ferulic Acid Correlates with Drought Tolerance. *Ann. Bot.* **2007**, *100*, 767–775. [CrossRef]
123. Sahay, S.; Khan, E.; Gupta, M. Nitric Oxide and Abscisic Acid Protects against PEG-Induced Drought Stress Differentially in Brassica Genotypes by Combining the Role of Stress Modulators, Markers and Antioxidants. *Nitric Oxide—Biol. Chem.* **2019**, *89*, 81–92. [CrossRef]
124. El-Beltagi, H.S.; Mohamed, H.I.; Sofy, M.R. Role of Ascorbic Acid, Glutathione and Proline Applied as Singly or in Sequence Combination in Improving Chickpea Plant through Physiological Change and Antioxidant Defense under Different Levels of Irrigation Intervals. *Molecules* **2020**, *25*, 1702. [CrossRef]
125. Ali, M.Y.; Sina, A.A.I.; Khandker, S.S.; Neesa, L.; Tanvir, E.M.; Kabir, A.; Khalil, M.I.; Gan, S.H. Nutritional Composition and Bioactive Compounds in Tomatoes and Their Impact on Human Health and Disease: A Review. *Foods* **2021**, *10*, 45. [CrossRef]
126. Bhardwaj, S.; Sharma, D.; Kapoor, D. Salicylic Acid Signaling and ROS Balance in Plants. *Salicylic Acid Contrib. Plant Biol. Chang. Environ.* **2021**, *2021*, 87–114.
127. Lin, K.H.; Chao, P.Y.; Yang, C.M.; Cheng, W.C.; Lo, H.F.; Chang, T.R. The Effects of Flooding and Drought Stresses on the Antioxidant Constituents in Sweet Potato Leaves. *Bot. Stud.* **2006**, *47*, 417–426.
128. Ahanger, M.A.; Qi, M.; Huang, Z.; Xu, X.; Begum, N.; Qin, C.; Zhang, C.; Ahmad, N.; Mustafa, N.S.; Ashraf, M.; et al. Improving Growth and Photosynthetic Performance of Drought Stressed Tomato by Application of Nano-Organic Fertilizer Involves up-Regulation of Nitrogen, Antioxidant and Osmolyte Metabolism. *Ecotoxicol. Environ. Saf.* **2021**, *216*, 112195. [CrossRef] [PubMed]

129. Guddimalli, R.; Somanaboina, A.K.; Palle, S.R.; Edupuganti, S.; Kummari, D.; Palakolanu, S.R.; Naravula, J.; Gandra, J.; Qureshi, I.A.; Marka, N.; et al. Overexpression of RNA-Binding Bacterial Chaperones in Rice Leads to Stay-Green Phenotype, Improved Yield and Tolerance to Salt and Drought Stresses. *Physiol. Plant.* **2021**, *173*, 1351–1368. [CrossRef] [PubMed]
130. Doneva, D.; Pál, M.; Brankova, L.; Szalai, G.; Tajti, J.; Khalil, R.; Ivanovska, B.; Velikova, V.; Misheva, S.; Janda, T.; et al. The Effects of Putrescine Pre-Treatment on Osmotic Stress Responses in Drought-Tolerant and Drought-Sensitive Wheat Seedlings. *Physiol. Plant.* **2021**, *171*, 200–216. [CrossRef]
131. Al-Ghzawi, A.L.A.; Khalaf, Y.B.; Al-Ajlouni, Z.I.; Al-Quraan, N.A.; Musallam, I.; Hani, N.B. The Effect of Supplemental Irrigation on Canopy Temperature Depression, Chlorophyll Content, and Water Use Efficiency in Three Wheat (*Triticum aestivum* L. and *T. durum* Desf.) Varieties Grown in Dry Regions of Jordan. *Agriculture* **2018**, *8*, 67. [CrossRef]
132. Khan, S.; Anwar, S.; Yu, S.; Sun, M.; Yang, Z.; Gao, Z.Q. Development of Drought-Tolerant Transgenic Wheat: Achievements and Limitations. *Int. J. Mol. Sci.* **2019**, *20*, 3350. [CrossRef]
133. Hassanein, R.A.; Amin, A.B.A.E.S.; Rashad, E.S.M.; Ali, H. Effect of Thiourea and Salicylic Acid on Antioxidant Defense of Wheat Plants under Drought Stress. *Int. J. Chem. Tech. Res.* **2015**, *7*, 346–354.
134. Li, J.; Luan, Y.; Liu, Z. SpWRKY1 Mediates Resistance to Phytophthora Infestans and Tolerance to Salt and Drought Stress by Modulating Reactive Oxygen Species Homeostasis and Expression of Defense-Related Genes in Tomato. *Plant Cell. Tissue Organ. Cult.* **2015**, *123*, 67–81. [CrossRef]
135. Rajput, V.D.; Harish; Singh, R.K.; Verma, K.K.; Sharma, L.; Quiroz-Figueroa, F.R.; Meena, M.; Gour, V.S.; Minkina, T.; Sushkova, S.; et al. Recent Developments in Enzymatic Antioxidant Defence Mechanism in Plants with Special Reference to Abiotic Stress. *Biology (Basel)* **2021**, *10*, 267. [CrossRef]
136. Gill, S.S.; Tuteja, N. Reactive Oxygen Species and Antioxidant Machinery in Abiotic Stress Tolerance in Crop Plants. *Plant Physiol. Biochem.* **2010**, *48*, 909–930. [CrossRef]
137. Xiong, H.; Hua, L.; Reyna-Llorens, I.; Shi, Y.; Chen, K.M.; Smirnoff, N.; Kromdijk, J.; Hibberd, J.M. Photosynthesis-Independent Production of Reactive Oxygen Species in the Rice Bundle Sheath during High Light Is Mediated by NADPH Oxidase. *Proc. Natl. Acad. Sci. USA* **2021**, *118*, e2022702118. [CrossRef]
138. Siddiqui, M.N.; León, J.; Naz, A.A.; Ballvora, A. Genetics and Genomics of Root System Variation in Adaptation to Drought Stress in Cereal Crops. *J. Exp. Bot.* **2021**, *72*, 1007–1019. [CrossRef] [PubMed]
139. Hsu, P.K.; Dubeaux, G.; Takahashi, Y.; Schroeder, J.I. Signaling Mechanisms in Abscisic Acid-Mediated Stomatal Closure. *Plant J.* **2021**, *105*, 307–321. [CrossRef] [PubMed]
140. Golldack, D.; Lüking, I.; Yang, O. Plant Tolerance to Drought and Salinity: Stress Regulating Transcription Factors and Their Functional Significance in the Cellular Transcriptional Network. *Plant Cell Rep.* **2011**, *30*, 1383–1391. [CrossRef] [PubMed]
141. Varoquaux, N.; Cole, B.; Gao, C.; Pierroz, G.; Baker, C.R.; Patel, D.; Madera, M.; Jeffers, T.; Hollingsworth, J.; Sievert, J.; et al. Transcriptomic Analysis of Field-Droughted Sorghum from Seedling to Maturity Reveals Biotic and Metabolic Responses. *Proc. Natl. Acad. Sci. USA* **2019**, *116*, 27124–27132. [CrossRef] [PubMed]
142. Hou, Z.; Yin, J.; Lu, Y.; Song, J.; Wang, S.; Wei, S.; Liu, Z.; Zhang, Y.; Fang, Z. Transcriptomic Analysis Reveals the Temporal and Spatial Changes in Physiological Process and Gene Expression in Common Buckwheat (Fagopyrum Esculentum Moench) Grown under Drought Stress. *Agronomy* **2019**, *9*, 569. [CrossRef]
143. Brozynska, M.; Furtado, A.; Henry, R.J. Genomics of Crop Wild Relatives: Expanding the Gene Pool for Crop Improvement. *Plant Biotechnol. J.* **2016**, *14*, 1070–1085. [CrossRef]
144. Zhu, M.; Monroe, J.G.; Suhail, Y.; Villiers, F.; Mullen, J.; Pater, D.; Hauser, F.; Jeon, B.W.; Bader, J.S.; Kwak, J.M.; et al. Molecular and Systems Approaches towards Drought-Tolerant Canola Crops. *New Phytol.* **2016**, *210*, 1169–1189. [CrossRef]
145. Nguyen, K.L.; Grondin, A.; Courtois, B.; Gantet, P. Next-Generation Sequencing Accelerates Crop Gene Discovery. *Trends Plant Sci.* **2019**, *24*, 263–274. [CrossRef]
146. Martineau, C.; Li, X.; Lalancette, C.; Perreault, T.; Fournier, E.; Tremblay, J.; Gonzales, M.; Yergeau, É.; Quach, C. Serratia Marcescens Outbreak in a Neonatal Intensive Care Unit: New Insights from next-Generation Sequencing Applications. *J. Clin. Microbiol.* **2018**, *56*, e00235-18. [CrossRef]
147. Zenda, T.; Liu, S.; Duan, H. Adapting Cereal Grain Crops to Drought Stress: 2020 and Beyond. *Abiotic Stress Plants* **2020**, *2020*, 1–30.
148. Kumar, A.; Saripalli, G.; Jan, I.; Kumar, K.; Sharma, P.K.; Balyan, H.S.; Gupta, P.K. Meta-QTL Analysis and Identification of Candidate Genes for Drought Tolerance in Bread Wheat (*Triticum aestivum* L.). *Physiol. Mol. Biol. Plants* **2020**, *26*, 1713–1725. [CrossRef] [PubMed]
149. Chen, K.; Wang, Y.; Zhang, R.; Zhang, H.; Gao, C. CRISPR/Cas Genome Editing and Precision Plant Breeding in Agriculture. *Annu. Rev. Plant Biol* **2019**, *70*, 667–697. [CrossRef] [PubMed]
150. Mahmood, T.; Khalid, S.; Abdullah, M.; Ahmed, Z.; Shah, M.K.N.; Ghafoor, A.; Du, X. Insights into Drought Stress Signaling in Plants and the Molecular Genetic Basis of Cotton Drought Tolerance. *Cells* **2020**, *9*, 105. [CrossRef] [PubMed]
151. Alagoz, Y.; Gurkok, T.; Zhang, B.; Unver, T. Manipulating the Biosynthesis of Bioactive Compound Alkaloids for Next-Generation Metabolic Engineering in Opium Poppy Using CRISPR-Cas 9 Genome Editing Technology. *Sci. Rep.* **2016**, *6*, 30910. [CrossRef]
152. Martignago, D.; Rico-Medina, A.; Blasco-Escámez, D.; Fontanet-Manzaneque, J.B.; Caño-Delgado, A.I. Drought Resistance by Engineering Plant Tissue-Specific Responses. *Front. Plant Sci.* **2020**, *10*, 1676. [CrossRef] [PubMed]

153. Muthusamy, M.; Kim, J.-H.; Kim, J.A.; Lee, S.-I. Plant RNA Binding Proteins as Critical Modulators in Drought, High Salinity, Heat, and Cold Stress Responses: An Updated Overview. *Int. J. Mol. Sci.* **2021**, *22*, 6731. [CrossRef]
154. Caddell, D.F.; Deng, S.; Coleman-Derr, D. Role of the Plant Root Microbiome in Abiotic Stress Tolerance. In *Seed Endophytes*; Springer: Berlin/Heidelberg, Germany, 2019; pp. 273–311.
155. Sattiraju, K.S.; Kotiyal, S.; Arora, A.; Maheshwari, M. Plant Growth-Promoting Microbes: Contribution to Stress Management in Plant Hosts. *Environ. Biotechnol. Sustain. Futur.* **2019**, *2019*, 199–236. [CrossRef]
156. Ullah, A.; Manghwar, H.; Shaban, M.; Khan, A.H.; Akbar, A.; Ali, U.; Ali, E.; Fahad, S. Phytohormones Enhanced Drought Tolerance in Plants: A Coping Strategy. *Environ. Sci. Pollut. Res.* **2018**, *25*, 33103–33118. [CrossRef]
157. Chumikina, L.V.; Arabova, L.I.; Kolpakova, V.V.; Topunov, A.F. The Role of Phytohormones in the Regulation of the Tolerance of Wheat, Rye, and Triticale Seeds to the Effect of Elevated Temperatures during Germination. *Appl. Biochem. Microbiol.* **2019**, *55*, 59–66. [CrossRef]
158. Jogawat, A.; Yadav, B.; Lakra, N.; Singh, A.K.; Narayan, O.P. Crosstalk between Phytohormones and Secondary Metabolites in the Drought Stress Tolerance of Crop Plants: A Review. *Physiol. Plant.* **2021**, *172*, 1106–1132. [CrossRef]
159. Jiang, K.; Asami, T. Chemical Regulators of Plant Hormones and Their Applications in Basic Research and Agriculture. *Biosci. Biotechnol. Biochem.* **2018**, *82*, 1265–1300. [CrossRef] [PubMed]
160. Yu, Z.; Duan, X.; Luo, L.; Dai, S.; Ding, Z.; Xia, G. How Plant Hormones Mediate Salt Stress Responses. *Trends Plant Sci.* **2020**, *25*, 1117–1130. [CrossRef] [PubMed]
161. Singh, P.; Dutta, P.; Chakrabarty, D. MiRNAs Play Critical Roles in Response to Abiotic Stress by Modulating Cross-Talk of Phytohormone Signaling. *Plant Cell Rep.* **2021**, *40*, 1617–1630. [CrossRef] [PubMed]
162. Raza, A.; Charagh, S.; Zahid, Z.; Mubarik, M.S.; Javed, R.; Siddiqui, M.H.; Hasanuzzaman, M. Jasmonic Acid: A Key Frontier in Conferring Abiotic Stress Tolerance in Plants. *Plant Cell Rep.* **2021**, *40*, 1513–1541. [CrossRef]
163. Yadav, A.N.; Yadav, N. Stress-Adaptive Microbes for Plant Growth Promotion and Alleviation of Drought Stress in Plants. *Acta Sci. Agric.* **2018**, *2*, 85–88.
164. Campos-Rivero, G.; Osorio-Montalvo, P.; Sánchez-Borges, R.; Us-Camas, R.; Duarte-Aké, F.; De-la-Peña, C. Plant Hormone Signaling in Flowering: An Epigenetic Point of View. *J. Plant Physiol.* **2017**, *214*, 16–27. [CrossRef]
165. Sheng, J.; Li, X.; Zhang, D. Gibberellins, Brassinolide, and Ethylene Signaling Were Involved in Flower Differentiation and Development in Nelumbo Nucifera. *Hortic. Plant J.* **2022**, *8*, 243–250. [CrossRef]
166. Waterman, P.G.; Mole, S. Extrinsic Factors Influencing Production of Secondary Metabolites in Plants. In *Insect-Plant Interactions*; CRC Press: Boca Raton, FL, USA, 2019; pp. 107–134. ISBN 0429290918.
167. Mir, R.A.; Bhat, B.A.; Yousuf, H.; Islam, S.T.; Raza, A.; Rizvi, M.A.; Charagh, S.; Albaqami, M.; Sofi, P.A.; Zargar, S.M. Multidimensional Role of Silicon to Activate Resilient Plant Growth and to Mitigate Abiotic Stress. *Front. Plant Sci.* **2022**, *13*, 819658. [CrossRef]
168. Iqbal, N.; Fatma, M.; Gautam, H.; Sehar, Z.; Rasheed, F.; Khan, M.I.R.; Sofo, A.; Khan, N.A. Salicylic Acid Increases Photosynthesis of Drought Grown Mustard Plants Effectively with Sufficient-N via Regulation of Ethylene, Abscisic Acid, and Nitrogen-Use Efficiency. *J. Plant Growth Regul.* **2022**, *2022*, 1–12. [CrossRef]
169. Yadav, B.; Jogawat, A.; Gnanasekaran, P.; Kumari, P.; Lakra, N.; Lal, S.K.; Pawar, J.; Narayan, O.P. An Overview of Recent Advancement in Phytohormones-Mediated Stress Management and Drought Tolerance in Crop Plants. *Plant Gene* **2021**, *25*, 100264.
170. Signorelli, S.; Tarkowski, Ł.P.; Van den Ende, W.; Bassham, D.C. Linking Autophagy to Abiotic and Biotic Stress Responses. *Trends Plant Sci.* **2019**, *24*, 413–430. [CrossRef] [PubMed]
171. Llanes, A.; Andrade, A.; Alemano, S.; Luna, V. Alterations of Endogenous Hormonal Levels in Plants under Drought and Salinity. *Am. J. Plant Sci.* **2016**, *7*, 1357–1371. [CrossRef]
172. Álvarez, S.; Gómez-Bellot, M.J.; Acosta-Motos, J.R.; Sánchez-Blanco, M.J. Application of Deficit Irrigation in Phillyrea Angustifolia for Landscaping Purposes. *Agric. Water Manag.* **2019**, *218*, 193–202. [CrossRef]
173. Shen, Q.; Liu, Y.; Naqvi, N.I. Fungal Effectors at the Crossroads of Phytohormone Signaling. *Curr. Opin. Microbiol.* **2018**, *46*, 1–6. [CrossRef]
174. La, V.H.; Lee, B.R.; Zhang, Q.; Park, S.H.; Islam, M.T.; Kim, T.H. Salicylic Acid Improves Drought-Stress Tolerance by Regulating the Redox Status and Proline Metabolism in Brassica Rapa. *Hortic. Environ. Biotechnol.* **2019**, *60*, 31–40. [CrossRef]
175. Castro, P.H.; Couto, D.; Freitas, S.; Verde, N.; Macho, A.P.; Huguet, S.; Botella, M.A.; Ruiz-Albert, J.; Tavares, R.M.; Bejarano, E.R.; et al. SUMO Proteases ULP1c and ULP1d Are Required for Development and Osmotic Stress Responses in Arabidopsis Thaliana. *Plant Mol. Biol.* **2016**, *92*, 143–159. [CrossRef]
176. Tiwari, R.K.; Lal, M.K.; Kumar, R.; Chourasia, K.N.; Naga, K.C.; Kumar, D.; Das, S.K.; Zinta, G. Mechanistic Insights on Melatonin-mediated Drought Stress Mitigation in Plants. *Physiol. Plant.* **2021**, *172*, 1212–1226. [CrossRef]
177. Khalvandi, M.; Siosemardeh, A.; Roohi, E.; Keramati, S. Salicylic Acid Alleviated the Effect of Drought Stress on Photosynthetic Characteristics and Leaf Protein Pattern in Winter Wheat. *Heliyon* **2021**, *7*, e05908. [CrossRef]
178. Garg, N.; Bharti, A. Salicylic Acid Improves Arbuscular Mycorrhizal Symbiosis, and Chickpea Growth and Yield by Modulating Carbohydrate Metabolism under Salt Stress. *Mycorrhiza* **2018**, *28*, 727–746. [CrossRef]
179. Bandurska, H. Salicylic Acid: An Update on Biosynthesis and Action in Plant Response to Water Deficit and Performance Under Drought. In *Salicylic Acid*; Springer: Berlin/Heidelberg, Germany, 2013; pp. 1–14.

180. Jahan, M.S.; Wang, Y.; Shu, S.; Zhong, M.; Chen, Z.; Wu, J.; Sun, J.; Guo, S. Exogenous Salicylic Acid Increases the Heat Tolerance in Tomato (*Solanum lycopersicum* L) by Enhancing Photosynthesis Efficiency and Improving Antioxidant Defense System through Scavenging of Reactive Oxygen Species. *Sci. Hortic. (Amsterdam)* **2019**, *247*, 421–429. [CrossRef]
181. Otálora, G.; Piñero, M.C.; Collado-González, J.; López-Marín, J.; Del Amor, F.M. Exogenous Salicylic Acid Modulates the Response to Combined Salinity-Temperature Stress in Pepper Plants (*Capsicum annuum* L. Var. Tamarin). *Plants* **2020**, *9*, 1790. [CrossRef] [PubMed]
182. Torun, H. Time-Course Analysis of Salicylic Acid Effects on ROS Regulation and Antioxidant Defense in Roots of Hulled and Hulless Barley under Combined Stress of Drought, Heat and Salinity. *Physiol. Plant.* **2019**, *165*, 169–182. [CrossRef] [PubMed]
183. Salem, K.F.M.; Saleh, M.M.; Abu-Ellail, F.F.B.; Aldahak, L.; Alkuddsi, Y.A. The Role of Salicylic Acid in Crops to Tolerate Abiotic Stresses. In *Salicylic Acid-A Versatile Plant Growth Regulator*; Springer: Berlin/Heidelberg, Germany, 2021; pp. 93–152.
184. Maghsoudi, K.; Emam, Y.; Ashraf, M.; Arvin, M.J. Alleviation of Field Water Stress in Wheat Cultivars by Using Silicon and Salicylic Acid Applied Separately or in Combination. *Crop Pasture Sci.* **2019**, *70*, 36–43. [CrossRef]
185. Munsif, F.; Shah, T.; Arif, M.; Jehangir, M.; Afridi, M.Z.; Ahmad, I.; Jan, B.L.; Alansi, S. Combined Effect of Salicylic Acid and Potassium Mitigates Drought Stress through the Modulation of Physio-Biochemical Attributes and Key Antioxidants in Wheat. *Saudi J. Biol. Sci.* **2022**, *29*, 103294. [CrossRef] [PubMed]
186. Saheri, F.; Barzin, G.; Pishkar, L.; Boojar, M.M.A.; Babaeekhou, L. Correction to: Foliar Spray of Salicylic Acid Induces Physiological and Biochemical Changes in Purslane (*Portulaca oleracea* L.) under Drought Stress (Biologia, (2020), 10.2478/S11756-020-00571-2). *Biologia (Bratisl)* **2020**, *75*, 2201. [CrossRef]
187. Seif El-Yazal, S.; Seif El-Yazal, M.; Dwidar, E.; Rady, M. Phytohormone Crosstalk Research: Cytokinin and Its Crosstalk with Other Phytohormones. *Curr. Protein Pept. Sci.* **2015**, *16*, 395–405. [CrossRef]
188. Vrabka, J.; Niehaus, E.-M.; Münsterkötter, M.; Proctor, R.H.; Brown, D.W.; Novák, O.; Pěnčik, A.; Tarkowská, D.; Hromadová, K.; Hradilová, M. Production and Role of Hormones during Interaction of Fusarium Species with Maize (*Zea mays* L.) Seedlings. *Front. Plant Sci.* **2019**, *9*, 1936. [CrossRef]
189. Tarkowská, D.; Strnad, M. Isoprenoid-Derived Plant Signaling Molecules: Biosynthesis and Biological Importance. *Planta* **2018**, *247*, 1051–1066. [CrossRef]
190. Yonekura-Sakakibara, K.; Kojima, M.; Yamaya, T.; Sakakibara, H. Molecular Characterization of Cytokinin-Responsive Histidine Kinases in Maize. Differential Ligand Preferences and Response to Cis-Zeatin. *Plant Physiol.* **2004**, *134*, 1654–1661. [CrossRef]
191. Bidon, B.; Kabbara, S.; Courdavault, V.; Glévarec, G.; Oudin, A.; Héricourt, F.; Carpin, S.; Spíchal, L.; Binder, B.M.; Cock, J.M. Cytokinin and Ethylene Cell Signaling Pathways from Prokaryotes to Eukaryotes. *Cells* **2020**, *9*, 2526. [CrossRef]
192. Borghi, L.; Kang, J.; de Brito Francisco, R. Filling the Gap: Functional Clustering of ABC Proteins for the Investigation of Hormonal Transport in Planta. *Front. Plant Sci.* **2019**, *10*, 422. [CrossRef] [PubMed]
193. Müller, M.; Munné-Bosch, S. Hormonal Impact on Photosynthesis and Photoprotection in Plants. *Plant Physiol.* **2021**, *185*, 1500–1522. [CrossRef] [PubMed]
194. Verma, S.; Negi, N.P.; Pareek, S.; Mudgal, G.; Kumar, D. Auxin Response Factors in Plant Adaptation to Drought and Salinity Stress. *Physiol. Plant.* **2022**, *2022*, e13714. [CrossRef] [PubMed]
195. Khosravi-nejad, F.; Khavari-nejad, R.A.; Moradi, F.; Najafi, F. Cytokinin and Abscisic Acid Alleviate Drought Stress through Changing Organic Acids Profile, Ion Immolation, and Fatty Acid Profile to Improve Yield of Wheat (*Triticum aestivum* L.) Cultivars. *Physiol. Mol. Biol. Plants* **2022**, *28*, 1119–1129. [CrossRef]
196. Farhangi-Abriz, S.; Torabian, S. Biochar Increased Plant Growth-Promoting Hormones and Helped to Alleviates Salt Stress in Common Bean Seedlings. *J. Plant Growth Regul.* **2018**, *37*, 591–601. [CrossRef]
197. Deng, Y.; Zhou, Q.; Wu, Y.; Chen, X.; Zhong, F. Properties and Mechanisms of Flavin-Dependent Monooxygenases and Their Applications in Natural Product Synthesis. *Int. J. Mol. Sci.* **2022**, *23*, 2622. [CrossRef]
198. Ronzan, M.; Piacentini, D.; Fattorini, L.; Caboni, E.; Eiche, E.; Ziegler, J.; Hause, B.; Riemann, M.; Betti, C.; Altamura, M.M. Auxin-Jasmonate Crosstalk in Oryza Sativa L. Root System Formation after Cadmium and/or Arsenic Exposure. *Environ. Exp. Bot.* **2019**, *165*, 59–69. [CrossRef]
199. Kalra, G.; Bhatla, S.C. Gibberellins. In *Plant Physiology, Development and Metabolism*; Springer: Berlin/Heidelberg, Germany, 2018; pp. 617–628.
200. Sedaghat, M.; Emam, Y.; Mokhtassi-Bidgoli, A.; Hazrati, S.; Lovisolo, C.; Visentin, I.; Cardinale, F.; Tahmasebi-Sarvestani, Z. The Potential of the Synthetic Strigolactone Analogue GR24 for the Maintenance of Photosynthesis and Yield in Winter Wheat under Drought: Investigations on the Mechanisms of Action and Delivery Modes. *Plants* **2021**, *10*, 1223. [CrossRef]
201. Rasheed, A.; Hassan, M.U.; Aamer, M.; Batool, M.; Sheng, F.; Ziming, W.U.; Huijie, L.I. A Critical Review on the Improvement of Drought Stress Tolerance in Rice (*Oryza sativa* L.). *Not. Bot. Horti Agrobot. Cluj-Napoca* **2020**, *48*, 1756–1788. [CrossRef]
202. Salvi, P.; Manna, M.; Kaur, H.; Thakur, T.; Gandass, N.; Bhatt, D.; Muthamilarasan, M. Phytohormone Signaling and Crosstalk in Regulating Drought Stress Response in Plants. *Plant Cell Rep.* **2021**, *40*, 1305–1329. [CrossRef]
203. Choudhary, P.; Pramitha, L.; Rana, S.; Verma, S.; Aggarwal, P.R.; Muthamilarasan, M. Hormonal Crosstalk in Regulating Salinity Stress Tolerance in Graminaceous Crops. *Physiol. Plant.* **2021**, *173*, 1587–1596. [CrossRef] [PubMed]
204. Emamverdian, A.; Ding, Y.; Mokhberdoran, F. The Role of Salicylic Acid and Gibberellin Signaling in Plant Responses to Abiotic Stress with an Emphasis on Heavy Metals. *Plant Signal. Behav.* **2020**, *15*, 1777372. [CrossRef] [PubMed]

205. Hedden, P. The Current Status of Research on Gibberellin Biosynthesis. *Plant Cell Physiol.* **2020**, *61*, 1832–1849. [CrossRef] [PubMed]
206. Vishal, B.; Kumar, P.P. Regulation of Seed Germination and Abiotic Stresses by Gibberellins and Abscisic Acid. *Front. Plant Sci.* **2018**, *9*, 838. [CrossRef] [PubMed]
207. Chen, W.; Cheng, Z.; Liu, L.; Wang, M.; You, X.; Wang, J.; Zhang, F.; Zhou, C.; Zhang, Z.; Zhang, H. Small Grain and Dwarf 2, Encoding an HD-Zip II Family Transcription Factor, Regulates Plant Development by Modulating Gibberellin Biosynthesis in Rice. *Plant Sci.* **2019**, *288*, 110208. [CrossRef]
208. Goldschmidt, E.E.; Sadka, A. Yield Alternation: Horticulture, Physiology, Molecular Biology, and Evolution. *Hortic. Rev. (Am. Soc. Hortic. Sci).* **2021**, *48*, 363–418.
209. Verbancic, J. Carbon Supply and the Regulation of Primary Cell Wall Synthesis in Arabidopsis Thaliana 2021. Doctoral Dissertation, Universität Potsdam, Potsdam, Germany, 2021.
210. Omena-Garcia, R.P.; Martins, A.O.; Medeiros, D.B.; Vallarino, J.G.; Ribeiro, D.M.; Fernie, A.R.; Araújo, W.L.; Nunes-Nesi, A. Growth and Metabolic Adjustments in Response to Gibberellin Deficiency in Drought Stressed Tomato Plants. *Environ. Exp. Bot.* **2019**, *159*, 95–107. [CrossRef]
211. Salazar-Cerezo, S.; Martínez-Montiel, N.; García-Sánchez, J.; Pérez-y-Terrón, R.; Martínez-Contreras, R.D. Gibberellin Biosynthesis and Metabolism: A Convergent Route for Plants, Fungi and Bacteria. *Microbiol. Res.* **2018**, *208*, 85–98. [CrossRef]
212. Kumar, B. Plant Bio-Regulators for Enhancing Grain Yield and Quality of Legumes: A Review. *Agric. Rev.* **2021**, *42*, 175–182. [CrossRef]
213. Draweel, M.M.; Soegianto, A.; Soetopo, L.; Kuswanto, K. Evaluation of Some Morphological Criteria to Drought Tolerance on Seedling of Bambara Groundnut [*Vigna subterranea* (L.) Verdc.] Using Polyethylene Glycol (Peg6000). *Legum. Res. Int. J.* **2021**, *1*, 10. [CrossRef]
214. Rademacher, W. Chemical Regulators of Gibberellin Status and Their Application in Plant Production. *Annu. Plant Rev. Online* **2018**, *2018*, 359–403.
215. Shohat, H.; Cheriker, H.; Kilambi, H.V.; Illouz Eliaz, N.; Blum, S.; Amsellem, Z.; Tarkowská, D.; Aharoni, A.; Eshed, Y.; Weiss, D. Inhibition of Gibberellin Accumulation by Water Deficiency Promotes Fast and Long-term 'Drought Avoidance'Responses in Tomato. *New Phytol.* **2021**, *232*, 1985–1998. [CrossRef] [PubMed]
216. Litvin, A.G.; van Iersel, M.W.; Malladi, A. Drought Stress Reduces Stem Elongation and Alters Gibberellin-Related Gene Expression during Vegetative Growth of Tomato. *J. Am. Soc. Hortic. Sci.* **2016**, *141*, 591–597. [CrossRef]
217. Binenbaum, J.; Weinstain, R.; Shani, E. Gibberellin Localization and Transport in Plants. *Trends Plant Sci.* **2018**, *23*, 410–421. [CrossRef] [PubMed]
218. Ali, S.; Hayat, K.; Iqbal, A.; Xie, L. Implications of Abscisic Acid in the Drought Stress Tolerance of Plants. *Agronomy* **2020**, *10*, 1323. [CrossRef]
219. Waterland, N.L.; Campbell, C.A.; Finer, J.J.; Jones, M.L. Abscisic Acid Application Enhances Drought Stress Tolerance in Bedding Plants. *HortScience* **2010**, *45*, 409–413. [CrossRef]
220. Ramachandran, P.; Wang, G.; Augstein, F.; de Vries, J.; Carlsbecker, A. Continuous Root Xylem Formation and Vascular Acclimation to Water Deficit Involves Endodermal ABA Signalling via MiR165. *Development* **2018**, *145*, dev159202. [CrossRef]
221. Fraudentali, I.; Ghuge, S.A.; Carucci, A.; Tavladoraki, P.; Angelini, R.; Rodrigues-Pousada, R.A.; Cona, A. Developmental, Hormone-and Stress-Modulated Expression Profiles of Four Members of the Arabidopsis Copper-Amine Oxidase Gene Family. *Plant Physiol. Biochem.* **2020**, *147*, 141–160. [CrossRef]
222. Yoon, Y.; Seo, D.H.; Shin, H.; Kim, H.J.; Kim, C.M.; Jang, G. The Role of Stress-Responsive Transcription Factors in Modulating Abiotic Stress Tolerance in Plants. *Agronomy* **2020**, *10*, 788. [CrossRef]
223. Chen, R.; Ma, J.; Luo, D.; Hou, X.; Ma, F.; Zhang, Y.; Meng, Y.; Zhang, H.; Guo, W. CaMADS, a MADS-Box Transcription Factor from Pepper, Plays an Important Role in the Response to Cold, Salt, and Osmotic Stress. *Plant Sci.* **2019**, *280*, 164–174. [CrossRef]
224. Khan, A.; Pan, X.; Najeeb, U.; Tan, D.K.Y.; Fahad, S.; Zahoor, R.; Luo, H. Coping with Drought: Stress and Adaptive Mechanisms, and Management through Cultural and Molecular Alternatives in Cotton as Vital Constituents for Plant Stress Resilience and Fitness. *Biol. Res.* **2018**, *51*. [CrossRef] [PubMed]
225. Haider, M.E. Advances in Transgenic Technology for Crop Cultivation and Stomatal Regulation as Potent Role in Agriculture. *Sch. Int. J. Biochem.* **2021**, *4*, 86–90.
226. Wang, N.-N.; Xu, S.-W.; Sun, Y.-L.; Liu, D.; Zhou, L.; Li, Y.; Li, X.-B. The Cotton WRKY Transcription Factor (GhWRKY33) Reduces Transgenic Arabidopsis Resistance to Drought Stress. *Sci. Rep.* **2019**, *9*, 724. [CrossRef] [PubMed]
227. Rehman, A.; Azhar, M.T.; Hinze, L.; Qayyum, A.; Li, H.; Peng, Z.; Qin, G.; Jia, Y.; Pan, Z.; He, S. Insight into Abscisic Acid Perception and Signaling to Increase Plant Tolerance to Abiotic Stress. *J. Plant Interact.* **2021**, *16*, 222–237. [CrossRef]
228. Ali, F.; Qanmber, G.; Li, F.; Wang, Z. Updated Role of ABA in Seed Maturation, Dormancy, and Germination. *J. Adv. Res.* **2022**, *35*, 199–214. [CrossRef]
229. Yang, W.H.; Lu, C.Z.; Chen, W.; Xu, H.Y. Reduction of Early Fruit Abscission by Main-Branch-Girdling in Macadamia Is Related to the Favorable Status of Carbohydrates and Endogenous Hormones. *HortScience* **2022**, *57*, 40–47. [CrossRef]
230. Razi, K.; Muneer, S. Drought Stress-Induced Physiological Mechanisms, Signaling Pathways and Molecular Response of Chloroplasts in Common Vegetable Crops. *Crit. Rev. Biotechnol.* **2021**, *41*, 669–691. [CrossRef]

231. Parveen, A.; Ahmar, S.; Kamran, M.; Malik, Z.; Ali, A.; Riaz, M.; Abbasi, G.H.; Khan, M.; Sohail, A.B.; Rizwan, M. Abscisic Acid Signaling Reduced Transpiration Flow, Regulated Na+ Ion Homeostasis and Antioxidant Enzyme Activities to Induce Salinity Tolerance in Wheat (*Triticum aestivum* L.) Seedlings. *Environ. Technol. Innov.* **2021**, *24*, 101808. [CrossRef]
232. Chen, K.; Li, G.J.; Bressan, R.A.; Song, C.P.; Zhu, J.K.; Zhao, Y. Abscisic Acid Dynamics, Signaling, and Functions in Plants. *J. Integr. Plant Biol.* **2020**, *62*, 25–54. [CrossRef]
233. Pál, M.; Tajti, J.; Szalai, G.; Peeva, V.; Végh, B.; Janda, T. Interaction of Polyamines, Abscisic Acid and Proline under Osmotic Stress in the Leaves of Wheat Plants. *Sci. Rep.* **2018**, *8*, 12839. [CrossRef]
234. Ouledali, S.; Ennajeh, M.; Ferrandino, A.; Khemira, H.; Schubert, A.; Secchi, F. Influence of Arbuscular Mycorrhizal Fungi Inoculation on the Control of Stomata Functioning by Abscisic Acid (ABA) in Drought-Stressed Olive Plants. *S. Afr. J. Bot.* **2019**, *121*, 152–158. [CrossRef]
235. El-Yazied, A.A.; Ibrahim, M.F.M.; Ibrahim, M.A.R.; Nasef, I.N.; Al-Qahtani, S.M.; Al-Harbi, N.A.; Alzuaibr, F.M.; Alaklabi, A.; Dessoky, E.S.; Alabdallah, N.M. Melatonin Mitigates Drought Induced Oxidative Stress in Potato Plants through Modulation of Osmolytes, Sugar Metabolism, ABA Homeostasis and Antioxidant Enzymes. *Plants* **2022**, *11*, 1151. [CrossRef] [PubMed]
236. Jiang, Z.; Zhu, H.; Zhu, H.; Tao, Y.; Liu, C.; Liu, J.; Yang, F.; Li, M. Exogenous ABA Enhances the Antioxidant Defense System of Maize by Regulating the AsA-GSH Cycle under Drought Stress. *Sustainability* **2022**, *14*, 3071. [CrossRef]
237. Xing, X.; Cao, C.; Xu, Z.; Qi, Y.; Fei, T.; Jiang, H.; Wang, X. Reduced Soybean Water Stress Tolerance by MiR393a-Mediated Repression of GmTIR1 and Abscisic Acid Accumulation. *J. Plant Growth Regul.* **2022**, *2022*, 1–17. [CrossRef]
238. Yari Kamrani, Y.; Shomali, A.; Aliniaeifard, S.; Lastochkina, O.; Moosavi-Nezhad, M.; Hajinajaf, N.; Talar, U. Regulatory Role of Circadian Clocks on ABA Production and Signaling, Stomatal Responses, and Water-Use Efficiency under Water-Deficit Conditions. *Cells* **2022**, *11*, 1154. [CrossRef]
239. Rehman, R.S.; Ali, M.; Zafar, S.A.; Hussain, M.; Pasha, A.; Naveed, M.S.; Ahmad, M.; Waseem, M. Abscisic Acid Mediated Abiotic Stress Tolerance in Plants. *Asian J. Res. C Sci.* **2022**, *7*, 1–17. [CrossRef]
240. Huang, T.-H.; Hsu, W.-H.; Mao, W.-T.; Yang, C.-H. The Oncidium Ethylene Synthesis Gene Oncidium 1-Aminocyclopropane-1 Carboxylic Acid Synthase 12 and Ethylene Receptor Gene Oncidium ETR1 Affect GA-DELLA and Jasmonic Acid Signaling in Regulating Flowering Time, Anther Dehiscence, and Flower Senescence In. *Front. Plant Sci.* **2022**, *13*, 785441. [CrossRef]
241. Patil, S.V.; Patil, C.D.; Mohite, B.V. Isolation and Screening of ACC Deaminase-Producing Microbes for Drought Stress Management in Crops. In *Practical Handbook on Agricultural Microbiology*; Springer: Berlin/Heidelberg, Germany, 2022; pp. 361–367.
242. Chandwani, S.; Amaresan, N. Role of ACC Deaminase Producing Bacteria for Abiotic Stress Management and Sustainable Agriculture Production. *Environ. Sci. Pollut. Res.* **2022**, *29*, 22843–22859. [CrossRef]
243. Gautam, H.; Fatma, M.; Sehar, Z.; Iqbal, N.; Albaqami, M.; Khan, N.A. Exogenously-Sourced Ethylene Positively Modulates Photosynthesis, Carbohydrate Metabolism, and Antioxidant Defense to Enhance Heat Tolerance in Rice. *Int. J. Mol. Sci.* **2022**, *23*, 1031. [CrossRef]
244. Costa, L.C.; Luz, L.M.; Nascimento, V.L.; Araujo, F.F.; Santos, M.N.S.; França, C.d.F.M.; Silva, T.P.; Fugate, K.K.; Finger, F.L. Selenium-Ethylene Interplay in Postharvest Life of Cut Flowers. *Front. Plant Sci.* **2020**, *11*, 2055. [CrossRef]
245. Meena, S.; Taria, S.; Nagar, S.; Yadav, S. Phytohormone Engineering: A Potential Approach for Inducing Abiotic Stress Tolerance in Crop Plants. *Multidisciplinary* **2022**, *2022*, 35.
246. Houben, M.; Van de Poel, B. 1-Aminocyclopropane-1-Carboxylic Acid Oxidase (ACO): The Enzyme That Makes the Plant Hormone Ethylene. *Front. Plant Sci.* **2019**, *10*, 695. [CrossRef] [PubMed]
247. Pérez-Pérez, J.G.; Puertolas, J.; Albacete, A.; Dodd, I.C. Alternation of Wet and Dry Sides during Partial Rootzone Drying Irrigation Enhances Leaf Ethylene Evolution. *Environ. Exp. Bot.* **2020**, *176*, 104095. [CrossRef]
248. Chung, K.M.; Demianski, A.J.; Harrison, G.A.; Laurie-Berry, N.; Mitsuda, N.; Kunkel, B.N. Jasmonate Hypersensitive 3 (JAH3) Negatively Regulates Both Jasmonate and Ethylene-Mediated Responses in Arabidopsis. *J. Exp. Bot.* **2022**. [CrossRef]
249. Kazan, K. Diverse Roles of Jasmonates and Ethylene in Abiotic Stress Tolerance. *Trends Plant Sci.* **2015**, *20*, 219–229. [CrossRef] [PubMed]
250. Siddiqi, K.S.; Husen, A. Plant Response to Jasmonates: Current Developments and Their Role in Changing Environment. *Bull. Natl. Res. Cent.* **2019**, *43*, 153. [CrossRef]
251. Ahmad, P.; Ahanger, M.A.; Alyemeni, M.N.; Wijaya, L.; Alam, P.; Ashraf, M. Mitigation of Sodium Chloride Toxicity in *Solanum lycopersicum* L. By Supplementation of Jasmonic Acid and Nitric Oxide. *J. Plant Interact.* **2018**, *13*, 64–72. [CrossRef]
252. Ghorbel, M.; Brini, F.; Sharma, A.; Landi, M. Role of Jasmonic Acid in Plants: The Molecular Point of View. *Plant Cell Rep.* **2021**, *40*, 1471–1494. [CrossRef]
253. Zaid, A.; Mohammad, F. Methyl Jasmonate and Nitrogen Interact to Alleviate Cadmium Stress in Mentha Arvensis by Regulating Physio-Biochemical Damages and ROS Detoxification. *J. Plant Growth Regul.* **2018**, *37*, 1331–1348. [CrossRef]
254. Ali, M.S.; Baek, K.H. Jasmonic Acid Signaling Pathway in Response to Abiotic Stresses in Plants. *Int. J. Mol. Sci.* **2020**, *21*, 621. [CrossRef]
255. Zamani, H.; Arvin, M.J.; Jahromi, A.A.; Abdossi, V.; Torkashvand, A.M. The Effect of Sodium Silicate and Methyl Jasmonate on Pigments and Antioxidant Activity of Tomato (*Solanum lycopersicum* L.) under Salinity Stress. *Tarim Bilim. Derg.* **2020**, *26*, 479–487. [CrossRef]
256. Per, T.S.; Khan, M.I.R.; Anjum, N.A.; Masood, A.; Hussain, S.J.; Khan, N.A. Jasmonates in Plants under Abiotic Stresses: Crosstalk with Other Phytohormones Matters. *Environ. Exp. Bot.* **2018**, *145*, 104–120. [CrossRef]

plants

Article

Foliar Application of Nano-Silicon Improves the Physiological and Biochemical Characteristics of 'Kalamata' Olive Subjected to Deficit Irrigation in a Semi-Arid Climate

Islam F. Hassan [1,*], Rahaf Ajaj [2], Maybelle S. Gaballah [1], Chukwuma C. Ogbaga [3,*], Hazem M. Kalaji [4,5], Harlene M. Hatterman-Valenti [6] and Shamel M. Alam-Eldein [7]

1 Water Relations and Field Irrigation Department, Agricultural and Biological Research Institute, National Research Center, Giza 12622, Egypt; msgaballa54@yahoo.com
2 Department of Environmental and Public Health, College of Health Sciences, Abu Dhabi University, Abu Dhabi 59911, United Arab Emirates; rahaf.ajaj@adu.ac.ae
3 Department of Microbiology and Biotechnology, Nile University of Nigeria, Abuja 900001, Nigeria
4 Department of Plant Physiology, Institute of Biology, Warsaw University of Life Sciences-SGGW, 02-776 Warsaw, Poland; hazem_kalaji@sggw.edu.pl or mohamed_kalaji@sggw.edu.pl
5 Institute of Technology and Life Sciences, National Research Institute, Falenty, Al.Hrabska 3, 05-090 Raszyn, Poland
6 Department of Plant Sciences, Department 7670, North Dakota State University, P.O. Box 6050, Fargo, ND 58108, USA; h.hatterman.valenti@ndsu.edu
7 Department of Horticulture, Faculty of Agriculture, Tanta University, Tanta 31527, Egypt; shamel.alameldein@agr.tanta.edu.eg or shamel@ufl.edu
* Correspondence: if.hassan@nrc.sci.eg (I.F.H.); chukwumaogbaga@gmail.com (C.C.O.)

Citation: Hassan, I.F.; Ajaj, R.; Gaballah, M.S.; Ogbaga, C.C.; Kalaji, H.M.; Hatterman-Valenti, H.M.; Alam-Eldein, S.M. Foliar Application of Nano-Silicon Improves the Physiological and Biochemical Characteristics of 'Kalamata' Olive Subjected to Deficit Irrigation in a Semi-Arid Climate. *Plants* **2022**, *11*, 1561. https://doi.org/10.3390/plants11121561

Academic Editors: Fulai Liu and Ferenc Fodor

Received: 3 May 2022
Accepted: 10 June 2022
Published: 13 June 2022

Abstract: In Egypt's arid and semi-arid lands where the main olive production zone is located, evapotranspiration is higher than rainfall during winter. Limited research has used nanomaterials, especially nano-silicon (nSi) to improve the growth, development, and productivity of drought-stressed fruit trees, amid the global water scarcity problem. To assess the role of nSi on drought-sensitive 'Kalamata' olive tree growth, and biochemical and physiological changes under drought conditions, a split-plot experiment was conducted in a randomized complete block design. The trees were foliar sprayed with nSi in the field using nine treatments (three replicates each) of 0, 150, and 200 $mg \cdot L^{-1}$ under different irrigation regimes (100, 90, and 80% irrigation water requirements 'IWR') during the 2020 and 2021 seasons. Drought negatively affected the trees, but both concentrations of nSi alleviated drought effects at reduced irrigation levels, compared to the non-stressed trees. Foliar spray of both concentrations of nSi at a moderate level (90% IWR) of drought resulted in improved yield and fruit weight and reduced fruit drop percentage, compared to 80% IWR. In addition, there were reduced levels of osmoprotectants such as proline, soluble sugars, and abscisic acid (ABA) with less membrane damage expressed as reduced levels of malondialdehyde (MDA), H_2O_2 and electrolyte leakage at 90% compared to 80% IWR. These results suggest that 'Kalamata' olive trees were severely stressed at 80% compared to 90% IWR, which was not surprising as it is classified as drought sensitive. Overall, the application of 200 $mg \cdot L^{-1}$ nSi was beneficial for the improvement of the mechanical resistance, growth, and productivity of moderately-stressed (90% IWR) 'Kalamata' olive trees under the Egyptian semi-arid conditions.

Keywords: abscisic acid; antioxidants; chloroplast degeneration; drought; malondialdehyde; nanoparticles; oxidative stress; photosynthesis

1. Introduction

The olive tree (*Olea Europaea* L.), family Oleaceae, is an evergreen tree and one of the oldest fruit trees believed to be indigenous to the entire Mediterranean Basin. Its cultivation was considerably developed in Syria and Palestine and spread to the island of

Crete towards Egypt [1,2]. The olive tree is a symbol of honor and culture and has been used as a prize for champions at the Olympic games [1]. Its global cultivated area is about 12,763,184 ha with a total annual production of approximately 23,640,307 t and an average yield of 1.85 t·ha^{-1}. In Egypt, olive cultivation increased during the past two decades and reached 100,826 ha with an annual production of about 932,927 t and an average yield of 9.25 t·ha^{-1} [2]. Egypt ranked eighth and third in global olive oil and table olive production, respectively. A total of 10% of the crop is planted for the purposes of producing olive oil, whilst 90% is for pickling [3]. The most important grown cultivars in Egypt are Toffahi, Aggizi, Picual, Manzanillo, Kroniaki, Coratina, Chemlali, and Kalamata.

As one of the most famous olive cultivars worldwide, 'Kalamata' olive was introduced to Egypt from Greece about three decades ago. The tree is drought-sensitive, cold intolerant, and fairly large, with distinguished large grey-green foliage, which grows to twice the size of other olive cultivars. The tree has inconspicuous white flowers that produce large deep reddish-purple fruit. The fruit is soft, almond-shaped, and slightly bitter in flavor. Fruit are hand-picked to avoid bruising, and they contain moderate oil levels, so growers produce them mainly for use as table olives (pickled) and some oil. The optimal harvest time is late fall, whilst other cultivars are usually picked earlier [4].

Egypt's Mediterranean climate of hot dry summers and warm humid winters is well suited for olive cultivation. In general, olives are drought-tolerant plants and can withstand hot and dry conditions; however, total annual production has been negatively affected in the past few years amidst climate change-induced water shortage and soil salinity problems [5]. Water scarcity has become a recent problem in Egypt and may translate to a limiting factor for the overall fruit industry in the future, particularly with the increased human demand for freshwater, which creates competition with agricultural activities [6]. Water resources and rainfall (20–200 mm annually) are limited, and the Nile River is the most important water resource [7]. Under such conditions, there is a need to reduce agricultural water demand and increase the economic productivity of water for the future expansion of olive agriculture in the water-scarce Mediterranean area [8]. Improving on-farm management of agricultural water through the utilization of advanced irrigation technology, (e.g., deficit irrigation) and improved irrigation scheduling, offer the prospect of significant increases in water productivity [9–13]. Deficit irrigation is a strategy where the amount of applied water is less than the full water requirements of a crop, and the resulting stress has minimal effects on crop yield [14]. Deficit irrigation effectively reduced water requirements, enhanced plant water use efficiency [15], and improved fruit quality of various deciduous and evergreen fruit trees [16–18], including olives, depending on the phenological stage when water shortage was applied, drought severity, and the cultivar [19–24].

The scenario of deficit irrigation is still under research for a few olive cultivars, such as 'Kalamata', which is classified as a drought-sensitive cultivar. Generally, drought mainly impacts plant morphology, physiology, and biochemistry [25]. Under such conditions, xylem vessels become susceptible to embolism or dysfunction, leading to lower hydraulic conductance and carbon intake, which in turn affect plant growth characteristics and productivity [26]. Drought stress causes a reduction in root and vegetative growth, number of leaves per branch, leaf area, and leaf water content [27]. It also causes photoinhibition of photosystem II (PS II), limits electron transfer from PSII to photosystem I (PSI) and induces stomatal closure. These eventually decrease CO_2 fixation in the chloroplast during the Calvin cycle [28–30]. Drought stress causes an increase in the formation of reactive oxygen species (ROS) negatively affecting plant metabolism through oxidative damage by lipids, proteins, and nucleic acids [31,32].

Various reports have shown that the foliar fertilization of micronutrients such as Zn, B, Cu, Mn, Si, Se, and Fe are effective and induce a very rapid plant response. Their application can improve the nutrient balance of a plant, resulting in increased fruit yield and quality, better disease resistance, and alleviation of the adverse effects of drought and salinity stress [33–36]. During the past three decades, many problems in different fields of science and industry have been resolved using nanotechnology. Materials that are smaller than 100 nm, at least in one dimension, are generally classified as nanomaterials [37]. Nanomaterials could be used for designing new fertilizers [38], to ensure the effective delivery of the required nutrients to the plant and a very rapid plant response [33] with only one-third of the required conventional counterparts added to the environment [39]. In the same context, it was reported that silicon is mostly toxic to plants in its bulk form, whereas silicon nanoparticles were beneficial for plants [40].

Silicon (Si) is the second most abundant element in the soil; however, it is not considered an essential element for plant growth, development, or productivity [41]. Recently, Si has gained global attention because it is safe for the environment, induces disease and pest resistance in plants, and can reduce doses of pesticides applied for plant protection [42]. Silicon is beneficial for alleviating the nutrient imbalance stress and improving the growth, development, and yield of various plants [43]. It improves organogenesis, embryogenesis, growth traits, and the morphological, anatomical, and physiological characteristics of leaves. It also enhances tolerance to chilling, freezing, salinity, and drought and protects cells against metal toxicity, oxidative stress, and phenolic browning [40,44,45]. Foliar application of Si is most powerful for plants under stressful conditions, such as salinity, drought, flood, heat, cold, and even biotic stress [46]. Silicon can potentially decrease the negative effect of oxidative stress and offer slight resistance to some abiotic and biotic plant stresses. A large number of genes are activated by stress, and several Si-produced proteins that participate in biochemical pathways lead to the enhancement of stress tolerance [47]. The application of nano-silicon (nSi) has been suggested to enhance plant tolerance to drought stress by reducing the production of reactive oxygen species (ROS) in barley, wheat, faba bean, feverfew, strawberry, Mahaleb cherry, and mango [34,48–54]. The primary mechanisms of Si-mediated abiotic stress reduction include plant antioxidant system activation, co-precipitation and immobilization of toxic metal ions in the growth medium, and metal absorption and separation within the plants [55–68].

Limited research findings have been reported on the effect of nanoparticles (NPs) on woody plants, especially fruit trees. Most reports focused on the role of NPs to alleviate the effect of stressful conditions on fruit tree seedlings, whilst some concentrated on using NPs to improve the growth, yield, and fruit quality of fruit trees growing under non-stressful conditions. To the best of the authors' knowledge, this is considered to be one of few reports to cover both goals and the first on a drought-sensitive olive cultivar. The aim of this research was to estimate the role of nano-silicon in alleviating the drastic effects of water stress on a drought-sensitive 'Kalamata' olive tree grown under semi-arid conditions. The hypothesis was that nano-silicon may enhance the physiological and biochemical characteristics of stressed 'Kalamata' trees by improving the resistance mechanisms, growth, and productivity.

2. Results

2.1. Yield

Olive yield differed slightly from one season to another (Figure 1a). In the 2020 season, the yield per tree increased with increased nSi concentration, regardless of the water stress level, with the exception of trees that received 80% IWR. The highest yield per tree was recorded in trees sprayed with 200 mg·L^{-1} nSi at 100% IWR, while the lowest yield was recorded in the control trees. In the 2021 season, total yield increased with increased nSi concentration, regardless of the water stress level, with the exception of trees that received 80% IWR. Like the first season, the highest yield was recorded in trees sprayed with 200 mg·L^{-1} nSi at 100% IWR, and the lowest for the control. In addition, trees that received 200 mg·L^{-1} nSi at 80% IWR showed an improved yield when compared to those that received 150 mg·L^{-1} nSi at 80% IWR. Results suggest that nSi application increased 'Kalamata' olive yield when trees were stressed.

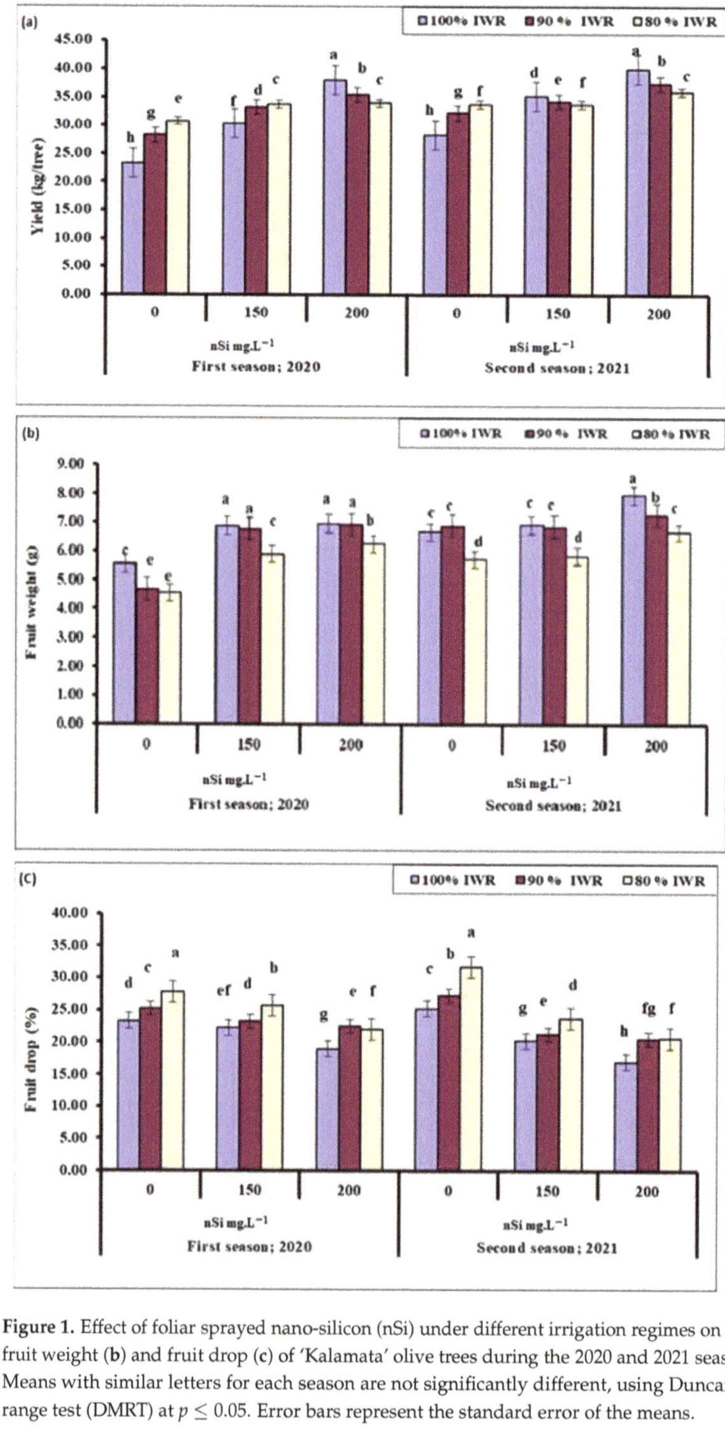

Figure 1. Effect of foliar sprayed nano-silicon (nSi) under different irrigation regimes on the yield (**a**), fruit weight (**b**) and fruit drop (**c**) of 'Kalamata' olive trees during the 2020 and 2021 seasons (n = 10). Means with similar letters for each season are not significantly different, using Duncan's multiple range test (DMRT) at $p \leq 0.05$. Error bars represent the standard error of the means.

2.2. Fruit Weight

Unlike total yield, average fruit weight was the highest at 100% or 90% IWR with the application of either 150 or 200 mg·L^{-1} nSi (Figure 1b) during the 2020 season. Average fruit weight was the least at 80% IWR combined with either 150 or 200 mg·L^{-1} nSi; however, fruit were still heavier than those that were not sprayed with nSi, regardless of the water regime. During the 2021 season, the highest fruit weight was recorded at 100% IWR, combined with 200 mg·L^{-1} nSi. At the same concentration, fruit weight decreased as water stress severity increased. Similar fruit weight was noticed at 100% or 90% IWR in combination with either distilled water or 150 mg·L^{-1} nSi.

2.3. Fruit Drop

The more severe the drought condition was, the higher percentage the fruit drop was at either distilled water or 150 mg·L^{-1} nSi spray during the 2020 season (Figure 1c). The lowest fruit drop was noticed at 200 mg·L^{-1} nSi for non-stressed trees. Increasing the concentration of nSi to 200 mg·L^{-1} nSi under drought stress conditions (either 90% or 80% IWR) reduced fruit drop in comparison to the control. Increased fruit drop percentages with the application of 150 mg·L^{-1} at 80% IWR, compared to that at 90% IWR indicated that plant nSi was not that effective during the severe drought. However, the application of 200 mg·L^{-1} nSi was more effective at 80% IWR during the first season only.

2.4. Biochemical Characteristics

2.4.1. Total Chlorophyll

Results indicated that generally, the higher the nSi concentration, the higher the leaf chlorophyll content was, regardless of the drought severity in both the 2020 and 2021 seasons. However, the difference between nSi (150 mg·L^{-1}) at 90% IWR and the non-treated plants at any irrigation level was insignificant (Figure 2a). Similar results were noticed in both seasons; however, chlorophyll content was significantly decreased in trees received 200 mg·L^{-1} nSi at 80% IWR compared to the nonstressed ones during the second season.

2.4.2. Proline

As osmoprotectant, proline increased with the increased levels of water stress. Unlike chlorophyll, leaf proline generally decreased with increased concentrations of nSi at all irrigation regimes during both seasons (Figure 2b). Trees subjected to 90% IWR and received nSi (200 mg·L^{-1}) were less stressed and had lower proline compared to those that received 150 mg·L^{-1} in both seasons confirming the role of nSi in mitigating the drought effects and suggesting that irrigation at 80% IWR was really stressful for olive trees.

2.4.3. Soluble Sugars

Like proline, leaf soluble sugars generally increased with increased levels of water stress, but they decreased with the increased concentrations of nSi in both seasons (Figure 2c). Trees sprayed with 200 mg·L^{-1} nSi at 80% IWR showed more soluble sugars than that of the control leaves during the first season, whereas they showed less soluble sugars during the second season, suggesting a cumulative effect of drought and nSi application from one season to another. nSi application mitigated the stress effect at 90%, compared to 80% IWR, suggesting that 80% IWR was really stressful to olive trees under the Egyptian semi-arid conditions.

2.5. Leaf Water Status and Membrane Damage Indicator

2.5.1. Relative Water Content (RWC)

The water contents of olive leaves were generally the highest in the control trees but decreased gradually with the severity of drought during both seasons. The application of nSi improved the RWC, which was the highest at 200 mg·L^{-1} (Figure 3a). The most pronounced effect was recorded for non-stressed trees that received nSi (200 mg·L^{-1}) in both seasons. Relative water content (RWC) was negatively affected in response to nSi and water stress levels, in comparison to proline and soluble sugars. Results suggest that the application of nSi 200 mg·L^{-1} effectively increased the leaf RWC of moderately stressed plants compared to the control for both seasons.

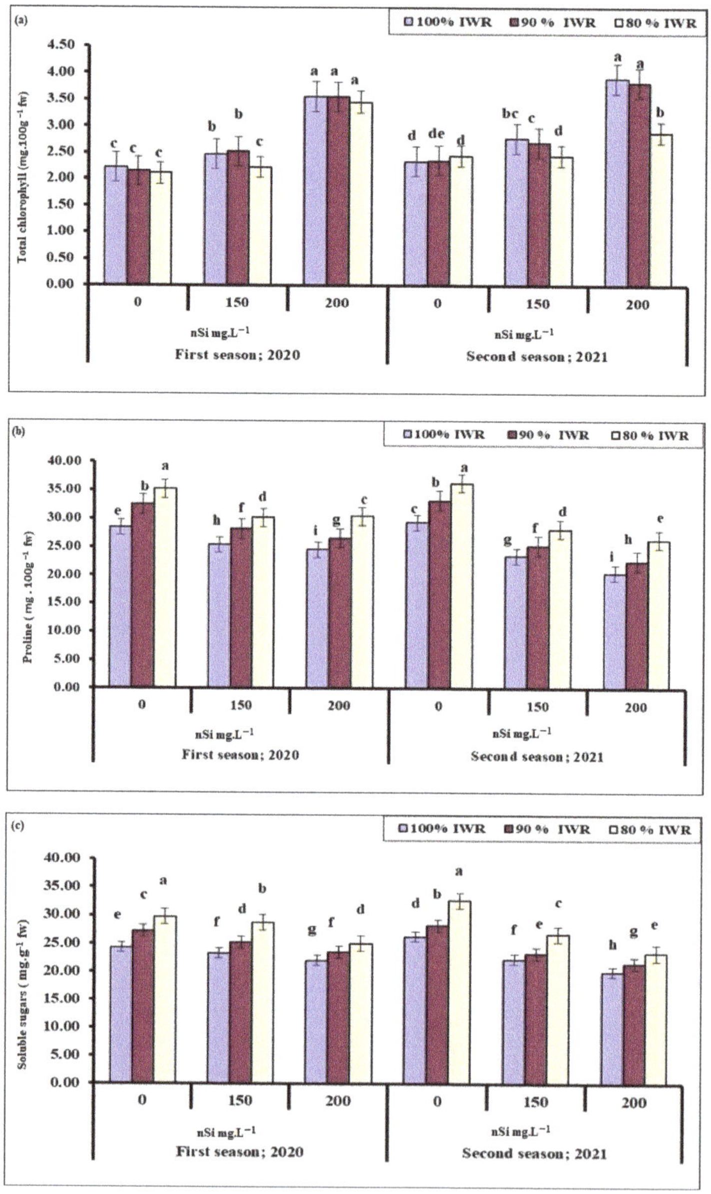

Figure 2. Effect of foliar sprayed nano-silicon (nSi) under different irrigation regimes on the leaf total chlorophyll (**a**), proline (**b**), and soluble sugars (**c**) of 'Kalamata' olive trees during the 2020 and 2021 seasons (n = 10). Means with similar letters for each season are not significantly different, using Duncan's multiple range test (DMRT) at $p \leq 0.05$. Error bars represent the standard error of the means.

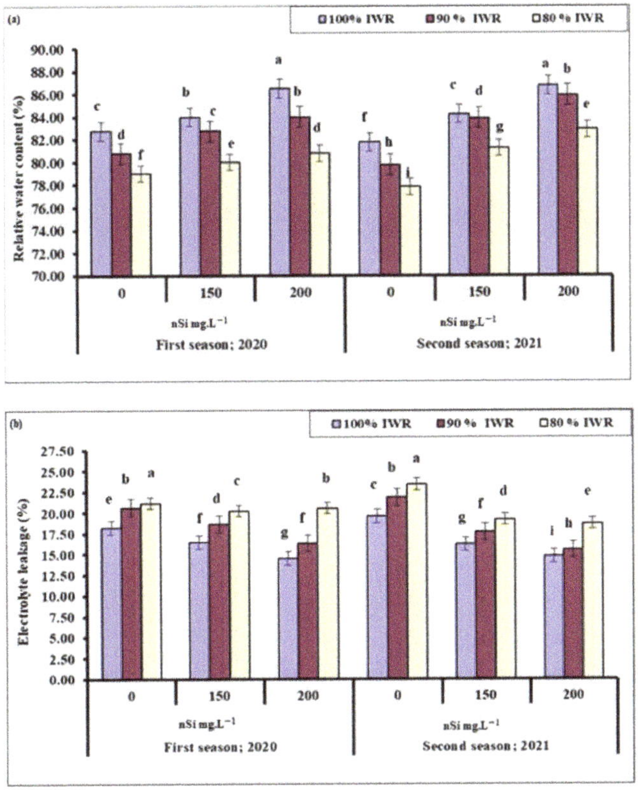

Figure 3. Effect of foliar sprayed nano-silicon (nSi) under different irrigation regimes on the leaf relative water content 'RWC' (**a**) and electrolyte leakage (**b**) of 'Kalamata' olive trees during the 2020 and 2021 seasons ($n = 10$). Means with similar letters for each season are not significantly different, using Duncan's multiple range test (DMRT) at $p \leq 0.05$. Error bars represent the standard error of the means.

2.5.2. Electrolyte Leakage

Like leaf proline and soluble sugar contents, the electrolyte leakage generally increased with the severity of drought, but gradually decreased with increased concentrations of nSi (Figure 3b). The most pronounced effect was recorded for trees sprayed with 200 mg·L^{-1} nSi at 100% IWR, followed by those at 90% IWR that were significantly different from non-stressed trees that received 150 mg·L^{-1} nSi during the second season only. Trees sprayed with 200 mg·L^{-1} nSi showed a significant reduction in electrolyte leakage compared to those that received distilled water for both seasons, which may suggest a cumulative nSi role.

2.6. Oxidative Stress Markers

2.6.1. Malondialdehyde (MDA)

The higher the plant stress level was, the higher the oxidative stress that the plant faced, represented by the increased levels of lipid peroxidation and the production of MDA, which significantly reduced with the increased nSi concentration (Figure 4a). With regards to the stressed olive trees, the most pronounced reduction in MDA was recorded for trees that received nSi (200 mg·L^{-1}) at 90% IWR, followed by those at 80% IWR during both seasons.

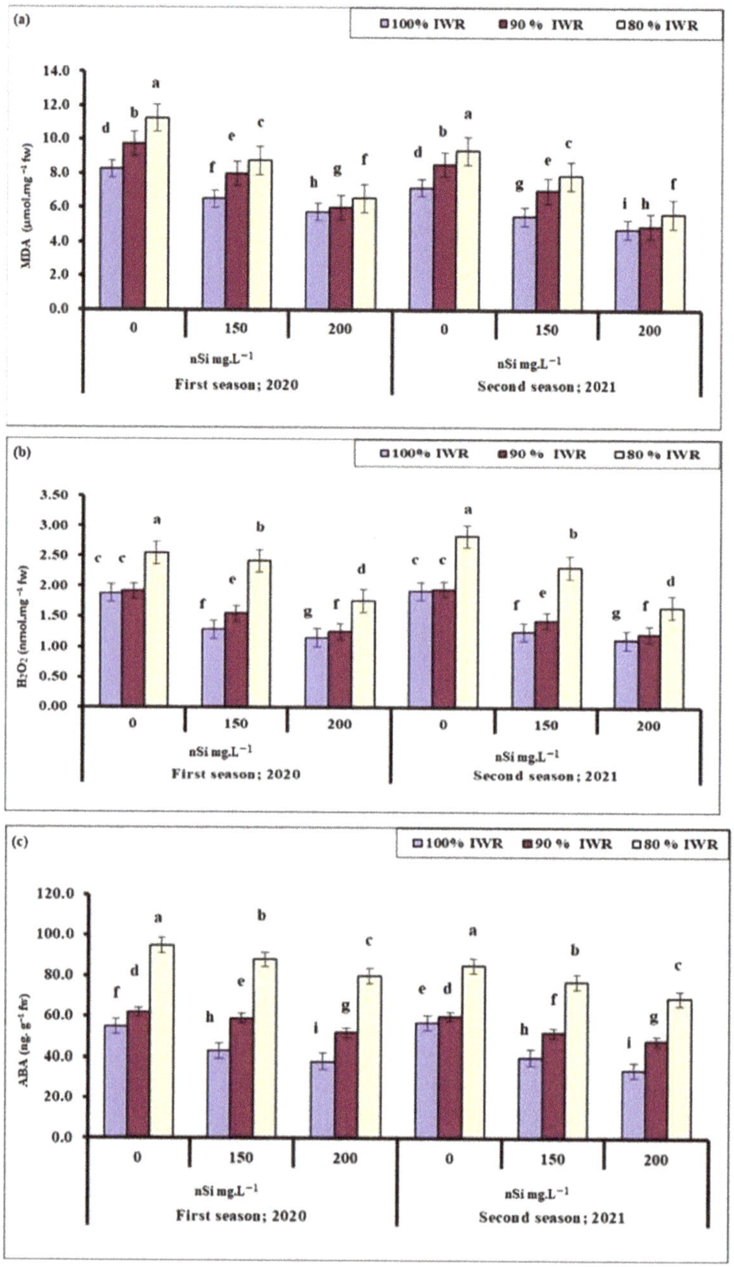

Figure 4. Effect of foliar sprayed nano-silicon (nSi) under different irrigation regimes on the leaf malondialdehyde 'MDA' (**a**), H_2O_2 (**b**), and abscisic acid 'ABA' (**c**) contents of 'Kalamata' olive trees during the 2020 and 2021 seasons (n = 10). Means with similar letters for each season are not significantly different, using Duncan's multiple range test (DMRT) at $p \leq 0.05$. Error bars represent the standard error of the means.

2.6.2. Hydrogen Peroxide (H_2O_2)

As another indicator of plant oxidative stress, H_2O_2 level increased significantly with drought severity, but the application of nSi significantly decreased its levels compared to the control, with the most conspicuous effect recorded at the highest nSi concentration (200 mg·L^{-1}), followed by those that received 150 mg·L^{-1}, for trees subjected to 90% and 80% IWR during both seasons (Figure 4b). Trees sprayed with nSi (200 mg·L^{-1}) at 80% IWR had higher H_2O_2 than those at 90 and 100% IWR during both seasons, suggesting that irrigation at 80% IWR was really stressful for olive trees.

2.6.3. Abscisic Acid (ABA)

The non-enzymatic antioxidant, ABA (Figure 4c), showed the same response as the organic osmolytes, (e.g., proline, soluble sugars) (Figure 2b,c) to drought and nSi application during both seasons. Plant response to stress, in terms of ABA biosynthesis, matched the oxidative stress markers (MDA, H_2O_2) (Figure 4a,b) produced during both seasons. Likewise, in other parameters, low ABA levels indicate less stressed trees. Thus, the most effective treatments were the application of 200 mg·L^{-1}, followed by 150 mg·L^{-1} nSi at 90% IWR, and then 200 mg·L^{-1}, followed by 150 mg·L^{-1} nSi at 80% IWR, suggesting that 80% IWR was really stressful for olive trees under semi-arid conditions during both seasons.

3. Discussion

Silicon nanoparticles have been found to play an important role in a plant's ability to overcome the adverse effects of environmental stresses [12,69–72]. Several researchers have reported the ability of nSi to improve a plant's resistance to drought stress [29,73]. However, this is the first time nSi has been sprayed on a mature woody plant to overcome water stress. Other research work utilized $nSiO_2$ to mitigate the drought effects on hawthorn (*Crataegus monogyna*) seedlings [74], and Mahaleb cherry [48].

The increase in olive yield in response to water stress was not expected (Figure 1) and contradicts the previous findings that indicated a negative effect on olive tree acclimation, and thereby yield, fruit dry mass, and oil content [75]. The total yield was also influenced by the percentage of fruit drop, which increased with the increased levels of water stress. Foliar sprayed nSi decreased the percentage of fruit drop, and lower percentages were achieved with the increase in nSi concentration (Figure 1). Trees' response to nSi (150 mg·L^{-1}), in terms of yield, varied from one season to another, while those that received 200 mg·L^{-1} showed a consistent response between seasons with the highest yield recorded for non-stressed trees, followed by those that received 90% IWR, and then trees at 80% IWR (Figure 1). Trees sprayed with nSi showed higher yield values compared to the non-treated stressed ones and the control. Foliar application of Si at tillering and anthesis stages increased the grain yield of the stressed and non-stressed plants [76]. The reduction in fruit weight, associated with increased fruit drop in stressed trees, resulted in reduced total yield (Figure 1), and was reported to make some improvement in fruit composition and oil quality [77–80]. However, average fruit weight was higher in nSi-treated trees (200 mg·L^{-1}) whether stressed or not during the 2020 season, while the treated trees at reduced water levels produced heavier fruit than the non-treated stressed ones during the 2021 season, suggesting that there is a positive role of nSi in combination with normal irrigation.

Water stress reduced the leaf's relative water content (RWC) in both years, but the application of nSi improved the RWC exponentially with increased concentration. Similar results were reported on faba beans [53], wheat [70], plums [81], and olives [13,82]. Increased level of ABA-induced stomatal closure had eventually increased RWC [83]. Chlorophyll content was the best at 200 mg·L^{-1} nSi for both stressed (90% IWR) and non-stressed trees during the 2021 season (Figure 2). Similar results were reported in faba bean, albeit a reduction in chlorophyll content was noticed with drought conditions [52]. Ma et al. [84] also reported that chlorophyll content in cucumber leaves decreased at moderate and severe drought stress. This effect has been mitigated with Si spray, which contributed to the protection of the chloroplasts. The variation in total chlorophyll content among the stressed faba bean [52], cucumber [83], and olive plants (Figure 2) could be attributed to differences in species, genotype-specific traits, growth stage, the method and concentration of nSi application, stress intensity and duration, and other environmental conditions [85]. The difference in chlorophyll content from one season to another could be due to the pre-exposure to stress that eventually improved the plant's acclimation [86].

Photosynthesis is needed for biomass productivity. Thus, increased chlorophyll content should also increase the photosynthetic capacity, as measured by leaf total soluble sugars [87], which unexpectedly decreased. However, chlorophyll contents increased with an increase in the levels of nSi and drought (Figure 2), perhaps due to an increase in leaf RWC (Figure 3). Iron oxide nanoparticles resulted in accumulated soluble sugar contents when plants were drought-stressed. However, iron also stimulates the redox process and chlorophyll biosynthesis in plants, thus should have helped to increase leaf chlorophyll and soluble sugar contents [88,89]. Similar results were reported in stressed cotton plants, but the used nSi concentration that resulted in the highest soluble sugar contents was 3200 $mg \cdot L^{-1}$ [90]; about 16-fold of what has been used in the present study.

Drought stress in olives is often associated with increased cellular levels of oxidative stress markers of lipids such as MDA and H_2O_2 [6]. When an imbalance between reactive oxygen species (ROS) production and the antioxidant defense system occurs, increased cellular membrane damage and electrolyte leakage occur [53,83]. It was also reported that when drought-stressed olive trees were re-watered, they still exhibited higher levels of H_2O_2, suggesting that this drought/wet rhythm is a possible means to keep the plant's antioxidative system on alert. It was reported that the enhanced level of H_2O_2 in drought-stressed leaves was accompanied by enhanced levels of MDA and electrolyte leakage [56]. In the current study, the content of H_2O_2 increased massively only when trees were subjected to 80% IWR (Figure 4), indicating that 90% IWR was mild water stress for olives, and the antioxidant defense system is little affected [91] since Kalamata olive is generally a drought-tolerant plant [5]. Trees sprayed with high nSi (200 $mg \cdot L^{-1}$) had lower leaf H_2O_2 when subjected to 80% IWR level (Figure 4), suggesting that the higher concentration of nSi was better at alleviating the H_2O_2-plant response to drought. Similar results were reported with soil application of nSi on drought-stressed barley seedlings [92]. In the current study, the electrolyte leakage (Figure 3) and MDA levels (Figure 4) also increased in response to water stress but decreased with increased nSi concentrations, which confirms the previous findings [93,94].

Proline has been described as an osmoprotectant that accumulates in response to abiotic stresses [95] and plays an essential role in the defense mechanisms of stressed plants through changes in key anatomical features of roots and leaves, the osmotic regulation of the cell sap, membrane and protein stability, enhanced enzyme activity, and scavenging the free radicals [96–98]. Enhanced endogenous proline levels improved leaf chlorophyll content, yield, fruit weight and diameter, and total soluble sugars (TSS) of non-stressed pomegranate [99] and orange [100], as well as salt-stressed mango [34] and tomato plants [101]. Leaf proline concentration increased with the severity of drought stress but decreased with elevated nSi concentrations (Figure 2). Increased levels of water deficit improved the biosynthesis of proline in castor bean, while foliar sprayed chitosan nanoparticles had no effect [102]. Additionally, higher concentrations of nFe mitigated the stress effects and reduced the accumulation of proline in drought-stressed wheat plants [88]. The difference in proline accumulation levels in the current study (Figure 2), compared to the previous findings could be due to the difference in water stress levels, as well as differences between wheat as an herbaceous plant and olive as a woody plant, as previously indicated [85]. However, the current results in Figure 2 contradict the previously reported findings on drought-stressed faba bean [30] and wheat [73].

The concentration of ABA increased with increased water stress levels, but decreased with increased concentration of nSi in both seasons (Figure 4), confirming the previous findings on wheat [103]. It was reported that the combined application of Si, B, Zn, and zeolite nanoparticles decreased the production of ABA in tomatoes under drought conditions [104]. Drought often leads to the formation of ROS, (e.g., H_2O_2, superoxide radical [$O^{\cdot-}{}_2$], singlet oxygen [$^1O_2$], and hydroxyl radicals [OH]), which are highly toxic and can react with proteins, lipids, and DNA, accelerating the aging process of chloroplasts, thus reducing photosynthetic capacity and decreasing plant growth and productivity [105]. Abiotic environmental stresses affect the plant through osmotic stress. Cell homeostasis is maintained against osmotic stress by the mechanism of osmotic adjustment, which is a primary stress-adaptive motor that positively correlates with plant production under drought conditions in various crops [106]. The mechanism of osmotic adjustment leads to the synthesis of organic osmolytes, (e.g., sugars, proline) [107], non-enzymatic antioxidants, (e.g., ascorbate, glutathione), and enzymatic (scavenger enzymes) antioxidants, (e.g., SOD, CAT, POD) [108] to balance the osmotic pressure of the cytosol and vacuole with that of the external environment [109]. The biosynthesis of ABA usually occurs in the roots under drought conditions and can increase up to 50 times, which is considered the highest change in any phytohormones under such conditions [110]. The ABA concentration has been shown to promote root growth and adjust shoot growth [83,111] via the regulation of transpiration and photosynthesis, as well as its potent effect on the production of primary and secondary metabolites, antioxidant enzymes, and lipoxygenase inhibitory activity [112–114]. A cross-talk mechanism between Ca^{+2} and ROS that originates from NADPH-oxidase in the ABA-induced antioxidant defense in the plant was also reported [115]. The role of ABA on sugar contents came through the accumulation of assimilates from the phloem into the fruit by strengthening sink capacity [116,117]. On the other hand, an increase in ABA has been reported as an indicator of the beginning of fruit senescence [118]. It was reported that the increased levels of ABA under drought conditions may trigger ethylene production, which promotes the activity of hydrolytic enzymes, such as endo-beta-glucan, cellulase, and polygalacturonase at the abscission zone of the fruit petiole inducing preharvest fruit abscission [119–121].

Overall, it could be said that the foliar application of nSi at 200 $mg{\cdot}L^{-1}$ followed by 150 $mg{\cdot}L^{-1}$ was effective in alleviating moderate drought effects (90% IWR) in 'Kalamata' olive trees compared to severe drought (80% IWR). During severe drought at 80% IWR, the tree's yield and average fruit weight reduced whilst fruit drop, along with levels of osmoprotectants increased with more membrane damage. The results of this study suggest that 'Kalamata' olive trees were severely stressed at 80% IWR compared to 90% IWR reinforcing its classification as drought sensitive [4] compared to most olive cultivars that are classified as drought tolerant [5].

4. Materials and Methods

4.1. Experiment

This research was carried out on nine-year-old 'Kalamata' olive trees (*Olea europaea* L.) grown in sandy soil in a private orchard located at Wadi El-Natrun, Beheira Governorate (30°46′98″ N, 30°27′43″ E, 23 m below the Mediterranean Sea level and 38 m below the Nile River), Egypt, during the 2020 and 2021 seasons. Ninety trees planted at a 6 m × 6 m spacing, similar in vigor and size, free from any symptoms of physiological disorders or nutrient deficiencies, were chosen for this experiment. Trees were subjected to drip irrigation from deep groundwater well and received other agricultural practices as the entire orchard during both seasons. Soil and water analysis were performed according to the methods described by Wilde et al. [122] and displayed in Table 1.

Irrigation treatments started in spring by full bloom (≈mid-March) and ceased at harvest by fall (≈late October). Trees were subjected to deficit irrigation based on differences in crop evapotranspiration (*ETc*), and three irrigation treatments–100% (control), 90%, and 80%–were used. The percentage of *ETc* was calculated based on the reference crop evapotranspiration (*ETo*) [$mm{\cdot}day^{-1}$] that is presented in Table 2 and crop coefficient factor (K_c) of olive, as suggested [123], using the following equations:

$$ETc = ETo \times K_c \quad (1)$$

Table 1. Soil and water analysis of the experimental site.

	Soil (0–40 cm)	Water
pH	8.22	7.01
Sand (%)	92.0	–
Silt (%)	5.0	–
Clay (%)	3.0	–
EC (dS/m)	1.82	1.56
$CaCO_3$ (%)	3.4	–
Ca^{2+} (meq·100 g^{-1})	8.6	9.4
Mg^{2+} (meq·100 g^{-1})	3.2	4.3
Na^{+} (meq·100 g^{-1})	6.9	9.80
K^{+} (meq·100 g^{-1})	1.5	0.22
Cl^{-} (meq·100 g^{-1})	8.2	6.46
$SO_4{}^{2-}$ (meq·100 g^{-1})	6.4	14.3
CO^{3-} (meq·100 g^{-1})	0.0	–
HCO^{3-} (meq·100 g^{-1})	5.6	3.0

Table 2. Average meteorological data of Wadi El Natrun area (2020 and 2021), source: own elaboration.

	Jan	Feb	Mar	Apr	May	Jun	Jul	Aug	Sept	Oct	Nov	Dec
Temp. mean Max (°C)	20.7	25.5	25.7	27.1	32.9	33.8	34.8	34.9	32.8	28	23.2	20.7
Temp. mean Min (°C)	9.1	8.98	11.1	13.7	16.7	19.5	20.2	22.7	20.2	17	10	9.3
Temp. average (°C)	14.9	17.24	19.4	20.4	24.3	26.65	27.45	28.5	26.4	23.4	20.05	15
Relative humidity (%)	65.1	62.5	62.56	58	58.1	59.2	58.8	59.9	63.1	62	65.1	65.2
Evaporation (mm·day^{-1})	6.2	7.7	9.8	12.5	13.8	15	14.3	12.7	10.5	8.6	6.1	5.1
ETo (mm·day^{-1})	2.80	3.30	4.0	4.80	5.30	5.80	6.10	5.40	4.40	3.10	3.02	2.90

Irrigation water requirements (IWR) for each irrigation regime (m^3·ha^{-1}·$season^{-1}$) during the entire season were determined using the following equation:

$$IWR = (A \times ETc \times Ii \times Kr)/(Ea \times 1000\ (1 - LR)) \quad (2)$$

where *ETC* expressed as m^3·ha^{-1} per irrigation time, *A* = cultivated area (ha), *ETc* = crop evapotranspiration, *Ii* = irrigation interval (day), *Kr* = reduction factor, *Ea* = irrigation efficiency, *LR* = leaching requirement = 10% of the total water amount delivered to the treatment.

The reduction factor was determined by the following equation:

$$Kr = (0.10 + GC) \leq 1 \quad (3)$$

where *GC* = the ground cover.

The leaching requirements (*LR*) were estimated according to the following equation:

$$LR = ECw/2ECe_{max} \quad (4)$$

where *ECw* = the electrical conductivity of the irrigation water (dS·m^{-1}), $2ECe_{max}$ = the maximum electrical conductivity of the soil saturated extract for a given crop.

The number of irrigation times varied among the three treatments with a frequency of 1–5 irrigation times per week, based on weather conditions (Table 2) and soil water content that was monitored weekly using soil tensiometer Model 64xx series (Spectrum Technologies Inc., Aurora, IL, USA). Two lateral lines of irrigation pipes (one on each side of the trees row) with 10 drippers per tree (8 L·h^{-1}·$dripper^{-1}$) were used for the control treatment (100% IWR), whereas 9 and 8 drippers were used for the 90% and 80% IWR treatments, which represent a total of 7007.14, 6306.43 and 5605.71 m^3·ha^{-1}·$season^{-1}$, respectively.

Foliar spray with nanoparticle chelate fertilizer of silicon (nSi = 5–15 nm) (Sigma-Aldrich, St. Louis, MO, USA) at 150 and 200 mg·L^{-1}, supplemented with Tween 20 as a surfactant (Sigma-Aldrich, St. Louis, MO, USA), was applied three times; by the onset of the vegetative growth in spring (≈late February), at full bloom (≈mid-March), and at fruit set (≈early April). Control trees were also sprayed at the same times with distilled water, supplemented with Tween 20 to avoid any effects between the treatments and the control, which could be related to the surfactant. Every tree received about 15 liters of the spray solution until dripping during the early morning period.

4.2. Yield and Average Fruit Weight

Total yield (kg·tree^{-1}) was recorded by harvest time in late October, and average fruit weight (g) was determined by weighing 90 randomly selected fruit samples from each replicate using a bench-top digital scale Model PC-500 (Doran scales, Inc., Batavia, IL, USA).

4.3. Fruit Drop

Fruit drop percentage was estimated per tree by randomly selecting four branches from the four directions (N, E, S, and W), and then branches were wrapped using net bags, and dropped fruit were collected and counted every 15 days until harvest. The number of the remaining fruit on the branches was recorded by the last observation time. The fruit drop percentage was calculated using the following equation:

$$\text{Fruit drop (\%)} = [(\text{Initial fruit number} - \text{Final fruit number})/\text{Initial fruit number}] \times 100 \quad (5)$$

4.4. Leaf Analysis

By the end of each harvest season, a sample of 50 mid-branch leaves was randomly collected, from the four directions (N, E, S, and W) and three levels (top, medium, and bottom) of the tree, for leaf analysis. All used chemicals were imported from Sigma-Aldrich, St. Louis, MO, USA.

Leaf chloroplasts were extracted in 85% acetone solution, and the absorbance of the aqueous phase of the extracted solution was estimated using a spectrophotometer Model UV-120-20 (Shimadzu, Kyoto, Japan) at λ = 663 and 645 nm [124]. Total chlorophyll was then calculated using the following equation:

$$\text{Total chlorophyll (mg·100 g}^{-1}\text{ fw)} = [(20.2 \times \text{OD 645 nm} + 8.02 \times \text{OD 663 nm}) \times \text{V}]/(\text{fw} \times 1000) \quad (6)$$

where: OD = optical density, V = the final volume of the solution (mL), and fw = tissue fresh weight (g).

Leaf proline (mg·100 g^{-1} fw) was extracted using 0.5 g of young leaves with sulfuric acid (3%), and the solution was quantified using ninhydrin reagent [125]. The solution was then mixed with toluene, and the absorbance of the toluene phase of the extracted solution was determined using the spectrophotometer at 520 nm.

The concentration of the soluble sugars (mg·100 g^{-1} fw) was determined using dried leaf samples [126]. Every dried leaf sample (150 mg) was extracted twice with 80% ethanol and centrifuged at 3500 rpm for 10 min and the volume of the supernatant was adjusted to 25 mL. The supernatant (1 mL) was then transferred to a test tube with the addition of 1 mL phenol (18%) and 5 mL sulfuric acid, and the mixture was shaken. The absorbance of the aqueous phase of the extracted solution was recorded at 490 nm using the spectrophotometer.

The RWC of the leaf was estimated using a fresh leaf sample (0.2 g) incubated in distilled water (50 mL) for 4 h. The turgid weight of the leaf sample was calculated, and then the sample was oven-dried at 60 °C for 48 h, followed by the determination of the dry weight [124]. Leaf RWC was calculated using the following equation:

$$\text{RWC (\%)} = [(\text{FW} - \text{DW})/(\text{TW} - \text{DW})] \times 100 \quad (7)$$

where FW, DW, and TW = fresh, dry, and turgid weights, respectively.

Ten discs per leaf (0.5 cm diameter) were collected from ten freshly expanded leaves and used to determine the electrolyte leakage of the membrane [124]. Leaf discs were washed three times with deionized water to remove dust, and then kept in closed tubes containing 10 mL of deionized water and shook for 30 min using a lab shaker, Model Bioshake 3000-T (Kobenhavn, NV, Danmark), and left in a dark at room temperature (≈22–23 °C) for 24 h. The initial electrical conductivity of the solution (EC1) was determined using an electrical conductivity meter, Model HI9032 (Hanna Instruments, Woonsocket, RI, USA). Samples were then kept in a 'Precision ™ General Purpose' water bath (ThermoFisher Scientific, Waltham, MA, USA) at 80 °C for 20 min to release all endogenous electrolytes. Afterward, the solution was cooled down to 25 °C, its final electrical conductivity (EC2) was estimated, and the percentage of EL was calculated using the following equation:

$$EL\ (\%) = (EC1/EC2) \times 100 \tag{8}$$

Lipid peroxidation of the membrane was determined with MDA concentration (μmol·mg^{-1} fw) using the thiobarbituric acid reactive substance assay (TBARS) [127]. A fresh leaf sample (100 mg) was extracted in 1% trichloroacetic acid (TCA) and then centrifuged for 10 min at 10,000× *g* using a benchtop general purpose centrifuge Model Allegra V-15R (Beckman Coulter Life Sciences, Indianapolis, IN, USA). The supernatant (1 mL) was mixed with 4 mL Thiobarbituric acid (TBA) [0.5%], heated for 30 min at 95 °C, and then cooled in an ice bath, followed by centrifugation at 5000× *g* for 5 min. The absorbance of the aqueous phase of the extracted solution was recorded at 532 and 600 nm using the spectrophotometer.

The non-radical H_2O_2 (nmol·g^{-1} fw) was determined by homogenizing a fresh leaf sample (100 mg) in 0.1% trichloroacetic acid (TCA). The homogenate was centrifuged at 12,000× *g* for 15 min, and a sample of the supernatant (0.5 mL) was mixed with 0.5 mL potassium phosphate buffer (10 mM, pH 7.0) and 1 mL potassium iodide (1 M). The absorbance of the aqueous phase of the extracted solution was recorded at 390 nm using the spectrophotometer, and a standard curve was used to calculate H_2O_2 content [128].

Leaf ABA content (ng·g^{-1} fw) was determined according to Koshioka et al. [129] using high-performance liquid chromatography (M5 Microflow HPLC system; SCIEX, Framingham, MA, USA).

4.5. Experimental Design and Statistical Analysis

The experimental design was in a randomized complete block system, as a split-plot experiment of three nano-silicon concentrations (main plots) and three irrigation regimes (sub-plots); a total of nine treatments, 10 replicates each. Each replicate was represented by one tree [130].

Data were analyzed using CoStat—Statistics Software (version 4.20) [131]. Data were first run for numerical normality and homogeneity of variance using the Shapiro–Wilk's and Levene's tests, respectively, and then the analysis of variance was performed, and means were compared using Duncan's multiple range tests (DMRT) at $p \leq 0.05$. Standard error bars were also added for mean comparisons in the figures [132,133].

5. Conclusions

Imposing water stress on a drought-sensitive 'Kalamata' olive trees induced oxidative stress, which was expressed as elevated H_2O_2, MDA, and electrolyte leakage with an increased fruit drop. The application of nSi generally improved fruit yield, fruit weight, leaf total chlorophyll, and RWC, and lowered fruit drop, leaf proline, soluble sugars, H_2O_2, electrolyte leakage, and ABA, with a more pronounced effect at moderate water stress (90% IWR). Increasing the nSi foliar concentration from 150 to 200 mg·L^{-1} improved the morphological, physiological, and biochemical characteristics of the olive trees, resulting in improved growth, development, and productivity under semi-arid conditions. Future research could study the molecular basis of olive defense mechanisms to enhance the drought tolerance of the 'Kalamata' cultivar in order to withstand more severe drought conditions, like other olive cultivars, amid the global water scarcity. Future research could also explore the cold tolerance mechanisms of this cultivar.

Author Contributions: I.F.H. and M.S.G.; methodology, I.F.H., R.A., M.S.G. and C.C.O.; validation, I.F.H., M.S.G., C.C.O. and H.M.K.; formal analysis, I.F.H., H.M.K., H.M.H.-V. and S.M.A.-E.; investigation, I.F.H., R.A., H.M.K. and S.M.A.-E.; resources, C.C.O., H.M.H.-V. and S.M.A.-E.; data curation, I.F.H., R.A. and H.M.K.; visualization, I.F.H. and M.S.G.; project administration, C.C.O. and H.M.H.-V.; writing—original draft preparation, R.A., C.C.O. and S.M.A.-E. writing—review and editing. All authors have read and agreed to the published version of the manuscript.

Funding: Authors declare no outside financial support.

Institutional Review Board Statement: Not applicable.

Informed Consent Statement: Not applicable.

Data Availability Statement: Not applicable.

Acknowledgments: The authors gratefully thank the owner of the olive orchard and his staff for providing all the required field materials to do this research. The authors' appreciation also extends to the staff of the Water Relations and Field Irrigation Department, Agricultural and Biological Research Institute, National Research Center, for their excellent technical assistance.

Conflicts of Interest: The authors declare no conflict of interest.

References

1. International Olive Council (IOC). *The Olive Tree*; International Olive Council (IOC): Madrid, Spain, 2022. Available online: https://www.internationaloliveoil.org/olive-world/olive-tree/ (accessed on 20 April 2022).
2. Food and Agriculture Organization of the United Nations (FAO). *FAO Statistics*; Food and Agriculture Organization of the United Nations (FAO): Rome, Italy, 2020. Available online: https://www.fao.org/faostat/en/#data/QCL (accessed on 20 April 2022).
3. Bayoumi, H.; Zaghloul, E.A.; Al-Gebaly, M.R.; Abd el-Ghani, S.S. An economic study of olive crop in North Sinai governorate. *Middle East J. Agric. Res.* **2014**, *3*, 994–1001.
4. Wiesman, Z. *Desert Olive Oil Cultivation: Advanced Biotechnologies*; Elsevier: New York, NY, USA, 2009; p. 147.
5. Yacout, D.A.; Soliman, N.F.; Zahran, H.F. Potentials of a Sustainable Olive Industry in Egypt. In Proceedings of the International Conference of Biotechnology and Environment (ICBE 2016), Alexandria, Egypt, 1–3 November 2016; p. 57.
6. Ogbaga, C.C.; Amir, M.; Bano, H.; Chater, C.C.; Jellason, N.P. Clarity on frequently asked questions about drought measurements in plant physiology. *Sci. Afr.* **2020**, *8*, e00405. [CrossRef]
7. Agricultural Statistics of Egypt. *Water Scarcity in Egypt: The Urgent Need for Regional Cooperation among the Nile Basin Countries. Report of the Ministry of Water Resources and Irrigation*; Government of Egypt: Cairo, Egypt, 2014; p. 5.
8. Ben Ahmed, C.; Ben Rouina, B.; Boukhris, M. Effects of water deficit on olive trees cv. Chemlali under field conditions in arid region in Tunisia. *Sci. Hortic.* **2007**, *113*, 267–277. [CrossRef]
9. Jury, W.A.; Vaux, J.H. The role of science in solving the world's emerging water problems. *Proc. Natl Acad. Sci. USA* **2005**, *102*, 15715–15720. [CrossRef] [PubMed]
10. Cosgrove, W.J.; Loucks, D.P. Water management: Current and future challenges and research directions. *Water Resour. Res.* **2015**, *51*, 4823–4839. [CrossRef]
11. Brito, C.; Dinis, L.-T.; Ferreira, H.; Moutinho-Pereira, J.; Correia, C. The role of nighttime water balance on Olea europaea plants subjected to contrasting water regimes. *J. Plant Physiol.* **2018**, *226*, 56–63. [CrossRef]
12. Farooq, M.; Wahid, A.; Kobayashi, N.; Fujita, D.; Basra, S.M.A. Plant drought stress: Effects, mechanisms and management. *Agron. Sustain. Dev.* **2009**, *29*, 185–212. [CrossRef]
13. Petridis, A.; Therios, I.; Samouris, G.; Koundouras, S.; Giannakoula, A. Effect of water deficit on leaf phenolic composition, gas exchange, oxidative damage and antioxidant activity of four Greek olive (*Olea europaea* L.) cultivars. *Plant Physiol. Biochem.* **2012**, *60*, 1–11. [CrossRef]
14. Abdallah, B.M.; Trupiano, D.; Polzella, A.; de Zio, E.; Sassi, M.; Scaloni, A.; Zarrouk, M.; Youssef, N.B.; Scippa, G.S. Unraveling physiological, biochemical and molecular mechanisms involved in olive (*Olea europaea* L. cv. Chétoui) tolerance to drought and salt stresses. *J. Plant Physiol.* **2018**, *220*, 83–95. [CrossRef]
15. English, M. Deficit irrigation. I. Analytical framework. *J. Irrig. Drain. Eng.* **1990**, *116*, 399–412. [CrossRef]
16. Costa, J.M.; Ortuno, M.F.; Chaves, M.M. Deficit irrigation as a strategy to save water: Physiology and potential application to horticulture. *J. Integr. Plant Biol.* **2007**, *49*, 1421–1434. [CrossRef]
17. Ferenes, E.; Goldhamer, D.A. Deciduous Fruit and Nut Trees. In *Irrigation of Agricultural Crops*, 1st ed.; Stewart, B.A., Nielsen, D.R., Eds.; American Society of Agronomy: Madison, WI, USA, 1990; pp. 987–1017.

18. Panigrahi, P.; Srivastava, A.K. Effective management of irrigation water in citrus orchards under a water scarce hot sub-humid region. *Sci. Hortic.* **2016**, *210*, 6–13. [CrossRef]
19. Zuazo, V.H.D.; Garcia-Tejero, I.F.; Rodriguez, B.C.; Tarifa, D.F.; Ruiz, B.G.; Sacristan, P.C. Deficit irrigation stratigies for subtropical mango farming. A review. *Agr. Sustain. Develop.* **2021**, *41*, 13. [CrossRef]
20. Rapoport, H.F.; Perez-Priego, O.; Orgaz, F.; Martins, P. Water deficiet effects during olive tree inflorescence and flower development. *Acta Hortic.* **2011**, *888*, 157–162. [CrossRef]
21. Connor, D.J.; Fereres, E. The physiology of adaptation and yield expression in olive. *Hortic. Rev.* **2010**, *31*, 155–229.
22. Sofo, A.; Manfreda, S.; Fiorentino, M.; Dichio, B.; Xiloyannis, C. The olive tree: A paradigm for drought tolerance in Mediterranean climates. *Hydrol. Earth Syst. Sci.* **2008**, *12*, 293–301. [CrossRef]
23. Shaheen, M.A.; Hegazi, A.A.; Hmmam, I.S. Effect of water stress on vegetative characteristics and leaves chemical constituents of some transplants olive cultivars. *Am-Eurasian J. Agric. Environ. Sci.* **2011**, *11*, 663–670.
24. Ahmed, C.B.; Rouina, B.B.; Sensoy, S.; Boukhris, M.; Abdallah, F.B. Changes in gas exchange, proline accumulation and antioxidative enzyme activities in three olive cultivars under contrasting water availability regimes. *Env. Exp. Bot.* **2009**, *67*, 345–352. [CrossRef]
25. Bolat, I.; Dikilitas, M.; Ikinci, A.; Ercisli, S.; Tonkaz, T. Morphological, physiological, biochemical characteristics and bud success responses of myrobolan 29 c plum rootstock subjected to water stress. *Can. J. Plant Sci.* **2015**, *96*, 485–493. [CrossRef]
26. Saiki, S.T.; Ishida, A.; Yoshimura, K.; Yazaki, K. Physiological mechanisms of drought-induced tree die-off in relation to carbon, hydraulic and respiratory stress in a drought-tolerant woody plants. *Sci. Rep.* **2017**, *7*, 2995. [CrossRef]
27. Tahir, F.M.; Ibrahim, M.; Hamid, K. Effect of drought stress on vegetative and reproductive growth behavior of mango (*Mangifera indica* L.). *Asian J. Plant Sci.* **2003**, *2*, 116–118. [CrossRef]
28. Tadina, N.; Germ, M.; Kreft, I.; Breznik, B.; Gaberščik, A. Effects of water deficit and selenium on common buckwheat (Fagopyrum esculentum Moench.) plants. *Photosynthetica* **2007**, *45*, 472–476. [CrossRef]
29. Djanaguiraman, M.; Prasad, P.V.; Seppanen, M. Selenium protects sorghum leaves from oxidative damage under high temperature stress by enhancing antioxidant defense system. *Plant Physiol. Biochem.* **2010**, *48*, 999–1007. [CrossRef] [PubMed]
30. Islam, F.H.; Abou Leila, B.; Gaballah, M.E.L.; Wakeel, H. Effect of antioxidants on Citrus leaf anatomical structure grown under saline irrigation water. *Plant Arch.* **2019**, *19*, 840–845.
31. Ahmadipour, S.; Arji, I.; Ebadi, A.; Abdossi, V. Physiological and biochemical responses of some olive cultivars (*Olea europaea* L.) to water stress. *Cell. Molec. Biol.* **2018**, *64*, 20–29. [CrossRef]
32. Boughalleb, F.; Hajlaoui, H. Physiological and anatomical changes induced by drought in two olive cultivars (cv Zalmati and Chemlali). *Acta Physiol Plant* **2011**, *33*, 53–65. [CrossRef]
33. Fernandez, V.; Sotiropoulos, T.; Brown, P.H. *Foliar Fertilization: Scientific Principles and Field Practices*, 1st ed.; International Fertilizer Industry Association (IFA): Paris, France, 2013; p. 140.
34. Elsheery, N.I.; Helaly, M.N.; El-Hoseiny, H.M.; Alam-Eldein, S.M. Zinc oxide and silicon nanoparticles to improve the resistance mechanism and annual productivity of salt-stressed mango trees. *Agronomy* **2020**, *10*, 558. [CrossRef]
35. Regni, L.; Del Buono, D.; Micheli, M.; Facchin, S.L.; Tolisano, C.; Proietti, P. Effects of Biogenic ZnO Nanoparticles on Growth, Physiological, Biochemical Traits and Antioxidants on Olive Tree In Vitro. *Horticulturae* **2022**, *8*, 161. [CrossRef]
36. Zahedi, S.M.; Hosseini, M.S.; Meybodi, N.D.H.; da Silva, J.A.T. Foliar application of selenium and nano-selenium affects pomegranate (*Punica granatum* cv. Malase Saveh) fruit yield and quality. *South Afr. J. Bot.* **2019**, *124*, 350–358. [CrossRef]
37. Scott, N.; Chen, H. *Nanoscale Science and Engineering for Agriculture and Food Systems. A Report Submitted to Cooperative State Research, Education and Extension Service*; National Planning Workshop, The United States Department of Agriculture (USDA): Washington, DC, USA, 2003; p. 62. Available online: http://www.nseafs.cornel.edu/web.roadmap.pdf (accessed on 25 April 2022).
38. Mosanna, R.; Khalilvand, B.E. Morpho-physiological response of maize (*Zea mays* L.) to zinc nano-chelate foliar and soil application at different growth stages. *J. New Biol. Rep.* **2015**, *4*, 46–50.
39. Boutchuen, A.; Zimmerman, D.; Aich, N.; Masud, A.M.; Arabshahi, A.; Palchoudhury, S. Increased Plant Growth with Hematite Nanoparticle Fertilizer Drop and Determining Nanoparticle Uptake in Plants Using Multimodal Approach. *J. Nanomater.* **2019**, *2019*, 6890572. [CrossRef]
40. Helaly, M.N.; El-Hoseiny, H.; El-Sheery, N.I.; Rastogi, A.; Kalaji, H.M. Regulation and physiological role of silicon in alleviating drought stress of mango. *Plant Physiol. Biochem.* **2017**, *118*, 31–44. [CrossRef] [PubMed]
41. Imtiaz, M.; Rizwan, M.S.; Mushtaq, M.A.; Ashraf, M.; Shahzad, S.M.; Yousaf, B.; Tu, S. Silicon occurrence, uptake, transport and mechanisms of heavy metals, minerals and salinity enhanced tolerance in plants with future prospects: A review. *J. Environ. Manag.* **2016**, *183*, 521–529. [CrossRef] [PubMed]
42. Ma, J.F.; Takahasi, E. *Soil, Fertilizer, and Plant Silicon Research in Japan*, 1st ed.; Elsevier: Amsterdam, The Netherlands, 2002; p. 294.
43. Ma, J.F.; Miyake, Y.; Takahashi, E. Silicon as a beneficial element for crop plants. In *Silicon in Agriculture. Studies in Plant Science, 8*; Datnoff, L.E., Snyder, G.H., Korndorfer, G.H., Eds.; Elsevier: Amsterdam, The Netherlands, 2001; pp. 17–39.
44. Laane, H.-M. The Effects of Foliar Sprays with Different Silicon Compounds. *Plants* **2018**, *7*, 45. [CrossRef] [PubMed]
45. Sivanesan, I.; Park, S.W. The role of silicon in plant tissue culture. *Front. Plant Sci.* **2014**, *5*, 571. [CrossRef]
46. Artyszak, A. Effect of Silicon Fertilization on Crop Yield Quantity and Quality—A Literature Review in Europe. *Plants* **2018**, *7*, 54. [CrossRef]

47. Balakhnina, T.; Borkowska, A. Effects of silicon on plant resistance to environmental stresses: Review. *Int. Agrophys.* **2013**, *27*, 225–232. [CrossRef]
48. Ashkavand, P.; Zarafshar, M.; Tabari, M.; Mirzaie, J.; Nikpour, A.; Bordbar, S.K.; Struve, D.; Striker, G.G. Application of SiO_2 nanoparticles as pretreatment alleviates the impact of drought on the physiological performance of *Prunus mahaleb* L. (Rosaceae). *Bol. Soc. Argent. Bot.* **2018**, *53*, 207–219. [CrossRef]
49. Behboudi, F.; Tahmasebi Sarvestani, Z.; Kassaee, M.Z.; Modares Sanavi, S.; Sorooshzadeh, A. Improving growth and yield of wheat under drought stress via application of SiO_2 nanoparticles. *J. Agric. Sci. Technol.* **2018**, *20*, 1479–1492.
50. Avestan, S.; Ghasemnezhad, M.; Esfahani, M.; Byrt, C.S. Application of nano-silicon dioxide improves salt stress tolerance in strawberry plants. *Agronomy* **2019**, *9*, 246. [CrossRef]
51. Hellala, F.; Amerb, A.K.; El-Sayeda, S.; El-Azab, K. Mitigation The negative effect of water stress on barley by nano silica application. *Plant Arch.* **2020**, *20*, 3224–3231.
52. Desoky, E.-S.M.; Mansour, E.; El-Sobky, E.-S.E.; Abdul-Hamid, M.I.; Taha, T.F.; Elakkad, H.A.; Arnaout, S.M.; Eid, R.S.; El-Tarabily, K.A.; Yasin, M.A. Physio-biochemical and agronomic responses of faba beans to exogenously applied nano-silicon under drought stress conditions. *Front. Plant. Sci.* **2021**, *12*, 637783. [CrossRef] [PubMed]
53. Bayati, P.; Karimmojeni, H.; Razmjoo, J.; Pucci, M.; Abate, G.; Baldwin, T.C.; Mastinu, A. Physiological, Biochemical, and Agronomic Trait Responses of *Nigella sativa* Genotypes to Water Stress. *Horticulturae* **2022**, *8*, 193. [CrossRef]
54. Elsheery, N.I.; Sunoj, V.; Wen, Y.; Zhu, J.; Muralidharan, G.; Cao, K. Foliar application of nanoparticles mitigates the chilling effect on photosynthesis and photoprotection in sugarcane. *Plant Physiol. Biochem.* **2020**, *149*, 50–60. [CrossRef]
55. Khot, L.R.; Sankaran, S.; Maja, J.M.; Ehsani, R.; Schuster, E.W. Applications of nanomaterials in agricultural production and crop protection: A review. *Crop Protec.* **2012**, *35*, 64–70. [CrossRef]
56. Matussin, S.; Harunsani, M.H.; Tan, A.L.; Khan, M.M. Plant-extract-mediated SnO_2 nanoparticles: Synthesis and applications. *ACS Sustain. Chem. Eng.* **2020**, *8*, 3040–3054. [CrossRef]
57. El-Dengawy, E.; EL-Abbasy, U.; El-Gobba, M.H. Influence of nano-silicon treatment on growth behavior of 'Sukkary'and 'Gahrawy'mango root-stocks under salinity stress. *J. Plant Prod.* **2021**, *12*, 49–61.
58. Al-Wasfy, M.M. Response of Sakkoti date palms to foliar application of royal jelly, silicon and vitamins B. *J. Amer. Sci.* **2013**, *9*, 315–321.
59. Zhang, W.; Xie, Z.; Wang, L.; Li, M.; Lang, D.; Zhang, X. Silicon alleviates salt and drought stress of Glycyrrhiza uralensis seedling by altering antioxidant metabolism and osmotic adjustment. *J. Plant. Res.* **2017**, *130*, 611–624. [CrossRef]
60. Ismail, L.M.; Soliman, M.I.; Abd El-Aziz, M.H.; Abdel-Aziz, H.M. Impact of Silica Ions and Nano Silica on Growth and Productivity of Pea Plants under Salinity Stress. *Plants* **2022**, *11*, 494. [CrossRef]
61. Mahmoud, L.M.; Dutt, M.; Shalan, A.M.; El-Kady, M.E.; El-Boray, M.S.; Shabana, Y.M.; Grosser, J.W. Silicon nanoparticles mitigate oxidative stress of in vitro-derived banana (Musa acuminata 'Grand Nain') under simulated water deficit or salinity stress. *S. Afr. J. Bot.* **2020**, *132*, 155–163. [CrossRef]
62. Molahoseini, H.; Feizian, M.; Mehdi Pour, E.; Davazdah Emami, S. Investigating the Effect of Coated Nanosilicon Oxide with Humic Acid on Yield, Ion Composition and Salinity Tolerance of Black Cumin (*Nigella sativa* L.). *Iran. J. Soil Water Res.* **2020**, *51*, 2711–2723.
63. Muhammad, H.M.D.; Abbas, A.; Ahmad, R. Fascinating Role of Silicon Nanoparticles to Mitigate Adverse Effects of Salinity in Fruit Trees: A Mechanistic Approach. *Silicon* **2022**, *9*, 1–8. [CrossRef]
64. Andreotti, C.; Rouphael, Y.; Colla, G.; Basile, B. Rate and Timing of Application of Biostimulant Substances to Enhance Fruit Tree Tolerance toward Environmental Stresses and Fruit Quality. *Agronomy* **2022**, *12*, 603. [CrossRef]
65. Attia, E.A.; Elhawat, N. Combined foliar and soil application of silica nanoparticles enhances the growth, flowering period and flower characteristics of marigold (*Tagetes erecta* L.). *Sci. Hortic.* **2021**, *282*, 110015. [CrossRef]
66. González-García, Y.; Cárdenas-Álvarez, C.; Cadenas-Pliego, G.; Benavides-Mendoza, A.; Cabrera-de-la-Fuente, M.; Sandoval-Rangel, A.; Valdés-Reyna, J.; Juárez-Maldonado, A. Effect of three nanoparticles (Se, Si and Cu) on the bioactive compounds of bell pepper fruits under saline stress. *Plants* **2021**, *10*, 217. [CrossRef] [PubMed]
67. González-Moscoso, M.; Martínez-Villegas, N.V.; Cadenas-Pliego, G.; Benavides-Mendoza, A.; Rivera-Cruz, M.d.C.; González-Morales, S.; Juárez-Maldonado, A. Impact of silicon nanoparticles on the antioxidant compounds of tomato fruits stressed by arsenic. *Foods* **2019**, *8*, 612. [CrossRef]
68. Mahmoud, L.M.; Shalan, A.M.; El-Boray, M.S.; Vincent, C.I.; El-Kady, M.E.; Grosser, J.W.; Dutt, M. Application of silicon nanoparticles enhances oxidative stress tolerance in salt stressed 'Valencia' sweet orange plants. *Sci. Hortic.* **2022**, *295*, 110856. [CrossRef]
69. Hussain, A.; Rizwan, M.; Ali, Q.; Ali, S. Seed priming with silicon nanoparticles improved the biomass and yield while reduced the oxidative stress and cadmium concentration in wheat grains. *Environ. Sci. Pollut. Res.* **2019**, *26*, 7579–7588. [CrossRef]
70. Fatemi, H.; Esmaiel Pour, B.; Rizwan, M. Foliar application of silicon nanoparticles affected the growth, vitamin C, flavonoid, and antioxidant enzyme activities of coriander (*Coriandrum sativum* L.) plants grown in lead (Pb)-spiked soil. *Environ. Sci. Pollut. Res.* **2021**, *28*, 1417–1425. [CrossRef]
71. Hassan, I.F.; Gaballah, M.S.; Ogbaga, C.C.; Murad, S.A.; Brysiewicz, A.; Bakr, B.M.; Mira, A.; Alam-Eldein, S.M. Does melatonin improve the yield attributes of field-droughted banana under Egyptian semi-arid conditions? *J. Water Land Dev.* **2022**, *52*, 221–231.

72. Helaly, M.N.; El-Hoseiny, H.M.; Elsheery, N.I.; Kalaji, H.M.; Santos-Villalobos, S.d.l.; Wróbel, J.; Hassan, I.F.; Gaballah, M.S.; Abdelrhman, L.A.; Mira, A.M. 5-Aminolevulinic acid and 24-epibrassinolide improve the drought stress resilience and productivity of banana plants. *Plants* **2022**, *11*, 743. [CrossRef] [PubMed]
73. Abdallah, D.B.; Tounsi, S.; Gharsallah, H.; Hammami, A.; Frikha-Gargouri, O. Lipopeptides from Bacillus amyloliquefaciens strain 32a as promising biocontrol compounds against the plant pathogen Agrobacterium tumefaciens. *Environ. Sci. Pollut. Res.* **2018**, *25*, 36518–36529. [CrossRef] [PubMed]
74. Ashkavand, P.; Tabari, M.; Zarafshar, M.; Tomásková, I.; Struve, D. Effect of SiO_2 nanoparticles on drought resistance in hawthorn seedlings. *Leśne Pr. Badaw.* **2015**, *76*, 350–359. [CrossRef]
75. Brito, C.; Dinis, L.-T.; Moutinho-Pereira, J.; Correia, C.M. Drought stress effects and olive tree acclimation under a changing climate. *Plants* **2019**, *8*, 232. [CrossRef]
76. Maghsoudi, K.; Emam, Y.; Pessarakli, M. Effect of silicon on photosynthetic gas exchange, photosynthetic pigments, cell membrane stability and relative water content of different wheat cultivars under drought stress conditions. *J. Plant Nutr.* **2016**, *39*, 1001–1015. [CrossRef]
77. Patumi, M.; d'Andria, R.; Marsilio, V.; Fontanazza, G.; Morelli, G.; Lanza, B. Olive and olive oil quality after intensive monocone olive growing (*Olea europaea* L., cv. Kalamata) in different irrigation regimes. *Food Chem.* **2002**, *77*, 27–34. [CrossRef]
78. Greven, M.; Neal, S.; Green, S.; Dichio, B.; Clothier, B. The effects of drought on the water use, fruit development and oil yield from young olive trees. *Agric. Water Manag.* **2009**, *96*, 1525–1531. [CrossRef]
79. Machado, M.; Felizardo, C.; Fernandes-Silva, A.A.; Nunes, F.M.; Barros, A. Polyphenolic compounds, antioxidant activity and l-phenylalanine ammonia-lyase activity during ripening of olive cv. "Cobrançosa" under different irrigation regimes. *Food Res. Int.* **2013**, *51*, 412–421. [CrossRef]
80. Caruso, G.; Gucci, R.; Urbani, S.; Esposto, S.; Taticchi, A.; Di Maio, I.; Selvaggini, R.; Servili, M. Effect of different irrigation volumes during fruit development on quality of virgin olive oil of cv. Frantoio. *Agric. Water Manag.* **2014**, *134*, 94–103. [CrossRef]
81. Hassan, I.F.; Gaballah, M.S.; El-Hoseiny, H.M.; El-Sharnouby, M.E.; Alam-Eldein, S.M. Deficit Irrigation to Enhance Fruit Quality of the 'African Rose'Plum under the Egyptian Semi-Arid Conditions. *Agronomy* **2021**, *11*, 1405. [CrossRef]
82. Karimi, S.; Rahemi, M.; Rostami, A.A.; Sedaghat, S. Drought effects on growth, water content and osmoprotectants in four olive cultivars with different drought tolerance. *Int. J. Fruit Sci.* **2018**, *18*, 254–267. [CrossRef]
83. Sharp, R.E.; LeNoble, M.E.; Else, M.A.; Thome, E.T.; Gherardi, F. Endogenous ABA maintains shoot growth in tomato independently of effects on plant water balance: Evidence for an interaction with ethylene. *J. Exp. Bot.* **2000**, *51*, 1575–1584. [CrossRef]
84. Ma, C.C.; Li, Q.F.; Gao, Y.B.; Xin, T.R. Effects of silicon application on drought resistance of cucumber plants. *Soil Sci. Plant Nutr.* **2004**, *50*, 623–632. [CrossRef]
85. Laxa, M.; Liebthal, M.; Telman, W.; Chibani, K.; Dietz, K.-J. The role of the plant antioxidant system in drought tolerance. *Antioxidants* **2019**, *8*, 94. [CrossRef]
86. Abdallah, M.B.; Methenni, K.; Nouairi, I.; Zarrouk, M.; Youssef, N.B. Drought priming improves subsequent more severe drought in a drought-sensitive cultivar of olive cv. Chétoui. *Sci. Hortic.* **2017**, *221*, 43–52. [CrossRef]
87. Ogbaga, C.C.; Stepien, P.; Johnson, G.N. Sorghum (*Sorghum bicolor*) varieties adopt strongly contrasting strategies in response to drought. *Phys. Plant.* **2014**, *152*, 389–401. [CrossRef]
88. Noor, R.; Yasmin, H.; Ilyas, N.; Nosheen, A.; Hassan, M.N.; Mumtaz, S.; Khan, N.; Ahmad, A.; Ahmad, P. Comparative analysis of iron oxide nanoparticles synthesized from ginger (*Zingiber officinale*) and cumin seeds (*Cuminum cyminum*) to induce resistance in wheat against drought stress. *Chemosphere* **2022**, *292*, 133201. [CrossRef]
89. Sreelakshmi, B.; Induja, S.; Adarsh, P.; Rahul, H.; Arya, S.; Aswana, S.; Haripriya, R.; Aswathy, B.; Manoj, P.; Vishnudasan, D. Drought stress amelioration in plants using green synthesised iron oxide nanoparticles. *Mater. Today Proc.* **2021**, *41*, 723–727. [CrossRef]
90. Shallan, M.A.; Hassan, H.M.; Namich, A.A.; Ibrahim, A.A. Biochemical and physiological effects of TiO_2 and SiO_2 nanoparticles on cotton plant under drought stress. *Res. J. Pharm. Biol. Chem. Sci.* **2016**, *7*, 1540–1551.
91. Ma, C.; Liu, H.; Guo, H.; Musante, C.; Coskun, S.H.; Nelson, B.C.; White, J.C.; Xing, B.; Dhankher, O.P. Defense mechanisms and nutrient displacement in *Arabifopsis thaliana* upon exposure to CeO_2 and In_2O_3 nanoparticles. *Environ. Sci. Nano* **2016**, *3*, 1369–1379. [CrossRef]
92. Ghorbanpour, M.; Mohammadi, H.; Kariman, K. Nanosilicon-based recovery of barley (Hordeum vulgare) plants subjected to drought stress. *Environ. Sci. Nano* **2020**, *7*, 443–461. [CrossRef]
93. Luyckx, M.; Hausman, J.-F.; Lutts, S.; Guerriero, G. Silicon and plants: Current knowledge and technological perspectives. *Front. Plant Sci.* **2017**, *8*, 411. [CrossRef]
94. Parveen, A.; Liu, W.; Hussain, S.; Asghar, J.; Perveen, S.; Xiong, Y. Silicon priming regulates morpho-physiological growth and oxidative metabolism in maize under drought stress. *Plants* **2019**, *8*, 431. [CrossRef] [PubMed]
95. Moradshahi, A.; Salehi, E.A.B.; Khold, B.B. Some Physiological Responses of Canola (*Brassica napus* L.) to Water Deficit Stress Under Laboratory Conditions. *Iran. J. Sci. Technol.* **2004**, *28*, 43–50.
96. Ghafoor, R.; Akram, N.A.; Rashid, M.; Ashraf, M.; Iqbal, M.; Lixin, Z. Exogenously applied proline induced changes in key anatomical features and physio-biochemical attributes in water stressed oat (*Avena sativa* L.) plants. *Physiol. Mol. Biol. Plants* **2019**, *25*, 1121–1135. [CrossRef]

97. Mansour, M.M.F. Nitrogen containing compounds and adaptation of plants to salinity stress. *Biol. Plant.* **2000**, *43*, 491–500. [CrossRef]
98. Meister, A. *Biochemistry of the Amino Acids*, 2nd ed.; Elsevier Science: Amsterdam, The Netherlands, 2012; ISBN 0323161472.
99. El Sayed, O.M.; El Gammal, O.H.M.; Salama, A.S.M. Effect of proline and tryptophan amino acids on yield and fruit quality of Manfalouty pomegranate variety. *Sci. Hortic.* **2014**, *169*, 1–5. [CrossRef]
100. Caronia, A.; Gugliuzza, G.; Inglese, P. Influence of L-proline on citrus sinensis (L.) ['new hall' and 'tarocco Sciré'] fruit quality. *Acta Hortic.* **2010**, *884*, 423–426. [CrossRef]
101. Kahlaoui, B.; Hachicha, M.; Misle, E.; Fidalgo, F.; Teixeira, J. Physiological and biochemical responses to the exogenous application of proline of tomato plants irrigated with saline water. *J. Saudi Soc. Agric. Sci.* **2018**, *17*, 17–23. [CrossRef]
102. Karimi, S.; Abbaspour, H.; Sinak, J.; Makarian, H. Evaluation of drought stress and foliar chitosan on biochemical characterices of castor bean (*Ricinus communis* L.). *Res. J. Biol. Sci.* **2012**, *7*, 117–122. [CrossRef]
103. Akhtar, N.; Ilyas, N.; Hayat, R.; Yasmin, H.; Noureldeen, A.; Ahmad, P. Synergistic effects of plant growth promoting rhizobacteria and silicon dioxide nano-particles for amelioration of drought stress in wheat. *Plant Physiol. Biochem.* **2021**, *166*, 160–176. [CrossRef] [PubMed]
104. Mahmoud, A.W.M.; Abdeldaym, E.A.; Abdelaziz, S.M.; El-Sawy, M.B.; Mottaleb, S.A. Synergetic effects of zinc, boron, silicon, and zeolite nanoparticles on confer tolerance in potato plants subjected to salinity. *Agronomy* **2019**, *10*, 19. [CrossRef]
105. Apel, K.; Hirt, H. Reactive oxygen species, metabolism, oxidative stress and signal transduction. *Ann. Rev. Plant Biol.* **2004**, *55*, 373–399. [CrossRef] [PubMed]
106. Blum, A. Osmotic adjustment is a prime drought stressadaptive engine in support of plant production. *Plant Cell Environ.* **2017**, *40*, 4–10. [CrossRef] [PubMed]
107. Munns, R. Comparative physiology of salt and water stress. *Plant Cell Environ.* **2002**, *25*, 239–250. [CrossRef]
108. Sen, A. Oxidative Stress Studies in Plant Tissue Culture. In *Antioxidant Enzyme; World's Largest Science, Technology and Medicine Open Access Book*; Chapter 3; El-Missiry, M.A., Ed.; INTECH: Rijeka, Croatia, 2012; pp. 59–88. Available online: https://doi.org/10.5772/2895 (accessed on 17 February 2021).
109. Gadallah, M.A.A. Effect of proline and glycinebetaine on Vicia faba responses to salt stress. *Biol. Plant.* **1999**, *42*, 249–257. [CrossRef]
110. Diaz-Mula, H.M.; Zapata, P.J.; Guillen, F.; Castilo, S.; Martinez-Romero, D.; Valero, D.; Serrano, M. Changes in phytochemical and nutritive parameters and bioactive compounds during development and on-tree ripening of eight plum cultivars: A comparative study. *J. Sci. Food Agric.* **2008**, *88*, 2499–2507. [CrossRef]
111. Norman, S.M.; Maier, V.P.; Pon, D.L. Abscisic acid accumulation and carotenoid and chlorophyll content in relation to water stress and leaf age of different types of citrus. *J. Agric. Food Chem.* **1990**, *38*, 1326–1334. [CrossRef]
112. Ibrahim, M.H.; Jaafar, H.Z.E. Abscisic acid induced changes in production of primary and secondary metabolites, photosynthetic cabacity, antioxidant capability, antioxidant enzymes and lipoxygenase inhibitory activity of *Orthosiphon stamineus* Benth. *Molecules* **2013**, *18*, 7957–7976. [CrossRef]
113. Shu, S.; Gao, P.; Li, L.; Yuan, Y.; Sun, J.; Guo, S. Abscisic acid-induced H_2O_2 accumulation enhances antioxidant capacity in pumpkin-grafted cucumber leaves under $Ca(NO_3)_2$ stress. *Front. Plant Sci.* **2016**, *7*, 1489. [CrossRef] [PubMed]
114. Ming-Yi, J.; Jian-Hua, Z. Abscisic acid and antioxidant defense in plant cells. *Acta Bot. Sin.* **2004**, *1*, 1–9.
115. Hegazi, E.S.; El-Motaium, R.A.; Yehia, T.A.; Hashem, M.E. Effect of foliar boron application on boron, chlorophyll, phenol, sugars and hormones concentration of olive (*Olea europea* L.) buds, leaves, and fruits. *J. Plant Nutr.* **2018**, *41*, 749–765. [CrossRef]
116. Johnson, R.S.; Handley, D.F.; Day, K.R. Postharvest water stress of an early maturing plum. *J. Hort. Sci.* **1994**, *69*, 1035–1041. [CrossRef]
117. Aung, L.H.; Houch, L.G.; Norman, S.M. The abscisic acid content of citrus with special references to lemon. *J. Exp. Bot.* **1991**, *42*, 1083–1088. [CrossRef]
118. Pinillos, V.; Ibáñez, S.; Cunha, J.M.; Hueso, J.J.; Cuevas, J. Postveraison deficit irrigation effects on fruit quality and yield of "Flame Seedless" table grape cultivated under greenhouse and net. *Plants* **2020**, *9*, 1437. [CrossRef]
119. Rasmussen, G.K. Gibberellin and cell-wall hydrolysis as related to the low response of 'Valencia' oranges to abscission chemicals. *HortScience* **1981**, *16*, 497–498.
120. Burns, J.K.; Lewandowski, D.J. *Genetics and Expression of Pectinmethylesterase, Endo-B-Glucanase and Polygalacturonase Genes in Valencia Oranges*; Proc. 1st Int.; Citrus Biotechnology Symposium: Eilat, Israel, 2000; pp. 65–80.
121. Kazokas, W.C.; Burns, J.K. Cellulase activity and gene expression in citrus fruit abscission zones during and after ethylene treatment. *J. Amer. Soc. Hort. Sci.* **1998**, *123*, 781–786. [CrossRef]
122. Wilde, S.A.; Corey, R.B.; Lyer, J.G.; Voight, G.K. *Soil and Plant Analysis for Tree Culture*, 3rd ed.; Oxford and IBH. Puplishing Co.: New Delhi, India, 1985; pp. 93–106.
123. Allen, R.G.; Pereira, L.S.; Raes, D.; Smith, M. FAO Irrigation and drainage paper No. 56. *Rome Food Agric. Organ. United Nations* **1998**, *56*, e156.
124. Saini, R. *Laboratory Manual of Analytical Techniques in Horticulture*; Agrobios: Jodhpur, India, 2001.
125. Bates, L.S.; Waldren, R.P.; Teare, I. Rapid determination of free proline for water-stress studies. *Plant Soil* **1973**, *39*, 205–207. [CrossRef]

126. Buysse, J.; Merckx, R. An improved colorimetric method to quantify sugar content of plant tissue. *J. Exp. Bot.* **1993**, *44*, 1627–1629. [CrossRef]
127. Heath, R.L.; Packer, L. Photoperoxidation in isolated chloroplasts: I. Kinetics and stoichiometry of fatty acid peroxidation. *Arch. Biochem. Biophys.* **1968**, *125*, 189–198. [CrossRef]
128. Sergiev, I.; Alexieva, V.; Karanov, E. Effect of spermine, atrazine and combination between them on some endogenous protective systems and stress markers in plants. *Compt. Rend. Acad. Bulg. Sci.* **1997**, *51*, 121–124.
129. Koshioka, M.; Harada, J.M.; Noma, T.; Sassa, T.; Ogiama, K.; Taylor, S.; Rood, S.B.; Legge, R.L.; Pharis, R.P.K. Reversed-phase C18 high performance liquid chromatography of acidic and conjugated gibberellins. *J. Chromatog.* **1983**, *256*, 101–115. [CrossRef]
130. Clarke, G.M. *Introduction to the Design and Analysis of Experiments*; Arnold: New York, NY, USA, 1997.
131. CoHort Software. Statistics Software. *CoStat*; Version 4.20; CoHort Software: Berkeley, CA, USA, 1990.
132. Snedecor, G.W.; Cochran, W.G. *Statistical Methods*, 7th ed.; Iowa State University Press: Ames, IA, USA, 1990; p. 593.
133. Duncan, D.B. Multiple ranges and multiple F. test. *Biometrics* **1955**, *11*, 1–42. [CrossRef]

plants

MDPI

Article

The Influence of Abiotic Stress Factors on the Morphophysiological and Phytochemical Aspects of the Acclimation of the Plant *Rhodiola semenowii* Boriss

Nina V. Terletskaya [1,2,*], Nazym K. Korbozova [1,2], Nataliya O. Kudrina [1,2,*], Tatyana N. Kobylina [2], Meruert S. Kurmanbayeva [1], Nataliya D. Meduntseva [1] and Tatyana G. Tolstikova [3]

1 Department of Biodiversity and Biological Resources, Faculty of Biology and Biotechnology, Al-Farabi Kazakh National University, Al-Farabi av., 71, Almaty 050040, Kazakhstan; naz-ik@mail.ru (N.K.K.); meruyert.kurmanbayeva@kaznu.kz (M.S.K.); nat.mdnt@gmail.com (N.D.M.)
2 Institute of Genetic and Physiology, Al-Farabi av., 93, Almaty 050040, Kazakhstan; kobylina.tatyana.n@mail.ru
3 N.N. Vorozhtsov Novosibirsk, Institute of Organic Chemistry, Siberian Branch of Russian Academy of Science, 630090 Siberia, Russia; tg_tolstikova@mail.ru
* Correspondence: teni02@mail.ru (N.V.T.); kudrina_nat@mail.ru (N.O.K.); Tel.: +7-(777)-299-3335 (N.V.T.); +7-(705)-181-1440 (N.O.K.)

Abstract: Plants of the *Crassulaceae* family are natural accumulators of many medicinal secondary metabolites (SM). This article describes the study of morphophysiological, anatomic and phytochemical responses of immature plants of *Rhodiolla semenovii* under water deficit and (or) cold-stress conditions. Changes in biomass production due to water content in plant tissues such as a decrease in water deficit and an increase in cold stress were revealed. A significant decrease in the efficiency of the photosynthetic apparatus under stress conditions was noted, based on the parameters quantum efficiency of Photosystem II and electron transport rate and energy dissipated in Photosystem II. The greatest decrease in efficiency was pointed out in conditions of water shortage. The anatomical modulations of root and shoot of *R. semenovii* under stress conditions were found. For the first time, a detailed study of the chemical composition of the ethanol extract of root and shoot of *R. semenovii* under stress was carried out using gas chromatography–mass spectrometry. The qualitative and quantitative composition of SM associated with acclimation to the effects of abiotic stresses was determined. Both nonspecific and specific phytochemical changes caused by the action of water deficiency and cold treatment were identified. It has been shown that the antioxidant system in plant tissues is complex, multicomponent, depending on a number of natural and climatic factors. Further research should be focused on the use of abiotic stressors for the targeted synthesis of bioactive SMs valuable for pharmaceutical use.

Keywords: water deficit; cold stress; water content; photosynthesis; anatomy; secondary metabolites

Citation: Terletskaya, N.V.; Korbozova, N.K.; Kudrina, N.O.; Kobylina, T.N.; Kurmanbayeva, M.S.; Meduntseva, N.D.; Tolstikova, T.G. The Influence of Abiotic Stress Factors on the Morphophysiological and Phytochemical Aspects of the Acclimation of the Plant *Rhodiola semenowii* Boriss. *Plants* **2021**, *10*, 1196. https://doi.org/10.3390/plants10061196

Academic Editors: Kinga Dziurka, Mateusz Labudda and Ewa Muszyńska

Received: 16 May 2021
Accepted: 8 June 2021
Published: 11 June 2021

1. Introduction

Nowadays dry and arid regions cover a vast territory. According to the Netherlands Environmental Assessment Agency [1], they make up about half of the earth's environment worldwide. Identification of the mechanisms of acclimation of plants to the habitat and resistance to water and temperature stress acquires particular importance in conjunction with the expected further global climate changes.

In addition to the agricultural aspect, the problem of resistance to abiotic stresses is of great natural and ecological importance, since the ability of plants to adapt to specific conditions in different parts of the planet is one of the factors that determine the distribution areas of wild species and the possibility of their introduction [2].

The changes caused by adverse environmental influences, including geo climatic and seasonal changes, external conditions of temperature, humidity, etc., negatively affect many metabolic processes in plants. Alteration of photosynthesis is one of the earliest processes to be affected by negative stress, long before stress-induced changes in plants become visible. For example, temperature fluctuations, a decrease in the availability of nutrients and water affect the parameters of chlorophyll (Chl) fluorescence, CO_2 assimilation, and a decrease in stomatal conductance and transpiration [3,4]. Among other things, abiotic stresses affect biomass production, often due to water content in plant tissues and PSM (plant secondary metabolism). Likewise, qualitatively and/or quantitatively on SM (secondary metabolites), which perform various functions important for physiological developmental processes, and the content of which can also change during growth and development in response to environmental changes [5,6]. Therefore, there is data that medicinal plants, the vegetation of which takes place against the background of exposure to abiotic stresses, usually exhibit significantly higher concentrations of SM than identical plants of the same species grown in favorable conditions. Perhaps, plant SMs which are synthesized in response to environmental stresses partly determine the ability of plants to survive and adapt to abiotic stress factors [7]. However, so far there is very little information about this well-known phenomenon [8].

There is no doubt that any plant, even those referred to by people as "medicinal", synthesizes its biologically active substances, first "for itself". In this way, it can be assumed that, depending on the nature of the stress effect (its intensity, duration, rate of exposure), biologically active structures or compounds allow the plant to flexibly and adequately respond to abiotic stress factors. In this case, the type and concentration of secondary metabolites produced under stress conditions can be determined by the genotype, specific features of physiology, and the stage of plant development. This also suggests that the production of secondary metabolites may be an indicator of a protective response [9–12].

Therefore, the protective role of plant secondary metabolism (PSM) in oxidative stress was established [13–16]. It was reported that the widely studied phenylpropanoids, in addition to the formation of structural components in plants (for example, the synthesis of lignin necessary for the formation of the cell wall), also participate in plant defense responses to abiotic stresses [17,18]. An increase in fatty acid unsaturation in the composition of membrane lipids in hypothermia was noted [19]. It was shown that seasonal climatic fluctuations influenced the production of sesquiterpenes, lactones, and phenols. The correlation of the content of these substances in plants with the amount of precipitation and temperature changes was established [20]. It was found that the effect of drought on plants promotes a higher production of such classes of secondary metabolites as terpenes, complex phenols and alkaloids [21–23]. The development and understanding of the contribution of individual stress-protective systems to the constitutive and induced resistance of plants was developed. In particular, the information on the role of such classical stress protectors as sugars [24] and proline [25] has expanded significantly in recent years. However, the formation of harmonious ideas about the functioning of plant protective systems under abiotic stresses is complicated. On the one hand, there is a variety of effects depending on the intensity and duration of exposure and the physiological state of plants at the time of exposure. On the other hand, the formation is compounded by a significant dependence of the nature of defense reactions on species characteristics.

The processes of protective stress reactions stimulate metabolic changes that can lead to the biosynthesis of biologically active compounds with pharmaceutical or nutritional value. For centuries, people in one way or another have been using the physiological acclimations of medicinal herbs as a source of improving the biosynthesis of useful bioactive compounds [13,26,27]. Nowadays, the physiology of stress associated with PSM is attracting more attention of researchers and more literary sources indicate the stabilizing role of metabolites [16,28–31]. Impulses are outlined for new practical approaches to improving product quality through the deliberate application of stress exposure during the cultivation of medicinal plants [8].

Therefore, the changes cause great interest in the synthesis of SM induced by stressful conditions. It causes the interest both from the point of view of understanding the biological processes of aclimation of a plant organism to unfavorable environmental conditions, and from the point of view of the introduction of wild valuable species and the development of pathways for the directed synthesis of plant biologically active substances valuable for pharmacy.

Plants of the *Crassulaceae* family are natural accumulators of many medicinal SM. *R. semenovii* (Regel and Herder) Boriss is a taproot, short-rhizome perennial that grows in moist, rocky soils and along riverbanks in the alpine belt up to 3500 m above sea level, preferring a sunny location. The studies carried out by phytochemists and pharmacologists show that plants of the genus *Rhodiola* and representatives of *R. semenovii*, contain proanthocyanidins, coumarins, flavone glycosides, and organic acids, tannins of the pyrogallic and pyrocatechol groups [32,33]. However, the metabolic profile of *R. semenovii* and possible changes in its metabolites, including drug-active and stress-resistant components, have not been illustrated in any way under different vegetation conditions yet.

The aim of this article was an experimental study of sudden water shortage or cold exposure on immature plants *R. semenovii*. The outcome measures will help to determine the effect of these stressors on changes in the physiological state and content of the main classes of secondary metabolites in the root and shoot of this member of the *Crassulaceae* family. Moreover, it will be useful both for understanding the mechanisms of defense against adverse conditions and for approaches to the targeted synthesis of valuable secondary metabolites.

2. Results

2.1. Morphophysiological Reactions of *R. semenowii* Immature Plants under Stress Conditions

Comparison of immature *R. semenowii* plants under stress and control conditions showed the absence of linear growth in stressed plants compared to control ones (Figure 1). At the same time, under conditions of water deficit, a significant decrease in biomass was noted, while under conditions of cold stress, its increase. The water content in the tissues of *R. semenowii* under stress conditions of water deficit was low and amounted to 46%, compared to 64% in the control plants. An opposite tendency to an increase in tissue hydration was observed under cold stress, the water content was 68%.

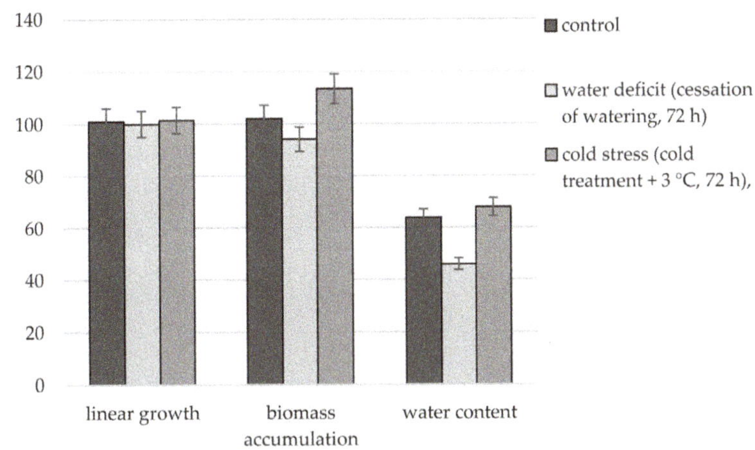

Figure 1. Changes in growth parameters, biomass accumulation and water content of *R. semenowii* under stress conditions.

The data presented in Figure 2 show that the stress of *R. semenowii* plants caused by both water deficiency and exposure to cold could be accompanied by a significant decrease in the efficiency of the photosynthetic apparatus. A decrease in the values of the maximum quantum yield of photosystem II (PSII) (ratio Fv/Fm) is shown. The rate of noncyclic electron transport through PSII (ETR) decreased. The values of the parameter of the quantum yield of the unregulated dissipation energy in PSII-Y(NO) increased. However, according to the level of decrease in the quantum yield of regulated energy dissipated in PSII Y(NPQ) and the value of the Fv/Fm index, as well as by the level of increase in Y(NO), damage to the photosynthetic apparatus was less under conditions of cold stress than under conditions of water deficiency.

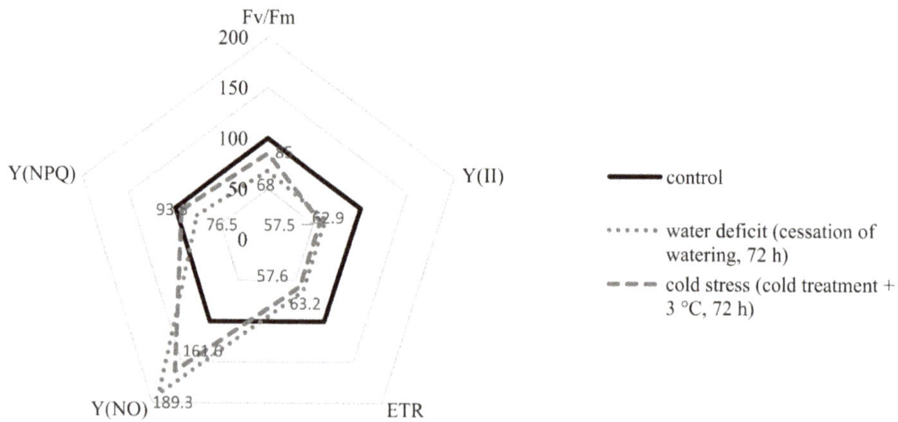

Figure 2. Changes in the activity of the photosynthetic apparatus of *R. semenowii* under stress conditions.

2.2. *Changes in the Aspects of the Anatomical Structure of* R. semenowii *under Stress Conditions*

The results of anatomical and histological examination are presented in Figures 3–5. It was determined that the rhizome of *R. semenowii* had a sparse cell structure in the control plots (Figure 3a). The cross section shows a three-layer periderm. The cell walls of the periderm are covered with suberin and are colored brown. There are three layers such as fellam, phellogen and phelloderm under the periderm. Moreover, there is a loose parenchyma with large cells with numerous small round and oval starch grains and unformed inclusions under the integumentary tissues. The cells of the parenchyma have a rounded-oblong shape.

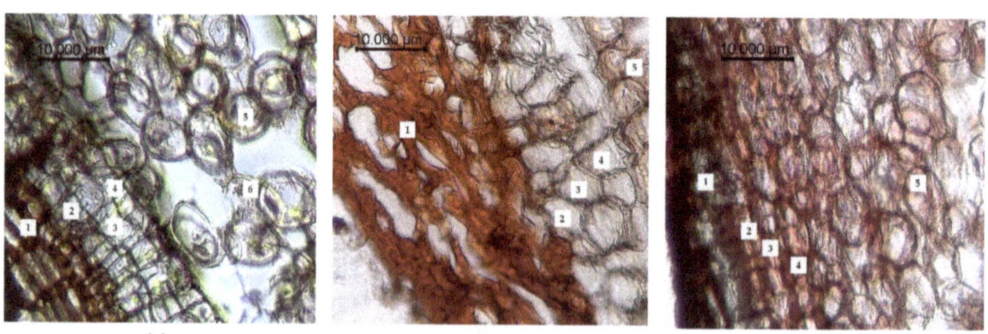

(a) (b) (c)

Figure 3. Changes in the anatomical structure of the root of *R. semenowii* under stress conditions: (**a**) control, (**b**) water deficit (cessation of watering, 72 h), (**c**) cold stress (cold treatment +3 °C, 72 h). 1—periderm; 2—fellam; 3—phellogen; 4—phelloderm; 5—parenchyma of the primary cortex; 6—starch grains; scale bar = 10 μm.

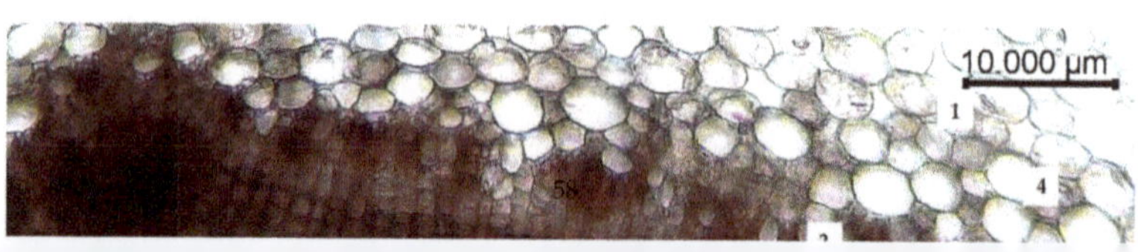

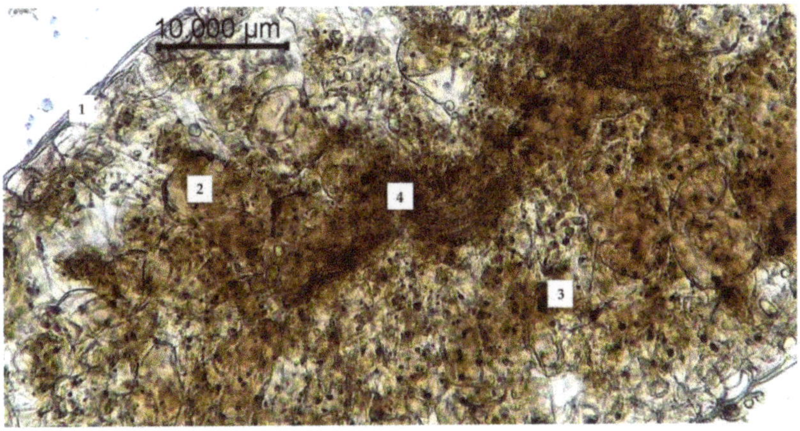

(a)

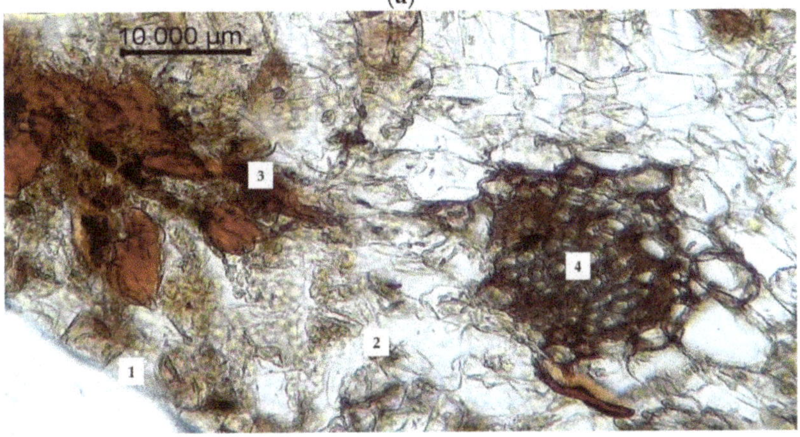

(b)

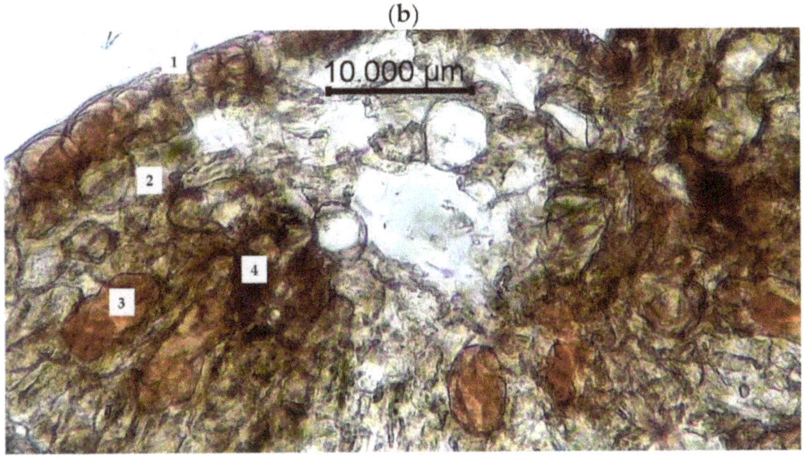

(c)

Figure 5. Changes in the anatomical structure of the leaf of *R. semenowii* under stress conditions: (**a**) control, (**b**) water deficit (cessation of watering, 72 h), (**c**) cold stress (cold treatment +3 °C, 72 h). 1—epidermis; 2—mesophyll, 3—inclusions; 4—conductive bundle; scale bar = 10 µm.

The conditions of water deficiency led to a significant deformation of the peridermal cells, their flattening and multiple ruptures. Turgor of parenchymal cells is reduced; starch grains are hydrolyzed, and the dye penetrates into cells (Figure 3b).

The cells become more hydrated under conditions of cold stress. In addition, the cells of the primary cortex acquire a more rounded shape, the periderm become denser and they are intensely stained (Figure 3c).

The anatomical study of the stem of *R. semenowii* noted features such as the cells of the epidermis are located in one row; the epiderm is characterized by the presence of a slightly thickened cuticle of the outer wall. The cells of the assimilation parenchyma, located in several rows under the layer of the epiderm, have a rounded-elongated shape and alternate with large intercellular spaces. Single inclusions are found among the cells of the primary cortex. The central cylinder is a conductive system of adjacent conductive bundles arranged in a circle. The pith consists of round or oval shaped parenchymal cells (Figure 4a).

Water deficiency led to changes in the cells of the primary cortex, which became multifaceted, flattened and stretched towards the central cylinder. Conducting bundles were "squeezed" by parenchymal cells, deformed, and destroyed. There were no inclusions. Turgor of parenchymal cells was reduced, and their deformation was noted (Figure 4b).

Inclusions of the primary cortex were displaced to the periphery of the stem and were located in the cells of the epiderm under conditions of cold stress. The cells of the parenchyma increased in size, becoming more rounded and more hydrated, the intercellular spaces were absent (Figure 4c).

The epidermal cells of the *R. semenowii* leaf are arranged in one row and have an oval shape, from the outside they are covered with a cuticle. Cells of the spongy type, the length of which decreases represent mesophyll from the outer to the inner layers. Palisade cells form 2–3 layers. Numerous point inclusions, as well as very large areas of biologically active compounds, were noted in leaf tissues. In the central part of the leaf, there are small collateral-type vascular bundles (Figure 5a).

The water deficiency conditions lead to a change in the shape of cells. Plasmolysis of cells develops; parenchymal cells flatten, stretch and shift towards the conducting beam. Conducting beams are deformed and disintegrated. Biologically active substances are concentrated in the middle of the leaf blade. The stomata plunge into the leaf mesophyll. Cells of the palisade parenchyma are destroyed (Figure 5b).

The leaf blade had acquired sparse cell structure under conditions of cold stress. Hydrated parenchymal cells were increased in size; being located in the central part of the leaf blade they took sharper outlines and the areas of accumulation of biologically active substances were more clearly distinguished (Figure 5c). Wilt and chlorosis of the leaf blades was observed when exposed to both water deficiency and cold stress.

2.3. Changes in Plant Secondary Metabolism of *R. semenowii* under Stress Conditions

GC-MS analysis of *R. semenowii* organs (shoots and roots) growing on control and stress backgrounds revealed the presence of up to 34 compounds (phytochemicals) in each of the studied variants, which can contribute to the medicinal qualities of the plant (Supplementary Table S1). The identification of phytochemicals was confirmed based on peak area, retention time and molecular formula. Analysis of mass spectra showed that stress conditions significantly alter the dominant spectrum of PSM shoots and roots of *R. semenowii* (Supplementary Table S2).

Analysis of the biologically active substances found in *R. semenowii* plants according to the classes of chemical compounds made it possible to reveal certain patterns.

Therefore, according to the data of the structural group composition in the ethanol extract of the root and stem of *R. semenowii* against stressful backgrounds, a change in the content of substances from the group of ubiquinones was illustrated. An increase in the shoot content of the γ-Tocopherol vitamin under stressful conditions and a decrease in its concentration in the root were noted (Figure 6). At the same time, an increase in the content in shoot 4,8,12,16-Tetramethylheptadecan-4-olide (Supplementary Tables S1 and S2) was shown against the stress background.

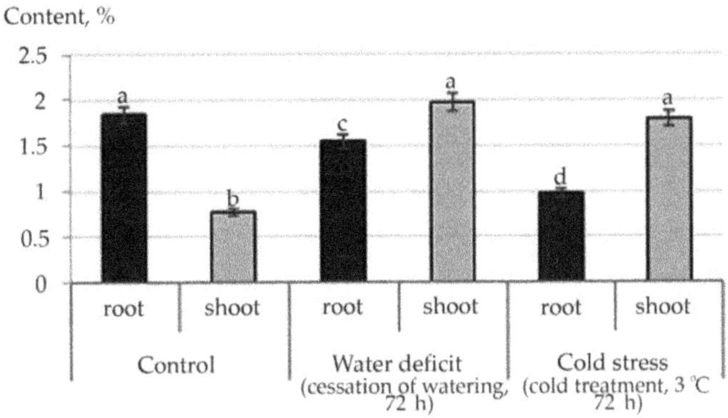

Figure 6. Change in the content of γ-Tocopherol in *R. semenowii* under stress conditions. Values presented are means (±SD). Different letters above the bars represent significant differences at $p \leq 0.05$, $n = 3$ plants in each of 3 replicates for all treatments.

Variations in the phytosterols content, exhibiting high biological activity in various physiological processes, were observed. (Supplementary Tables S1 and S2). An increase in the content of β-Sitosterol in root during cold stress was shown, the formation of in root against stressful backgrounds γ-Sitosterol with a higher content during cold treatment than with water deficiency.

The fatty acids were represented by 17-Octadecynoic acid and Propanoic acid, 3-(acetylthio)-2-methyl-; the first one was identified in shoot under control conditions, and the second in root under water deficit conditions (Supplementary Tables S1 and S2).

An increase in the content of all detected fatty acid esters in shoot under the conditions of the studied abiotic stresses, and a decrease in almost all of them, except for Ethyl oleate under water stress conditions in root, were noted (Supplementary Tables S1 and S2). This trend is illustrated by the example of linolenic acid ethyl ester, a plant metabolite with antioxidant activity (Figure 7).

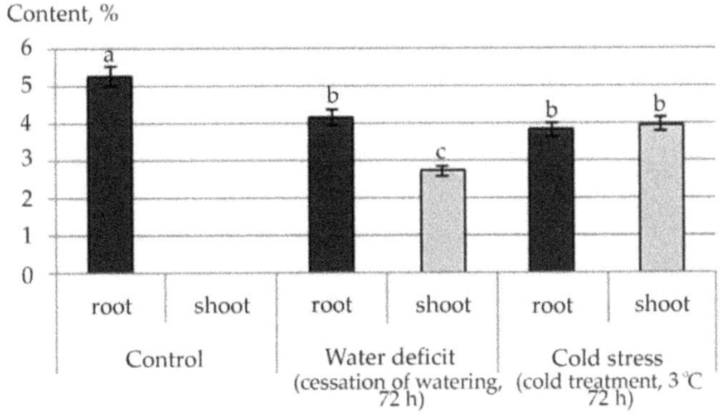

Figure 7. Changes in the content of fatty acid esters in *R. semenowii* under stress conditions by the example of Ethyl 9,12,15-octadecatrienoate (Ethyl 9α-linolenate, linolenic acid ethyl ester). Values presented are means (±SD). Different letters above the bars represent significant differences at $p \leq 0.05$, $n = 3$ plants in each of 3 replicates for all treatments.

The pharmacological action of biologically active substances *R. semenowii* is also determined by the content in their composition of aldehydes, glycosides, alcohols, hydrocarbons, amino acids and their derivatives, which have a rather complex structure, the content of which in plant tissues also changed under the influence of stress factors. For example, the content of Tetracosyl acetate (Wax monoesters) was not detected experimentally in shoot of *R. semenowii*. However, in root, we noted a significant change in the content of Tetracosyl acetate under stress conditions: a decrease under water deficit and an increase under cold stress (Figure 8).

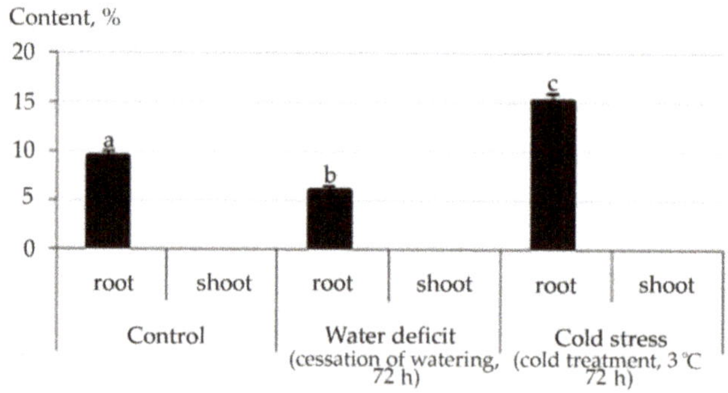

Figure 8. Change in the content of Tetracosyl acetate in root of *R. semenowii* under stress conditions. Values presented are means (±SD). Different letters above the bars represent significant differences at $p \leq 0.05$, $n = 3$ plants in each of 3 replicates for all treatments.

A decrease in Cyclopropyl carbinol in root under cold stress, an increase in the content of 1-Docosanol acetate under cold stress, and Cyclopropyl carbinol in root under osmotic stress were demonstrated (Supplementary Tables S1 and S2).

A change in the content of cyclic five-membered ketones, lactones and their derivatives under stress conditions was revealed (Supplementary Tables S1 and S2).

The tendency to an increase in the content of ketones in root was noted under water deficiency, while cold stress caused a decrease in the content of such detected ketones as 4-Cyclopentene-1,3-dione (shoot), 2-Propanone, 1-(acetyloxy)-(root), 2-Cyclopenten-1-one, 2-hydroxy-(root and shoot), 1,2-Cyclopentanedione, 3-methyl-(root and shoot) (Supplementary Tables S1 and S2). An increase in the concentration of psychoactive oxybutyrate 2-Hydroxy-gamma-butyrolactone in root under water deficit conditions and a decrease in it under conditions of cold stress with complete absence in stress conditions in shoot were revealed (Figure 9).

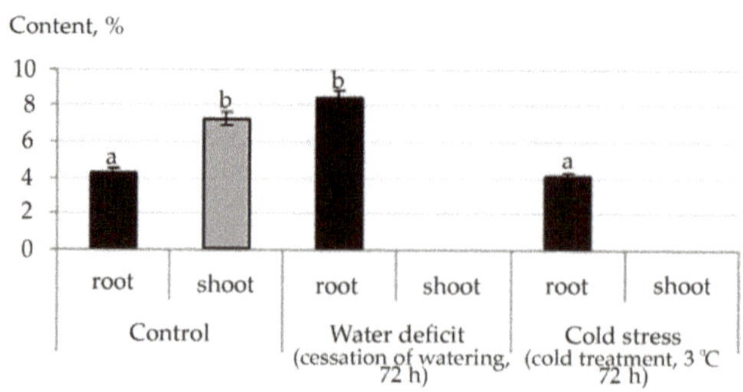

Figure 9. Change in the content of 2-Hydroxy-gamma-butyrolactone in *R. semenowii* under stress conditions. Values presented are means (±SD). Different letters above the bars represent significant differences at $p \leq 0.05$, $n = 3$ plants in each of 3 replicates for all treatments.

Furan and Pyran derivatives containing active alcohol, aldehyde and ketone groups, which also exhibit high biological activity, are of particular importance in the formation of the direction and specificity of the pharmacological action of *R. semenowii* preparations (Supplementary Tables S1 and S2, Figure 10).

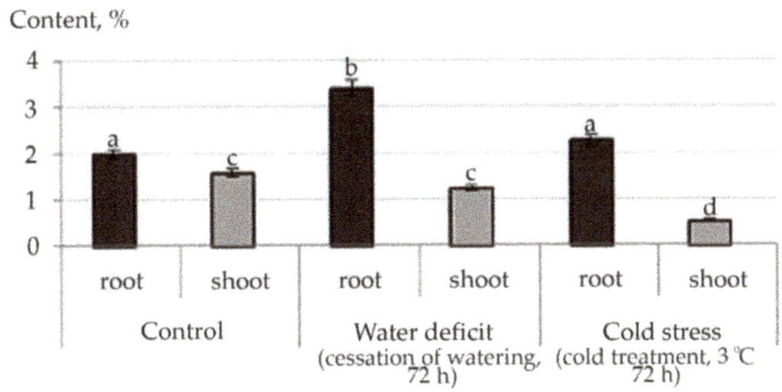

Figure 10. Changes in the content of furan derivatives in root and in shoot using the example of 2 (5H)-Furanone in *R. semenowii* under stress conditions. Values presented are means (±SD). Different letters above the bars represent significant differences at $p \leq 0.05$, $n = 3$ plants in each of 3 replicates for all treatments.

Phenols were presented only in shoot under control conditions and in the root against the background of water deficiency (Supplementary Tables S1 and S2).

The ethanol extract of *R. semenowii* contains derivatives (esters) of Benzoic acid (aromatic carboxylic acid). At the same time, an increase in the concentration of Benzoic acid, pentadecyl ester under water and cold stresses in root and disappearance in shoot were revealed, while the content of Benzoic acid, tridecyl ester under conditions of both osmotic and cold stress increased in both shoot and root. The content of Benzoic acid, heptyl ester, increased under conditions of cold stress in root and was not detected in shoot, while under conditions of water deficit it tended to decrease in all the studied organs. However, under conditions of water deficit, neoplasms of Benzoic acid, tetradecyl ester in root were noted (Supplementary Tables S1 and S2).

The change in the content of esters of saturated monobasic acids by the example of an ester of palmitic (hexadecanoic) acid is illustrated in Figure 11.

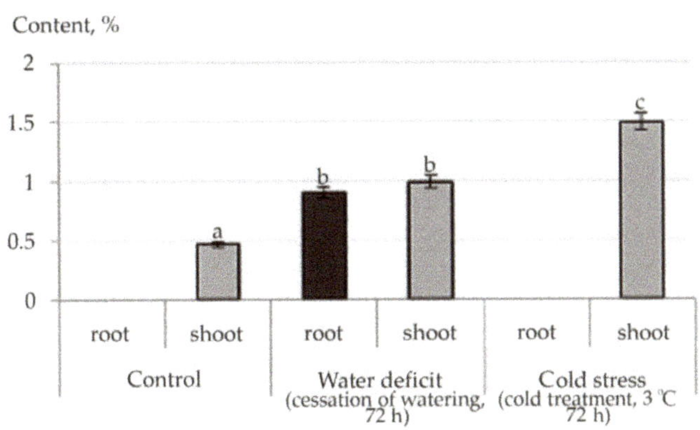

Figure 11. Change in the content of Hexadecanoic acid, 1-(hydroxymethyl)-1,2-ethanediyl ester in *R. semenowii* under stress conditions. Values presented are means (±SD). Different letters above the bars represent significant differences at $p \leq 0.05$, $n = 3$ plants in each of 3 replicates for all treatments.

A significant increase in the content of Hexadecanoic acid, 1- (hydroxymethyl) -1,2-ethanediyl ester under in shoot stresses and the appearance under conditions of water deficit in root was revealed. The appearance under the influence of osmotic stress of Formic acid, 2,6-dimethoxyphenyl ester in shoot (Supplementary Tables S1 and S2) was shown.

The content of esters of lower and middle carboxylic acids is presented in Supplementary Tables S1 and S2. An increase in the concentration of Diisooctyl Phthalate (DIOP) was noted under the action of the studied abiotic stressors, both in root and in shoot (Figure 12).

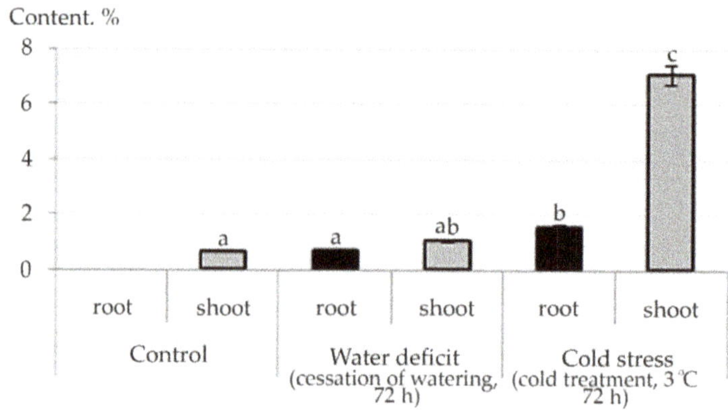

Figure 12. Change in concentration content of Diisooctyl phthalate (DIOP) in *R. semenowii* under stress conditions. Values presented are means (±SD). Different letters above the bars represent significant differences at $p \leq 0.05$, $n = 3$ plants in each of 3 replicates for all treatments.

A significant increase in the content of Phosphoric acid, diethyloctyl ester was shown under conditions of osmotic stress, and Phosphoric acid, diethylnonyl ester was found under conditions of cold stress in root (Supplementary Tables S1 and S2).

Abiotic stresses caused an increase in the content of diterpene (phytols) in shoot (Supplementary Tables S1 and S2) and triterpene (Squalen) hydrocarbons in both root and shoot (Figure 13).

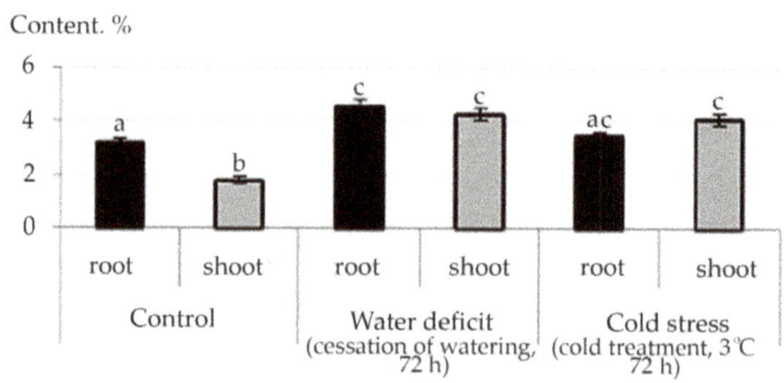

Figure 13. Change in content of Squalen in *R. semenowii* under stress conditions. Values presented are means (±SD). Different letters above the bars represent significant differences at $p \leq 0.05$, $n = 3$ plants in each of 3 replicates for all treatments.

3. Discussion

As a rule, the plant at different sensory levels perceives both water deficiency and cold. A decrease in biomass under water deficit is due to the loss of water by plant tissues and can serve as one of the indicators of water stress [34,35]. Generally, hypothermia also limits the growth and development of plants, although it does not always reduce the accumulation of biomass, preserving or increasing the water content of plant tissues. Nevertheless, it has such effects as a violation of the stability of proteins or protein complexes and a decrease in enzymatic activity [36]. Both water stress and cold lead to photoinhibition and disruption of photosynthesis, as well as significant membrane damage. The results of the study with water deficiency indicate damage to chlorophyll-bearing tissues, an increase in membrane permeability (the dye intensively penetrates into cells), a decrease in the content of free water in plant tissues, and a deterioration in the functioning of the conducting system in both root and shoot. In this case, water deficiency leads to a decrease in turgor, the development of plasmolysis and an increase in the concentration of cell juice and cytosol [37,38]. The high density of plant tissues of the leaf and the stomata immersed in the mesophyll are adaptive features that ensure a decrease in water loss during transpiration. Cold stress also causes damage to chlorophyll-bearing tissues, provoking numerous violations of the ultrastructure of cell membranes. The most common is an increase in the viscosity of their lipid part. The electron density of the cytoplasm decreases; structural changes occur (disintegration of granules, accumulation of lipid droplets, and the disappearance of starch grains). Therefore, membrane-bound processes such as photosynthesis and respiration are sensitive to cold stress [39]. In this way, the changes in the anatomical parameters of the leaf and stem of *R. semenowii* under stress conditions can also indicate damage to the photosynthetic apparatus.

A reduction in the values of the maximum quantum yield of photosystem II (PSII) is an indicator of photoinhibition and a decrease in the performance of PSII reaction centers [40]. At the same time, a decrease in the rate of noncyclic electron transport through PSII (ETR) indicates the activation of non-photochemical mechanisms of quenching [41], which points out a certain disruption in the functioning of the photosynthetic apparatus of *R. semenowii* under the created stress conditions. Any increase in Y (NPQ) under stress is an attempt to dissipate excess energy. Moreover, an increase in the quantum yield of uncontrolled heat dissipation and fluorescence Y (NO) means that the excess energy flows are out of control. High Y (NO) values at a relatively low Y (NPQ) in the studied plants indicate serious issues with the redistribution of excess light energy entering PSII [42,43]. This means it indicates possible damage to the photosynthetic apparatus of young plants of *R. semenowii*, caused by both water and cold stresses.

During their evolution, plants have adapted to drought conditions by accumulating SMs, however, increased SM accumulation is usually accompanied by reduced biomass [12]. Changes in the morphophysiological characteristics of plant organs growing under stress conditions are considered important acclimatization indicators [44]. However, the synthesis of SM under stress is also a significant component that is involved in defense reactions in response to biotic and abiotic stresses [45].

Oxidative stress is considered as one of the major causes of plant damage from abiotic stress, which is associated primarily with impaired electron transport in electron transport chains, caused by a change in the state of lipids.

A change in the parameters of the water balance or temperature causes a sharp increase in the generation of reactive oxygen species in the plant cell, and the cell needs antioxidant protection. The inhibition of free radicals in the plant is carried out by the antioxidant system. Due to the fact that SM has strong antioxidant properties, they may be associated with a mechanism for combating the harmful effects of reactive oxygen species (ROS), most of SM participating in maintaining redox balance by ROS scavenging (which also confers stress tolerance in plants) [7,13]. Antioxidants can be divided into two classes such as enzymatic antioxidants and nonenzymatic PSM [46,47], which we consider in this work.

The initial stage of autooxidation in membranes is inhibited by ubiquinones (tocopherol), polyphenols and superoxide dismutase. The ability to inhibit lipid peroxidation reactions is inherent only in the reduced forms of natural antioxidants and is associated with the presence of a hydroxyl group in the molecule. Natural antioxidants have a labile hydroxyl group and react relatively easily with hydrocarbon peroxide radicals [45].

For ubiquinones, the quinone form is the most stable. Tocopherols are found in lipids mainly in cyclic form which is both in the form of free tocopherol and in the form of its esters. In the body, the ester forms are easily hydrolyzed to free tocopherol with the help of enzymes. The increased content of γ-Tocopherol in shoot under stress conditions both water deficit and cold, confirms that tocopherols function as antioxidants in oxidative reactions. They form phenoxyl radicals by reacting with peroxide radicals, which are then converted into quinones, dimers, trimers. The study of lipid extracts from stress photodegradable cells demonstrates that enhanced autooxidation of vitamin E occurs with the production of 4, 8, 12, 16-tetramethylheptadecan-4-olide [48]. In our experiment, this phenomenon was found in shoot of *R. semenowii* against the background of stressful influences. Therefore, given the decrease in the photosynthetic ability of the plant when exposed to water deficit or low temperature, it is possibly also associated with photodegradation of cells.

Both the direction of the pharmacological action and its specificity are necessarily determined by glycosides, which are formed by sterols in reaction with carbohydrates, significant amounts of which have been identified in the ethanol extract. Phytosterols are bioactive compounds that are an important structural component of plant cell membranes in nature and play a vital role in the regulation of membrane fluidity and permeability. Moreover, phytosterols have a chemical structure similar to cholesterol obtained from mammalian cells. Among the various phytosterols, beta-sitosterol (SIT) and its stereoisomer gamma-sitosterol is the main compound found in abundance in plants. The primary precursor of sterol biosynthesis is cycloartenol, formed from squalene by cycloartenol synthase (CAS) [49]. Therefore, an increase in the level of SIT in root under stressful conditions both water deficit and cold stress, indirectly indicates its role in strengthening the membranes of plant cells.

The major building blocks of botanical membranes are phospholipids and glycolipids, both of which contain a glycerol core linked to two tails derived from fatty acids (FAs). As a result, FAs have a strong influence on the properties of membranes [50].

Both exposure to cold and water stress can cause disturbances in the conductivity of the biomembrane. It is proven by deactivation of proteins and ion leakage [51]. Reducing the fluidity of the cell membrane after exposure to stress is considered the first line of defense [52]. It has been shown in the literature that the fluidity of the plasma membrane correlates with the proportion of unsaturated fatty acids (UFAs) [53]. We experimentally revealed a decrease in the proportion of unsaturated FA in shoot both under water deficit and cold stress conditions. Furthermore, it has been shown that under stress with an increased content of UFAs in the lipids of the inner membranes of chloroplasts and mitochondria, the weakening of PSII photoinhibition may be associated [54,55]. Linoleic and linolenic fatty acids are essential and included in the composition of cell membranes, regulating their microviscosity, permeability, electrical properties, reducing excitability, forming the corresponding lipid environment of membrane proteins and enzymes [56]. In plants, fatty acids (FAs) can be present not only in a free state, but in the form of their methyl, ethyl, and other esters as well [57]. The nature of lipid components can vary greatly and include waxes, hydrocarbons (including squalene), sterol esters, aliphatic aldehydes, primary and secondary alcohols, 1,2-, 2,3- and α, ω-diols, ketones, β-diketones, triacylglycerols and numerous others. It was experimentally revealed that under stress conditions of both water deficit and cold, the content of FA in shoot esters significantly increases.

The interaction between plants and the environment is provided for the aerial organs by epicuticular waxes, which typically contain esters of long chain fatty alcohol esters with long chain fatty acids. However, little is known about the nature, biosynthesis and role of waxes at the root–rhizosphere interface. In this context, the observed changes in the content of Wax monoesters in the root of *R. semenowii* under abiotic stresses stimulated the interest. Li et al. [58] suggest a direct metabolic relationship between some root waxes and suberin. The direct physical connection between wax and suberin implies extracellular location under the primary cell wall, since suberin is deposited outside the plasma membrane. It is possible that most root waxes can be embedded deeper into the peridermal cell walls, where suberization occurs [58].

Chemicals from fatty acid metabolism can act as important chemical signals. Superoxide radical anion and hydrogen peroxide, which are produced under oxidative stress, can directly oxidize lipids or be converted to a hydroxyl radical via the Fenton and Haber-Weiss reactions [59]. The hydroxyl radical easily initiates the peroxidation of polyunsaturated, mainly linoleic and linolenic fatty acids. Spontaneous rearrangements of oxidized polyunsaturated fatty acids (PUFAs) lead to the formation of various phytoprostanes and aldehydes and other reactive electrophile species (RES), which are often toxic to plants [60]. Reactive oxygen species, especially singlet oxygen, formed in chloroplasts under stressful conditions, can also oxidize carotenoids, which also leads to the formation of such oxidized products as aldehydes, ketones, endoperoxides, and lactones [61]. As an example of this in our experiment is the detection of Benzeneacetaldehyde in shoot under cold stress as well as an increase in the concentration of 2-Hydroxy-gamma-butyrolactone in root under water deficit conditions and a decrease in it under conditions of cold stress with complete absence in stress conditions in shoot.

Perhaps, stress conditions that negatively influence the fragile and sensitive shoot are able to stimulate the internal solvation of the critical transition state by the neighboring hydroxyl group due to the binding and/or orientation of water molecules as a result of stress. The biological effects of linoleic and linolenic acids are realized at the cellular and organ levels. In response to abiotic stress in plants, lipases can be activated, which release unsaturated fatty acids and trigger the synthesis of a number of oxylipins with different functions [62]. Some of them have direct antimicrobial functions, while others are powerful regulators of defense mechanisms. Oxylipins are involved in plant acclimation to abiotic stresses as well. They are part of complex interactive networks of phytohormones, including salicylic acid, ethylene, auxin, brassinosteroids, gibberellic acid and abscisic acid, which control all aspects of plant growth and development and how plants adapt to their environment. These are signaling molecules formed from the group of polyunsaturated fatty acids, which are involved in the formation of the body's responses to signals from the external environment [62].

Compared with other kinds of raw materials, the content of fatty acids in plants is low. Saturated and unsaturated fatty acids are part of the acyl lipids of plant tissue. Lipids, in turn, actively change metabolism and increase plant resistance, particularly to low temperatures [63,64]. Perhaps, it is the oxidative stress caused by the action of abiotic stressors that initiates the increase in the plant tissues of *R. semenowii* of such furan compounds as 2 (5H)-Furanone and Benzofuran. The appearance in the phytochemical spectrum of 2,5-Dimethyl-4-hydroxy-3 (2H)-furanone in root under cold stress also stimulated the interest.

It is noteworthy that the pyrrolidine ring is one of the most frequent heterocycles in the structure of medicines. This structural fragment is a part of many biologically active natural compounds (alkaloids nicotine, hygrin, the amino acid proline, etc.), among which there are osmolytes actively accumulated in plant tissues as a result of stress [65]. It is possible that the functions of osmolytes can be performed in this case by Methylglucoside (appearance in root under conditions of water deficiency).

In determining the directions of the potential pharmacological action of drugs based on *R. semenowii*, special attention should be paid to essential oils, which are based on various structures and in the presence of certain functional groups, terpenes, phenols, polyunsaturated carboxylic acids containing 1–4 double and triple bonds, ethers, alcohols [65]. Vegetable essential oils are intended to mediate the plant's attitude to abiotic and biotic stress factors of various nature and to increase antioxidant activity [66–68]. In this context, free volatile substances are glycosylated, which are stored in the cell vacuoles and the internal swelling of the cells reduces the stress effect, particularly, from water deficiency [69]. This might be related to a decrease in leaf area [70].

Therefore, phenols exhibit antimicrobial and antioxidant (membrane stabilizing, cytoprotective) action. The antioxidant effect of phenols is responsible for stabilizing the cell membrane; phenols prevent mitochondrial autolysis. Moreover, they are involved in the suppression or blocking of free radicals, the most characteristic reaction of lipid peroxidation (LPO), and generally have a cytoprotective effect [71]. It is possible that the results of this experiment indicate precisely the antioxidant effect of phenols for plants under osmotic stress caused by water deficiency.

In the metabolism of higher plants, Benzoic acid (BA) influences their growth, anatomy, morphology, and stress resistance [72]. In plants, BA is a precursor of a wide range of primary and secondary metabolites, including various esters. Various modifications in BA molecules affect the volatility, permeability of substances in various cell compartments, their solubility and activity, and are crucial for their transport and functioning. Senaratna et al. [73] suggest that the structural part of benzoic acid is most likely the main functional molecular structure that makes plants resistant to stress. In this regard, the revealed changes in the concentrations of various BA esters in *R. semenowii* in shoot and root tissues under water deficit and cold stressful conditions arouse the interest for further research.

Unsaturated fatty acids such as palmitic acid are required to maintain a certain level of membrane fluidity. In the case of formic acid, in addition to acidic properties, it also exhibits some properties of aldehydes, notably, reducing properties. Unsaturated fatty acids are present in plant organisms in the form of esters. There is information about their connection with the acclimation of plants to low temperatures [74]. The data of our experiment also testify to their adaptive role to water deficit.

Esters of lower and medium carboxylic acids are constituents of the essential oils of many plants. The known phosphoric acid esters are extremely numerous. The majority of them play a central role in life processes and therefore they are of direct biological interest. In addition, there are crop improvers, plant growth regulators, ripening agents and others. Generally, in small quantities, they are involved in various processes taking place in a living organism and are aroma-forming components. We detected experimentally Diisooctylphthalate (DIOP) (diester) of phthalic acid (1,2-benzenedicarboxylic acid) which is the simplest representative of dibasic aromatic carboxylic acids. Commonly, the presence of phthalic acid is not typical of plant matter. In the case of detection of dialkyl phthalates in various objects and the interpretation of the data obtained is considered a serious problem [75]. However, it was noted that in response to stress induction by phytophages or phytopathogens, the content of diethyl phthalate and diisooctyl phthalate in plants could increase by more than five times [76]. For those in this experiment in conditions of both water deficiency and cold in shoot and in root, an incompletely understood stress response takes place.

Terpenes like other volatile components of green leaves (aldehydes, alcohols, and ethers) can serve as a signal of stress transmitted from plant to plant [77]. Diterpene hydrocarbons phytols are part of chlorophyll; therefore, there is no doubt about the relationship of their accumulation in the plant *R. semenowii* under stressful backgrounds, when the photosynthetic activity changes significantly. In this regard, the detected accumulation of phytols in shoot against the background of cold stress is of interest.

Squalen is a naturally occurring triterpene hydrocarbon. It belongs to the group of carotenoids. The accumulation of carotenoids in shoot is also closely related to stress acclimation processes. The decay products of hexoses serve as the basis for the synthesis of carotenoids and terpenes in a plant. In that case, carotenoids play an active part in the absorption of light energy shoot and its transfer to the reaction centers of the photosystem and serve as photoprotectors. Carotenoids protect the photosynthetic apparatus from photooxidizing damage by quenching the triplet of chlorophyll molecules [62]. Carotenoids are potent scavengers of reactive oxygen species that protect pigments and unsaturated fatty acids from lipids from oxidative damage [78]. They can react with free radicals by electron transfer, transfer of a hydrogen atom or its addition, as well as by modulating the physical properties of photosynthetic membranes with the participation of the xanthophyllic cycle in this process [79]. The positive effect of carotenoids on root growth was proven experimentally. Physical and physicochemical studies of the reactivity of carotenoids in redox processes are of priority today [80]. Therefore, the accumulation of Squalen in both shoot and root can be considered an adaptive stress response of the species *R. semenowii*.

Therefore, we can talk about the nonspecificity of many physiological and phytochemical reactions of plant tissues of *R. semenowii* to the effect of sudden cold or water stress. These reactions are expressed in a decrease in the level of photosynthetic activity, an increase in the content of such SM as γ-Tocopherol in shoot, SIT in root, a decrease in the proportion of unsaturated FA and an increase in the content of FA in shoot esters, in the accumulation of Squalen under stress in both shoot and root and others.

Nevertheless, specific adaptive mechanisms and nuances of reacting both shoot and root of *R. semenowii* to the action of each of the studied abiotic stresses were also noted. These are, for example, a change in the concentrations of various BA esters in the tissues of *R. semenowii*, the accumulation of phytols in shoot under cold stress conditions, as well as the pathways for the formation of such oxidized products as aldehydes, ketones and lactones and their derivatives, esters of lower and medium carboxylic acids, etc.

The variety of mechanisms of acclimation of *R. semenowii* plants to the abiotic stresses action by changing the SM content stimulates the interest for further experiments. In general, further research should be focused on the use of abiotic stressors for the targeted synthesis of bioactive SMs valuable for pharmaceutical use.

4. Materials and Methods

4.1. Plant Material and Growing Conditions

Plants *R. semenovii* grown in vegetation pots were studied. The plants had already lost their juvenile characteristics but had not entered the generative period of ontogenesis yet during the immature period of development, when an intensive growth of the shoot was observed.

Plants at the time of the experiment were divided into three groups: (1) a control group were grown under 26 ± 3 °C at day and 20 ± 3 °C at night, with average air humidity 37% and optimal irrigation (up to 60% of full moisture capacity); (2) a group subjected to sudden cold treatment +3 °C in refrigerator cabinet with lighting ("Polair", Moscow, Pussia) under circadian illumination (using commercial fluorescent white light tubes): 16 h light/8 h darkness regime [200 μmol m^{-2} s^{-1} PAR, light metre LI-205 (Li-Cor, Lincoln, NE, USA)]) and (3) a group subjected to water deficiency (cessation of watering). The duration of the stress exposure was 72 h.

The calculation of the soil moisture was carried out according to the formula: W= (a x 100): b (%), where (W)—the soil moisture, in % of the dry soil mass; (a)—the mass of water in the soil sample, g; (b)—dry soil mass, g.

Growth parameters were determined by measurements before and on the third day after the onset of stress exposure [81]. The water content (WC) in plant tissues was calculated using the formula:

$$WC = ((a - б)/a) \times 100\%, \quad (1)$$

where a is the initial mass, mg; b is the mass after drying at 105 °C, mg.

4.2. Photosynthetic Activity Determination

Photosynthetic activity parameters were estimated by determination of fluorescence levels. Rapid light curves (RLCs) were recorded using Junior-PAM ("Heinz WalzGmbH", Effeltrich, Germany) under actinic illumination of 450 nm. The RLC for each sample was recorded after quasi-darkness to assess the effect of actinic light absence, while complete darkness is difficult to achieve under field conditions [82]. For each measurement the fluorometer provided eight saturation light pulses of 10,000 μmol/m^{2}s every 20 s, while actinic light increased from 0 to 625 μmol/m^{2}s gradually. For comparison, the data obtained from the last pulse of the light curve were taken [83]. The following parameters were calculated using WinControl-3.29 (Walz, Effeltrich, Germany) software: Fv/Fm: maximum quantum yield of PSII photochemistry; Y(II): effective photochemical quantum yield of PSII; Y(NPQ): quantum yield of non-photochemical energy conversion in PSII due to downregulation of the light-harvesting function; Y(NO): quantum yield of non-photochemical energy conversion in PSII that caused by downregulation of the light-harvesting function; PSII relative electron transport (ETR). In the experiment, each time the region of the middle third of the active leaf was selected. All measurements were performed on a sunny day from 09:00 to 11:00 a.m.

4.3. Analysis of Changes in the Elements of the Anatomical Structure

Fixation of roots was performed in 70% ethanol and preservative fluid was a Strasburger-Flemming's mixture: 96% ethanol:glycerol:water in ratio of 1:1:1 [84]. The material was infused for a 24 h. Anatomical specimens were prepared with a microtome MZP-01 ("Technom", Ekaterinburg, Russia) with a freezing unit OL-ZSO 30 ("Inmedprom", Yaroslavl, Russia). The thickness of anatomical sections varied between 10 and 15 microns. Sudan IV-stained sections were placed on a glass slide in a drop of pure glycerin and covered with a cover slip to obtain a temporary preparation. Micrographs of anatomic sections were made on a microscope with Micro Opix MX 700 (T) (West Medica, Brown Boveri-Straße 6, B17-1 2351 Wiener Neudorf, Austria), CAM V1200C HD-camera (West Medica, Brown Boveri-Straße 6, B17-1 2351 Wiener Neudorf, Austria). All anatomical data were obtained in 3–5 replicates (5 plants in each) with a 40× objective.

4.4. Determination of Organic Compounds in Extracts

Analysis methods: Gas chromatography with mass spectrometric detection (Agilent 6890N/5973N, Santa Clara, CA, USA). Sample volume 1.0 μL, sample injection temperature 260 °C, without flow division. Separation was carried out using a chromatographic capillary column DB-35MS with a length of 30 m, an inner diameter of 0.25 mm, and a film thickness of 0.25 μm at a constant carrier gas (helium) velocity of 1 mL/min. The chromatographic temperature was programmed from 40 (exposure 0 min) to 150 °C with a heating rate of 10 °C/min (exposure 0 min) and up to 300 °C with a heating rate of 5 °C/min (exposure 10 min). Detection was carried out in the SCANm/z 34-850 mode. AgilentMSDChemStation software (version 1701EA) (Santa Clara, CA, USA) was used to control the gas chromatography system, register and process the obtained results and data. Data processing included determination of retention times, peak areas, as well as processing, spectral information obtained using a mass spectrometric detector. The Wiley 7th edition and NIST'02 libraries were used to decode the obtained mass spectra (the total number of spectra in the libraries is more than 550 thousand).

All experiments were done in three replicates. The processing of data and graphing was performed using Microsoft Excel (Microsoft Corp., Redmond, Washington, DC, USA). Atypical values were excluded from the data based on t-tests, the standard error of the average sample was calculated. Differences were considered significant at $p < 0.05$.

5. Conclusions

The results of this study demonstrated that under water deficit and cold stress conditions, morphophysiological responses and elements of anatomical structure of organs of *R. semenovii* changed at some degrees.

For the first time, a detailed study of the chemical composition of the ethanol extract of root and shoot of *R. semenovii* under stress was carried out using gas chromatography–mass spectrometry, which made it possible to state that the antioxidant system in plant tissues is multicomponent and includes SM. All components are in functional interaction and are due to the body's adaptive stress responses.

The results obtained for SM, which are medicinal biologically active substances, are useful both for understanding the mechanisms of protection against adverse conditions and for approaches to the targeted synthesis of secondary metabolites valuable for pharmaceutical applications.

Supplementary Materials: The following are available online at https://www.mdpi.com/article/10.3390/plants10061196/s1, Table S1: Change in content of SM in shoot under stress conditions, Table S2: Change in content of SM in root under stress conditions.

Author Contributions: Conceptualization, N.V.T.; methodology, N.V.T., T.G.T. and N.O.K.; formal analysis, N.V.T., N.O.K. and M.S.K.; investigation, N.K.K., M.S.K. and T.N.K.; data curation, N.V.T. and N.O.K.; writing—original draft preparation, N.V.T., N.O.K. and N.D.M.; writing—review and editing N.V.T. and T.G.T.; visualization, N.K.K.; supervision, N.V.T. and N.O.K.; resource, M.S.K.; project administration, N.V.T.; funding acquisition, N.V.T. All authors have read and agreed to the published version of the manuscript.

Funding: This research was carried out in the framework of the project AP08855699 "The impact of abiotic stresses on the morphophysiological and phytochemical aspects of adaptation and biological activity of Kazakhstan plant *Rhodiola semenovii* Boriss." (2020–2022). This grant is funded by the Ministry of Education and Science of the Republic of Kazakhstan.

Institutional Review Board Statement: In this the study did not involve humans or animals.

Informed Consent Statement: Not applicable.

Data Availability Statement: Not applicable.

Acknowledgments: Authors are grateful to Center of Physico-Chemical Methods of Research and Analysis (Almaty) for consultations in the SM identification.

Conflicts of Interest: The authors declare that they have no conflict of interest.

References

1. Van der Esch, S.; ten Brink, B.; Stehfest, E.; Bakkenes, M.; Sewell, A.; Bouwman, A.; Meijer, J.; Westhoek, H.; Van den Berg, M.; Van den Born, G.J. *Exploring Future Changes in Land Use and Land Condition and the Impacts on Food, Water, Climate Change and Biodiversity: Scenarios for the UNCCD Global Land Outlook*; Policy Report; PBL Netherlands Environmental Assessment Agency: The Hague, The Netherlands, 2017; p. 115.
2. El-Sherbeny, G.A.; Dakhil, M.A.; Eid, E.M.; Abdelaal, M. Structural and Chemical Adaptations of Artemisia monosperma Delile and *Limbarda crithmoides* (L.) Dumort. in Response to Arid Coastal Environments along the Mediterranean Coast of Egypt. *Plants* **2021**, *10*, 481. [CrossRef] [PubMed]
3. Lawlor, D.W.; Tezara, W. Cause of decreased photosynthetic rate and metabolic capacity in water-deficient leaf cells: A critical evaluation of mechanisms and integration of processes. *Ann. Bot.* **2009**, *103*, 561–579. [CrossRef] [PubMed]
4. Del Pozo, A.; Méndez-Espinoza, A.M.; Romero-Bravo, S.; Garriga, M.; Estrada, F.; Alcaíno, M.; Camargo-Rodriguez, A.V.; Corke, F.M.; Doonan, J.H.; Lobos, G.A. Genotypic variations in leaf and whole-plant water use efficiencies are closely related in bread wheat genotypes under well-watered and water-limited conditions during grain filling. *Sci. Rep.* **2020**, *10*, 460. [CrossRef] [PubMed]
5. Bartwal, A.; Mall, R.; Lohani, P.; Guru, S.K.; Arora, S. Role of Secondary Metabolites and Brassinosteroids in Plant Defense Against Environmental Stresses. *J. Plant Growth Regul.* **2013**, *32*, 216–232. [CrossRef]
6. Berini, J.L.; Brockman, S.A.; Hegeman, A.D.; Reich, P.B.; Muthukrishnan, R.; Montgomery, R.A.; Forester, J.D. Combinations of abiotic factors differentially alter production of plant secondary metabolites in five woody plant species in the boreal-temperate transition zone. *Front. Plant Sci.* **2018**, *9*, 1257. [CrossRef]
7. Jan, R.; Asaf, S.; Numan, M.; Lubna; Kim, K.-M. Plant Secondary Metabolite Biosynthesis and Transcriptional Regulation in Response to Biotic and Abiotic Stress Conditions. *Agronomy* **2021**, *11*, 968. [CrossRef]
8. Selmar, D.; Kleinwächter, M.; Abouzeid, S.; Yahyazadeh, M.; Nowak, M. The Impact of Drought Stress on the Quality of Spice and Medicinal Plants. In *Medicinal Plants and Environmental Challenges*; Ghorbanpour, M., Varma, A., Eds.; Springer: Cham, Switzerland, 2017. [CrossRef]
9. Shulaev, V.; Cortes, D.; Miller, G.; Mittler, R. Metabolomics for plant stress response. *Physiol. Plant* **2008**, *132*, 199–208. [CrossRef]
10. Edreva, A.; Velikova, V.; Tsonev, T.; Dagnon, S.; Gürel, A.L.; Aktas, L.Y. Stress-protective role of secondary metabolites: Diversity of functions and mechanisms. *Gen. Appl. Plant Physiol.* **2008**, *34*, 67–78.
11. Wink, M. Evolution of Secondary Metabolites from an Ecological and Molecular Phylogenetic Perspective. *Phytochemistry* **2003**, *64*, 3–19. [CrossRef]

12. Agrawal, A.A.; Konno, K. Latex: A Model for Understanding Mechanisms, Ecology, and Evolution of Plant Defense against Herbivory. *Annu. Rev. Ecol. Evol. Syst.* **2009**, *40*, 311–331. [CrossRef]
13. Isah, T. Stress and defense responses in plant secondary metabolites production. *Biol. Res.* **2019**, *52*. [CrossRef]
14. Foyer, C.H.; Noctor, G. Redox regulation in photosynthetic organisms: Signaling, acclimation, and practical implications. *Antioxid. Redox Signal.* **2009**, *11*, 861–905. [CrossRef]
15. Goyal, S.; Lambert, C.; Cluzet, S.; Mérillon, J.M.; Ramawat, K.G. Secondary metabolites and plant defense. *In Plant Defence: Biological Control; Mérillon, J.M., Ramawat, K.G., Eds.; Springer, Dordrecht.* **2012**, *12*, 109–138. [CrossRef]
16. Selmar, D.; Kleinwächter, M. Stress enhances the synthesis of secondary plant products: The impact of stress-related over-reduction on the accumulation of natural products. *Plant Cell Physiol.* **2013**, *54*, 817–826. [CrossRef]
17. Zhao, J.; Davis, L.C.; Verpoorte, R. Elicitor signal transductions leading to the production of plant secondary metabolite. *Biotechnol. Adv.* **2005**, *23*, 283–333. [CrossRef]
18. Sato, H.; Tanaka, S.; Tabata, M. Kinetics of alkaloid uptake by cultured cells of Coptis japonica. *Phytochemistry* **1993**, *34*, 697–701. [CrossRef]
19. Gupta, B.; Huang, B. Mechanism of salinity tolerance in plants: Physiological, biochemical, and molecular characterization. *Intl. J. Genomics* **2014**. [CrossRef]
20. Theocharis, A.; Clément, C.; Barka, E.A. Physiological and molecular changes in plants grown at low temperatures. *Planta* **2012**, *235*, 1091–1105. [CrossRef]
21. Sampaio, B.L.; Edrada-Ebel, R.; Da Costa, F.B. Effect of the environment on the secondary metabolic profile of Tithonia diversifolia: A model for environmental metabolomics of plants. *Sci. Rep.* **2016**, *6*, 29265. [CrossRef]
22. Niinemets, Ü. Uncovering the hidden facets of drought stress: Secondary metabolites make the difference. *Tree Physiol.* **2015**, *36*, 129–132. [CrossRef]
23. Afzal, S.F.; Yar, A.K.; Ullah, R.H.; Ali, B.G.; Ali, J.S.; Ahmad, J.S.; Fu, S. Impact of drought stress on active secondary metabolite production in Cichorium intybus roots. *J. Appl. Environ. Biol. Sci.* **2017**, *7*, 39–43.
24. Piasecka, A.; Sawikowska, A.; Kuczyńska, A.; Ogrodowicz, P.; Mikołajczak, K.; Krystkowiak, K.; Gudyś, K.; Guzy-Wróbelska, J.; Krajewski, P.; Kachlicki, P. Drought-related secondary metabolites of barley (*Hordeum vulgare* L.) leaves and their metabolomic quantitative trait loci. *Plant J.* **2017**, *89*, 898–913. [CrossRef]
25. Ramel, F.; Sulmon, C.; Bogard, M.; Couée, I.; Gouesbet, G. Differential patterns of reactive oxygen species and antioxidative mechanisms during atrazine injury and sucrose-induced tolerance in Arabidopsis thaliana plantlets. *BMC Plant Biol.* **2009**, *9*. [CrossRef]
26. Liang, X.; Zhang, L.; Natarajan, S.K.; Becker, D.F. Proline mechanisms of stress survival. *Antioxid. Redox Signal.* **2013**, *19*, 998–1011. [CrossRef]
27. De Luca, V.; Salim, V.; Atsumi, S.M.; Yu, F. Mining the biodiversity of plants: A revolution in the making. *Science* **2012**, *336*, 1658–1661. [CrossRef]
28. Wurtzel, E.T.; Kutchan, T.M. Plant metabolism, the diverse chemistry set of the future. *Science* **2016**, *353*, 1232–1236. [CrossRef]
29. Ramakrishna, A.; Ravishankar, G.A. Influences of abiotic stress signals on secondary metabolites in plants. *Plant Signal. Behav.* **2011**, *6*, 1720–1731. [CrossRef]
30. Rejeb, I.B.; Pastor, V.; Mauch-Mani, B. Plant responses to simultaneous biotic and abiotic stress: Molecular mechanisms. *Plants* **2014**, *3*, 458–475. [CrossRef]
31. Szabó, K.; Radácsi, P.; Rajhárt, P.; Ladányi, M.; Németh, É. Stress-induced changes of growth, yield and bioactive compounds in lemon balm cultivars. *Plant Physiol. Biochem.* **2017**, *119*, 170–177. [CrossRef]
32. Kuliev, Z.A.; Kim, K.K.; Vdovin, A.D.; Abdullaev, N.D.; Khushbaktova, Z.A.; Syrov, V.N. Oligomeric proanthocyanidin glycosides of Clementsia semenovii and their biological activity. *Chem. Nat. Compd.* **2000**, *36*, 60. [CrossRef]
33. Kuliev, R.Z.; Akhmedov, U.; Khalmatov, K.K. Dimeric Proanthocyanidins from Rhodiola semenovii. *Chem. Nat. Compd.* **2004**, *40*, 94. [CrossRef]
34. Pour-Aboughadareh, A.; Omidi, M.; Naghavi, M.R.; Etminan, A.; Mehrabi, A.A.; Poczai, P.; Bayat, H. Effect of Water Deficit Stress on Seedling Biomass and Physio-Chemical Characteristics in Different Species of Wheat Possessing the D Genome. *Agronomy* **2019**, *9*, 522. [CrossRef]
35. Wang, X.; Wang, X.; Sun, X.; Berlyn, J.P.; Rehim, A. Effect of pavement and water deficit on biomass allocation and whole-tree transpiration in two contrasting urban tree species. *Urban Ecosyst.* **2020**, *23*, 893–904. [CrossRef]
36. Orvar, B.L.; Sangwan, V.; Omann, F.; Dhindsa, R. Early steps in cold sensing by plant cells: The role of actin cytoskeleton and membrane fluidity. *Plant J.* **2000**, *23*, 785–794. [CrossRef]
37. Terletskaya, N.; Duisenbayeva, U.; Rysbekova, A.; Kurmanbayeva, M.; Blavachinskaya, I. Architectural traits in response to salinity of wheat primary roots. *Acta Physiol. Plant.* **2019**, *41*, 157. [CrossRef]
38. Terletskaya, N.V.; Lee, T.E.; Altayeva, N.A.; Kudrina, N.O.; Blavachinskaya, I.V.; Erezhetova, U. Some Mechanisms Modulating Root Growth of Various Wheat Species under Osmotic-Stress Conditions. *Plants* **2020**, *9*, 1545. [CrossRef]
39. Kolupaev, Y.E.; Gorelova, E.I.; Yastreb, T.O. Mechanizms of plant adaptation to hypothermia: Role of antioxidant system. *Вісник ХНАУ* **2018**, *1*, 6–33.
40. Guidi, L.; Piccolo, E.L.; Landi, M. Chlorophyll Fluorescence, Photoinhibition and Abiotic Stress: Does it Make Any Difference the Fact to Be a C3 or C4 Species? *Front. Plant Sci.* **2019**. [CrossRef]

41. Ralph, P.J.; Gademann, R. Rapid light curves: A powerful tool to assess photosynthetic activity. *Aquat. Bot.* **2005**, *82*, 222–237. [CrossRef]
42. Ruban, A.V.; Johnson, M.P.; Duffy, C.D.P. The photoprotective molecular switch in the photosystem II antenna. *Biochim. Biophys. Acta* **2012**, *1817*, 167–181. [CrossRef]
43. Sperdouli, I.; Moustakas, M. Spatio-temporal heterogeneity in Arabidopsis thaliana leaves under drought stress. *Plant Biol.* **2012**, *14*, 118–128. [CrossRef]
44. Ashraf, M.; Harris, P.J.C. Photosynthesis under Stressful Environments: An Overview. *Photosynthetica* **2013**, *51*, 163–190. [CrossRef]
45. Edreva, A.M.; Velikova, V.B.; Tsonev, T.D. Phenylamides in Plants. *Russ. J. Plant Physiol.* **2007**, *54*, 287–301. [CrossRef]
46. Selmar, D. Influencing the Product Quality by Deliberately Applying Drought Stress during the Cultivation of Medicinal Plants. *Ind. Crops Prod.* **2013**, *42*, 558–566. [CrossRef]
47. Hameed, A.; Gulzar, S.; Aziz, I.; Hussain, T.; Gul, B.; Khan, M.A. Effects of Salinity and Ascorbic Acid on Growth, Water Status and Antioxidant System in a Perennial Halophyte. *AoB Plants* **2015**, *7*. [CrossRef]
48. Rontani, J.-F.; Nassiry, M.; Mouzdahir, A. Free radical oxidation (autoxidation) of α-tocopherol (vitamin E): A potential source of 4,8,12,16-tetramethylheptadecan-4-olide in the environment. *Org. Geochem.* **2007**, *38*, 37–47. [CrossRef]
49. Shyamaladevi, B.; Selvaraj, J. An update on β-sitosterol: A potential herbal nutraceutical for diabetic management. *Biomed. Pharmacother.* **2020**, *131*, 110702. [CrossRef]
50. He, M.; He, C.-Q.; Ding, N.-Z. Abiotic Stresses: General Defenses of Land Plants and Chances for Engineering Multistress Tolerance. *Front. Plant Sci.* **2018**, *9*, 1771. [CrossRef]
51. Hazel, J.R. Thermal adaptation in biological membranes: Is homeoviscous adaptation the explanation? *Annu. Rev. Physiol.* **1995**, *57*, 19–42. [CrossRef]
52. Sangwan, N.; Farooqi, A.; Shabih, F.; Sangwan, R.S. Regulation of essential oil production in plants. *Plant Growth Regul.* **2001**, *34*, 3–21. [CrossRef]
53. Martiniere, A.; Gayral, P.; Hawes, C.; Runions, J. Building bridges: Formin1 of Arabidopsis forms a connection between the cell wall and the actin cytoskeleton. *Plant J.* **2011**, *66*, 354–365. [CrossRef] [PubMed]
54. Sui, N.; Han, G.L. Salt-induced photoinhibition of PSII is alleviated in halophyte Thellungiella halophila by increases of unsaturated fatty acids in membrane lipids. *Acta Physiol. Plant.* **2014**, *36*, 983–992. [CrossRef]
55. Liu, S.S.; Wang, W.Q.; Li, M.; Wan, S.B.; Sui, N. Antioxidants and unsaturated fatty acids are involved in salt tolerance in peanut. *Acta Physiol. Plant.* **2017**, *39*, 207. [CrossRef]
56. Matistov, N.V.; Valuyskikh, O.E.; Shirshova, T.I. Qualitative characteristics and content of micronutrients in berries of *Rubus chamaemorus* L. *in European northeast of Russia. Bull. Komi Sci. Center URO RAS* **2012**, *1*, 41–45. (In Russian)
57. Lyutikova, M.N.; Turov, Y.P.; Botirov, E.H. Application of chromatography-mass spectrometry to determine free and esterified fatty acids in their combined presence in plant raw materials. *Fine Chem. Technol.* **2013**, *8*, 52–57. (In Russian)
58. Li, Y.; Beisson, F.; Ohlrogge, J.; Pollard, M. Monoacylglycerols Are Components of Root Waxes and Can Be Produced in the Aerial Cuticle by Ectopic Expression of a Suberin-Associated Acyltransferase. *Plant Physiol.* **2007**, *144*, 1267–1277. [CrossRef]
59. Tola, A.J.; Jaballi, A.; Germain, H.; Missihoun, T.D. Recent Development on Plant Aldehyde Dehydrogenase Enzymes and Their Functions in Plant Development and Stress Signaling. *Genes* **2021**, *12*, 51. [CrossRef]
60. Farmer, E.E.; Davoine, C. Reactive electrophile species. *Curr. Opin. Plant Biol.* **2007**, *10*, 380–386. [CrossRef]
61. Havaux, M. Carotenoid oxidation products as stress signals in plants. *Plant J.* **2013**, *79*, 597–606. [CrossRef]
62. Pushkina, N.V. Fatty acids and oxylipins accumulation in *Zea Maize*, L. seedlings under the influence of the electromagnetic field of the super high-frequency range. *Chem. Plant Raw. Mater.* **2020**, *2*, 93–99. [CrossRef]
63. Dylenova, E.P.; Zhigzhitzhapova, S.V.; Randalova, T.E.; Tykheev, Z.A.; Imikhenova, E.I.; Radnaeva, L.D. Composition of Lipid Fraction from the Aerial Part of Artemisia frigida. *Chem. Nat. Compd.* **2018**, *54*, 339–341. [CrossRef]
64. Novitskaya, G.V.; Suvorova, T.A.; Trunova, T.I. Lipid Composition of Tomato Leaves as Related to Plant Cold Tolerance. *Russ. J. Plant Physiol.* **2000**, *47*, 728–733. [CrossRef]
65. Melyashova, A.S.; Smolobochkin, A.V.; Gazizov, A.S.; Voronina, J.K.; Burilov, A.R.; Pudovik, M.A. Convenient Synthesisof 2-(Het) ArylpyrrolidinesviaStable 1-PyrroliniumSalts. *Tetrahedron* **2019**, *75*, 130681. [CrossRef]
66. Del Rio, D.; Rodriguez-Mateos, A.; Spencer, J.P.E.; Tognolini, M.; Borges, G.; Crozier, A. Dietary (Poly) Phenolics in Human Health: Structures, Bioavailability, and Evidence of Protective Effects Against Chronic Diseases. *Antiox Redox Signal.* **2013**, *18*, 1818–1892. [CrossRef]
67. Verma, N.; Shukla, S. Impact of Various Factors Responsible for Fluctuation in Plant Secondary Metabolites. *J. Appl. Res. Med. Aromat. Plants* **2015**, *2*, 105–113. [CrossRef]
68. Yang, L.; Wen, K.S.; Ruan, X.; Zhao, Y.X.; Wei, F.; Wang, Q. Response of Plant Secondary Metabolites to Environmental Factors. *Molecules* **2018**, *23*, 762. [CrossRef]
69. Pichersky, E.; Noel, J.P.; Dudareva, N. Biosynthesis of Plant Volatiles: Nature's Diversity and Ingenuity. *Science* **2006**, *311*, 808–811. [CrossRef]
70. Dunford, N.T.; Vazquez, R.S. Effect of Water Stress on Plant Growth and Thymol and Carvacrol Concentrations in Mexican Oregano Grown under Controlled Conditions. *J. Appl. Hortic.* **2005**, *7*, 20–22. [CrossRef]

71. Platonov, V.V.; Khadartsev, A.A.; Sukhikh, G.T.; Dunaeva, I.V.; Volochaeva, M.V. Chromato-mass-spectrometry of the ethanol extract of liquorice root (*Glycyrrhiza glabra*, L.; Fabaceae family). *J. New Med. Technol.* **2020**, *3*, 137–142. [CrossRef]
72. Sepúlveda, L.V.; González-Morales, S.; Mendoza, A.B. Benzoic acid: Biosynthesis, modification and function in plants. *Rev. Mex. Cienc. Agríc.* **2015**, *6*, 1667–1678. [CrossRef]
73. Senaratna, T.; Merritt, D.J.; Dixon, K.; Bunn, E.; Touchell, D.H.; Sivasithamparam, K. Benzoic acid may act as the functional group in salicylic acid and derivatives in the induction of multiple stress tolerance in plants. *Plant Growth Regul.* **2003**, *39*, 77–81. [CrossRef]
74. Los, D.A.; Mironov, K.S.; Allakhverdiev, S.I. Regulatory role of membrane fluidity in gene expression and physiological functions. *Photosynth. Res.* **2013**, *2–3*, 489–509. [CrossRef]
75. Zenkevich, I.G.; Rotaru, K.I.; Selivanov, S.I.; Kostikov, R.R. Determination of dialkyl phthalates in different objects (Problems for discussion). *St. Petersbg. State Univ. Bull.* **2015**, *2*, 386–394.
76. Stepanycheva, E.A.; Petrova, M.O.; Shchenikova, A.V.; Chermenskaya, T.D. Allelochemics: An interaction between phytophages and Pseudomonas syringae pv. tomato on tomato Solanum lycopersicum plants. *Agric. Biol.* **2016**, *51*, 731–738. [CrossRef]
77. Farmer, E.E. Surface-to-air signals. *Nature* **2001**, *411*, 854–856. [CrossRef]
78. Edge, R.; McGarvey, D.J.; Truscott, T.G. The Carotenoids as Antioxidants. *Photochem. Photobiol.* **1997**, *4*, 1189–1200. [CrossRef]
79. Gruszecki, W.I.; Strzałka, K. Does the Xanthophyll Cycle Take Part in the Regulation of Fluidity of the Thylakoid Membrane? *Biochim. Biophys. Acta* **1991**, *1060*, 310–314. [CrossRef]
80. Polyakov, N.E.; Leshina, T.V. Certain aspects of the reactivity of carotenoids. Redox processes and complexation. *Russ. Chem. Rev.* **2006**, *75*, 1049–1064. [CrossRef]
81. Udovenko, G.V. Diagnostics stability of plant resistance to stress. In *Methodical Guidance*; VIR: Leningrad, Russia, 1988; p. 89.
82. Rascher, U.; Liebig, M.; Lüttge, U. Evaluation of instant light-response curves of chlorophyll fluorescence parameters obtained with a portable chlorophyll fluorometer on site in the field. *Plant Cell Environ.* **2000**, *23*, 1397–1405. [CrossRef]
83. Terletskaya, N.V.; Stupko, Y.U.; Altayeva, N.A.; Kudrina, N.O.; Blavachinskaya, I.V.; Kurmanbayeva, M.S.; Erezhetova, U. Photosynthetic activity of Triticum dicoccum × Triticum aestivum alloplasmic lines during vegetation in connection with productivity traits under varying moister conditions. *Photosynthetica* **2021**, *59*, 1–11. [CrossRef]
84. Barykina, R.P.; Veselova, T.D.; Devyatov, A.G.; Dzhalilova , Kh.Kh.; Ilyina, G.M.; Chubatova, N.V. Guide on Botanical Microtechique. In *Base and Methods*; MSU: Moscow, Russia, 2004; p. 312.

plants

MDPI

Article

The Variability for the Biochemical Indicators at the Winter Wheat Assortment and Identifying the Sources with a High Antioxidant Activity

Ramona Aida Paunescu [1], Elena Bonciu [2,*], Elena Rosculete [3,*], Gabriela Paunescu [4], Catalin Aurelian Rosculete [2] and Cristina Babeanu [5]

[1] Syngenta Agro Romania, 73-81 Bucuresti-Ploiesti Street, 013685 Bucharest, Romania; aida.paunescu@yahoo.com
[2] Department of Agricultural and Forestry Technology, Faculty of Agronomy, University of Craiova, 13 A.I. Cuza Street, 200585 Craiova, Romania; catalin_rosculete@yahoo.com
[3] Department of Land Measurement, Management, Mechanization, Faculty of Agronomy, University of Craiova, 13 A.I. Cuza Street, 200585 Craiova, Romania
[4] SCDA Caracal, University of Craiova, 106 Vasile Alecsandri Street, 235200 Caracal, Romania; paunescucraiova@yahoo.com
[5] Department of Chemistry, Faculty of Sciences, University of Craiova, 13 A.I. Cuza Street, 200585 Craiova, Romania; cbabeanu@yahoo.com
* Correspondence: elena.agro@gmail.com (E.B.); rosculeta2000@yahoo.com (E.R.); Tel.: +40-722996073 (E.B.); +40-729589490 (E.R.)

Abstract: This study presents the variability of some biochemical indicators in the winter wheat assortments tested in south-western Oltenia (Romania) and identification of the sources showing a high antioxidant activity. The peroxidase activity has intensified as the stress induced by treatment with PEG of different concentrations and in different doses increased. Regarding the peroxidase content, among the varieties treated with PEG 10,000 25%, the majority of the Romanian varieties tested showed higher values of the PEG/control treatment ratio, which suggests tolerance to drought. In reverse, the activity of ascorbate peroxidase is lower in tolerant varieties. The varieties with a subunit report have been noted. Among them are the Izvor variety, known as the drought-tolerant variety, as well as other Romanian varieties: Alex, Delabrad, Lovrin 34, etc. An increased activity of catalase was present in most varieties, so there is the possibility of drought tolerance. Among the varieties highlighted are Romanian varieties (Dropia, Trivale, Nikifor, etc.) but also foreign varieties (Kristina, GH Hattyu, Karlygash, etc.). However, the correlation between yield index in the limited assortment and the antioxidant enzyme content ratios between PEG and control treatments does not exist, suggesting that none of these biochemical indicators are a selective indicator for drought tolerance under the experimental condition.

Keywords: wheat; peroxidase; ascorbate peroxidase; catalase; yield index

Citation: Paunescu, R.A.; Bonciu, E.; Rosculete, E.; Paunescu, G.; Rosculete, C.A.; Babeanu, C. The Variability for the Biochemical Indicators at the Winter Wheat Assortment and Identifying the Sources with a High Antioxidant Activity. *Plants* **2021**, *10*, 2443. https://doi.org/10.3390/plants10112443

Academic Editors: Kinga Dziurka, Mateusz Labudda and Ewa Muszyńska

Received: 19 October 2021
Accepted: 10 November 2021
Published: 12 November 2021

1. Introduction

Wheat is a cheap source of essential amino acids (which are not synthesized in the body), good quality minerals, vitamins, and vital dietary fibres to the human diet [1]. Besides this, it is also considered a natural source of both enzymatic and non-enzymatic antioxidants [2]. The enzymatic antioxidants include superoxide dismutase (SOD), glutathione reductase (GR), and ascorbate peroxidase (APX), catalase (CAT), and peroxidase (POD), while non-enzymatic antioxidants include vitamin C (tocopherols and tocotrienols), vitamin E, and carotenoids [3].

Drought tolerance is a complex trait that refers to the degree to which a plant is adapted to arid or drought conditions that lead to different morphological and physiological changes. Adaptation processes to drought stress conditions involve the genetics at different molecular, physiological, biochemical, and biological levels and processes [4,5].

Under drought conditions, oxidative degradation products occur at the cellular level, leading to oxidative stress. Numerous experiments on the study of wheat drought resistance showed cell-based induction of enzyme oxidative stress protection systems [6–8]. A drought-tolerant genotype had the highest activity of peroxidase and catalase ascorbate and high ascorbic acid content and showed the lowest accumulation of hydrogen peroxide and lipid peroxidase, compared to a sensitive genotype that had the lowest activity of antioxidant enzymes and ascorbic acid content and the highest content of hydrogen peroxide and lipid peroxidase [7,8]. Reactive oxygen species (ROS) are generated in plants upon exposure to stressful conditions [9]. ROS are byproducts of numerous enzymatic reactions in various cell compartments [10].

The combating ROS in plants during stressful conditions is maintained by the enzymatic components comprising of the superoxide dismutase (SOD), ascorbate peroxidase (APX), guaiacol peroxidase (GPX), glutathione-S-transferase (GST), and catalase (CAT). The omnipresent nature of these enzymatic components underlies the necessity of the detoxification of ROS for cellular survival [9,10].

One of the most important goals of plant breeding is to produce new wheat cultivars with a high degree of drought tolerance. Thus, the first step is to select the potential germplasm that contains genotypic differences for drought tolerance [11].

Some authors [12] studied *Triticum* genotypes with three levels of polyploidy: hexaploid, tetraploid, and diploid, submitted to a stress of 4, 8, and 12 days, respectively. In general, catalase showed an increase or maintenance in the early stages of drought and then a decrease as the magnitude of stress increased. In contrast, peroxidase increased to water stress.

According to [13], who studied the role of the plant's antioxidant system in stress tolerance, drought induced in two different stages after anthesis resulted in increased accumulation of oxygenated water and decreased ascorbic acid content. Antioxidant enzymes such as ascorbate peroxidase and catalase have increased under water stress conditions. A drought-tolerant genotype had the highest activity of ascorbate peroxidase and catalase and high ascorbic acid content and also showed the lowest accumulation of oxygenated water and lipid peroxidase. By comparison, a sensitive genotype had the lowest antioxidant enzyme activity and ascorbic acid content and the highest content of oxygenated water and lipid peroxidase.

A solution of polyethylene glycol (PEG) can be used to induce drought stress that is measured using a timescale of days after treating the seeds with the PEG solution. There are many different concentrations of PEG; therefore, it is essential to test a wide range of concentrations. In germination experiments using PEG, the seeds of genotypes are tested to different concentration [14,15].

Some authors [16] analysed five wheat cultivars submitted to 3, 6, and 9 days water stress, respectively. The activity of peroxidase and ascorbate peroxidase showed an initial increase. In cultivars that were found to be more stress-tolerant than others, peroxidase activity increased with increasing stress duration while ascorbate peroxidase activity decreased. The study conducted by [17] on an assortment of wheat cultivars and lines created at Șimnic regarding the phenol content and the activity of the antioxidants revealed that they are significantly influenced by genotype and environment. The Dropia variety was superior to the Boema variety and to the lines tested by point of view of phenol content.

Results reported by some authors suggest that water stress alters the balance between free radical production and enzymatic protection mechanisms in wheat plants [18]. Studies conducted by [19] investigated the effects of salicylic acid (SA) and cold on apoplastic protein levels and activities of apoplastic catalase (CAT), peroxidase (POX), and polyphenol oxidase (PPO) in winter wheat (*Triticum aestivum* cv. Dogu-88) leaves. When the activities with cold + SA treatment are compared to their cold treatments, CAT and POX activities were decreased while PPO activity was increased by SA.

The activities of antioxidant enzyme defence system depended on wheat cultivar, duration of drought, and the stage of leaf development [20].

Drought-tolerant genotypes also kept higher ascorbate compared with sensitive genotypes under stress and non-stress conditions, while peroxidase activity was not affected by drought stress [21]. Antioxidants and stress markers can be efficiently and economically used as biochemical indices to screen or to enrich wheat germoplasm for drought tolerance at the early seedling stage [21]. Plant materials should be phenotyped accurately using an appropriate assay and trait that has a direct relation to drought tolerance. Single-trait evaluation for drought tolerance to distinguish between tolerant and susceptible genotypes is not recommendable [22,23].

This study aimed at investigating the variability of some biochemical indicators (peroxidase, ascorbate peroxidase, and catalase) at the winter wheat assortments and identification of the sources showing a high antioxidant activity.

2. Results

In our experiment in south-western Oltenia, the activity of peroxidase increased as the stress induced by PEG treatment of different concentrations and at different doses increased. The varieties that showed an increase in the activity of the peroxidase are shown in Table 1.

Table 1. The activity of peroxidase in an assortment of 50 wheat cultivars analysed.

No.	Cultivar	PEROXIDASE (ΔA/1 min/1 gsp)				
		In Normal Conditions (Water–Control)	In Water-Stress-Induced Conditions (25% PEG 10,000)	In Water-Stress-Induced Conditions (40% PEG 4000)	Ratio PEG (25%/Control)	Ratio PEG (40%/Control)
35.	KRISTINA	57.60	211.43	319.02	3.671	5.539
39.	MOLDAU	91.69	308.8	0	3.368	0.000
40.	MV PALMA	66.21	220.78	0	3.335	0.000
50.	TRIVALE	23.27	71.31	0	3.064	0.000
31.	GK HATTYU	104.88	293	431.14	2.794	4.111
36.	LADA	72.94	188.09	141.2	2.579	1.936
48.	SHOHAM	71.46	177.44	0	2.483	0.000
30.	GRUIA	75.75	176.8	200.6	2.334	2.648
47.	ROMULUS	75.16	168.75	0	2.245	0.000
34.	KARLYGASH	72.01	158.66	275.62	2.203	3.828
41.	NIKIFOR	85.89	180.38	100.54	2.100	1.171
29.	GK GOBE	119.08	238.71	174.45	2.005	1.465
43.	ORQUAL	69.33	134.46	0	1.939	0.000
25.	GIAVA	111.83	216	321.96	1.932	2.879
33.	IZVOR	114.15	216.16	190.17	1.894	1.666
27.	GK ELET	105.89	199.7	387	1.886	3.655
21.	EXOTIC	76.19	141	332	1.851	4.358
42.	ORATORIO	72.12	128.97	0	1.788	0.000
18.	ELIANA	85.96	125.55	87.59	1.461	1.019
26.	GK DAVID	207.61	301	543.27	1.450	2.617
46.	ROMANSA	111.45	157.97	76.76	1.417	0.689
11.	CUBUS	108.57	151.5	219.64	1.395	2.023
1.	AGRON	134.4	176	218.3	1.310	1.624
5.	BITOP	140.21	176.25	162.7	1.257	1.160
38.	LOVRIN 34	126.92	144.34	0	1.137	0.000
6.	BOEMA	151.28	171.53	165	1.134	1.091
13.	DELABRAD	122.7	137.88	166.65	1.124	1.358
28.	GLOSA	137.53	151.6	372	1.102	2.705
2.	ALEX	130.9	141.28	191.75	1.079	1.465
23.	FLAMURA 85	171	176	316	1.029	1.848
3.	AZTEC	138	140.77	161	1.020	1.167
32.	MIRANDA	90.6	92.25	170.25	1.018	1.879
22.	FAUR	107.55	108	184.48	1.004	1.715
14.	DEMETRA	129.66	128.76	215.12	0.993	1.659
7.	SIMNIC 50	121.1	113.92	255.1	0.941	2.107
16.	DROPIA	120.4	111.57	140.68	0.927	1.168
19.	ENESCO	132.89	121.96	321.83	0.918	2.422
24.	GABRIELA	149.67	135.62	265	0.906	1.771
20.	ESQUISIT	199.74	168.06	219.21	0.841	1.097
12.	DARIEL	132.97	110.44	163.74	0.831	1.231
4.	BEZOSTAIA	162.83	130.92	175	0.804	1.075
15.	DOR	160.66	126.42	166.11	0.787	1.034
17.	DUNAI	164.75	125.78	211.99	0.763	1.287
49.	SIMNIC 30	59.58	43.96	0	0.738	0.000
10.	CRINA	133.22	84.7	138.87	0.636	1.042
44.	JULIUS	207.74	121.65	0	0.586	0.000
8.	CAPO	179.5	103.76	182.28	0.578	1.015
37.	LITERA	139.04	77.74	0	0.559	0.000
45.	NATHAN	238.33	113.04	0	0.474	0.000
9.	CAROLINA	126.33	56.63	163.25	0.448	1.292

The cultivars from the limited assortment were yellow highlighted (winter wheat cultivars tested for 14 years, year by year). It is observed that they are distributed everywhere in the range of the results obtained by the 50 cultivars, this fact indicating that the varieties that are part of this assortment and the obtained results could have a high probability of identifying the sources of stress tolerance for the south-western Oltenia drought conditions.

Among the varieties treated with PEG 10,000 25%, most of the Romanian varieties tested, components of the limited assortment, Gruia, Izvor, Lovrin 34, Boema, Delabrad, Glosa, Alex, Miranda, and Faur, presented super unitary values of the ratio between PEG/control treatment, which suggests drought tolerance.

Conversely, the activity of ascorbate peroxidase is lower in the tolerant varieties, so in Table 2, the varieties with subunit ratio were noted. Among them are the Izvor variety, known as the drought-tolerant variety, as well as other Romanian varieties: Alex, Delabrad, Lovrin 34, Miranda, Dor, and Romulus (Table 2).

Table 2. The activity of ascorbat peroxidase in an assortment of 50 wheat cultivars analysed.

No.	Cultivar	ASCORBAT PEROXIDAZE (µgAsA/1 min/1 gsp)				
		In Normal Conditions—Water-Control	In Water-Stress-Induced Conditions—25% PEG 10,000	In Water-Stress-Induced Conditions—40% PEG 4000	Ratio PEG 25%/Control	Ratio PEG 40%/Control
50.	TRIVALE	4693	40,522	0	8.635	0.000
49.	SIMNIC 30	4754	35,528	0	7.473	0.000
29.	GK GOBE	2990	21,994	3661	7.356	1.224
8.	CAPO	16,313	57,206	4550	3.507	0.279
9.	CAROLINA	5834	17,921	11,101	3.072	1.903
7.	SIMNIC 50	9829	29,888	14,349	3.041	1.460
11.	CUBUS	4606	11,865	13,154	2.576	2.856
40.	MV PALMA	52,089	129,907	0	2.494	0.000
10.	CRINA	13,795	33,881	30,699	2.456	2.225
41.	NIKIFOR	75,225	179,104	41,893	2.381	0.557
37.	LITERA	26,825	62,474	0	2.329	0.000
14.	DEMETRA	9350	20,147	6711	2.155	0.718
48.	SHOHAM	34,285	70,888	0	2.068	0.000
23.	FLAMURA 85	31,788	60,423	80,790	1.901	2.542
5.	BITOP	13,523	24,456	20,302	1.808	1.501
6.	BOEMA	11,930	20,898	0	1.752	0.000
22.	FAUR	29,801	51,107	34,830	1.715	1.169
24.	GABRIELA	33,376	51,546	43,200	1.544	1.294
30.	GRUIA	18,476	28,177	13,396	1.525	0.725
21.	EXOTIC	20,906	31,348	52,677	1.499	2.520
3.	AZTEC	12,420	17,802	4565	1.433	0.368
28.	GLOSA	28,813	37,094	14,970	1.287	0.520
39.	MOLDAU	63,595	70,754	0	1.113	0.000
43.	ORQUAL	35,074	38,866	0	1.108	0.000
16.	DROPIA	23,299	25,689	22,446	1.103	0.963
17.	DUNAI	28,531	31,389	18,868	1.100	0.661
36.	LADA	13,827	13,470	16,152	0.974	1.168
12.	DARIEL	7299	6995	12,564	0.958	1.721
27.	GK ELET	25,027	23,328	7278	0.932	0.291
2.	ALEX	9663	9001	5758	0.931	0.596
25.	GIAVA	56,518	50,899	61,391	0.901	1.086
45.	NATHAN	34,463	29,747	0	0.863	0.000
13.	DELABRAD	20,398	17,253	20,925	0.846	1.026
15.	DOR	20,618	17,412	39,525	0.845	1.917
35.	KRISTINA	37,092	29,585	0	0.798	0.000
47.	ROMULUS	46,875	35,744	0	0.763	0.000
38.	LOVRIN 34	29,747	21,739	0	0.731	0.000
33.	IZVOR	43,547	29,441	26,106	0.676	0.599
1.	AGRON	10,847	6428	6007	0.593	0.554
46.	PKB ROMANSA	70,721	39,318	60,942	0.556	0.862
32.	MIRANDA	51,011	27,675	15,796	0543	0.310
26.	GK DAVID	35,090	18,654	0	0.532	0.000
4.	BEZOSTAIA	9221	45,85	5098	0.497	0.553
42.	ORATORIO	73,625	31,888	0	0.433	0.000
44.	JULIUS	80,042	30,550	0	0.382	0.000
19.	ENESCO	65,789	24,314	12,227	0.370	0.186
18.	ELIANA	36,772	13,413	29,207	0.365	0.794
31.	GK HATTYU	17,537	6002	12,975	0.342	0.740
34.	KARLYGASH	34,110	11,376	6852	0.334	0.201
20.	ESQUISIT	10,981	2838	20,284	0.258	1.847

Increased activity of catalase has been present in most varieties, so that the possibility of drought tolerance exists. Among the varieties listed in Table 3 are Romanian varieties: Dropia, Trivale, Nikifor, Simnic 30, Simnic 50, Faur, Glosa, Gruia, Miranda, and Flamura 85. Of the foreign varieties that maintain a high activity of catalase at higher dose stress intensification, we can see: Kristina, GH Hattyu, Karlygash, Esquisit, Lada, Enesco, GK Elet, Cubus, and GK Gobe (Table 3).

Table 3. The activity of catalase in an assortment of 50 wheat cultivars.

No.	Cultivar	CATALASE				
		In Normal Conditions—Water-Control	**In Water-Stress-Induced Conditions—25% PEG 10,000**	**In Water-Stress-Induced Conditions—40% PEG 4000**	**Ratio PEG 25%/Control**	**Ratio PEG 40%/Control**
48.	SHOHAM	656	3318	-	5.058	-
16.	DROPIA	298	1407	1492	4.721	5.007
35.	KRISTINA	896.8	3155.81	5534.08	3.519	6.171
4.	BEZOSTAIA	1536.98	4471	1087.7	2.909	0.708
50.	TRIVALE	987	2765	-	2.801	-
42.	ORATORIO	1652	4578	-	2.771	-
31.	GK HATTYU	1945.44	5378.13	5536.33	2.764	2.846
18.	ELIANA	1046	2862	290	2.736	0.277
41.	NIKIFOR	1288	3422	3422	2.657	2.657
9.	CAROLINA	435.73	1142.05	-	2.621	-
34.	KARLYGASH	1819.21	3519.15	2557.91	1.934	1.406
49.	SIMNIC 30	982	1876	-	1.910	-
20.	ESQUISIT	1749	3309	3137	1.892	1.794
36.	LADA	1482.57	2753.87	4020.1	1.857	2.712
7.	SIMNIC 50	366.97	680.13	-	1.853	-
33.	IZVOR	929	1685.36	1021.05	1.814	1.099
22.	FAUR	763	1362.8	-	1.786	-
46.	PKB ROMANSA	1128	1864	-	1.652	-
19.	ENESCO	701	1141	4565	1.628	6.512
28.	GLOSA	1782.55	2868.62	3832.3	1.609	2.150
27.	GK ELET	1993.32	3110.42	4885.5	1.560	2.451
43.	ORQUAL	1808	2784	-	1.540	-
30.	GRUIA	1839.45	2604	2703.61	1.416	1.470
11.	CUBUS	1253	1654	1497	1.320	1.195
45.	NATHAN	1387	1829	-	1.319	-
32.	MIRANDA	1432.37	1668.02	1904.12	1.165	1.329
26.	GK DAVID	3992.5	4158.68	1923.07	1.042	0.482
29.	GK GOBE	3472.56	3501.76	3645.2	1.008	1.050
12.	DARIEL	1370	1380	965	1.007	0.704
39.	MOLDAU	1208	1211	-	1.002	-
23.	FLAMURA 85	5086	5086	5086	1.000	1.000
24.	GABRIELA	5086	5086	5086	1.000	1.000
21.	EXOTIC	1115	1103.4	1274	0.990	1.143
2.	ALEX	1020.46	921.75	3071.01	0.903	3.009
40.	MV PALMA	2200	1877	-	0.853	-
15.	DOR	1173	990	1265	0.844	1.078
8.	CAPO	1020	854.3	-	0.838	-
44.	JULIUS	2100	1685	-	0.802	-
38.	LOVRIN 34	1625	1265	-	0.778	-
1.	AGRON	3010	2285.71	4605.66	0.759	1.530
3.	AZTEC	2700	1898.88	2207.8	0.703	0.818
37.	LITERA	987	674	-	0.683	-
10.	CRINA	820.55	542.1	382.04	0.661	0.466
13.	DELABRAD	1450	938	-	0.647	-
5.	BITOP	3300.23	2086.96	1082.78	0.632	0.328
47.	ROMULUS	4562	2647	-	0.580	-
25.	GIAVA	3014.31	1266.82	3274.2	0.420	1.086
14.	DEMETRA	1662	286	1577	0.172	0.949
6.	BOEMA	3800	334.38	-	0.088	-
17.	DUNAI	3570	312	1342	0.087	0.376

However, the correlation between YI in the limited assortment and the ratio of antioxidant enzyme content between PEG and control treatments does not exist, suggesting that none of these biochemical indicators represent a selection indicator for drought tolerance under the conditions of south-western Oltenia (Table 4).

Table 4. The correlation between the YI index and the antioxidant activity for the limited assortment.

Cultivar	YI	Ratio PEROX PEG 25%/ct	Ratio PEROX PEG 40%/ct	Ratio ASC PEG 25%/ct	Ratio ASC PEG 40%/ct	Ratio CAT PEG 25%/ct	Ratio CAT PEG 40%/ct
GLOSA	1.252	1.102	2.705	1.287	0.520	1.609	2.150
GRUIA	1.220	2.334	2.648	1.525	0.725	1.416	1.470
IZVOR	1.219	1.894	1.666	0.676	0.599	1.814	10.991
FAUR	1.153	1.004	1.715	1.715	1.169	1.786	0.000
DELABRAD	1.116	1.124	1.358	0.846	1.026	0.647	0.000
CRINA	0.996	0.636	1.042	2.456	2.225	0.000	0.000
ALEX	0.955	1.079	1.465	0.931	0.596	0.903	3.009
DROPIA	0.941	0.927	1.168	1.103	0.963	4.721	5.007
SIMNIC 30	0.895	0.738	1.029	7.473	5.972	1.910	-
BEZOSTAIA	0.854	0.804	1.075	0.497	0.553	2.909	0.708
BOEMA	0.847	1.134	1.091	1.752	1.400	0.000	0.000
ROMULUS	0.800	2.245	3.130	0.763	0.609	0.580	0.000
LOVRIN 34	0.753	1.137	1.585	0.731	0.584	0.000	0.000
Corelation with YI		0.25	0.31	−0.11	−0.18	0.14	0.42

From previous studies [24,25], it was clear that there is a close correlation between the YI index, which expresses drought tolerance, and the ratio between the stem length at PEG treatment 20% and the length of the stem at the control of 15 days from the sowing date (T1) or on average at three moments of determination (15, 24, and 35 days from sowing) and the ratio of the weight of the stem to PEG treatment 20% against the weight of the stem on the control, on the other hand.

In turn, these reports were correlated, as follows (Table 5):

The ratio of stem length to 20% PEG treatment and control on average of three determination times (Med) was correlated with:

- Ratio between peroxidase content after 25% PEG treatment and peroxidase content in the control-significant correlation.

The ratio of the stem weight to PEG treatment 20% and the weight of the stem to the control was correlated with:

- Ratio between catalase content at 25% PEG treatment and control catalase content—distinctly significant correlation;
- Ratio between catalase content at 40% PEG treatment and control catalase content—significant correlation.

Directly or indirectly, all these determinations mentioned above can be used as selection indicators for drought tolerance, and, depending on the results obtained from the varieties tested, certain parents may be suggested.

The biochemical indicators were correlated, as follows:

The ratio of peroxidase content to 25% PEG treatment to peroxidase content in the control was significantly correlated with:

1. The ratio between the stem length at 15% PEG treatment and the length of the stem at the control after 24 days from sowing (T2);
2. The ratio of stem length to 15% PEG treatment and control stem length on average on determinations made at 15, 24, and 35 days;
3. The ratio of stem length to 20% PEG treatment and control stem length on average on determinations made at 15, 24, and 35 days;
4. The ratio of stem weight to the 15% PEG treatment and the weight of the stem to the control.

When the PEG dose is increased to 40%, another correlation appears:

- Distinctly significant positive with the ratio of peroxidase content to 20% PEG treatment and peroxidase content to control.

The ratio of ascorbate peroxidase content to 25% PEG treatment and ascorbate peroxidase content control and ratio of ascorbate peroxidase to 40% PEG treatment and ascorbat peroxidase determined in control were significantly negatively correlated with reduction of seedling length in 40% PEG treatment compared to the control (treated with water).

Table 5. Correlations between the characters determined in south-western Oltenia on a limited assortment of wheat varieties.

Legend: 0.600 - 0.740: + * (green); - 0 (red). > 0.740: + ** (yellow); - 00 (blue).

		Ratio Length Stem PEG 15%/ct T_1	Ratio Length Stem PEG 15%/ct T_2	Ratio Length Stem PEG 15%/ct T_3	Ratio Length Stem PEG 15%/ct aver.	Ratio Length Stem PEG 20%/ct T_1	Ratio Length Stem PEG 20%/ct T_2	Ratio Length Stem PEG 20%/ct T_3	Ratio Length Stem PEG 20%/ct Aver.	Ratio Weight Root PEG 15%/ct	Ratio Weight Root PEG 20%/ct	Ratio Weight Stem PEG 15%/ct	Ratio Weight Stem PEG 20%/ct	Ratio Root/Stem PEG 15%/Root/Stem/ct	Ratio Root/Stem PEG 20%/Root/Stem/ct	Seed Germ. Stress/Normal Var (%)
		1	2	3	4	5	6	7	8	9	10	11	12	13	14	15
Peroxidase PEG 25%/per ct	23	0.491	0.727	0.380	0.710	0.469	0.589	0.296	0.636	0.540	0.033	0.702	0.150	0.095	−0.131	0.224
Peroxidase PEG 40%/per ct	24	0.463	0.668	0.412	0.684	0.321	0.479	0.322	0.515	0.436	−0.114	0.689	−0.081	0.109	−0.001	0.269
Asc PEG 25%/asc ct	25	−0.090	−0.383	0.388	−0.062	−0.108	−0.055	0.208	0.002	−0.186	0.076	−0.223	−0.248	−0.007	0.402	−0.407
Asc PEG 40%/asc ct	26	−0.231	−0.440	0.379	−0.173	−0.205	−0.095	0.219	−0.066	−0.220	0.091	−0.242	−0.231	−0.042	0.384	−0.467
Cat PEG 25%/cat ct	27	0.169	0.352	0.127	0.280	0.183	0.740	0.357	0.561	0.250	0.796	0.223	0.762	−0.105	0.204	−0.406
Cat PEG 40%/cat ct	28	0.279	0.326	0.003	0.294	0.236	0.656	0.170	0.479	0.218	0.734	0.181	0.700	−0.084	0.189	−0.259

Legend: 0.600 - 0.740: + * (green); - 0 (red). > 0.740: + ** (yellow); - 00 (blue).

		Red. Coleoptil Length (%)	Ratio Length pl PEG 25%/Length pl. ct	Ratio Length pl PEG 40%/Length pl. ct	Initial Water Cont. (%)	Water Loss after 4 h (%)	Water Loss after 20 h (%)	Water Loss after 24 h (%)	Peroxidase PEG 25%/per ct	Peroxidase PEG 40%/per ct	Asc per PEG 25%/Asc per ct	Asc per PEG 40%/Asc per ct	Catalase PEG 25%/Cat ct	Catalase PEG 40%/Cat ct
		16	17	18	19	20	21	22	23	24	25	26	27	28
Peroxidase PEG 25%/per ct	23	−0.393	0.045	0.111	0.204	0.590	−0.295	0.114	1					
Peroxidase PEG 40%/per ct	24	−0.781	0.128	0.115	0.067	0.156	−0.351	−0.194	0.771	1				
Ascorbat peroxidase PEG 25%/asc ct	25	0.235	−0.468	−0.863	−0.055	0.156	−0.200	−0.131	−0.336	−0.298	1			
Ascorbat peroxidase PEG 40%/asc ct	26	0.338	−0.450	−0.843	−0.043	0.027	−0.127	−0.088	−0.397	−0.404	0.985	1		
Catalase PEG 25%/cat ct	27	0.202	−0.255	0.049	0.242	0.410	0.009	0.255	0.290	−0.042	−0.147	−0.130	1	
Catalase PEG 40%/cat ct	28	0.109	−0.115	0.160	0.189	0.446	0.006	0.274	0.289	0.005	−0.258	−0.260	0.932	1

3. Discussion

Drought (water deficit) one the emerging threat worldwide and adversely affects the morpho-physiology and biochemical activity of plants, finally leading to a decrease in the grain yield of wheat [26,27]. Additionally, [28] showed that drought stress is one of the main threats that negatively affected the morphological, physiological, and biochemical behaviours of plants than other abiotic stresses. Drought adversely deteriorated the plant metabolic process by affecting the photosynthesis and water relations of the plant and also the uptake of nutrients [29].

At the biochemical indices, there is a change of meaning of the classification because a high ratio of peroxidase and catalase but a lower ratio of ascorbate peroxidase are desirable. In our study, the best of this pattern is the folded Kristina cultivar. On the other hand, the Izvor cultivar, known for its drought tolerance, is not noticeable by very good values for the enzymatic reaction possibly involved in drought conditions. Our results suggest ways to improve the Izvor cultivar performance under water stress conditions by hybridizing it with Kristina or Dropia varieties. Thus, descendants can be selected that accumulate good growth in the presence of water stress (simulated by PEG treatment), with a better reaction of the enzyme apparatus.

The results of many authors suggest that ascorbate peroxidase, a central enzyme for ROS scavenging in plants, can be induced under abiotic and biotic stresses [30–34]. Thus, the antioxidant enzymes APX, SOD, POD, and CAT are produced under different environmental stresses (such as drought, salt, etc.) for scavenging the activity of ROS in plants [35,36].

Additionally, the peroxidase activity has intensified as stress induced by PEG treatment of different concentrations and in different doses increased. Among the varieties treated with PEG 10,000 25%, most of the Romanian varieties tested presented super unit values of the PEG/control ratio, suggesting tolerance to drought. In reverse, the activity of ascorbate peroxidase is lower in tolerant varieties. Among the varieties evidenced by the increased activity of catalase were the Romanian varieties: Dropia, Trivale, Nikifor, Simnic 30, Simnic 50, Faur, Glosa, Gruia, Miranda, and Flamura 85. Among the foreign varieties that have maintained a high catase activity to increase stress through a higher dose of PEG were Kristina, GH Hattyu, Karlygash, Esquisit, Lada, Enesco, GK Elet, Cubus, and GK Gobe.

Drought stress can occur at any growth stage and depends on the local environment. According to [11], genotypes may be tested for their drought tolerance at relevant and often different growth stages because some genotypes may tolerate drought at the germination or seedling stage, but these may be very sensitive to drought at the flowering stage or vice versa. The ability of seeds and young seedlings to cope with oxidative stress during early vegetative growth and biotic (attachment of the soil and seed-borne pathogens) and abiotic stresses (drought, salinity, heat, and chilling) is vital for crop performance and production [37]. High activities of antioxidant enzymes such as superoxide dismutase (SOD), catalase (CAT), peroxidase (POD), and ascorbate peroxidase (APX) have been recorded during seed germination, early growth, and biotic and abiotic stresses [38–40]. APX plays a considerable role in wheat drought tolerance by detoxifying plants from the accumulation of H_2O_2 [41]

There are many studies that suggest that the yield index is correlated to antioxidant activity [38–42]. Generally, the genotypes respond differently to drought tolerance at different growth stages [11]. Some wheat genotypes exhibited a similar pattern of stress response, comprising proline accumulation, rise in hydrogen peroxide content, oxidative damage to membrane lipids, and increase in total antioxidant and antiradical activities, phenolic and flavonoid content, ascorbate and glutathione pools, and mobilization of superoxide dismutase (SOD), catalase (CAT), and peroxidase (POX) enzyme isoforms [42]. Other results reveal the important role of certain chloroplast chaperone proteins in drought stress response and different strategies of stress adaptation depending on the wheat genotype [43]. According to [44], the combination of lower N supply and water deprivation (osmotic stress induced by polyethylene glycol treatment) led to greater damage of the photosynthetic efficiency and a higher degree of oxidative stress than the individually applied stresses. Plant materials should be phenotyped accurately using an appropriate assay and trait that has a direct relation to drought tolerance [22].

Reactive oxygen species (ROS) plays an important signalling role in plants, controlling processes such as growth, development, and especially response to biotic and abiotic environmental stressors [45,46]. However, ROS are unable to cause damage, as they are being scavenged by different antioxidant mechanisms [47–49].

Plants treated with herbicides, similarly to those grown under various abiotic stress conditions, are subjected to enhanced attacks by ROS. According to [50], the stress markers, enzymatic and non-enzymatic antioxidant defence, were additionally increased during the stress period after the combined herbicide and drought treatment.

4. Materials and Methods

The experiments were located in the South-West area of Oltenia region (Romania). This belongs to the temperate climate zone, with Mediterranean influences due to its south-western position. The position and the depressional feature of the land it occupies, close to the curvature of the Carpathian–Balkan mountain range, determine, on the whole, a warmer climate than in the central and northern part of the country, with an annual average of 10–11.5 °C.

4.1. Plant Materials

Fifty wheat varieties of various origins were tested in the laboratory to detect differences between biochemical indicators: peroxidase, ascorbate peroxidase, and catalase. Fresh tissue, necessary for enzymatic analysis, was collected after 30 days in which the seedlings were grown in a controlled environment, under three experimental conditions (H_2O, 25% PEG 10,000, and 40% PEG 4000).

From 50 varieties, 13 varieties were also field tested for the period 2002–2015, and the YI-specific drought tolerance index was calculated from yield data from the field, taking into account the average yield of years with the most severe drought 2002 and 2003 as Ys. This is the limited assortment.

Field experiments were placed in triple balanced grid without repeating the basic scheme.

4.2. Laboratory Researche Methods

Plants can be protected by antioxidant synthesis and by increasing the activity of antioxidant enzymes (peroxidase, superoxide dismutase, and catalase). The response of plants to exposure to water stress can be determined by different mechanisms, including the ability to maintain high levels of antioxidants and to regenerate them.

Peroxidase is the most extensively used enzyme as a biochemical marker of plant growth and development processes. The implications of peroxidases in plant physiology are multiple, but the most intensively studied refer to participation in the control of cell growth.

The laboratory analyses were carried out in 2017. The soil used in the planting pots had the same origin and was subsequently dried and brought to a uniform humidity.

Fifty variants were sown in plant pots containing the same amount of soil. After sprouting, six seedlings were kept in each vegetation pot. These were placed in the Sanyo growth chamber, previously adjusted to the optimal temperature, light, and atmospheric humidity parameters for proper growth of wheat plants (Table 6).

Table 6. Optimal parameters for proper growth of wheat plants.

Parameters	Values									
Time (h)	0:00	3:00	6:00	9:00	11:00	13:00	15:00	17:00	19:00	22:00
Temperature (°C)	15.0	14.0	15.0	18.0	20.0	25.0	25.0	20.0	18.0	16.0
Light (lux)	0	0	1	2	4	5	4	3	1	0
Humidity (%)	60	65	65	60	55	55	55	55	55	55

The enzymatic determinations were performed according to the working methodology specific to the biochemical analyses. Fresh tissue was homogenized with 0.1 M phosphate buffer (pH = 7.0) containing 0.1 mM ascorbic acid and 0.1 mM EDTA. The homogenate was centrifuged 20 min at 10,000 r.p.m. (rotation per minute), and the supernatant was used for enzyme assays.

- The activity of peroxidase (guaiacol-peroxidase type E.C.1.11.1.7) was determined colorimetrically at λ = 470 nm and was expressed as the variation in absorbance per minute due to the oxidation of guaiacol from the extract of one gram of fresh substance [51].
- The activity of catalase (E.C.1.11.1.6) was determined by the Sinha method by colorimetrically determining the amount of H_2O_2 decomposed for 1 min by the enzyme in 1 g fresh substance. The method is based on the fact that potassium chromate in acidic medium is reduced by hydrogen peroxide to chromic acetate, which can be colorimetric at 570 nm [52,53].

4.3. Statistical Analysis

The paper contains the YI computation and the calculation of correlation coefficients. Yield index [54],

$$\mathrm{YI} = \frac{Ys}{\bar{Y}s}$$

where Ys is the production in the dry year and $\bar{Y}s$ is the average of the production in the dry year, best reflects the behaviour under stress conditions, compared to the average of all varieties. It is not influenced by other conditions and therefore seems the most adequate to characterize the ability of indirect methods to describe the drought resistance. The correlations between yield index and the antioxidant activity for the limited assortment were calculated.

The correlations were performed with the Pearson correlation test after [55].

5. Conclusions

Peroxidase activity has intensified as stress induced by PEG treatment of different concentrations and in different doses increased. Among the varieties treated with PEG 10,000 25%, most of the Romanian varieties tested (Gruia, Izvor, Lovrin 34, Boema, Delabrad, Glosa, Alex, Miranda, and Faur) presented super unit values of the PEG/control ratio, suggesting tolerance to drought. In reverse, the activity of ascorbate peroxidase is lower in tolerant varieties. The varieties with a subunit report were remarked upon. Among them are the Izvor variety, known as the drought-tolerant variety, as well as other Romanian varieties: Alex, Delabrad, Lovrin 34, Miranda, Dor, and Romulus.

Among the varieties evidenced by the increased activity of catalase were the Romanian varieties: Dropia, Trivale, Nikifor, Simnic 30, Simnic 50, Faur, Glosa, Gruia, Miranda, and Flamura 85. Among the foreign varieties that have maintained a high catalase activity to increase stress through a higher dose of PEG were Kristina, GH Hattyu, Karlygash, Esquisit, Lada, Enesco, GK Elet, Cubus, and GK Gobe.

There was no significant correlation between field behaviour to stress, expressed by the YI index on the limited assortment and ratios of antioxidant enzyme content between PEG and control treatments, suggesting that none of these biochemical indicators individually represent a selection marker for drought tolerance under the conditions of south-western Oltenia.

From the improver's point of view, thus, given that the Izvor cultivar does not stand out for very good values for the enzyme (activity of catalase—1814 ratio PEG 10,000 25%/control and 1099 ratio PEG 4000 40%/control) device possibly involved in drought behaviour, one can expect that from its hybridization with the Kristina (activity of catalase —3519 ratio PEG 10,000 25%/control and 6171 ratio PEG 4000 40%/control) or Dropia (activity of catalase—4721 ratio PEG 10,000 25%/control and 5.007 ratio PEG 4000 40%/control cultivars), it is possible to select progeny that will cumulate good growth in the presence of water stress (simulated by PEG treatment), causing a better device enzymatic reaction.

Author Contributions: Conceptualization, R.A.P. and G.P.; methodology, C.B.; validation, E.R., E.B., and C.A.R.; formal analysis, E.R.; investigation, R.A.P.; writing—original draft preparation, G.P.; writing—review and editing, E.B.; visualization, G.P.; supervision, C.A.R. All authors have read and agreed to the published version of the manuscript.

Funding: This research received no external funding.

Institutional Review Board Statement: Not applicable.

Informed Consent Statement: Not applicable.

Data Availability Statement: All the data supporting this article were included in the main text.

Conflicts of Interest: The authors declare no conflict of interest.

References

1. Okarter, N.; Liu, C.S.; Sorrells, M.E.; Liu, R.H. Phytochemical content and antioxidant activity of six diverse varieties of whole wheat. *Food Chem.* **2010**, *119*, 249–257. [CrossRef]
2. Mughal, I.; Shah, Y.; Tahir, S.; Haider, W.; Fayyaz, M.; Yasmin, T.; Ilyas, M.; Farrakh, S. Protein quantification and enzyme activity estimation of Pakistani wheat landraces. *PLoS ONE* **2020**, *15*, e0239375. [CrossRef]
3. Narwal, S.; Thakur, V.; Sheoran, S.; Dahiya, S.; Jaswal, S.; Gupta, R.K. Antioxidant activity and phenolic content of the Indian wheat varieties. *J. Plant Biochem. Biotechnol.* **2014**, *23*, 11–17. [CrossRef]
4. Hu, H.; Xiong, L. Genetic engineering and breeding of drought-resistant crops. *Annu. Rev. Plant Biol.* **2014**, *65*, 715–741. [CrossRef]
5. Tricker, P.J.; ElHabti, A.; Schmidt, J.; Fleury, D. The physiological and genetic basis of combined drought and heat tolerance in wheat. *J. Exp. Bot.* **2018**, *69*, 3195–3210. [CrossRef]
6. Khanna-Chopra, R.; Selote, D.S. Acclimation to drought stress generates oxidative stress tolerance in drought-resistant than-susceptible wheat cultivar under field conditions. Environ. *Expt. Bot.* **2007**, *60*, 276–283. [CrossRef]
7. Osakabe, Y.; Osakabe, K.; Shinozaki, K.; Tran, L.S.P. Response of plants to water stress. *Front. Plant Sci.* **2014**, *5*, 86. [CrossRef]
8. Zhang, G.; Zhang, M.; Zhao, Z.; Ren, Y.; Li, Q.; Wang, W. Wheat TaPUB1 modulates plant drought stress resistance by improving antioxidant capability. *Sci. Rep.* **2017**, *7*, 7549. [CrossRef] [PubMed]
9. Apel, K.; Hirt, H. Reactive oxygen species: Metabolism, oxidative stress, and signal transduction. *Annu. Rev. Plant Biol.* **2004**, *55*, 373–399. [CrossRef] [PubMed]
10. Dietz, K.J. Thiol-based peroxidases and ascorbate peroxidases: Why plants rely on multiple peroxidase systems in the photosynthesizing chloroplast? *Mol. Cells* **2016**, *39*, 20–25.
11. Sallam, A.; Alqudah, A.M.; Dawood, M.F.A.; Baenziger, P.S.; Börner, A. Drought Stress Tolerance in Wheat and Barley: Advances in Physiology, Breeding and Genetics Research. *Int. J. Mol. Sci.* **2019**, *20*, 3137. [CrossRef]
12. Zhang, J.N.; Kirkha, M.B. Antioxidant responses to drought in sunflower and sorghum seedlings. *New Phytol.* **1996**, *132*, 361–373. [CrossRef]
13. Sairam, R.K.; Deshmukh, P.S.; Saxena, D.C. Role of antioxidant systems in wheat genotypes tolerance to water stress. *Biol. Plant.* **1998**, *41*, 384–394. [CrossRef]
14. Kido, É.A.; Ferreira-Neto, J.R.C.; Pandolfi, V.; de Melo Souza, A.C.; Benko-Iseppon, A.M. Drought Stress Tolerance in Plants: Insights from Transcriptomic Studies. In *Drought Stress Tolerance in Plants*; Springer: Cham, Switzerland, 2016; Volume 2, pp. 153–185.
15. Khazayi, H.; Kafi, M.; Masumi, A. Physiological effects of stress induced by polyethylene glycol on germination of chickpea genotypes. *J. Agron. Res. Iran* **2008**, *2*, 453.
16. Chakraborty, U.; Pradhan, B. Oxidative stress in five wheat varieties (*Triticum aestivum* L.) exposed to water stress and study of their antioxidant enzyme defense system, water stress responsive metabolites and H2O2 accumulation. *Braz. J. Plant Physiol.* **2012**, *24*, 17–130. [CrossRef]

17. Păunescu, A.; Dodocioiu, A.M.; Băbeanu, C.; Păunescu, G.; Buzatu, G.D. Total phenols content and antioxidant activity of whole grain flours from some wheat lines tested in the south-west of Romania. *SGEM* **2016**, *1*, 845–852.
18. Zhang, Y.; Shih, D.S. Isolation of an osmotion like protein genefron strawerry and analysis of the response of this gene to abiotic stresses. *J. Plant Physiol.* **2007**, *164*, 68–77. [CrossRef]
19. Taşgin, E.; Okkeş, A.; Nalbantoglu, B.; Petrova Popova, L. Effects of salicylic acid and cold treatments on protein levels and on the activities of antioxidant enzymes in the apoplast of winter wheat leaves. *Phytochemistry* **2006**, *67*, 71–75. [CrossRef]
20. Nikolaeva, M.K.; Maevskaya, S.N.; Shugaev, A.G.; Bukhov, N.G. Effect of drought on chlorophyll content and antioxidant enzyme activities in leaves of the wheat cultivars varying in productivity. *Russ. J. Plant Physiol.* **2010**, *57*, 87–95. [CrossRef]
21. Hameed, A.; Goher, M.; Iqbal, N. Biochemicals Indices of Drought Tolerance in Wheat (*Triticum aestivum* L.) at Early Seedling Stage. *Philipp. Agric. Sci.* **2014**, *97*, 236–242.
22. Sallam, A.; Amro, A.; EL-Akhdar, A.; Dawood, M.F.A.; Kumamaru, T.; Stephen Baenziger, P. Genetic diversity and genetic variation in morpho-physiological traits to improve heat tolerance in Spring barley. *Mol. Biol. Rep.* **2018**, *45*, 2441–2453. [CrossRef] [PubMed]
23. Al Abdallat, A.M.; Ayad, J.Y.; Abu Elenein, J.M.; Al Ajlouni, Z.; Harwood, W.A. Overexpression of the transcription factor HvSNAC1 improves drought tolerance in barley (*Hordeum vulgare* L.). *Mol. Breed.* **2014**, *33*, 401–414. [CrossRef]
24. Păunescu, G.; Păunescu, A.R. Identification of wheat varieties tolerant to water stress based on ratio between the stem growth measured in seedlings after 20% PEG treatment and the stem growth measured after water treatment 15 days after sowing. *EWAC* **2018**, *17*, 91–97.
25. Păunescu, R.A. The stem growth measured in seedlings after 20% PEG treatment 15 days after sowing is significantly correled with field response to drought in the field. *Rom. Agric. Res.* **2018**, *35*, 29–37.
26. Zahoor, A.; Waraich, E.A.; Barutçular, C.; Hossain, A.; Erman, M.; Çiğ, F.; Gharib, H. El Sabagh, A. Enhancing drought tolerance in wheat through improving morphophysiological and antioxidants activities of plants by the supplementation of foliar silicon. *Phyton* **2020**, *89*, 529–539.
27. Wang, J.Y.; Xiong, Y.C.; Li, F.M.; Siddique, K.H.M.; Turner, N.C. Effects of drought stress on morphophysiological traits, biochemical characteristics, yield and yield components in different ploidy wheat: A meta-analysis. *Adv. Agron.* **2017**, *143*, 139–173.
28. El Sabagh, A.; Hossain, A.; Barutcular, C.; Islam, M.S.; Awan, S.I.; Galal, A.; Iqbal, M.A.; Sytar, O.; Yildirim, M.; Meena, R.S.; et al. Wheat (*Triticum aestivum* L.) production under drought and heat stress-adverse effects, mechanisms and mitigation: A review. *Appl. Ecol. Environ. Res.* **2019**, *17*, 8307–8332. [CrossRef]
29. Zhu, Y.; Gong, H. Beneficial effects of silicon on salt and drought tolerance in plants. *Agron. Sustain. Dev.* **2014**, *34*, 455–472. [CrossRef]
30. Agrawal, G.K.; Jwa, N.S.; Iwahashi, H.; Rakwal, R. Importance of ascorbate peroxidases OsAPX1 and OsAPX2 in the rice pathogen response pathways and growth and reproduction revealed by their transcriptional profiling. *Gene* **2003**, *322*, 93–103. [CrossRef]
31. Baxter, A.; Mittler, R.; Suzuki, N. ROS as key players in plant stress signalling. *J. Exp. Bot.* **2014**, *65*, 1229–1240. [CrossRef] [PubMed]
32. Caverzan, A.; Passaia, G.; Barcellos Rosa, R.; Werner Ribeiro, C.; Lazzarotto, F.; Margis-Pinheiro, M. Plant responses to stresses: Role of ascorbate peroxidase in the antioxidant protection. *Genet. Mol. Biol.* **2012**, *35*, 1011–1019. [CrossRef] [PubMed]
33. Chen, Y.; Cai, J.; Yang, F.X.; Zhou, B.; Zhou, L.R. Ascorbate peroxidase from *Jatropha curcas* enhances salt tolerance in transgenic *Arabidopsis*. *Genet. Mol. Res.* **2015**, *14*, 4879–4889. [CrossRef]
34. Fryer, M.J.; Ball, L.; Oxborough, K.; Karpinski, S.; Mullineaux, P.M.; Baker, N.R. Control of Ascorbate Peroxidase 2 expression by hydrogen peroxide and leaf water status during excess light stress reveals a functional organisation of Arabidopsis leaves. *Plant J.* **2003**, *33*, 691–705. [CrossRef] [PubMed]
35. Abdelaal, K.A.A.; Omara, I.R.; Hafez, M.Y.; Samar, M.E.; El Sabagh, A. Anatomical, biochemical and physiological changes in some Egyptian wheat cultivars inoculated with *Puccinia gramini* f. sp. tritici. *Fresenius Environ. Bull.* **2018**, *27*, 296–305.
36. Shen, X.; Zhou, Y.; Duan, L.; Li, Z.; Eneji, A.E.; Li, J. Silicon effects on photosynthesis and antioxidant parameters of soybean seedlings under drought and ultraviolet-B radiation. *J. Plant Physiol.* **2010**, *167*, 1248–1252. [CrossRef]
37. Yu, Q.; Rengel, Z. Drought and salinity differentially influence activities of superoxide dismutase in narrow-leafed lupins. *Plant Sci.* **1999**, *142*, 1–11. [CrossRef]
38. Giannopolitis, C.N.; Ries, S.K. Superoxide dismutases: I. Occurrence in higher plants. *Plant Physiol.* **1977**, *59*, 309–314. [CrossRef]
39. Prasad, T.K. Mechanisms of chilling-induced oxidative stress injury and tolerance in developing maize seedlings: Changes in antioxidant system, oxidation of proteins and lipids, and protease activities. *Plant J.* **1996**, *10*, 1017–1026. [CrossRef]
40. Gupta, V.K.; Sharma, S.K. Plants as natural antioxidants. *Nat. Prod. Radiance* **2006**, *5*, 326–334.
41. Jallouli, S.; Ayadi, S.; Landi, S.; Capasso, G.; Santini, G.; Chamekh, Z.; Zouari, I.; Ben Azaiez, F.E.; Trifa, Y.; Esposito, S. Physiological and Molecular Osmotic Stress Responses in Three Durum Wheat (*Triticum Turgidum* ssp *Durum*) Genotypes. *Agronomy* **2019**, *9*, 550. [CrossRef]
42. Kirova, E.; Pecheva, D.; Simova-Stoilova, L. Drought response in winter wheat: Protection from oxidative stress and mutagenesis effect. *Acta Physiol. Plant.* **2021**, *43*, 8. [CrossRef]
43. Simova-Stoilova, L.; Kirova, E.; Pecheva, D. Drought stress response in winter wheat varieties–changes in leaf proteins and proteolytic activities. *Acta Bot. Croat.* **2020**, *79*, 121–130. [CrossRef]

44. Kartseva, T.; Dobrikova, A.; Kocheva, K.; Alexandrov, V.; Georgiev, G.; Brestič, M.; Misheva, S. Optimal Nitrogen Supply Ameliorates the Performance of Wheat Seedlings under Osmotic Stress in Genotype-Specific Manner. *Plants* **2021**, *10*, 493. [CrossRef] [PubMed]
45. Das, K.; Roychoudhury, A. Reactive oxygen species (ROS) and response of antioxidants as ROS-scavengers during environmental stress in plants. *Front. Environ. Sci.* **2014**, *2*, 53. [CrossRef]
46. Shao, H.B.; Liang, Z.S.; Shao, M.A.; Sun, Q. Dynamic changes of antioxidative enzymes of 10 wheat genotypes at soil water deficits. *Colloids Surf. B Biointerfaces* **2005**, *42*, 187–195. [CrossRef]
47. Bhattacharjee, S. Reactive oxygen species and oxidative burst: Roles in stress, senescence and signal. *Curr. Sci.* **2005**, *89*, 1113–1121.
48. Farnese, F.S.; Menezes-Silva, P.E.; Gusman, G.S.; Oliveira, J.A. When bad guys become good ones: The key role of reactive oxygen species and nitric oxide in the plant responses to abiotic stress. *Front. Plant Sci.* **2016**, *7*, 471. [CrossRef]
49. Foyer, C.H.; Noctor, G. Redox homeostasis and antioxidant signaling: A metabolic interface between stress perception and physiological responses. *Plant Cell* **2005**, *17*, 1866–1875. [CrossRef]
50. Todorova, D.; Sergiev, I.; Katerova, Z.; Shopova, E.; Dimitrova, L.; Brankova, L. Assessment of the Biochemical Responses of Wheat Seedlings to Soil Drought after Application of Selective Herbicide. *Plants* **2021**, *10*, 733. [CrossRef]
51. Putter, J. Peroxidase. In *Methods of Enzymatic Analysis*; Bergmeyer, H.U., Ed.; Verlag Chemie: Weinhan, Germany, 1974; pp. 685–690.
52. Sinha, A.K. Colorimetric assay of catalase. *Anal. Biochem.* **1972**, *47*, 389–391. [CrossRef]
53. Paunescu, R.A. Identification of Resistant Phenotypes in order to Improve the Behaviour of Wheat in Drought Conditions on the Luvisoil from Simnic. Ph.D. Thesis, University of Agriculture and Veterinary Medicine, Bucharest, Romania, 2017.
54. Gavuzzi, P.; Rizza, F.; Palumbo, M.; Campaline, R.G.; Ricciardi, G.L.; Borghi, B. Evaluation of field and laboratory predictors of drought and heat tolerance in winter cereals. *Can. J. Plant Sci.* **1997**, *77*, 523–553. [CrossRef]
55. Hawkins, D. *Biomeasurement: A Student's Guide to Biological Statistics*, 2rd ed.; Oxford University Press: Oxford, UK, 2009; pp. 10–360.

plants

MDPI

Article

Aluminum Stress Induces Irreversible Proteomic Changes in the Roots of the Sensitive but Not the Tolerant Genotype of Triticale Seedlings

Agnieszka Niedziela [1,*], Lucyna Domżalska [2], Wioletta M. Dynkowska [1], Markéta Pernisová [3,4] and Krystyna Rybka [1,*]

1 Department of Biochemistry and Biotechnology, Plant Breeding and Acclimatization Institute-National Research Institute in Radzików, 05-870 Blonie, Poland; w.dynkowska@ihar.edu.pl
2 Center for Biological Diversity Conservation in Powsin, Polish Academy of Sciences Botanical Garden, Prawdziwka 2, 02-973 Warsaw, Poland; lucyna.domzalska@gmail.com
3 Plant Sciences Core Facility, Mendel Centre for Plant Genomics and Proteomics, Central European Institute of Technology (CEITEC), Masaryk University, Kamenice 5, 62500 Brno, Czech Republic; marketa.pernisova@ceitec.muni.cz
4 Laboratory of Functional Genomics and Proteomics, National Centre for Biomolecular Research, Faculty of Science, Masaryk University, Kamenice 5, 62500 Brno, Czech Republic
* Correspondence: a.niedziela@ihar.edu.pl (A.N.); k.rybka@ihar.edu.pl (K.R.); Tel.: +48-227-334-535 (A.N.); +48-227-334-537 (K.R.)

Citation: Niedziela, A.; Domżalska, L.; Dynkowska, W.M.; Pernisová, M.; Rybka, K. Aluminum Stress Induces Irreversible Proteomic Changes in the Roots of the Sensitive but Not the Tolerant Genotype of Triticale Seedlings. *Plants* **2022**, *11*, 165. https://doi.org/10.3390/plants11020165

Academic Editors: Ewa Muszyńska, Kinga Dziurka and Mateusz Labudda

Received: 18 November 2021
Accepted: 5 January 2022
Published: 8 January 2022

Abstract: Triticale is a wheat–rye hybrid with a higher abiotic stress tolerance than wheat and is better adapted for cultivation in light-type soils, where aluminum ions are present as Al-complexes that are harmful to plants. The roots are the first plant organs to contact these ions and the inhibition of root growth is one of the first plant reactions. The proteomes of the root apices in Al-tolerant and -sensitive plants were investigated to compare their regeneration effects following stress. The materials used in this study consisted of seedlings of three triticale lines differing in Al^{3+} tolerance, first subjected to aluminum ion stress and then recovered. Two-dimensional electrophoresis (2-DE) was used for seedling root protein separation followed by differential spot analysis using liquid chromatography coupled to tandem mass spectrometry (LC-MS-MS/MS). The plants' tolerance to the stress was evaluated based on biometric screening of seedling root regrowth upon regeneration. Our results suggest that the Al-tolerant genotype can recover, without differentiation of proteome profiles, after stress relief, contrary to Al-sensitive genotypes that maintain the proteome modifications caused by unfavorable environments.

Keywords: acidic soils; abiotic stress tolerance; proteomic studies; two dimensional electrophoresis (2-DE); × *Triticosecale* Wittmack

1. Introduction

Aluminum is the third most abundant element on earth, after oxygen and silicon. Its toxic effect in plants results from the physicochemical properties of common aluminum minerals, presented in the lithosphere as, for example: gibbsite and bauxite (hydroxylated Al-ions), kaolinite, or muscovite (hydrated complexes of aluminum and potassium). All minerals containing aluminum are insoluble at a neutral pH (6.5–7.0); hence, aluminum ions in such soils are biologically passive, non-available and thus non-harmful to plants. In acidic (pH 5.0–6.0) or very acidic (pH 4.0–5.5) soils, aluminum containing minerals can become soluble, releasing hydroxyl complexes of Al-ions in trivalent cationic forms, which are complexed in humus soils but picked up by plant roots from acidic sandy soils [1]. Since acidic soils constitute 30–40% of the world's arable land, with a constantly growing share due to anthropogenic impact, crop plants' aluminum tolerance is one of the features that affect higher/stable yielding in changing environments [2].

Tolerance to Al-ions relies on the inhibition of Al uptake by the roots (external tolerance) and/or on the inhibition of transport to the aerial parts (internal tolerance). In fact, most of the tolerant crop plants are Al-exuders, avoiding the stress by prevention of ion intake into the symplast [3]. Their basic mechanisms are citrate, malate, and/or oxalate secretion into the rhizosphere for chelating Al-ions into non-absorbable complexes as well as a pH increase in the rhizosphere as an effect of plasma membrane H^+-ATPase activity, higher in the roots of tolerant plants [4]. The base of the internal tolerance is Al-cation binding by negatively-charged carboxyl groups of the plant cell wall pectins in the root apex. Pectin content and the degree of methylation differentiate the Al tolerance response [5]. Another mechanism for Al tolerance is ion chelation in the cytosol and relocation to leaf vacuoles. Al-ions are present mostly in hydroxylated forms in the apoplast, whereas in symplast, they form complexes with sulfate, phosphate, and organic ligands [2]. When the tolerance mechanisms fail, the Al-ions gradually move inside the cells, misbalancing and eventually blocking ion channels, alternating lipid fluidity, and inducing changes in the cytoskeleton structure by binding to G-proteins and their substrates as well as to ATP-ases and nucleotide polyphosphate groups. Finally, the disruption of DNA synthesis and cell division in the root apex and the lateral roots is accompanied by increased rigidity of the cell walls and DNA double helix, which leads to rapid inhibition (within an hour) of root growth, even with micromolar Al^{3+} concentrations [2,6]. The biochemical consequences of aluminum intake are an increase in reactive oxygen species, increased fatty acid peroxidation, and inhibition of proton adenosine triphosphatase H^+-ATPases. The displacement of Ca^{2+} in cell membranes contributes to Al^{3+} accumulation in the apoplast, stimulates callose synthesis, and finally inhibits intercellular transport. The decreasing concentration of Ca^{2+} in the cytosol alters the pH balance, which in turn interferes with sugar phosphorylases and the deposition of cell wall polysaccharides [2,6,7].

The genetic bases of Al tolerance concern ion chelation and ion transport. The gene families responsible for Al-ion exudation are ALMT (aluminum-activated malate transporter), responsible for malate, and MATE (multidrug and toxin efflux) for citrate exudation [8–10]. The transcriptional expression of ALMT and MATE is controlled by a master zinc-finger transcription factor, STOP1 (sensitive to proton rhizotoxicity 1) [9,11]. Internal Al tolerance is less well characterized. The NRAT1 (Nramp aluminum transporter 1) transporter has been identified as a putative Al transporter involved in rice's internal resistance mechanism, which lowers Al-ion concentrations in the root cell wall, transporting the ions inside the root cells for sequestration in vacuoles [12]. Those genes are conserved in numerous plant crop genomes [13]. The STOP1 transcription factor, in addition to several phytohormones, hydrogen peroxide, and other reactive oxygen species (ROS), take part in the upregulation of genes of the ALMT family or root growth inhibition. In addition, ROS detoxifying enzymes are activated to respond to aluminum ion stress, activating the gene network towards the induction of aluminum tolerance in plant tissues [13].

Hexaploid triticale, a rye–wheat hybrid species of wheat (AABB genomes) and rye (RR genome), is characterized by an intermediate tolerance between wheat considered as an Al-sensitive and rye as an Al-tolerant parent. The beneficial influence of the rye genome was confirmed by Quantitative Trait Loci (QTL) localization on the 7R chromosome, explaining up to 36% of the phenotypic variance, including the malate transporter gene [14]. The other loci were found on chromosomes 3R [15,16], 4R, and 6R [15], without a recognized function in the triticale genome. Triticale's tolerance to Al-ions is less than that of rye, suggesting a suppressive effect of the wheat genome on the expression of rye genes.

In the present work, two-dimensional electrophoresis (2-DE) with immobilized pH gradients (IPGs), combined with protein identification by mass spectrometry (MS), was used to detect changes in the proteomes of triticale root tips after Al stress removal. This method has been successfully used to identify proteins involved in various stress responses in plants [17,18]. Al-responsive proteins identified in both monocots [19–22] and dicots [23] were functionally associated with cell division and structure, carbohydrate metabolism, antioxidant system, amino acid metabolism, protein degradation, signal transduction, and transporters [21,24,25]. The upregulation of the enzymes involved in cysteine and methionine metabolism, such as cysteine synthase, S-adenosylmethionine synthase, and O-methyltransferase, is a common response despite Al-ion tolerance [21–23,25,26]. These enzymes are required to maintain methyl cycling and glutathione metabolism, which are important mechanisms that lead Al detoxification [27]. The recent proteomic studies of soybean plasma membrane changes in response to Al-ions revealed about fifty different membrane traffic and transporter proteins [28].

To our knowledge, there are no reports concerning protein identification in triticale roots in the context of aluminum tolerance; however, the triticale root proteome response to drought was studied by Grębosz et al. [29]. For our comparative proteomic studies, we used the seedling root tips of triticale plants differing in Al tolerance subjected to recovery after the stress treatment.

2. Results

2.1. Biometric and Biochemical Evaluation of Tested Materials

The materials used in the studies were selected based on a biometric screening of 232 triticale lines in earlier published experiments [15] (Figure S1). One Al-tolerant line (L198, spring form) and two Al-sensitive lines (L17, spring form and L444, winter form) were chosen for the present studies. Root growth in response to 24 h of Al treatment was inhibited and in sensitive lines the root regrowth remained suppressed 48 h after the stress release, contrary to the roots of tolerant line, which resumed root growth after the stress release (Figure 1). Eriochrome cyanine R dye penetrates damaged or partially damaged root tips. The root regrowth ranged from 0.3 to 2.5 cm for individual seedlings of the tolerant line (Table 1). Additionally, the root apex redox potential of the seedlings released from Al stress was assessed in comparison with the control seedlings as a measure of the dynamics of the response to the Al stress removal [30]. Antioxidant capacity was determined using 2,2-diphenyl-1-picrylhydrazyl anion radical ($DPPH^{*-}$), and the cation radical 2,2′-azino-bis(3-ethyl benzothiazoline-6-sulphonic acid ($ABTS^{*+}$) [31]. According to one-way analysis of variance (ANOVA), no significant differences (with $p \leq 0.05$) were detected between the control and stress treated seedlings in the case of reactions with $DPPH^{*-}$ anion radical, despite a 20% difference in the L444 line between the control and Al-treated samples (Table 1). In the case of reactions with $ABTS^{*+}$, 13% higher activity, with statistical importance at $p \leq 0.05$, was detected after the stress release in the case of the L444 line, and 10% (with no statistical importance) in the case of the L17 line. There were no differences in redox potential between the control and treated roots of the tolerant line L198 (Table 1).

Table 1. Root regrowth (cm) and antioxidant potential of triticale root tips from control seedlings and from seedlings 48 h after 16 (ppm) Al treatment. The total antioxidant capacity of tolerant (L198) and sensitive (L444 and L17) genotypes was expressed as the μmol Trolox equivalent antioxidant capacity (TEAC) per mg of root tip tissue.

Line	Control/Stress	Root Regrowth (cm)	$DPPH^{*-}$ (μmol TEAC/mg)	$ABTS^{*+}$ (μmol TEAC/mg)
L198	control	-	15.264 ± 1.17 a	8.706 ± 0.39 ab
L198	stress (16ppm Al)	0.3–2.5	15.786 ± 1.29 a	9.170 ± 0.39 a
L444	control	-	13.854 ± 1.00 a	8.072 ± 0.11 b
L444	stress (16ppm Al)	no regrowth	16.643 ± 3.34 a	9.124 ± 0.25 a
L17	control	-	12.544 ± 2.05 a	8.086 ± 0.58 b
L17	stress (16ppm Al)	no regrowth	12.602 ± 0.9 a	8.870 ± 0.10 ab

a,b—statistically different mean values ($p < 0.05$); $DPPH^{*-}$—(2,2-diphenyl-1-picrylhydrazyl); $ABTS^{*+}$—(2,2′-azinobis-(3-ethylbenzothiazoline-6-sulfonic acid).

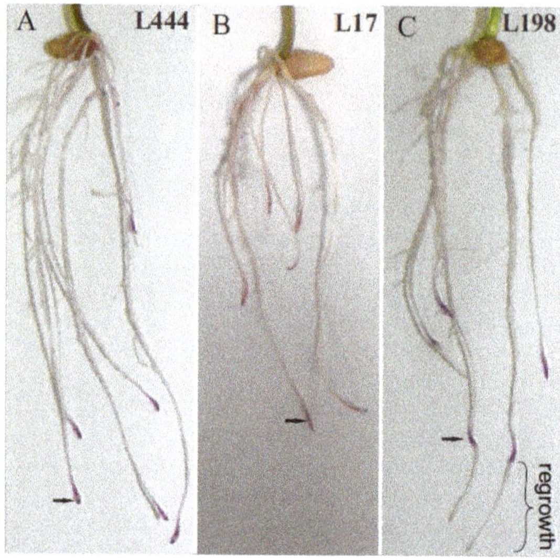

Figure 1. Damaged regions of triticale seedling roots (denoted by arrows) stained with Eriochrome cyanine R after Al-ion treatment prior to recovery. After 48 h recovery, the purple root tips and no rooth regrowth were visible in case of the Al-sensitive lines (L444 and L17 (**A**,**B**), respectively), whereas the dark purple bands on regrown roots were detected in the case of the Al-tolerant line (L198 (**C**)).

2.2. *Two-Dimensional Electrophoresis (2-DE)*

The analysis of 2-DE gels revealed approximately 590 spots in each experimental and biological replication of both the control samples and those stress released after the Al-treatment (Table 2). Ninety-five percent of the protein spots were matched and quantified. Isoelectrophocusing using IPG strips of pH 3–10 revealed the protein spots at pI values in the range of 4.0–8.5 and protein masses between 6.5 and 95 kDa (Figure S2A–F, Table 3). In the tolerant line (L198), the proteomes of stress-subjected and control roots were not differentiated according to the established criteria ($p \leq 0.01$ and difference in spot intensity $\geq$ two-fold). When the criterion of probability was weakened from 99% to 95%, only four differential protein spots were found (three spots as downregulated and one spot as upregulated). On the other hand, in the root tip proteomes of the Al-sensitive triticale lines, a higher number of differentiated protein spots were found. Regardless of the protein spot intensity, with a $p \leq 0.01$ probability criterion, in total seventy-one differential protein spots were found in the L17 proteome (23 upregulated, 21 downregulated, 15 induced, and 12 silenced) and forty-three in the L444 proteome (23 upregulated, eight downregulated, three induced, and nine silenced). When the criterion of two-fold difference in spot intensity was added, 14 upregulated and eight downregulated spots were found in the L17 proteome. In L444, upon a double criterion (probability and spot intensity), 18 spots of upregulated proteins, exclusively, were found. For induced or silenced protein spots, a second criterion of spot intensity ≥ 0.2 was decided. In the proteome of L17, nine spots had a relative signal intensity > 0.2 among induced proteins, and three among those silenced, whereas in the proteome of L444 root tips, three induced and nine silenced protein spots were found (Table 2, Figure S3).

Table 2. The number of identified protein spots on 2-DE gels from analyzed root tip proteomes of tolerant (L198) and susceptible (L17 and L44) triticale lines, in accordance with the t-test with probability $p \leq 0.01$. In parenthesis are the number of spots selected with double criterion: high probability ($p \leq 0.01$) and at least two-fold difference in spot intensity ($\geq 2\times$). For induced and silenced protein spots, the second criterion was spot relative intensity ≥ 0.2.

Line	Tolerant	Sensitive		L17-L444
Spots Characteristic	L198	L17	L444	Common Spots (with $p \leq 0.01$)
• Total number	579	584	602	
• Upregulated	0	23 ($14 \geq 2\times$)	23 ($18 \geq 2\times$)	13
• Downregulated	0	21 ($8 \geq 2\times$)	8 ($0 \geq 2\times$)	2
• Silenced upon Al^{3+}	0	12 ($3 \geq 0.2$)	9 ($1 \geq 0.2$)	3
• Induced upon Al^{3+}	0	15 ($9 \geq 0.2$)	3 ($0 \geq 0.2$)	2

The highest, nearly 8-fold, difference in upregulated protein spot intensity was detected for spot #6, and a 5.2-fold decline for #24 were detected in the L17 proteome (Table 3, Figure S2B). In the L444 proteome, a spot identical to spot #6 was induced de novo in response to Al stress with a relative intensity of 0.26, and was numbered as spot #31 (Table 4, Figure S2D). Furthermore, seven of the differential proteins (numbered: 2, 4, 9, 12, 16, 17, and 18 in the proteome of L17 and 3, 8, 13, 14, 15, 19, and 21 in the proteome of L444) showed a more than two-fold increase in signal intensity in the L17 proteome and about a two-fold increase in the L444 proteome (Table 3, Figure S2B,D). Ten spots revealed significant differentiation (with $p \leq 0.01$) for one of the two lines. Out of the eight protein spots of L17 that were downregulated with two-fold or higher intensities, none differed in this intensity in L444 (Table 3). This number of spots was counted according to the acute cut-off double criterion (probability $p < 0.01$ and intensity difference > 2); however, with weakened criteria and a single cut-off (probability $p < 0.05$), more than 80% of the protein spots were common for L17 and L444 lines. Since the spot identification was performed according to a cut-off by strong, double criteria, the weaker criteria data are not shown in detail (Tables 3 and 4).

2.3. Identification of Differential Proteins

The identification of differential proteins was carried out using liquid chromatography coupled to tandem mass spectrometry LC-MS-MS/MS system for 25 selected spots, with at least a two-fold change in intensity for at least one of the Al-sensitive lines (Table 3). Differential proteins from the spots marked as identical by the gel analysis software (Image Master 2D Platinum 7.0) were analyzed, except three upregulated protein spots (#2, #16, and #17) that represented both sensitive lines L17 and L444. The spots were extracted from the 2-DE gels followed by the separation of L17 and L444 root tip proteomes. Moreover, six induced and five silenced protein spots, with a signal intensity ≥ 0.2, were identified. In total, out of 36 protein spots, thirty-two represented 23 differentially expressed proteins, whereas three were unassigned (#24 and #25 from L17, and #36 from L444) (Tables 3 and 4).

Table 3. Identification of differential protein spots from 2-DE gels obtained by separation of seedling root tip proteins extracted from Al-sensitive triticale lines, L17 and L444, after stress release. Protein spots were chosen according to the double criterion of $p \leq 0.01$ and difference in spot intensity ≥ 2, and for induced or silenced proteins the criterion of relative spot intensity ≥ 0.2 was decided. Protein spots were detected using Image Master 2D Platinum 7.0 software, followed by MS-MS separation and further identification, characterization, and quantitation using Mascot Distiller v. 2.3 software.

Spot No.	Pathway/Protein Name	UniProt ID	Mascot Score	Mass	pI	[1] MP	Fold Changed	
							L17	L444
	Cell signaling							
1	Calmodulin	P04464	25	16,893	4.9	1	+2.52	* n.s.
	Metabolic pathway							
2	ATP synthase subunit alpha, mitochondrial	P12862	201/96	55,515	6.6	4	+3.54	+2.73
3	Adenosylhomocysteinase	P32112	34	54,086	7.85	1	−2.00	−0.83
4	Phosphoglycerate kinase, cytosolic	P12783	372	42,153	5.6	7	+3.13	+5.00
5	Fructose-1,6-bisphosphatase	P09195	24	44,703	7.1	1	+2.03	* n.s.
	Metabolic pathway/Flavonoid metabolism							
6	Flavone O-methyltransferase 1	Q84N28	1053	39,177	5.7	22	+7.61	see Table 3
	Methyl cycle							
7	S-adenosylmethionine synthase	B0LXM0	448	43,609	5.51	7	+2.91	n.s.
	Protease inhibitor							
8	Serpin-Z1C	Q9ST58	185	42,969	5.45	4	+2.03	+1.72
	Protein degradation/cell signaling							
9	Ubiquitin	P69326	42	8648	7.2	1	+2.01	+3.6
10	Ubiquitin	P69326	38	8648	6.79	1	+2.05	n.s.
11	Ubiquitin	P69326	55	8648	7.25	1	+2.43	n.s.
	Protein synthesis							
12	Protein disulfide-isomerase	P52589	113	56,726	4.9	4	+2.51	+3.61
13	Eukaryotic initiation factor 4A	P41378	114	47,183	5.25	2	+4.12	+1.92
14	Protein disulfide-isomerase	P52589	81	56,726	5.11	3	+2.65	+1.78
	Stress related							
15	Dehydrin COR410	P46524	50	28,166	6.9	1	+2.50	+1.34
16	Oxalate oxidase	P26759	341/394	23,711	6.35	5	+2.75	+5.00
17	Glutathione S-transferase	O04437	218/474	24,022	6,2	11	+2.85	+4.52
18	1-Cys peroxiredoxin	Q6W8Q2	298	24178	6	5	+2.87	+3.51
	Transcription control							
19	Splicing factor U2af large subunit B	Q2QKB4	334	60,720	5.2	4	+2.01	+1.51
20	DNA-directed RNA polymerase subunit beta	Q9XPS9	14	170,794	6.25	1	+2.05	n.s.
	Transport							
21	Mitochondrial outer membrane porin	P46274	16	28,944	6.5	1	+2.33	+1.62
	Lignin synthesis							
22	[2] DIMBOA1b, chloroplastic	Q1XH05	56	64,898	5.25	2	n.s.	+2.87
23	[2] DIMBOA 1c, chloroplastic	Q1XH04	80	64,980	5.4	2	n.s.	+2.16
	Unassigned peptides							
24	Unassigned peptide	-	-	-	-	-	−5.2	n.s.
25	Unassigned peptide	-	-	-	-	-	+2.65	n.s.

[1] MP—matched peptides; [2] DIMBOA—4-hydroxy-7-methoxy-3,4-dihydro-2H-1,4-benzoxazin-2yl beta glucosidase; * n.s.—protein spot not significantly changed.

Table 4. Aluminum ion-responsive proteins from seedling roots of Al-sensitive triticale lines, L17 and L444, present in roots of control (proteins silenced upon Al^{3+}) or in roots after the stress removal (proteins induced upon Al^{3+}). Double cut-off criterion $p \leq 0.01$ and relative spot intensity ≥ 0.2 were used. The relative intensity of protein spots on the gels are shown.

Spot No.	Pathway/Protein Name	UniProt/String (MLOC) ID	Mascot Score	Mass	pI	[1] MP	[2] Spot Intensity	
							L17	L444
	Cell division/Cytoskeleton							
26	Tubulin beta-3 chain	Q9ZRB0	791	50,555	4.9	18	−0.21	* n.s.
27	Tubulin alpha chain	Q9ZRB7	2061	50,396	4.95	26	−0.20	n.s.
	Lignin synthesis							
28	[3] DIMBOA 1b, chloroplastic	Q1XH05	328	64,898	5.55	7	+0.21	n.s.
29	[3] DIMBOA 1b, chloroplastic	Q1XH05	215	64,898	5.45	4	+0.23	n.s.
	Metabolic pathway							
30	Phosphomannomutase	Q1W374	505	28,405	6	11	n.s.	−0.20
	Metabolic pathway/Flavonoid metabolism							
31	Flavone O-methyltransferase 1	Q84N28	1053	39,177	5.7	22	see Table 2	+0.26
	Methyl cycle							
32	Adenosylhomocysteinase	P32112	137	54,086	6.8	3	−0.20	n.s.
	Protease inhibitor							
33	Ubiquitin	P69326	67	8648	7.6	1	+0.21	n.s.
34	Ubiquitin	P69326	70	8648	6.45	1	+0.22	n.s.
35	Ubiquitin	P69326	40	8648	7.25	1	+0.28	+0.26
	Unassigned peptides							
36	Unassigned peptide	-	-	-	-		n.s.	−0.24

[1] MP—matched peptides; [2] relative spot intensity on the gel; (−) silenced; (+) induced; [3] DIMBOA—4-hydroxy-7-methoxy-3,4-dihydro-2H-1,4-benzoxazin-2yl beta glucosidase; * n.s.—protein spot visible on the gel but not changed in a significant manner.

The identified proteins represented ten protein functional groups involved in cell division, protein folding, protein synthesis, stress-related response, metabolic pathways, lignin synthesis, transcription control, protease inhibition, protein degradation, and transport (Tables 3 and 4). The highest upregulation was detected for the protein folding (protein disulfide-isomerase), stress-related response (glutathione S-transferase, oxalate oxidase, 1-Cys peroxiredoxin PER1), and metabolic pathways (flavone O-methyltransferase 1). On the contrary, the downregulated proteins belonged to the cell division (tubulin) and metabolic pathways associated with amino acid metabolism and methylation control (adenosylhomocysteinase) as well as ascorbic acid biosynthesis (phosphomannomutase). The Protein-Protein Interaction Networks analysis (using STRING database) [32] revealed a functional network containing 19 nodes (flavone O-methyltransferase 1, oxalate oxidase, and serpin-Z1C were not connected) with 39 edges (vs. 34 expected) (Figure 2). We discovered two major proteins (1-Cys peroxiredoxin and phosphoglycerate kinase) with nine interactions in the network. Moreover, seven interactions were detected for ubiquitin.

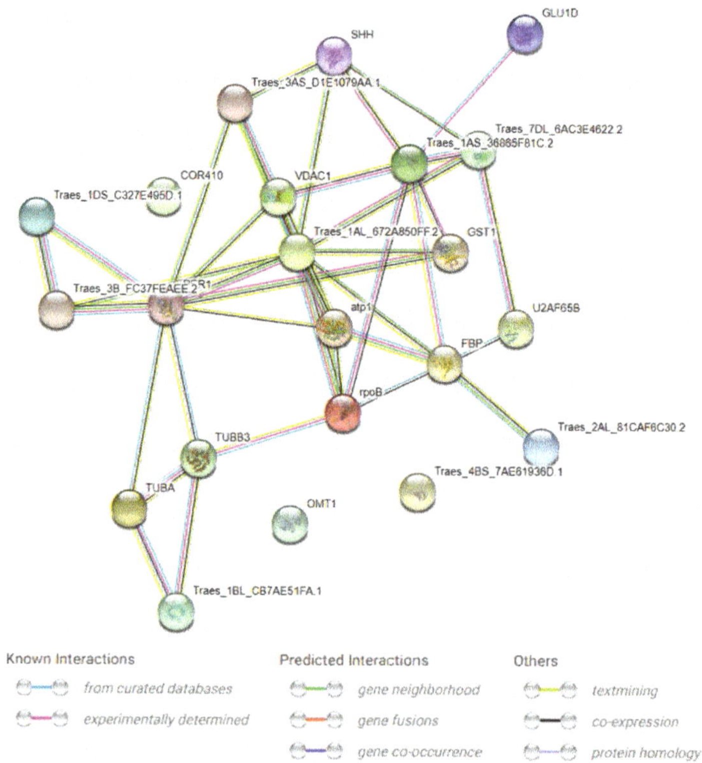

Figure 2. Computational prediction of the functional network between differential proteins. The proteins used for analysis are presented in Tables 3 and 4. The number of connecting lines is in proportion to the amount of information about the protein interactions available. The line color indicates the type of interaction evidence. The explanation of symbols used in STRING database annotations are as follows: Traes_4BS_7AE61936D.1—oxalate oxidase; GST1—glutathione S-transferase; atp1— ATP synthase subunit alpha, mitochondrial; PER1—1-Cys peroxiredoxin; Traes_3B_FC37FEAEE.2—protein disulfide-isomerase; U2AF65B—splicing factor U2af large subunit B; Traes_1DS_C327E495D.1—serpin-Z1C; Traes_7DL_6AC3E4622.2—eukaryotic initiation factor 4A; Traes_1BL_CB7AE51FA.1—calmodulin; Traes_1AS_36865F81C.2—ubiquitin; SHH—adenosylhomocysteinase; Traes_1AL_672A850FF.2—phosphoglycerate kinase, cytosolic; GLUD1—DIMBOA 1b, chloroplastic; TUBB3—tubulin beta-3 chain; TUBA—tubulin alpha chain; Traes_3AS_D1E1079AA1—S-adenosylmethionine synthase; VDAC1—mitochondrial outer membrane porin; FBP—fructose-1,6-bisphosphatase; COR410—dehydrin COR410; rpoB—DNA-directed RNA polymerase subunit beta; OMT1—flavone O-methyltransferase 1; Traes_2AL81CAF6C30.2—phosphomannomutase.

3. Discussion

3.1. Evaluation of the Al Stress Response of Tolerant (L198) and Sensitive (L17, L444) Triticale Lines

The earliest symptoms of Al toxicity concern the meristematic zone in the root apex [6,7,33], which results in the inhibition of root growth and, finally, in declined crop yield. Such root damages, confirmed by microscopic studies, have been described for many plant species [34,35]. Our experiment, performed in the frame of Al tolerance biometric phenotyping [36] developed for breeding selection purposes, also showed the inhibition of seedling root regrowth (Figure 1). This method enables one to distinguish the Al-tolerant genotypes, without completely damaging the root meristems [37]. The test developed by Aniol [37] is broadly used in the breeding selection of cereal crops. Depending on the plant species, different concentrations of Al-ions are used [36]. For triticale, which is less tolerant than rye but much more tolerant than wheat, a 16 ppm ion concentration is common and allows for clear differentiation between tolerant, intermediate, and sensitive lines, among which the tolerant forms are in minority [14,36,38]. Recent studies by Szewińska et al. [35], performed in line with the biometric phenotyping method, revealed that after the Al stress release, the epidermal cells of root tips in tolerant rye and triticale seedlings are replaced by new cells and the root growth is maintained [35], and this process is independent on organic acid exudation [39]. In our studies, the roots of Al-tolerant L198 triticale line regrew, and the proteomic data showed identity between the control and root tips recovered after the stress release. In contrast, the root tips of susceptible lines were permanently damaged and roots did not regrow, along with evidenced alterations between the control and stress-treated proteomes, which is in line with the literature data [24].

Additionally, the detected differences in the redox balance in root tips showed complete recovery in the case of the tolerant line after the stress release, and differentiation between the control and stress treated seedlings in the case of the susceptible lines [30]; however, these differences in the present experiment carried out according to the biometric test protocol were found only in the case of line L444 root extract reactions with the $ABTS^{*+}$ cation radical (Table 1). The proteomic data were analyzed with double, strong cut-off criterion (probability $p \leq 0.01$ and difference in intensity ≥ 2), which influenced the small number of identities between the proteomes of L17 and L444. When the criterion was weakened and the intensity ratio was neglected, more than 80% of proteome patterns were common for L17 and L 444 lines (Table 2).

3.2. Annotation of Protein Spots

The flavone O-methyltransferase 1 (OMT1), with homology to the *Triticum aestivum* enzyme, was the strongest upregulated protein found in triticale root tips of the L17 sensitive line, whereas in L444 it was synthesized de novo. This enzyme catalyzes the sequential O-methylation of tricetin to mono-, di-, or trimethylated derivatives, with tricin, a dimethyl derivative, as a component found in monocotyledonous lignins [40]. Cell wall lignification along with hemicellulose deposition is an important and well-documented mechanism of plant tissue protection against harmful Al-ions, is positively correlated with the inhibition of root elongation, is more strongly expressed in sensitive genotypes [41], and was also detected in our experiment. *Ta*OMT1 was initially considered as a putative caffeic acid O-methyltransferase [42] involved in lignin biosynthesis; however, Zhou et al. [43] documented its low activity in methylation of lignin precursors such as caffeic and 5-hydroxyferulic acids. The next intensity difference was assigned to oxalate oxidase (OXO). Despite the fact that oxalate synthesis and degradation in plant cell walls is not clearly understood [44], it is specified as the enzyme oxidizing the oxalate to CO_2 and H_2O_2. It was reported that an Al-induced increase in OXO was correlated with Al uptake, growth inhibition, damage of the plasma membrane, and disruption of membrane permeability in barley seedling roots [45]. The increased activity of OXO has been observed in the roots of barley [45] and wheat [21]. The increased concentration of H_2O_2 disturbs cell redox homeostasis, leading to the activation of stress response pathways on the one hand and apoptosis on the other. Delisle et al. [46] concluded that a high level of OXO expression may support trapping the Al-ions in the root cells rather than induction of H_2O_2-dependent cell death, which was observed in wheat epidermal cells after only 8 h exposure to Al. The other antioxidant enzymes, glutathione-S-transferase (GSH) and 1-Cys peroxiredoxin (PER1), were also found. GSH is known as a universal antioxidant and detoxifier, induced in response to various stresses [47]. It is one of the most common enzymes identified in protein and transcript analyses of different plant species exposed to Al stress, with increased activity in both Al-tolerant and -sensitive genotypes of soybean [23], flax [48], maize [49], *Arabidopsis* [27], and pea roots [50]. However, an unexpected suppression of GSH protein was also observed in tomato (−1.56-fold) and wheat (−2.5-fold) [21,43].

The identified proteins S-adenosylmethionine synthase and adenosylhomocysteinase are enzymes of methyl cycling. S-adenosylmethionine synthase (SAMS) catalyzes the formation of S-adenosylmethionine (SAM) from methionine and ATP. Adenosylhomocysteinase may play a key role in the control of methylations via regulation of the intracellular concentration of adenosylhomocysteine, an inhibitor of SAM-dependent methyl transferase reactions. In earlier proteomic studies, a dynamic induction of SAMS in Al-treated roots of wheat [21], tomato [26], and rice [25] was found. The same analysis showed downregulation of adenosylhomocysteinase in wheat [21]. It was proposed [51] that stimulation of SAM synthesis could be involved in the alteration of the cell wall and polymer structures in roots and/or ethylene-mediated inhibition of root growth. S-adenosylmethionine (SAM) may also serve as an important methyl donor for O-methyltrasferases (OMT), involved in lignin synthesis [52]. Due to the fact, that the synthesis of DIMBOA-Glc requires O-methylation catalyzed by O-methyltransferases with the presence of SAM as methyl donor, we speculate that methyl cycling also plays an important role in DIMBOA synthesis in triticale plants exposed to Al stress. The increase or de novo synthesis of DIMBOA (2,4- dihydroxy-7- methoxy-1,4- benzoxazin-3-one) glucosidases, GLU1b and GLU1c, in protein extracts from Al-sensitive root tips suggests that the hydrolysis of terminal, nonreducing beta-D-glucosyl residues releases the DIMBOA benzoxazinoid, a key defense compound, along with the DIBOA (2,4-dihydroxy-1,4-benzoxazin-3-one), present in major agricultural crops, such as maize and wheat, and biologically active in both the aboveground and underground parts of plants. Poschenrieder et al. [53] documented its role in maize root tip protection by chelating Al-ions in the rhizosphere, and Neal et al. [54] found their attractive function for *Pseudomonas putida* in the maize. The inhibition of root growth entails enhanced cell wall rigidity [7] and changes in the organization of cortical microtubules [55]. A significant decrease in α- and β-tubulins, the main components of microtubules, was observed in proteomes of both sensitive lines. Similar results were obtained for Al-sensitive maize [56] and rice [20] under Al stress. Interestingly, different subunits of tubulin were differentially expressed and changed dynamically in the Al-sensitive soybean [23].

A significant induction of several proteins involved in protein synthesis and degradation was observed as well. Among them, DNA-directed RNA polymerase subunit beta, which catalyzes the transcription of DNA into RNA, as well as splicing factor U2af large subunit B, necessary for the splicing of pre-mRNA, were upregulated. Moreover, we found a high induction of the eukaryotic initiation factor 4A (eIF4A), an ATP-dependent RNA helicase that is a subunit of the eukaryotic translation initiation factor 4F (eIF4F) complex involved in cap recognition, required for mRNA binding to ribosomes [57]. As aluminum stress affects the cellular gene expression machinery, it is evident that molecules involved in nucleic acid processing, including helicases, are likely to be affected in root tips [58]. The two eIF4As/helicases from pea have been shown to play a role in abiotic stress tolerance, especially for salinity and cold stress- [59–61]. The expression of *Pennisetum glaucum* eukaryotic translational initiation factor 4A exhibited superior growth performance and higher chlorophyll retention under simulated drought and salinity stresses compared to the control plants. Abiotic stress usually leads to protein unfolding, misfolding, and aggregation [62]. Protein disulfide isomerase-like proteins (PDIs) catalyze protein disulfide bonds, inhibit aggregation of misfolded proteins, and function in isomerization during protein folding in the endoplasmic reticulum and responses during abiotic stresses [63]. In triticale plants affected by aluminum, PDIs were found to be upregulated in both susceptible lines. PDIs from *Brachypodium distachyon* L., *Brassica rapa* ssp. *pekinensis*, and *Arabidopsis thaliana* were upregulated under abiotic stresses, such as drought or salt, as well as under the influence of abscisic acid (ABA), and hydrogen peroxide (H_2O_2) an reactive oxygen species, suggesting their involvement in multiple stress responses [62,64,65]. The activation of ubiquitin enzymes illustrates the proteolytic activity in response to Al stress. Ubiquitination plays a critical role in protein inactivation, the degradation of damaged proteins, and the regulation of several mechanisms related to abiotic stress responses [66]. The enzymes of glycolysis and the tricarboxylic acid TCA cycle were activated as well, showing the influence of Al-ions on these main biochemical pathways. The phosphoglycerate kinase, which catalyzes the ADP-dependent dephosphorylation of 1,3-bisphospho glycerate to 3-bispsphoglycerete in glycolysis, was activated. We also observed upregulation of fructose-1,6-bisphosphatase, a key metabolic enzyme that catalyzes the reversible aldol cleavage of fructose-1,6-bisphosphate into glyceraldehyde-3-phosphate, either in glycolysis or gluconeogenesis and in the Calvin–Benson cycle [67]. Stimulating glycolysis in Al-treated plants may accelerate pyruvate and acetyl CoA production for organic acid synthesis, such as citrate or malate, which serve as Al chelators in the tolerant genotypes [7]. Moreover, acetyl-CoA may be used for the synthesis of malonyl-CoA, an essential substrate of fatty-acid synthesis [68]. The regulation of lipid membrane composition and modification of membrane fluidity by changes in unsaturated fatty acid levels is an efficient barrier that prevents metals from entering to the symplasm [69]. The mechanisms of Al tolerance based on increasing the plasma membrane (PM) permeability by binding Al to negative sites on the PM surface of root cells have been well documented for numerous plant species [69–72]. A similar response of sensitive plants in the regeneration phase may suggest that plants still attempt to eliminate aluminum accumulated in root tips.

The other proteins, such as mitochondrial ATP synthase, mitochondrial outer membrane porin, calmodulin, and dehydrin COR410 were upregulated in susceptible triticale lines at 48h after Al treatment, which suggest their important role in the response to Al toxicity. Mitochondrial ATP synthase subunit alpha produces the energy storage molecule adenosine triphosphate (ATP), which is suggested to provide energy for active Al efflux and detoxification [22]. Mitochondrial outer membrane porin is responsible for forming a channel through the cell membrane that allows the passage of small molecules. This protein was upregulated in Al-susceptible triticale lines. The abundance change of these mitochondrion transport-related proteins under Al stress indicates that the ion/metabolite exchange between the mitochondria and cytosol was modulated in the roots to cope with the stress. It was also observed that Al induces calmodulin synthesis, a major sensory molecule that decodes Ca^{2+} signals in the presence of different biotic and abiotic stresses [73]. Dehydrins (DHNs) play an important protective role in plant cells during dehydration [74]; however, those containing relatively large amounts of reactive residues on their surface exhibit also reactive oxygen species (ROS) scavenging and metal ion binding properties. However, the role of dehydrin in Al stress has not been explained so far, though its documented properties may suggest a positive correlation with Al tolerance.

Our results indicate that seedlings of Al-tolerant genotypes can recover after 16 ppm Al3+ stress relief without differentiation of proteome profiles (according to criteria: $p \leq 0.01$ and difference in spot intensity $\geq$ two-fold), contrary to seedlings of Al-sensitive genotypes that maintain the proteome modifications caused by unfavorable environments. Enzymes involved in cell wall lignification were highly induced whereas proteins involved in cell division were strongly downregulated.

4. Materials and Methods

4.1. Plant Materials

The experiments were performed using triticale inbred lines differing in aluminum (Al^{3+}) stress tolerance: one Al-tolerant line, L198 (MAH3405 (Milewo) × Matejko), spring form, and two sensitive lines L17 (Gabo × 6944/97), spring form and L444 (MAH3198 × CHD2807/98-7-1), winter form. Seeds were obtained from Plant Breeding Strzelce Ltd., Experimental Station Małyszyn (Poland). The lines were highly homozygotic (F10 generation) and screened for Al tolerance annually in line with our previous and present projects [15,75].

The research was carried out using the common Al tolerance detection method developed by Anioł [37]. Seeds sterilized and germinated for one day to form a 3 mm sprout were sown on polyethylene nets floated in a tray filled with a base medium of 2.0 $CaCl_2$, 3.25 KNO_3, 1.25 $MgCl_2$, 0.5 $(NH_4)_2SO_4$, and 0.2 NH_4NO_3, in mM concentrations, and a final pH of 4.5. After three days, the seedlings were transferred for 24 h onto the same medium containing 16 ppm Al^{3+} ions in the form of $AlCl_3$. Next, after washing of Al^{3+} ions, seedlings were placed again into the base solution for 48 h to induce root regrowth. The roots of tolerant forms regrow in the opposite to the roots of sensitive forms. The control in this experiment was seedlings grown in medium without Al-ions [37]. The experiment was run in a growth chamber (Pol-Eko Aparatura, ST500 B40 FOT10) at 25 °C with a 12 h day/night photoperiod and a light intensity of 40 $W \cdot m^{-2}$. The seedlings' aluminum tolerance was assessed on the basis of the regrowth rate of roots stained prior to evaluation in 0.1% Eriochrome cyanine R within 10 min (Figure 1). For antioxidant activity estimation and proteomic analysis, the root tips (0.3–0.4 cm) from 7-day old days seedlings, both exposed and non-exposed to Al^{3+} ions, were excised. The root staining was omitted in this case. The results were based on four independent biological experiments.

4.2. Antioxidant Potential Determination

Antioxidant potential was determined using two radicals, stable anion radical $DPPH^{*-}$ (2,2-diphenyl-1-picrylhydrazyl radical) and cation radical $ABTS^{*+}$ (2,2′-azino-bis (3-ethylbenzo thiazoline-6-sulphonic acid radical) (Sigma-Aldrich Ltd., Poznań, Poland), and was expressed in (μmol/mg) of Trolox equivalents (6-hydroxy-2,5,7,8-tetramethychroman-2-carboxylic acid) (Sigma-Aldrich Ltd., Poznań, Poland) [31]. A UV-2101PC UV-Vis scanning spectrophotometer (Shimadzu, Kioto, Japan) was used for absorbance measurements. The root tips were mashed into powder in liquid nitrogen, extracted in 80% methanol (MetOH) (100 mg/1 mL) at room temperature for 2h, and centrifuged. The reaction mixture consisted of 200 (μL) of the root extract and 3.2 (mL) of $DPPH^{*-}$ in 80% MetOH (10 mg/25 mL). Absorbance was measured at 515 nm after 20 min. The $ABTS^{*+}$ cation radical was prepared by oxidation of 7 mM ABTS water solution by 2.45 mM potassium persulfate overnight (16 h), at room temperature in the dark, and then dilution with 80% methanol to absorbance ca. 0.70 at 734 nm. The reaction mixture consisted of 50 μL of the root extract and 3.7 mL ABTS $^{*+}$, and the measurement was performed at 734 nm after 6 min.

One-way analysis of variance (ANOVA) was performed with Addinsoft 2020 XL-STAT (New York, NY, USA. https://www.xlstat.com, accessed on 20 December 2021). A Tukey HSD (honestly significant difference) multiple comparison test was used to identify statistically homogeneous subsets at $\alpha = 0.05$.

4.3. Proteomic Studies

Phenol-SDS Buffer Extraction with Sonication (PSWS)

The phenol extraction of proteins was carried out as described by Hurkman and Tanaka [76]. Root tissue (300 mg) was ground in a mortar in the presence of liquid nitrogen and transferred to a 1.5 mL Eppendorf tube. Proteins were extracted with 3 mL of SDS buffer (30% sucrose, 2% SDS, 0.1 M Tris-Cl, 5% β-mercaptoethanol, and 1 mM phenyl-methylsulfonyl fluoride (PMSF), pH 8.0) by triple sonication for 15 s at 60 amps. After sonication, 0.8 mL of Tris buffered phenol was added to the mixture and vortexed for 10 mins at 4 °C. The set was centrifuged at 14,000× *g* for 5 min at 4 °C, and the phenolic phase was collected and re-extracted with 0.8 mL SDS buffer and shaken for 5 min. Centrifugation was further repeated using the same settings, with the phenolic phase collected and precipitated overnight with four volumes of 0.1 M ammonium acetate in methanol at −20 °C. The precipitate obtained by centrifugation at 14,000× *g* for 10 min at 4 °C was washed thrice with cold 0.1 M ammonium acetate and finally with cold 80% acetone. The pellet was dried and resuspended in 100 μL of sample buffer (Biorad) and used for further analyses. Protein concentrations were quantified using the Bradford protein assay method, using BSA as a standard.

4.4. Two-Dimensional Electrophoresis (2-DE)

IPG strips (ReadyStripTMIPG, pH = 3–10, 17cm, Biorad) were passively rehydrated overnight with rehydration sample buffer (7M urea, 2 M thiourea, 4% CHAPS, 0.5% IPG Buffer, 20 mM DTT, 0.002% bromophenol blue) containing 250 μg of isolated protein. First-dimension

Isoelectric focusing (IEF) was conducted using the following parameters: step 1-gradient volt, 1000 V for 60 mins, step 2-gradient volt, 12,000 V for 60 min, step 3-constant volt, 12,000 V for 25,000 volt hours, and step 4-constant volt, 1000 V for 60 min. All steps were performed at 20 °C using IEF 100 (Hoefer Scientific Instruments, San Francisco, CA, USA). Following IEF, the strips were reduced with 130 mM DTT in 10 mL of equilibration buffer (29.3% glycerol, 75 mM Tris-Cl, 6 M urea, 2% SDS, pH 8.8) for 15 min and alkylated with 135 mM iodoacetamide in 10 mL equilibration buffer for 15 min. The 2-DE was performed according to the Laemmli [77] protocol in lab cast 1.5 mm 12.5% (w/v) polyacrylamide gels using a Hoefer SE 600 Chroma Vertical Electrophoresis System (Hoefer Scientific Instruments, San Francisco, CA, USA). The following program was implemented: 15 mA/gel for 15 min and 30 mA/gel for 90 min in Tris glycine-SDS running buffer. Three gels, one from each independent biological replication, were used for the identification of differential proteins. The gels were stained with 0.1% (w/v) Coomassie brilliant blue R-250 (Sigma-Aldrich Ltd., Poznań, Poland) overnight, destained, and stored in 5% acetic acid at 4 °C for further analysis [78].

4.5. Analysis of 2D PAGE Gel Images

Stained gels were digitalized, annotated, and analyzed using Image Master 2D Platinum 7.0 software (GE Healthcare). Data were normalized by expressing abundance as relative volume (% vol). A difference in protein expression was accepted when the Student's *t*-test was at a significance level of 99% ($p \leq 0.01$). Spots were only accepted as present or absent if they were present or missing in all four gels from control or treated material/groups. Moreover, in the case of spots appearing only in the control (silenced) or stressed (induced) roots, only those with a signal intensity value >0.2 were considered as significant. The gels obtained for both NT lines were compared visually for identification of the identical spots showing the highest signal intensity changes.

4.6. Protein Identification by Mass Spectrometry and Database Search

To identify the protein content in interesting spots, gel pieces were manually cut out and subjected to a standard procedure during which proteins were reduced with DTT, alkylated with iodoacetamide, and digested overnight with trypsin (Sequencing Grade Modified Trypsin, Promega, Madison, WI, USA). The analyses were made by the Mass Spectrometry Laboratory, Institute of Biochemistry and Biophysics Polish Academy of Science (MS Lab IBB-PAN, Warsaw, Poland). The peptide mixtures were analyzed by liquid chromatography coupled to tandem mass spectrometry (LC-MS-MS/MS) with a classic mass spectrometer and LTQ (linear trap quadrupole ion trap-Orbitrap) (Thermo Electron Corporation, San Jose, CA, USA). Briefly, the peptide mixture was applied to an RP-18 precolumn (nanoACQUITY Symmetry® C18, Waters, Milford, CT, USA) using water containing 0.1% (*m/v*) formic acid (FA) as a mobile phase and then transferred to a nano-HPLC RP-18 column (nanoACQUITY BEH C18, Waters) using an acetonitrile gradient (0–60%, *v/v*, in 120 min) in the presence of 0.05% (*m/v*) formic acid with a flow rate of 0.25 $mm^3\ min^{-1}$. The column outlet was directly coupled with the ion source of the spectrometer working in the regime of data dependent MS to MS/MS switch.

After pre-processing the raw data with the Mascot Distiller v. 2.6.1.0 software (Matrix Science, London, UK), the obtained peak lists were used to search the non-redundant protein database of the National Centre for Biotechnology Information (NCBI) using the Mascot search engine (v. 2.5.1, Matrix Science). The taxonomic category selected was *Triticum aestivum*. Only peptides passing a Mascot-defined expectation value of 0.05 were considered as positive identifications [78]. The functional networks of differentially expressed proteins were constructed using the STRING database [32].

Supplementary Materials: The following supporting information can be downloaded at: https://www.mdpi.com/article/10.3390/plants11020165/s1, Figure S1 The stages of the experiment: (A) triticale seeds germinating on the polyethylene grid tray; (B) 5th days old triticale seedlings; (C) the growth chamber view; Figure S2 Protein separation by 2-DE on gels stained in Coomassie Brilliant Blue (A-F). Proteome of L17 Al-sensitive line control (A) and 48h after Al stress treated (B); proteome of L444 Al-sensitive line control (C) and 48h after Al stress treated (D); proteome of L198 Al-tolerant line control (E) and 48h after Al stress treated (F). The differential protein spots, which were common for both studied Al-sensitive lines, are marked in red on gel pictures of L17 and L444 line (control vs. Al-treated). The differential protein spots, which were characteristic only for one studied Al-sensitive lines, are marked in green on gel pictures of L17 and L444 line (control vs. Al-treated). The Image Master 2D Platinum 7.0 software was used for differential spots identification; Figure S3 Comparison in: (A) number of up/down-regulated and silenced/induced proteins and (B) number of common protein spots according to established criterions ($p \leq 0.01$ and difference in spot intensity $\geq$ 2-fold or 0.2 relative intensity of silenced/induced proteins).

Author Contributions: Conceptualization, A.N.; methodology, A.N., L.D., W.M.D. and K.R.; investigation, A.N. and L.D.; writing—original draft preparation, A.N.; writing—review and editing, K.R. and M.P.; visualization, A.N, L.D, W.M.D., M.P. and K.R. All authors have read and agreed to the published version of the manuscript.

Funding: This research was funded by the Ministry of Agriculture and Rural Development, Poland, as statutory grant of IHAR-PIB N° 1-1-03-2-01 and the European Regional Development Fund-Project "SINGING PLANT" (No. CZ.02.1.01/0.0/0.0/16_026/0008446), which received a financial contribution from the Ministry of Education, Youths and Sports of the Czech Republic in the form of special support through the National Programme for Sustainability II funds.

Institutional Review Board Statement: Not applicable.

Informed Consent Statement: Not applicable.

Data Availability Statement: Data available upon request.

Acknowledgments: The laboratory equipment was purchased in frame of the grant founded by the National Centre of Research and Development, Poland: grant number NCBR-PBS3/B8/19/2015.

Conflicts of Interest: The authors declare no conflict of interest. The funders had no role in the design of the study; in the collection, analyses, or interpretation of data; in the writing of the manuscript, or in the decision to publish the results.

References

1. Lee, S.S.; Schmidt, M.; Sturchio, N.; Nagy, K.; Fenter, P. Effect of pH on the Formation of Gibbsite-Layer Films at the Muscovite (001)−Water Interface. *J. Phys. Chem. C* **2019**, *123*, 6560–6571. [CrossRef]
2. Kochian, L.V.; Piñeros, M.A.; Hoekenga, O.A. The physiology, genetics and molecular biology of plant aluminum resistance and toxicity. In *Root Physiology: From Gene to Function*; Lambers, H., Colmer, T.D., Eds.; Springer: Dordrecht, The Netherlands, 2005; pp. 175–195.
3. Barceló, J.; Poschenrieder, C. Fast root growth responses, root exudates, and internal detoxification as clues to the mechanisms of aluminium toxicity and resistance: A review. *Environ. Exp. Bot.* **2002**, *48*, 75–92. [CrossRef]
4. Yang, Y.; Wang, Q.L.; Geng, M.J.; Guo, Z.H.; Zhao, Z. Rhizosphere pH difference regulated by plasma membrane H^+-ATPase is related to differential Al tolerance of two wheat cultivars. *Plant Soil Envion.* **2011**, *57*, 201–206. [CrossRef]
5. Blamey, F.P.C. The Role of the Root Cell Wall in Aluminum Toxicity. In *Plant Nutrient Acquisition*; Ae, N., Arihara, J., Okada, K., Srinivasan, A., Eds.; Springer: Tokyo, Japan, 2001; pp. 201–226.
6. Matsumoto, H. Cell biology of aluminium toxicity and tolerancein higher plants. *Int. Rev. Cytol.* **2000**, *200*, 1–46. [CrossRef]
7. Kochian, L.V. Cellular mechanisms of aluminium toxicity and tolerance in plants. *Annu. Rev. Plant Physiol. Plant Mol. Biol.* **1995**, *46*, 237–260. [CrossRef]
8. Fontecha, G.; Silva-Navas, J.; Benito, C.; Mestres, M.A.; Espino, F.J.; Hernández-Riquer, M.V.; Gallego, F.J. Candidate gene identification of an aluminum-activated organic acid transporter gene at the Alt4 locus for aluminum tolerance in rye (*Secale cereale* L.). *Theor. Appl. Genet.* **2007**, *114*, 249–260. [CrossRef]

9. Silva-Navas, J.; Benito, C.; Téllez-Robledo, B.; Abd El-Moneim, D.; Gallego, F.J. The ScAACT1 gene at the Qalt5locus as a candidate for increased aluminum tolerance in rye (*Secale cereale* L.). *Mol. Breed.* **2012**, *30*, 845–856. [CrossRef]
10. Ryan, P.R.; Raman, H.; Gupta, S.; Horst, W.J.; Delhaize, E. A Second Mechanism for Aluminum Resistance in Wheat Relies on the Constitutive Efflux of Citrate from Roots. *Plant Physiol.* **2009**, *149*, 340–351. [CrossRef] [PubMed]
11. Garcia-Oliveira, A.L.; Benito, C.; Prieto, P.; de Andrade Menezes, R.; Rodrigues-Pousada, C.; Guedes-Pinto, H.; Martins-Lopes, P. Molecular characterization of TaSTOP1 homoeologues and their response to aluminium and proton (H^+) toxicity in bread wheat (*Triticum aestivum* L.). *BMC Plant Biol.* **2013**, *13*, 134. [CrossRef]
12. Li, J.; Liu, J.; Dong, D.; Jia, X.; McCouch, S.; Kochian, L. Natural variation underlies alterations in Nramp aluminum transporter (NRAT1) expression and function that play a key role in rice aluminum tolerance. *Proc. Natl. Acad. Sci. USA* **2014**, *111*, 6503–6508. [CrossRef] [PubMed]
13. Tyagi, W.; Yumnam, J.S.; Sen, D.; Rai, M. Root transcriptome reveals efficient cell signaling and energy conservation key to aluminum toxicity tolerance in acidic soil adapted rice genotype. *Sci. Rep.* **2020**, *10*, 4580. [CrossRef]
14. Niedziela, A.; Bednarek, P.T.; Labudda, M.; Mańkowski, D.R.; Anioł, A. Genetic mapping of a 7R Al tolerance QTL in triticale (× *Triticosecale* Wittmack). *J. Appl. Genet.* **2014**, *55*, 1–14. [CrossRef] [PubMed]
15. Niedziela, A.; Bednarek, P.T.; Cichy, H.; Budzianowski, G.; Kilian, A.; Anioł, A. Aluminum tolerance association mapping in triticale. *BMC Genom.* **2012**, *13*, 67. [CrossRef] [PubMed]
16. Budzianowski, G.; Woś, H. The effect of single D-genome chromosomes on aluminum tolerance of triticale. *Euphytica* **2004**, *137*, 165–172. [CrossRef]
17. Kosová, K.; Vítámvás, P.; Urban, M.O.; Klíma, M.; Roy, A.; Tom Prášil, I. Biological networks underlying abiotic stress tolerance in temperate crops-a proteomic perspective. *Int. J. Mol. Sci.* **2015**, *16*, 20913–20942. [CrossRef]
18. Kosová, K.; Vítámvás, P.; Prášil, I.T. Proteomics of stress responses in wheat and barley—Search for potential protein markers of stress tolerance. *Front. Plant Sci.* **2014**, *5*, 711. [CrossRef] [PubMed]
19. Yang, J.L.; Li, Y.Y.; Zhang, Y.J.; Zhang, S.S.; Wu, Y.R.; Wu, P.; Zheng, S.J. Cell Wall Polysaccharides Are Specifically Involved in the Exclusion of Aluminum from the Rice Root Apex. *Plant Physiol.* **2008**, *146*, 602–611. [CrossRef]
20. Wang, C.; Shen, R.F.; Wang, W. Root protein profile changes induced by Al exposure in two rice cultivars differing in Al tolerance. *J. Proteom.* **2012**, *78*, 281–293. [CrossRef]
21. Oh, M.W.; Roy, S.K.; Kamal, A.H.M.; Cho, K.; Cho, S.-W.; Park, C.-S.; Choi, J.-S.; Komatsu, S.; Woo, S.-H. Proteome analysis of roots of wheat seedlings under aluminum stress. *Mol. Biol. Rep.* **2014**, *41*, 671–681. [CrossRef]
22. Dai, J.; Bai, G.; Zhang, D.; Hong, D. Validation of quantitative trait loci for aluminum tolerance in Chinese wheat landrace FSW. *Euphytica* **2013**, *192*, 171–179. [CrossRef]
23. Duressa, D.; Soliman, K.; Taylor, R.; Senwo, Z. Proteomic Analysis of Soybean Roots under Aluminum Stress. *Int. J. Plant Genom.* **2011**, *2011*, 282531. [CrossRef] [PubMed]
24. Zheng, L.; Lan, P.; Shen, R.F.; Li, W.F. Proteomics of aluminum tolerance in plants. *Proteimics* **2014**, *14*, 566–578. [CrossRef] [PubMed]
25. Fukuda, T.; Saito, A.; Wasaki, J.; Shinano, T.; Osaki, M. Metabolic alterations proposed by proteome in rice roots grown under low P and high Al concentration under low pH. *Plant Sci.* **2007**, *172*, 1157–1165. [CrossRef]
26. Zhou, S.; Sauvé, R.; Thannhauser, T.W. Proteome changes induced by aluminium stress in tomato roots. *J. Exp. Bot.* **2009**, *60*, 1849–1857. [CrossRef]
27. Ezaki, B.; Katsuhara, M.; Kawamura, M.; Matsumoto, H. Different Mechanisms of Four Aluminum (Al)-Resistant Transgenes for Al Toxicity in *Arabidopsis*. *Plant Physiol.* **2001**, *127*, 918–927. [CrossRef]
28. Wei, Y.; Jiang, C.; Han, R.; Xie, Y.; Liu, L.; Yu, Y. Plasma membrane proteomic analysis by TMT-PRM provides insight into mechanisms of aluminum resistance in tamba black soybean roots tips. *PeerJ* **2020**, *8*, e9312. [CrossRef] [PubMed]
29. Grębosz, J.; Badowiec, A.; Weidner, S. Changes in the root proteome of Triticosecale grains germinating under osmotic stress. *Acta Physiol. Plant.* **2014**, *36*, 825–835. [CrossRef]
30. Meriño-Gergichevich, C.; Ondrasek, G.; Zovko, M.; Šamec, D.; Alberdi, M.; Reyes-Díaz, M. Comparative study of methodologies to determine the antioxidant capacity of Al-toxified blueberry amended with calcium sulfate. *J. Soil Sci. Plant Nutr.* **2015**, *15*, 965–978. [CrossRef]
31. Arnao, M.B. Some methodological problems in the determination of antioxidant activity using chromogen radicals: A practical case. *Trends Food Sci. Technol.* **2000**, *11*, 419–421. [CrossRef]
32. Szklarczyk, D.; Morris, J.H.; Cook, H.; Kuhn, M.; Wyder, S.; Simonovic, M.; Santos, A.; Doncheva, N.T.; Roth, A.; Bork, P.; et al. The STRING database in 2017: Quality-controlled protein–protein association networks, made broadly accessible. *Nucleic Acids Res.* **2017**, *45*, D362–D368. [CrossRef]
33. Matsumoto, H.; Senoo, Y.; Kasai, M.; Maeshima, M. Response of the plant root to aluminum stress: Analysis of the inhibition of the root elongation and changes in membrane function. *J. Plant Res.* **1996**, *109*, 99–105. [CrossRef]
34. Vartapetian, B.B.; Andeeva, I.N.; Generozova, I.P.; Polyakova, L.I.; Maslova, I.P.; Dogikh, Y.I.; Stepanova, A.Y. Functional Electron Microscopy in Studies of Plant response and adaptation to Anaerobic Stress. *Ann. Bot.* **2003**, *91*, 155–172. [CrossRef] [PubMed]
35. Szewińska, J.; Różańska, E.; Papierowska, E.; Labudda, M. Proteolytic and Structural Changes in Rye and Triticale Roots under Aluminum Stress. *Cells* **2021**, *10*, 3046. [CrossRef] [PubMed]

36. Aniol, A.; Gustafson, J. Chromosome location of genes controlling aluminium tolerance in wheat, rye, and triticale. *Genome* **1984**, *26*, 701–705. [CrossRef]
37. Aniol, A. Induction of Aluminum Tolerance in Wheat Seedlings by Low Doses of Aluminum in the Nutrient Solution. *Plant Physiol.* **1984**, *76*, 551–555. [CrossRef]
38. Kim, B.Y.; Baier, A.C.; Somers, D.J.; Gustafson, J.P. Aluminum tolerance in triticale, wheat and rye. *Euphytica* **2001**, *120*, 329–337. [CrossRef]
39. Wang, Y.; Cai, Y.; Cao, Y.; Liu, J. Aluminum-activated root malate and citrate exudation is independent of NIP1;2-facilitated root-cell-wall aluminum removal in *Arabidopsis*. *Plant Signal. Behav.* **2018**, *13*, e1422469. [CrossRef]
40. Li, M.; Pu, Y.; Yoo, C.G.; Ragauskas, A. The occurrence of tricin and its derivatives in plants. *Green Chem.* **2016**, *18*, 1439–1454. [CrossRef]
41. Chandran, D.; Sharopova, N.; Ivashuta, S.; Gantt, J.S.; VandenBosch, K.A.; Samac, D.A. Transcriptome profiling identified novel genes associated with aluminum toxicity, resistance and tolerance in *Medicago truncatula*. *Planta* **2008**, *228*, 151–166. [CrossRef]
42. Jung, J.; Hong, M.; Kim, D.; Kim, J.; Heo, H.; Kim, T.; Jang, C.; Seo, Y.W. Structural and expressional divergence of genes encoding O-methyltransferase in wheat. *Genome* **2008**, *51*, 856–869. [CrossRef]
43. Zhou, J.-M.; Seo, Y.W.; Ibrahim, R.K. Biochemical characterization of a putative wheat caffeic acid O-methyltransferase. *Plant Physiol. Biochem.* **2009**, *47*, 322–326. [CrossRef] [PubMed]
44. Cai, X.; Ge, C.; Xu, C.; Wang, X.; Wang, S.; Wang, Q. Expression Analysis of Oxalate Metabolic Pathway Genes Reveals Oxalate Regulation Patterns in Spinach. *Molecules* **2018**, *23*, 1286. [CrossRef]
45. Tamás, L.; Budíková, S.; Huttová, J.; Mistrík, I.; Simonovicová, M.; Siroká, B. Aluminum-induced cell death of barley-root border cells is correlated with peroxidase- and oxalate oxidase-mediated hydrogen peroxide production. *Plant Cell Rep.* **2005**, *24*, 189–194. [CrossRef]
46. Delisle, G.; Champoux, M.; Houde, M. Characterization of Oxalate Oxidase and Cell Death in Al-Sensitive and Tolerant Wheat Roots. *Plant Cell Physiol.* **2001**, *42*, 324–333. [CrossRef] [PubMed]
47. Noctor, G.; Arisi, A.; Jouanin, L.; Kunert, K.; Rennenberg, H.; Foyer, C. Review article. Glutathione: Biosynthesis, metabolism and relationship to stress tolerance explored in transformed plants. *J. Exp. Bot.* **1998**, *49*, 623–647. [CrossRef]
48. Dmitriev, A.A.; Krasnov, G.S.; Rozhmina, T.A.; Kishlyan, N.V.; Zyablitsin, A.V.; Sadritdinova, A.F.; Snezhkina, A.V.; Fedorova, M.S.; Yurkevich, O.Y.; Muravenko, O.V.; et al. Glutathione S-transferases and UDP-glycosyltransferases Are Involved in Response to Aluminum Stress in Flax. *Front. Plant Sci.* **2016**, *7*, 1920. [CrossRef]
49. Cancado, G.M.A.; De Rosa, V.E.; Fernandez, J.H.; Maron, L.G.; Jorge, R.A.; Menossi, M. Glutathione S-transferase and aluminum toxicity in maize. *Funct. Plant Biol.* **2005**, *32*, 1045–1055. [CrossRef] [PubMed]
50. Panda, S.K.; Matsumoto, H. Changes in antioxidant gene expression and induction of oxidative stress in pea (*Pisum sativum* L.) under Al stress. *Biometals* **2010**, *23*, 753–762. [CrossRef] [PubMed]
51. Fukuda, H.; Hirakawa, Y.; Sawa, S. Peptide signaling in vascular development. *Curr. Opin. Plant Biol.* **2007**, *10*, 477–482. [CrossRef]
52. Kim, B.-G.; Sung, S.H.; Chong, Y.; Lim, Y.; Ahn, J.-H. Plant Flavonoid O-Methyltransferases: Substrate Specificity and Application. *J. Plant Biol.* **2010**, *53*, 321–329. [CrossRef]
53. Poschenrieder, C.; Tolrà, R.; Barceló, J. A role for cyclic hydroxamates in aluminum resistance in maize? *J. Inorg. Biochem.* **2005**, *99*, 1830–1836. [CrossRef]
54. Neal, A.L.; Ahmad, S.; Gordon-Weeks, R.; Ton, J. Benzoxazinoids in Root Exudates of Maize Attract *Pseudomonas putida* to the Rhizosphere. *PLoS ONE* **2012**, *7*, e35498. [CrossRef] [PubMed]
55. Fujita, S.; Pytela, J.; Hotta, T.; Kato, T.; Hamada, T.; Akamatsu, R.; Ishida, Y.; Kutsuna, N.; Hasezawa, S.; Nomura, Y.; et al. An Atypical Tubulin Kinase Mediates Stress-Induced Microtubule Depolymerization in *Arabidopsis*. *Curr. Biol.* **2013**, *23*, 1969–1978. [CrossRef] [PubMed]
56. Sivaguru, M.; Baluška, F.; Volkmann, D.; Felle, H.H.; Horst, W.J. Impacts of Aluminum on the Cytoskeleton of the Maize Root Apex. Short-Term Effects on the Distal Part of the Transition Zone. *Plant Physiol.* **1999**, *119*, 1073–1082. [CrossRef] [PubMed]
57. Bi, X.; Ren, J.; Goss, D.J. Wheat Germ Translation Initiation Factor eIF4B Affects eIF4A and eIFiso4F Helicase Activity by Increasing the ATP Binding Affinity of eIF4A. *Biochemistry* **2000**, *39*, 5758–5765. [CrossRef] [PubMed]
58. Tuteja, N.; Vashisht, A.; Tuteja, R. Translation initiation factor 4A: A prototype member of dead-box protein family. *Physiol. Mol. Biol. Plants Int. J. Funct. Plant Biol.* **2008**, *14*, 101–107. [CrossRef]
59. Pham, X.H.; Reddy, M.K.; Ehtesham, N.Z.; Matta, B.; Tuteja, N. A DNA helicase from *Pisum sativum* is homologous to translation initiation factor and stimulates topoisomerase I activity. *Plant J.* **2000**, *24*, 219–229. [CrossRef]
60. Vashisht, A.A.; Pradhan, A.; Tuteja, R.; Tuteja, N. Cold- and salinity stress-induced bipolar pea DNA helicase 47 is involved in protein synthesis and stimulated by phosphorylation with protein kinase C. *Plant J.* **2005**, *44*, 76–87. [CrossRef]
61. Santosh, R.B.R.T.; Vijaya Naresh, J.; Sudhakar, R.P.; Reddy, M.K.; Mallikarjuna, G. Expression of *Pennisetum glaucum* Eukaryotic Translational Initiation Factor 4A (PgeIF4A) Confers Improved Drought, Salinity, and Oxidative Stress Tolerance in Groundnut. *Front. Plant Sci.* **2017**, *8*, 453. [CrossRef]
62. Zhang, Z.; Liu, X.; Li, R.; Yuan, L.; Dai, Y.; Wang, X. Identification and Functional Analysis of a Protein Disulfide Isomerase (AtPDI1) in *Arabidopsis thaliana*. *Front. Plant Sci.* **2018**, *9*, 913. [CrossRef]

63. Khan, R.; Siddiqui, M.; Salahuddin, P. Protein Disulfide Isomerase: Structure, Mechanism of Oxidative Protein Folding and Multiple Functional Roles. *J. Biochem. Mol. Biol. Res.* **2016**, *2*, 173–179. [CrossRef]
64. Zhu, C.; Luo, N.; He, M.; Chen, G.; Zhu, J.; Yin, G.; Li, X.; Hu, Y.; Li, J.; Yan, Y. Molecular Characterization and Expression Profiling of the Protein Disulfide Isomerase Gene Family in *Brachypodium distachyon* L. *PLoS ONE* **2014**, *9*, e94704. [CrossRef]
65. Kayum, M.A.; Park, J.-I.; Nath, U.K.; Saha, G.; Biswas, M.K.; Kim, H.-T.; Nou, I.-S. Genome-wide characterization and expression profiling of PDI family gene reveals function as abiotic and biotic stress tolerance in Chinese cabbage (*Brassica rapa* ssp. pekinensis). *BMC Genom.* **2017**, *18*, 885. [CrossRef]
66. Lyzenga, W.J.; Stone, S.L. Abiotic stress tolerance mediated by protein ubiquitination. *J. Exp. Bot.* **2011**, *63*, 599–616. [CrossRef] [PubMed]
67. Lv, G.-Y.; Guo, X.-G.; Xie, L.-P.; Xie, C.-G.; Zhang, X.-H.; Yang, Y.; Xiao, L.; Tang, Y.-Y.; Pan, X.-L.; Guo, A.-G.; et al. Molecular Characterization, Gene Evolution, and Expression Analysis of the Fructose-1, 6-bisphosphate Aldolase (FBA) Gene Family in Wheat (*Triticum aestivum* L.). *Front. Plant Sci.* **2017**, *8*, 1030. [CrossRef] [PubMed]
68. Foster, D.W. Malonyl-CoA: The regulator of fatty acid synthesis and oxidation. *J. Clin. Investig.* **2012**, *122*, 1958–1959. [CrossRef]
69. Wagatsuma, T. The membrane lipid bilayer as a regulated barrier to cope with detrimental ionic conditions: Making new tolerant plant lines with altered membrane lipid bilayer. *Soil Sci. Plant Nutr.* **2017**, *63*, 507–516. [CrossRef]
70. Lindberg, S.; Griffiths, G. Aluminium Effects on ATPase Activity and Lipid Composition of Plasma Membranes in Sugar Beet Roots. *J. Exp. Bot.* **1993**, *44*, 1543–1550. [CrossRef]
71. Zhang, G.; Slaski, J.J.; Archambault, D.J.; Taylor, G.J. Alternation of plasma membrane lipids in aluminum-resistant and aluminum-sensitive wheat genotypes in response to aluminum stress. *Physiol. Plant.* **1997**, *99*, 302–308. [CrossRef]
72. Maejima, E.; Watanabe, T. Proportion of phospholipids in the plasma membrane is an important factor in Al tolerance. *Plant Signal. Behav.* **2014**, *9*, e29277. [CrossRef]
73. Zeng, H.; Xu, L.; Singh, A.; Wang, H.; Du, L.; Poovaiah, B.W. Involvement of calmodulin and calmodulin-like proteins in plant responses to abiotic stresses. *Front. Plant Sci.* **2015**, *6*, 600. [CrossRef] [PubMed]
74. Hanin, M.; Brini, F.; Ebel, C.; Toda, Y.; Takeda, S.; Masmoudi, K. Plant dehydrins and stress tolerance. *Plant Signal. Behav.* **2011**, *6*, 1503–1509. [CrossRef] [PubMed]
75. Niedziela, A. The influence of Al^{3+} on DNA methylation and sequence changes in the triticale (× *Triticosecale* Wittmack) genome. *J. Appl. Genet.* **2018**, *59*, 405–417. [CrossRef] [PubMed]
76. Hurkman, W.J.; Tanaka, C.K. Solubilization of Plant Membrane Proteins for Analysis by Two-Dimensional Gel Electrophoresis. *Plant Physiol.* **1986**, *81*, 802–806. [CrossRef]
77. Laemmli, U. Cleavage of structural protein during the assembly of the head of bacteriophage T4. *Nature* **1970**, *227*, 680–685. [CrossRef] [PubMed]
78. Nykiel, M.; Lisik, P.; Dębski, J.; Florea, B.I.; Rybka, K. Chl *a* fluorescence and proteomics reveal the protection of photosynthetic apparatus in tolerant but not susceptible to dehydration wheat cultivar. *Biol. Plant.* **2019**, *63*, 287–297. [CrossRef]

plants

Article

Copper Tolerance and Accumulation on *Pelargonium graveolens* L'Hér. Grown in Hydroponic Culture

Antonios Chrysargyris [1], Rita Maggini [2], Luca Incrocci [2,*], Alberto Pardossi [2] and Nikolaos Tzortzakis [1,*]

1 Department of Agricultural Sciences, Biotechnology and Food Science, Cyprus University of Technology, Limassol 3603, Cyprus; a.chrysargyris@cut.ac.cy

2 Department of Agriculture, Food and Environment, University of Pisa, 56124 Pisa, Italy; rita.maggini@unipi.it (R.M.); alberto.pardossi@unipi.it (A.P.)

* Correspondence: luca.incrocci@unipi.it (L.I.); nikolaos.tzortzakis@cut.ac.cy (N.T.)

Abstract: Heavy metal contamination is a major health issue concerning the commercial production of medicinal and aromatic plants (MAPs) that are used for the extraction of bioactive molecules. Copper (Cu) is an anthropogenic contaminant that, at toxic levels, can accumulate in plant tissues, affecting plant growth and development. On the other hand, plant response to metal-induced stress may involve the synthesis and accumulation of beneficial secondary metabolites. In this study, hydroponically grown *Pelargonium graveolens* plants were exposed to different Cu concentrations in a nutrient solution (4, 25, 50, 100 μM) to evaluate the effects Cu toxicity on plant growth, mineral uptake and distribution in plants, some stress indicators, and the accumulation of bioactive secondary metabolites in leaf tissues. *P. graveolens* resulted in moderately tolerant Cu toxicity. At Cu concentrations up to 100 μM, biomass production was preserved and was accompanied by an increase in phenolics and antioxidant capacity. The metal contaminant was accumulated mainly in the roots. The leaf tissues of Cu-treated *P. graveolens* may be safely used for the extraction of bioactive molecules.

Keywords: antioxidants; bioaccumulation; copper toxicity; hydroponics; translocation factor

Citation: Chrysargyris, A.; Maggini, R.; Incrocci, L.; Pardossi, A.; Tzortzakis, N. Copper Tolerance and Accumulation on *Pelargonium graveolens* L'Hér. Grown in Hydroponic Culture. *Plants* **2021**, *10*, 1663. https://doi.org/10.3390/plants10081663

Academic Editors: Ewa Muszyńska, Kinga Dziurka and Mateusz Labudda

Received: 13 July 2021
Accepted: 6 August 2021
Published: 12 August 2021

1. Introduction

Copper (Cu) is an abundant transition metal of the lithosphere that is considered a relevant anthropogenic contaminant, as large amounts of this element have been released into the environment over the past decades [1,2]. In addition to the environmental impact of mining and smelting operations, the extensive application of Cu-containing fertilizers, pesticides and fungicides in agricultural practices has contributed to water body and soil contamination [1–4]; therefore, agricultural soils are particularly exposed to pollution by this contaminant. For example, Chen et al. [5] reported that in China, over 16% of agricultural soil is contaminated by heavy metals, and 2% is polluted by Cu only. Among heavy metals, Cu is often the only contaminant in vineyards, where it is extensively used against downy mildew [1–4]. According to the European Council Directive 86/278/EEC [6] on the protection of the environment, the permitted Cu concentration in agricultural soils amended with sewage sludge is 50–140 mg kg^{-1} for pH values in the range 6–7. For uncontaminated soils, Kabata-Pendias and Szteke [7] indicated a Cu concentration range of 1–140 mg kg^{-1}, depending on soil texture; the same authors reported that soil Cu concentrations in the range 25–40 mg kg^{-1} may be toxic to plants below pH 5.5, as Cu availability increases with soil acidity.

In nature, Cu commonly exists in the elemental metal form or as Cu^{+} or Cu^{2+} ions, although the oxidation states +3 and +4 can also be found [8]. Due to its redox properties, Cu at low concentration has a fundamental biological role for all living organisms, taking part in several metabolic reactions [9,10]. In higher plants, Cu is an essential micronutrient that is necessary for normal growth and development [11], being involved in mineral nutrition and electron transfer reactions that occur in vital processes, such as respiration and photosynthesis, chlorophyll and primary metabolites biosynthesis, or the scavenging of radicals [4]. The Cu concentration range that is considered normal for plants ranges from 2–5 to 30 mg kg^{-1} dry weight (DW) [12,13], while at higher concentrations Cu can cause toxicity symptoms [14]. Toxic levels of Cu in plants can impair biochemical reactions, affect gas exchanges, reduce plant growth [15,16]. Plants grown in Cu-polluted soils undergo oxidative stress and accumulate reactive oxygen species (ROS) [17], which induces the activation of antioxidant enzymes and the biosynthesis of antioxidant molecules. Cu toxicity has also been associated with an increased content of proline in plant tissues [18,19].

To counteract the effects of metal toxicity, plants have developed tolerance mechanisms such as metal complexation, storage in vacuoles, precipitation in cell walls, and down-regulation of metal transporters via the plasma membrane [20,21]. On the other hand, plant species capable of effectively up taking heavy metals from the soil and accumulating these contaminants in tissues generally show a high translocation rate from roots to shoots. These species can be profitably used for phytoremediation through the removal of toxic metals from the soil, which represents a green and cost-effective strategy for the amelioration of marginal lands [22,23]. According to recent literature [2], about 500 species are currently used for the phytoremediation of metal polluted soils. Conversely, the accumulation of toxic metals by plant species that are employed for human usage represents a serious threat for the consumers safety and has become a health concern worldwide. Medicinal and aromatic plants (MAPs) are typically used in the food, pharmaceutical and cosmetic industries as a natural source of biologically active compounds [24], and are increasingly cultivated on a commercial scale to sustain the expansion of the market demand. Contamination of the plant material is a major health issue concerning the commercial production of MAPs [25]. On the other hand, the physiological markers of plant response to metal-induced stress are often beneficial bioactive secondary metabolites, mainly antioxidants such as phenolic compounds or essential oils constituents [26–29].

The *Pelargonium* genus in the family Geraniaceae comprises several hundreds of aromatic species, distributed worldwide in subtropical and temperate regions [30]. The essential oil from *Pelargonium* spp. is among the top 20 essential oils used all over the world [31] due to its well-known bioactive properties [32–34]. In addition, the pharmacological activity of *Pelargonium* spp. is attributed also to phenolic constituents such as flavonoids and hydroxycinnamic acid-derivatives [35]. *Pelargomium* spp. are tolerant to toxicity by heavy metals and have been successfully applied as hyperaccumulators for several metal contaminants [36], including Cu [37]. Particularly, *Pelargomium graveolens* L'Hér., popularly known as rose-scented geranium, has been reported by several authors as a good candidate for phytoremediation; in addition, the effect of heavy metals on the yield and quality of its essential oil has been widely investigated [38–40]. However, in recent years the pharmacological activity of *P. graveolens* has been increasingly linked also to the leaf content and composition of the pool of antioxidant phenolics [27,41–44], and much less is known about the influence of heavy metals on the concentration of these compounds. Therefore, the aim of the present study was to verify the tolerance of *P. graveolens* to Cu toxicity and to test the hypothesis that Cu-induced stress could stimulate the synthesis of antioxidant phenolic constituents, thus improving the medicinal properties of this species. With these objectives, we evaluated the effects of Cu exposure in *P. graveolens*, in terms of metal translocation to different plant organs, plant growth, and synthesis of bioactive phenolic metabolites.

2. Results

2.1. Visible Injury and Plant Growth

The plants appeared healthy during the whole growing cycle. Typical toxicity symptoms, such as leaf chlorosis and necrosis, were not observed in Cu-treated plants. Two-way ANOVA revealed that sampling date (D) significantly ($p < 0.05$; $p < 0.001$) affected the number of leaves produced and total upper fresh biomass, while neither Cu nor the interaction of date x Cu (D x Cu) affected the plant height, leaf number and total upper fresh biomass and dry matter content (Table 1). Copper concentration in the nutrient solution affected pelargonium growth parameters (Table 1). Plants grown with ≥25 μM Cu in the nutrient solution produced lower number of leaves at 35 DAT (days after transplanting), but this effect did not persist at 49 DAT. Total upper fresh biomass (including leaves, petioles and stems) decreased at the highest Cu (100 μM Cu) levels compared with the plants grown at 25 μM or 50 μM Cu after 35 DAT. Dry matter content at 49 DAT increased in plants grown in ≥50 μM Cu compared to 25 μM Cu and control treatment.

Table 1. Effect of increasing copper (Cu) concentration (4–25–50–100 μM Cu^{+2}) in the nutrient solution and sampling date after transplanting (DAT, 35 days and 49 days) on plant height (cm), leaf number, total upper fresh biomass (g $plant^{-1}$), and biomass dry matter content (%) in pelargonium plants grown hydroponically in perlite.

DAT	Cu^{2+} (μM)	Plant Height	Leaf Number	Total Upper Fresh Biomass	Total Upper Biomass Dry Matter Content
35 days	**4**	43.16 ± 2.19 [Y]	32.50 ± 4.66 a	57.14 ± 3.78 ab	12.19 ± 0.48
	25	44.33 ± 1.72	24.66 ± 1.17 b	68.39 ± 2.75 a	12.19 ± 0.28
	50	42.33 ± 2.67	25.50 ± 3.52 b	70.02 ± 4.69 a	12.16 ± 0.91
	100	42.83 ± 2.15	22.33 ± 2.84 b	50.72 ± 6.10 b	13.16 ± 0.91
49 days	**4**	53.00 ± 4.57	58.00 ± 6.29	160.04 ± 0.80	12.89 ± 0.07 b
	25	50.00 ± 5.19	41.83 ± 6.87	152.92 ± 21.32	12.68 ± 0.15 b
	50	54.83 ± 9.22	42.33 ± 8.36	156.89 ± 10.77	13.77 ± 0.08 a
	100	56.40 ± 2.22	52.80 ± 4.66	133.55 ± 20.87	13.72 ± 0.19 a
Significance					
Days (D)		*ns*	*	***	*ns*
Copper (Cu)		*ns*	*ns*	*ns*	*ns*
D x Cu		*ns*	*ns*	*ns*	*ns*

[Y] At each sampling date, values ($n = 6$) in columns followed by different letters are significantly different, $p < 0.05$, for each plant growth stage. *ns*, * and *** indicate non-significant or significant differences at $p < 5\%$, and 0.1%, respectively, following two-way ANOVA.

Looking at the fresh and dry biomass of individual plant organs, it was found that leaves and stems were increased at 35 DAT at 25–50 μM Cu compared to 100 μM Cu (Figure 1). Petiole fresh weight (FW) was also increased at 25–50 μM Cu compared with control or 100 μM Cu. Copper levels did not affect the root FW at 35 DAT. Following 49 DAT, leaf, stem, petiole and root FW were at similar levels (averages of 64.24, 26.13, 60.48 and 14.44 g, respectively), independent of the Cu concentration in the nutrient solution. Petiole dry matter content increased in 25–50 μM Cu, compared to the control treatment at 35 DAT. However, root dry matter content increased in 100 μM Cu compared to 25 μM Cu at 49 DAT.

2.2. *Effects on Plant Physiology Attributes*

Plants grown at the high Cu concentration of 100 μM Cu revealed higher stomatal resistance at both 35 and 49 DAT, compared to the control treatment (Table 2). Contrarily, chlorophyll fluorescence as measured by Fv/Fm (representing the maximum quantum yield of PSII), decreased at 100 μM Cu when compared to control and/or 25 μM Cu at 35 DAT and 49 DAT. The content of chlorophylls, as measured by chlorophyll a, chlorophyll b and total chlorophylls, did not change among the treatments at 35 DAT, but decreased at higher Cu levels (i.e., 100 μM Cu) at 49 DAT compared to the control and/or 25 μM Cu.

Two-way ANOVA revealed that sampling date (D) significantly affected stomatal conductivity and chlorophyll fluorescence ($p < 0.001$); copper levels significantly affected stomatal conductance and chlorophyll a ($p < 0.05$), while the interaction of sampling date and copper (D x Cu) did not affect the examined physiological parameters (Table 2).

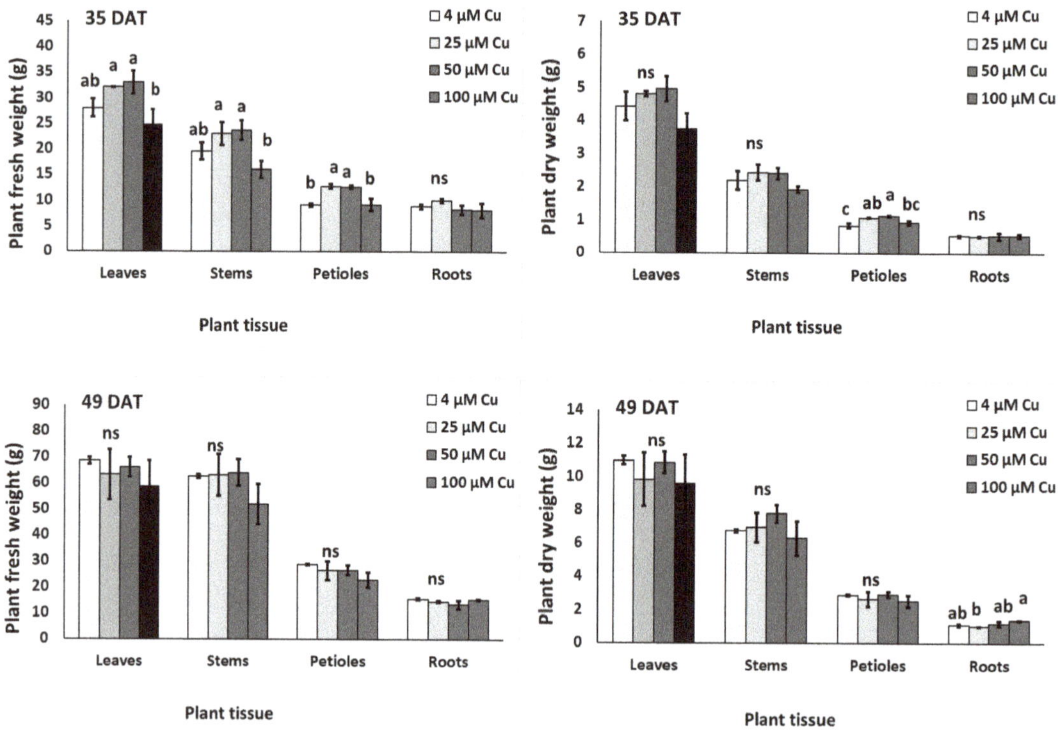

Figure 1. Effect of increasing copper (Cu) concentration (4–25–50–100 µM Cu^{2+}) in the nutrient solution and sampling date after transplanting (DAT, 35 days and 49 days) on the fresh (FW; g plant^{-1}) and dry weight (DW; g plant^{-1}) of leaves, stems, petioles and roots respectively, of pelargonium plants grown hydroponically in perlite. Significant differences ($p < 0.05$) among Cu concentrations for each plant tissue are indicated by different letters; ns indicates non-significant. Error bars show SE ($n = 6$).

2.3. *Effects on Total Phenols, Flavonoids and Antioxidant Activity*

Two-way ANOVA revealed that sampling dates (35 vs. 49 DAT) significantly effected total phenols and DPPH (2,2-diphenyl-1-picrylhydrazyl) ($p < 0.01$), Cu levels significantly effected ABTS (2,2′-azino-bis(3-ethylbenzothiazoline-6-sulphonic acid) ($p < 0.05$) and flavonoids ($p < 0.01$), while the interaction of the sampling date x Cu effected ABTS and flavonoids ($p < 0.05$) and total phenolics ($p < 0.01$). The content of flavonoids and antioxidant activity (as assayed by ferric reducing antioxidant power; FRAP, DPPH, ABTS) revealed their highest values at 50 µM Cu, when compared with control and 100 µM of Cu, and differed significantly also from 25 µM Cu in the case of flavonoids and DPPH at 35 DAT (Figure 2). The content of total phenols and flavonoids, as well as antioxidant activity as assayed by FRAP and ABTS, revealed an increased trend as the Cu level increased at 49 DAT, with significant differences at the high Cu levels compared to the control treatment (Figure 2A–C,E).

Table 2. Effect of increasing copper (Cu) concentration (4–25–50–100 μM Cu^{2+}) in the nutrient solution and sampling date after transplanting-DAT (35 days and 49 days) on leaf stomatal resistance (cm s^{-1}), chlorophyll fluorescence (Fv/Fm), chlorophylls (Chl a, Chl b, Total Chl) content (μg g^{-1} fresh weight) in pelargonium plants grown hydroponically in perlite.

DAT	Cu^{2+} (μM)	Stomatal Resistance	Fv/Fm	Chl a	Chl b	Total Chl
35 days	**4**	0.90 ± 0.13 b [Y]	0.83 ± 0.003 a	23.47 ± 0.35	36.32 ± 0.77	59.77 ± 1.04
	25	1.33 ± 0.08 ab	0.83 ± 0.001 a	23.80 ± 2.21	35.12 ± 2.57	58.90 ± 4.77
	50	1.34 ± 0.15 ab	0.83 ± 0.050 ab	23.97 ± 0.39	35.59 ± 0.71	59.54 ± 0.83
	100	1.61 ± 0.19 a	0.82 ± 0.003 b	23.49 ± 2.57	34.56 ± 2.15	58.03 ± 5.72
49 days	**4**	6.75 ± 0.50 b	0.80 ± 0.006 a	28.14 ± 1.10 a	39.76 ± 0.96 a	66.55 ± 1.39 a
	25	9.11 ± 1.20 ab	0.78 ± 0.005 ab	26.81 ± 0.43 ab	36.62 ± 0.14 ab	66.81 ± 2.60 a
	50	9.05 ± 0.78 ab	0.78 ± 0.007 ab	24.79 ± 0.85 b	38.68 ± 1.52 ab	66.81 ± 2.60 ab
	100	10.58 ± 1.38 a	0.75 ± 0.019 b	21.46 ± 0.91 c	35.22 ± 1.30 b	56.66 ± 2.17 b
Significance						
Days (D)		***	***	*ns*	*ns*	*ns*
Copper (Cu)		*	*ns*	*	*ns*	*ns*
D x Cu		*ns*	*ns*	*ns*	*ns*	*ns*

[Y] At each sampling date, values (n = 6) in columns followed by different letters are significantly different, $p < 0.05$, for each plant growth stage. *ns*, * and *** indicate non-significant or significant differences at $p < 5\%$, and 0.1%, respectively, following two-way ANOVA.

2.4. Plant Stress Indices

Two-way ANOVA revealed that sampling dates (35 vs. 49 DAT), Cu levels and their interactions significantly affected hydrogen peroxide (H_2O_2) and malondialdehyde (MDA) levels ($p < 0.01$, $p < 0.001$). Hydrogen peroxide levels increased at 25 μM Cu in comparison to the 50–100 μM Cu, but did not differ from the control at 35 DAT (Figure 3A). Following 49 DAT, H_2O_2 increased at 100 μM Cu compared to lower Cu levels and/or control. Lipid peroxidation (as assayed by MDA) increased at 50 μM Cu in comparison to higher or lower Cu levels at 35 DAT, while MDA decreased with ≥25 μM Cu compared to the control treatment (at 4 μM Cu) at 49 DAT (Figure 3B).

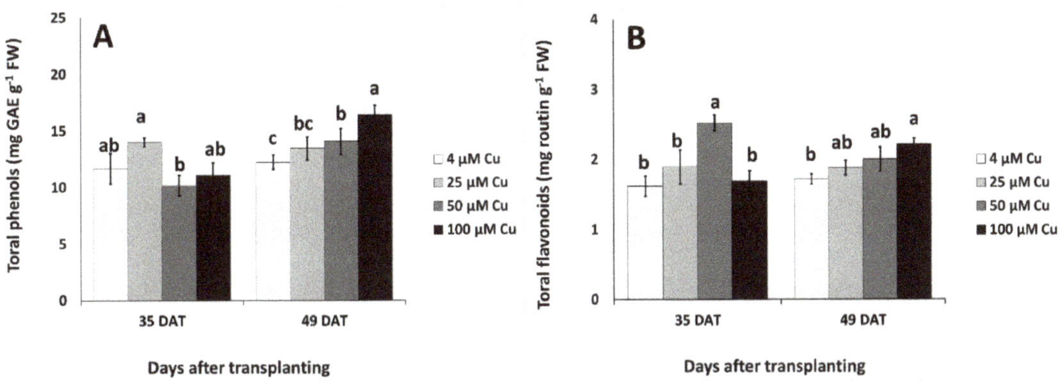

Figure 2. *Cont.*

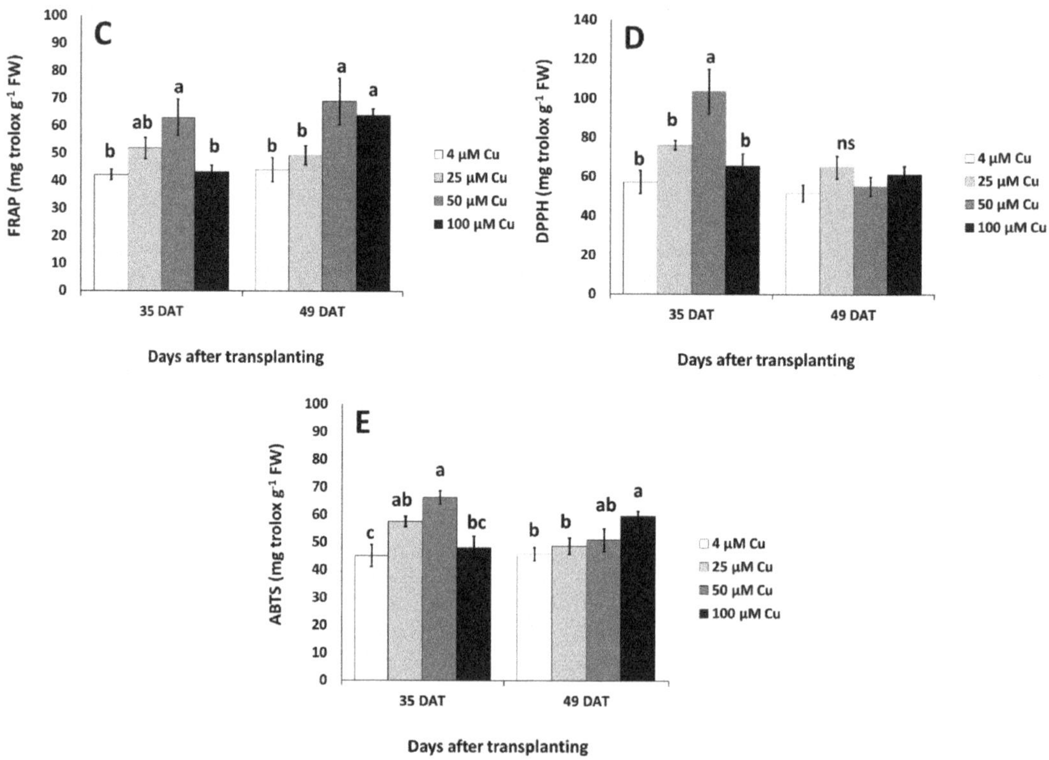

Figure 2. Effect of increasing copper (Cu) concentration (4–25–50–100 µM Cu^{2+}) in the nutrient solution and sampling date after transplanting (DAT, 35 days and 49 days) on the leaf content of total phenols, total flavonoids and antioxidant activity in pelargonium plants grown hydroponically in perlite. (**A**) Total phenols, (**B**) total flavonoids, (**C**) FRAP (**D**) DPPH, and (**E**) ABTS. Significant differences ($p < 0.05$) among Cu concentrations at each sampling date are indicated by different letters; ns indicates non-significant. Error bars show SE ($n = 4$).

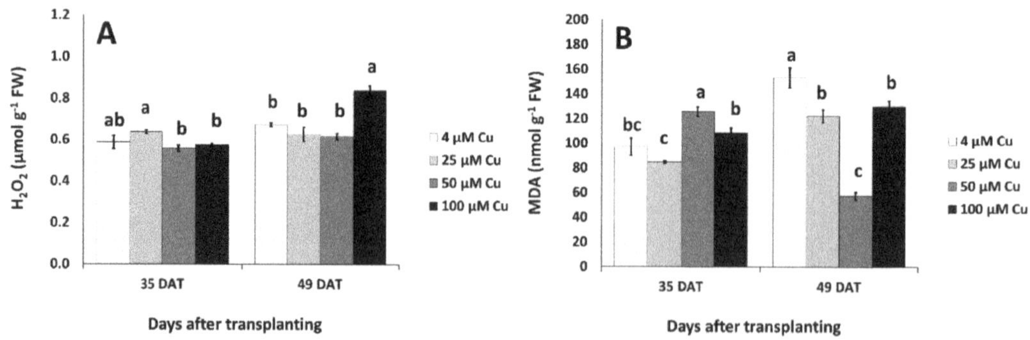

Figure 3. Effect of increasing copper (Cu) concentration (4–25–50–100 µM Cu^{2+}) in the nutrient solution and sampling date after transplanting (DAT, 35 days and 49 days) on the leaf content of hydrogen peroxide (H_2O_2; (**A**)) and malondialdehyde (MDA; (**B**)) in pelargonium plants grown hydroponically in perlite. Significant differences ($p < 0.05$) among Cu concentrations at each sampling date are indicated by different letters. Error bars show SE ($n = 4$).

2.5. Copper Content in Plant Tissues

Two-way ANOVA revealed that sampling date (D) significantly affected AR, BAC-roots, BAC-stems, TF-leaves, TF-stems, TF-petioles ($p < 0.001$), and BAC-petioles ($p < 0.05$); Cu levels significantly affected BAC-roots, BAC-leaves, BAC-stems, BAC-petioles, TF-leaves, TF-stems, and TF-petioles ($p < 0.001$), while the interaction of sampling date and Cu (D x Cu) affected BAC-roots, BAC-stems, BAC-petioles, TF-leaves, TF-stems and TF-petioles ($p < 0.001$) (Table 3). The copper accumulation rate increased at 50 µM of Cu compared with the control at 35 DAT. All bioaccumulation coefficients and translocation factors for leaves, stems, petioles and roots were significantly decreased with ≥25 µM of Cu in the nutrient solution at 35 and 49 DAT (Table 3).

Regarding tolerance index, two-way ANOVA revealed that sampling date (D) significantly affected TI-petiole FW and TI-petiole DW ($p < 0.001$); copper levels significantly affected TI-total biomass, TI-stem FW and TI-petiole FW ($p < 0.05$); while the interaction of sampling date and Cu (D x Cu) affected only the TI-petiole FW ($p < 0.05$) (Table 4). Tolerance index values of plant growth were affected at 35 DAT, as TI increased at 25–50 µM Cu for leaf, stem and petiole FW and as a consequence of the plant total biomass when compared with 100 µM Cu in the nutrient solution (Table 4). Similarly, TI-leaf DW and TI-petiole DW were also increased at 20–50 µM Cu. The TI of leaf number was decreased with ≥25 µM Cu in the nutrient solution at 35 DAT. At 49 DAT, the TI for total biomass was increased with 25 µM Cu in the nutrient solution, while TI-root DW increased at 100 µM Cu when compared to ≤25 µM Cu (Table 4).

2.6. Responses of Other Nutrients

The accumulation of nutrients in different plant organs (leaves, stems, petioles and roots) under different Cu levels at two sampling periods (35 and 49 DAT) is described in Figures 4 and 5. At 35 DAT, the leaf content of N and K increased in 25 µM Cu and decreased or remained unaffected in ≥50 µM Cu compared to the control (Figure 4A,C). Stem N decreased at 100 µM Cu when compared with 25 µM Cu, however the N level in petioles and roots remained similar in plants grown with different Cu levels in the nutrient solution (Figure 4A). Leaf and stem N levels were similar at 49 DAT in all examined Cu levels in the nutrient solution (Figure 4B). Petiole K increased in 25–50 µM Cu compared to the control (Figure 4C). Phosphorus content in leaves, stems and petioles was unaffected by the Cu levels in the nutrient solution, while P in roots decreased at 50 µM Cu and increased at 100 µM Cu compared to the control treatment (Figure 4E). Increased P levels were found at 49 DAT in roots at 50 µM Cu (Figure 4F). Sodium accumulated more in petioles, compared to leaves, stems and roots, while Na content decreased at high Cu levels (Figure 4G,H).

Copper accumulated in stems, petioles and roots as the Cu concentration increased in the nutrient solution; at 35 DAT, greater effects were observed in roots (2.2-fold increase at 100 µM Cu compared to the control treatment) (Figure 5A). A similar trend was found in Cu accumulation even at 49 DAT, but the increment in roots at 50–100 µM Cu was 6.9-fold greater compared with the control treatment (Figure 5B). Zinc accumulated more in leaves, stems, and petioles at 35 DAT as Cu levels increased in the nutrient solution, whereas Zn content decreased in roots with increasing Cu concentration in the nutrient solution (Figure 5C). However, the reverse was evidenced at 49 DAT, as Zn accumulated in roots following increases of Cu levels in the nutrient solution (Figure 5D).

Table 3. Accumulation rate (AR, mg kg^{-1} DW day^{-1}), bioaccumulation coefficient (BAC), and translocation factor (TF) for Cu after 35 and 49 DAT in pelargonium plants grown hydroponically in perlite.

DAT	Cu^{2+} (μM)	Accumulation Rate-AR (mg kg^{-1} DW day^{-1})	Bioaccumulation Coefficient (BAC)				Translocation Factor (TF)		
			Leaves	Stems	Petioles	Roots	Leaves	Stems	Petioles
35 days	4	94.94 ± 18.78 b [Y]	140.66 ± 13.50 a	124.48 ± 4.71 a	104.66 ± 1.17 a	575.99 ± 90.69 a	0.25 ± 0.03 a	0.22 ± 0.03 a	0.19 ± 0.03 a
	25	145.97 ± 9.78 ab	24.03 ± 0.66 b	21.66 ± 0.64 b	19.34 ± 0.45 b	193.77 ± 15.76 b	0.12 ± 0.01 b	0.11 ± 0.01 b	0.10 ± 0.01 b
	50	165.14 ± 30.56 a	12.40 ± 0.14 b	11.49 ± 0.41 c	11.89 ± 0.24 c	84.51 ± 14.58 b	0.15 ± 0.03 b	0.14 ± 0.02 b	0.11 ± 0.03 b
	100	124.05 ± 12.80 ab	6.27 ± 0.14 b	5.89 ± 0.11 c	6.67 ± 0.13 d	73.67 ± 6.63 b	0.08 ± 0.01 b	0.08 ± 0.00 b	0.09 ± 0.01 b
49 days	4	549.13 ± 23.12	133.81 ± 6.98 a	104.82 ± 2.14 a	124.29 ± 6.70 a	212.23 ± 3.35 a	0.63 ± 0.02 a	0.49 ± 0.00 a	0.58 ± 0.03 a
	25	676.26 ± 170.78	23.60 ± 0.71 b	18.66 ± 2.62 b	19.89 ± 0.66 b	137.88 ± 4.70 b	0.17 ± 0.01 b	0.13 ± 0.02 b	0.14 ± 0.00 b
	50	965.19 ± 42.52	11.31 ± 0.72 c	9.96 ± 0.77 c	8.58 ± 0.62 c	124.42 ± 21.48 b	0.09 ± 0.02 c	0.08 ± 0.01 c	0.07 ± 0.01 c
	100	955.28 ± 260.27	6.01 ± 0.35 c	6.49 ± 0.45 c	4.37 ± 0.06 c	67.25 ± 0.92 c	0.09 ± 0.00 c	0.09 ± 0.01 c	0.06 ± 0.00 c
Significance									
Days (D)		***	*ns*	***	*	***	***	***	***
Copper (Cu)		*ns*	***	***	***	***	***	***	***
D x Cu		*ns*	*ns*	***	***	***	***	***	***

[Y] At each sampling date, values (n = 6) in columns followed by different letters are significantly different, $p < 0.05$, for each plant growth stage. *ns*, *, and *** indicate non-significant or significant differences at $p < 5\%$, and 0.1%, respectively, following two-way ANOVA.

Table 4. Tolerance indices (TI (%)) for Cu after 35 and 49 DAT in pelargonium plants grown hydroponically in perlite.

DAT	Cu^{2+} (μM)	Tolerance Indices-TI (%)										
		Total Biomass	Plant height	Leaf No	Leaf FW	Stem FW	Petiole FW	Root FW	Leaf DW	Stem DW	Petiole DW	Root DW
35 days	4	100.00 ± 0.00 ab [Y]	100.00 ± 0.00	100.00 ± 0.00 a	100.00 ± 0.00 ab	100.00 ± 0.00 ab	100.00 ± 0.00 b	100.00 ± 0.00	100.00 ± 0.00 ab	100.00 ± 0.00	100.00 ± 0.00 b	100.00 ± 0.00
	25	111.30 ± 1.93 a	102.70 ± 3.98	75.89 ± 3.61 b	114.85 ± 0.12 a	117.05 ± 11.33 a	139.71 ± 4.48 a	113.49 ± 4.23	108.33 ± 1.82 a	110.93 ± 10.65	127.97 ± 1.03 a	98.97 ± 6.01
	50	113.84 ± 7.53 a	98.07 ± 6.20	78.46 ± 10.85 b	117.98 ± 8.13 a	120.51 ± 9.77 a	138.56 ± 6.94 a	93.54 ± 9.51	111.93 ± 8.57 a	109.79 ± 7.88	134.52 ± 2.06 a	117.89 ± 10.14
	100	88.31 ± 8.01 b	99.22 ± 4.98	68.71 ± 8.76 b	88.34 ± 10.83 b	81.90 ± 8.49 b	100.82 ± 23.47 b	91.53 ± 15.81	84.16 ± 10.72 b	87.92 ± 3.94	111.31 ± 7.21 b	93.33 ± 12.07
49 days	4	100.00 ± 0.00 b	100.00 ± 0.00	100.00 ± 0.00	100.00 ± 0.00	100.00 ± 0.00	100.00 ± 0.00	100.00 ± 0.00	100.00 ± 0.00	100.00 ± 0.00	100.00 ± 0.00	100.00 ± 0.00 b
	25	219.37 ± 77.36 a	94.33 ± 9.80	72.12 ± 11.84	92.36 ± 14.09	100.74 ± 12.76	91.83 ± 12.71	94.11 ± 3.02	89.62 ± 14.77	103.03 ± 13.16	91.55 ± 14.43	92.56 ± 1.19 b
	50	104.63 ± 6.51 b	103.45 ± 17.39	72.98 ± 14.42	96.47 ± 5.56	102.19 ± 8.24	92.66 ± 6.20	86.56 ± 10.47	98.95 ± 5.91	115.32 ± 7.77	101.20 ± 5.87	106.55 ± 13.72 ab
	100	89.36 ± 15.10 b	106.41 ± 4.20	91.03 ± 8.03	85.44 ± 14.69	83.03 ± 12.44	79.57 ± 10.38	98.23 ± 1.84	87.38 ± 15.64	93.41 ± 15.21	87.41 ± 12.84	124.10 ± 1.03 a
Significance												
Days (D)		*ns*	*ns*	*ns*	*ns*	*ns*	***	*ns*	*ns*	*ns*	***	*ns*
Copper (Cu)		*	*ns*	*ns*	*ns*	*	*	*ns*	*ns*	*ns*	*ns*	*ns*
D x Cu		*ns*	*ns*	*ns*	*ns*	*ns*	*	*ns*	*ns*	*ns*	*ns*	*ns*

[Y] At each sampling date, values (n = 6) in columns followed by different letters are significantly different, $p < 0.05$, for each plant growth stage. *ns*, * and *** indicate non-significant or significant differences at $p < 5\%$, and 0.1%, respectively, following two-way ANOVA.

2.7. Regression Analysis

Pearson's correlation coefficients were determined between individual pairs of parameters associated with Cu uptake (leaf and root Cu concentrations), leaf antioxidant systems (content of total phenols and flavonoids, antioxidant capacity according to FRAP, DPPH and ABTS assays), and oxidative stress (H_2O_2 and MDA) (Table 5).

Leaf Cu concentration was positively correlated to the root content of the element at both sampling dates. The correlation coefficients between leaf or root Cu and the biochemical parameters were generally higher at 49 than 35 DAT, and in older plants all correlations were positive except those involving MDA content.

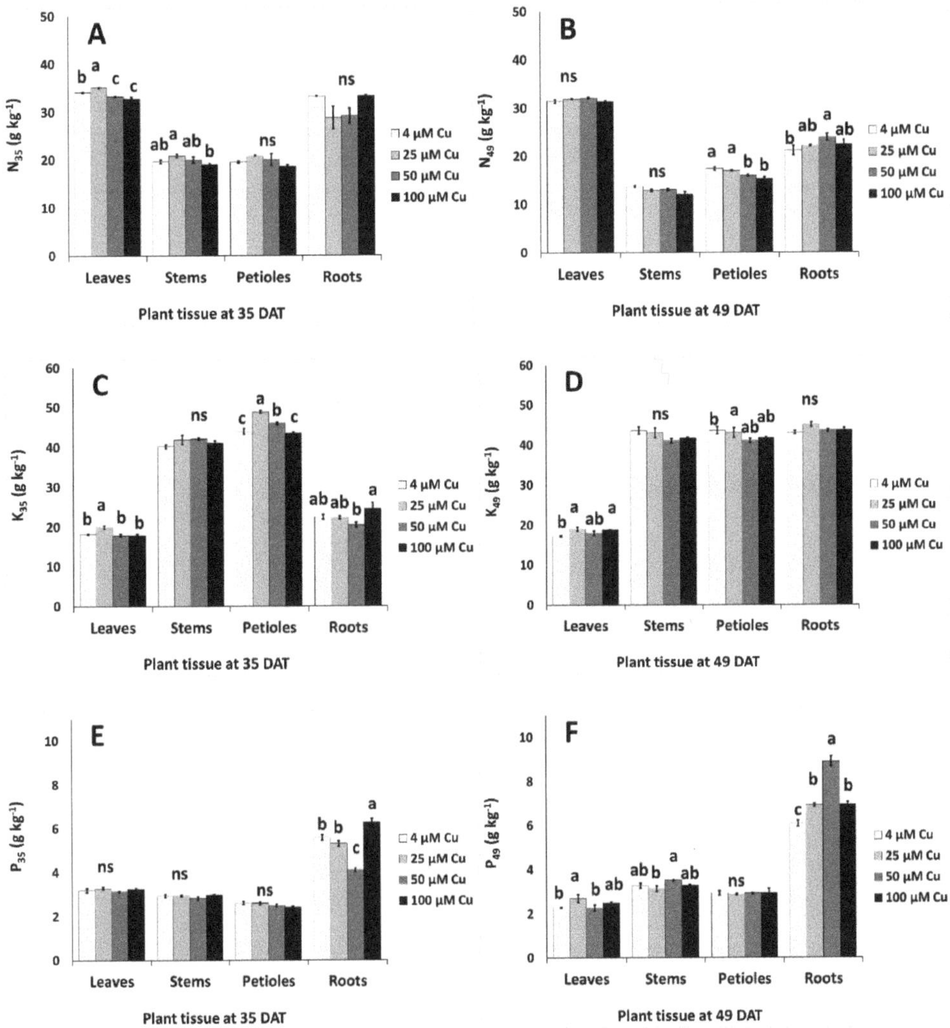

Figure 4. *Cont.*

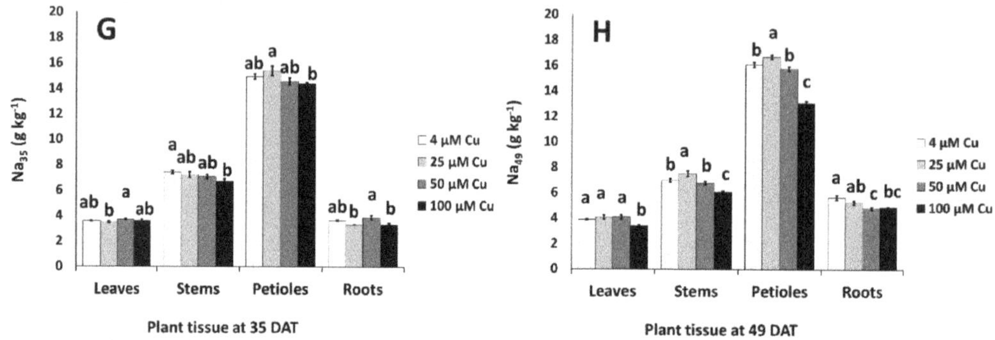

Figure 4. Effect of increasing copper (Cu) concentration (4–25–50–100 µM Cu^{2+}) in the nutriment solution and sampling date after transplanting (DAT, 35 days and 49 days) on the content of macronutrients and sodium in different organs of pelargonium plants grown hydroponically in perlite. (**A**,**B**) Nitrogen–N, (**C**,**D**) potassium–K, (**E**,**F**) phosphorus–P, (**G**,**H**) sodium–Na. Significant differences ($p < 0.05$) among Cu concentrations at each sampling date are indicated by different letters; ns indicates non-significant. Error bars show SE ($n = 4$).

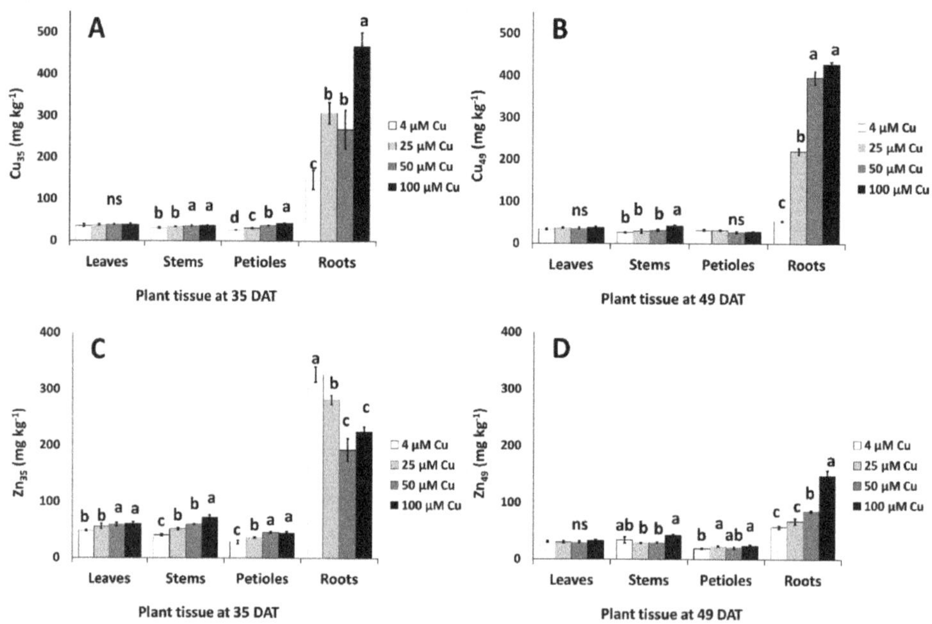

Figure 5. Effect of increasing copper (Cu) concentration (4–25–50–100 µM Cu^{2+}) in the nutriment solution and sampling date after transplanting (DAT, 35 days and 49 days) on the content of micronutrients in different organs of pelargonium plants grown hydroponically in perlite. (**A**,**B**) Copper–Cu, and (**C**,**D**) zinc–Zn. Significant differences ($p < 0.05$) among Cu concentrations at each sampling date are indicated by different letters; ns indicates non-significant. Error bars show SE ($n = 4$).

Table 5. Pearson's correlation table for leaf and root content of copper (Cu), leaf content of total phenols, flavonoids, H_2O_2 and malondialdehyde (MDA), and leaf antioxidant capacity determined using FRAP, DPPH or ABTS assays, in pelargonium plants grown hydroponically in perlite and exposed to four different Cu concentrations (4–25–50–100 µM Cu^{2+}) in the nutrient solution, sampled at 35 and 49 days after transplanting (DAT).

	Leaf Cu	*Root Cu*	*Phenols*	*Flavonoids*	*FRAP*	*DPPH*	*ABTS*	*H_2O_2*	*MDA*
35 DAT									
Leaf Cu	1								
Root Cu	0.8406	1							
Phenols	−0.3210	−0.0500	1						
Flavonoids	0.4770	−0.0714	−0.4142	1					
FRAP	0.4453	−0.0852	−0.2426	0.9835	1				
DPPH	0.5612	0.0339	−0.3562	0.9920	0.9866	1			
ABTS	0.5018	0.0037	−0.1406	0.9551	0.9901	0.9750	1		
H_2O_2	−0.3121	−0.0544	0.9996	−0.3893	−0.2161	−0.3310	−0.1139	1	
MDA	0.5334	0.1429	−0.9396	0.6778	0.5357	0.6432	0.4571	−0.9308	1
49 DAT									
Leaf Cu	1								
Root Cu	0.7078	1							
Phenols	0.8042	0.8743	1						
Flavonoids	0.7933	0.9364	0.9890	1					
FRAP	0.4427	0.9454	0.7185	0.8086	1				
DPPH	0.8973	0.3775	0.4626	0.4440	0.0855	1			
ABTS	0.7591	0.8380	0.9954	0.9751	0.6881	0.4034	1		
H_2O_2	0.4695	0.3948	0.7745	0.6815	0.2355	0.1705	0.8267	1	
MDA	−0.1303	−0.6349	−0.1807	−0.3241	−0.7818	0.0074	−0.1173	0.4215	1

3. Discussion

Plant growth and development are regulated by plant physiology, which in turn is tightly linked to both environmental conditions and, in hydroponic cultivation, to the composition of the nutrient solution that is supplied to the plants [45,46]. For evaluation of the effects of Cu toxicity, the choice of appropriate Cu levels depends on the species being tested, the duration of the Cu treatment, and other growing parameters such as the pH of the nutrient solution. In this study, hydroponically grown plants of *P. graveolens* were exposed to Cu concentrations of up to 100 µM. Similar concentrations were tested in several species including Moso bamboo [5], maize [15], or *Carthamus tinctorius* L. [1]. Although higher Cu levels have been reported for turfgrass (120 µM) [16], and particularly for tomato (250 or 350 µM) [47,48], lower concentrations (up to 40 µM) were employed for both tree species [49] or vegetable crops [46].

In our experiments, all growth parameters were, of course, significantly higher at 49 than 35 DAT, with the only exceptions being plant height and total upper dry matter percentage, which were not affected by plant age. At 35 DAT, despite the lower number of leaves in Cu-treated plants than in the control, Cu concentrations up to 50 µM in the nutrient solution increased the fresh biomass production of the aerial part (Table 1); however, a relevant increase in dry matter was observed only for stem tissues, suggesting that the overall effect was partially due to increased water absorption. In contrast, at 49 DAT the fresh weight of the distinct aboveground plant organs did not change across treatments (Figure 1), while the increase in dry mass percentage above 50 µM Cu (Table 1) indicated a lower water content in those tissues. An increase in the percentage of dry matter was observed also in the leaf tissues of tomato plants grown in hydroponics, after 15 days exposure to 100–350 µM Cu concentrations [48]. The detrimental effect of high Cu levels in the root zone on biomass production was observed in several food crops [4], and in MAPs such as *Carthamus tinctorius* [1]. Similarly, in this work both fresh (Table 1) and dry (Figure 1) biomass production were tendentially lower in the 100 µM Cu treatment than the control at both sampling dates, although the difference was never significant. These results showed that *P. graveolens* is a species tolerant to Cu toxicity of up to 100 µM concentration, consistent with the tolerance indices toward metal stress reported in Table 4. For each plant organ, the values of the latter parameters were generally similar to those of the corresponding control. Toxic Cu concentrations well below 100 μM have been reported in the literature for several species. For example, Reichman et al. [50] reported that the highest Cu concentration in the nutrient solution without negative effects on plant growth was 35 μM for Cu-tolerant populations of *Silene cucubalus*; the no-effect threshold was about 5 μM for Cu-sensitive cultivars of mung bean, sweet potato and wheat, and was below 1 μM in Australian tree species such as ironbark, *Acacia holosericea*, and *Melaleuca leucadendra*.

Plant age significantly influenced both stomatal resistance, which was much higher at 49 than 35 DAT for all the Cu treatments, and photosynthetic efficiency, expressed as chlorophyll fluorescence Fv/Fm, which was lower in older plants. At both sampling dates, increasing Cu concentrations in the nutrient solution interfered with the process of photosynthesis by increasing stomatal resistance and decreasing leaf chlorophyll fluorescence (Table 2). Along with the determination of stomatal resistance, the assessment of chlorophyll fluorescence is a key parameter in the rapid detection of response to physiological stress in higher plants; specifically, the Fv/Fm ratio is a physiological marker of photoinhibition of photosystem II (PSII) induced by stress conditions [51]. It has been reported that excess Cu can impair photosynthetic electron transport particularly at the PSII level, and Cu toxicity has been associated with quenching of variable fluorescence Fv [52]. The values of Fv/Fm reported in Table 2 remained within the typical range for healthy plants, that is 0.75–0.85 [53], and suggested that, despite a significant decline of the indicator at the 100 μM Cu concentration at both sampling dates, the function of the PSII reaction centers was preserved with all Cu treatments. Although plant age did not affect the content of chlorophylls, a significant decrease with increasing Cu concentration was observed at 49 DAT for these pigments. The above data could be reasonably interpreted as early indicators of Cu toxicity that became more severe with the duration of exposure; despite the effects of a possible photosynthetic imbalance this did not translate into a significant biomass decrease, or in the typical visible symptoms of toxicity such as leaf chlorosis [54]. Under impaired photosynthesis, plant metabolism is affected and one possible biochemical process that can be activated is the Mehler reaction, with formation of oxygenated molecules such as H_2O_2 [55]. This process is consistent with the significant increase in H_2O_2 concentration that was observed at 49 DAT in plants treated with 100 μM Cu (Figure 3A).

It is generally acknowledged that excess Cu causes oxidative stress in plants [1,2,13,17]. However, due to the time course of the antioxidant response, the levels of stress indicators in plant tissues may undergo fluctuations. This could account for the significant effect of sampling date, Cu concentration, and their interaction on the observed contents of both H_2O_2 and MDA, and could also explain the contrasting results reported in the literature concerning the levels of ROS or MDA in several plant species [4]. The antioxidant activity of pelargonium at 35 DAT showed the same behavior across the Cu treatments (Figure 2C–E), regardless of the assay used for the determination (FRAP, DPPH or ABTS) and the stimulation of the plant antioxidant response was strictly related to the occurrence of lipid peroxidation, since a similar pattern was observed also for the concentration of MDA (Figure 3B). The content of total flavonoids followed the same trend (Figure 2B), suggesting that this class of compounds could play a key role in the pool of antioxidant molecules of *P. graveolens* that are involved in plant response to excess Cu in the early stages of exposure. Interestingly, the highest values of antioxidant power and flavonoid concentration were obtained with the 50 μM Cu treatment, which also resulted in the highest rate of Cu accumulation in plant tissues (Table 3). On the other hand, a different behavior was observed for total phenols (Figure 2A), whose amount, unlike the content of flavonoids, was affected also by the sampling date. These dissimilarities indicated that, along with flavonoids, other classes of phenolic compounds could contribute significantly to the pool of phenolics of this species.

At 49 DAT, all parameters except DPPH scavenging capacity increased with Cu concentration (Figure 2), suggesting the occurrence of an effective dose-dependent response of the antioxidant system to excess Cu and a central role of phenolic compounds in the plant tolerance to Cu toxicity. These findings showed the effectiveness of high Cu concentrations in the nutrient solution in stimulating the synthesis and accumulation of beneficial antioxidant molecules in the plant tissues of *P. graveolens*. In addition to phenolics, other non-enzymatic antioxidant compounds could have an impact on plant response to Cu toxicity; for example, an increased content of proline has been observed in different species exposed to excess Cu [18,19]. In our experiments, the level of MDA decreased with Cu concentration in the nutrient solution up to 50 μM, showing that Cu-exposed *P. graveolens* could well counteract lipid peroxidation. Likewise, the leaf concentration of H_2O_2 was effectively controlled in up to 50 μM Cu. In contrast, the increase in the content of H_2O_2 at 100 μM Cu may indicate a less efficient plant response at this high concentration of the element (Figure 3A,B).

The relationships among the indicators linked to Cu uptake and antioxidant response to Cu treatments are confirmed in the Pearson's correlation table (Table 5). Leaf Cu content followed root content at both sampling dates, and, in younger plants, the content of flavonoids rather than the level of total phenols was strongly correlated with the antioxidant capacity as obtained using either the FRAP, DPPH or ABTS assays. However, at 35 DAT, Cu exposure did not elicit a marked response, as extremely weak relationships were evidenced between Cu levels in the tissues and all other biochemical parameters. The correlation coefficients between the concentrations of Cu and those of H_2O_2 and MDA also remained low at 49 DAT, suggesting that even older plants could efficiently prevent oxidative stress. On the other hand, higher values of the correlation coefficients were generally evidenced between root or leaf Cu levels and the other biochemical parameters (phenols or flavonoids content, or antioxidant capacity). Therefore, the results of the regression analysis are consistent with an initial antioxidant response to Cu toxicity at 49 DAT.

The TF and BAC indexes are important parameters for the evaluation of plant phytoremediation potential. In hyperaccumulator species, both parameters are greater than 1; in contrast, in our study only the BAC factor was higher than 1, indicating that *P. graveolens* acted as a Cu excluder. According to Saleem et al. [13]. Cu excluders, which have a low potential for metal extraction, could be effectively employed for phytostabilization. The BAC and TF indexes showed significant decreases with increasing Cu concentration in the nutrient solution at both sampling dates (Table 3). At low concentrations (up to 25 μM), young plants accumulated more Cu in root tissues and showed a much larger variation of the root BAC index among the Cu treatments than those sampled at 49 DAT. However, the decrease of root BAC values at both sampling dates indicated a strong inhibition of Cu uptake as the concentration of the element in the nutrient solution increased. A much larger variation was observed for the BAC index of the aerial parts, which decreased more than 10-fold in all aboveground tissues during the whole growing cycle, suggesting a synergistic effect of reduced Cu uptake and reduced element translocation in Cu-treated plants. This outcome was confirmed by lower TF values in the Cu treatments as compared to the control. Although in the latter treatment the TF in the aboveground tissues was higher in older plants, Cu translocation was markedly limited with increasing Cu concentration at both sampling dates. Therefore, the observed tolerance of *P. graveolens* to Cu toxicity was both due to the plant's ability to exclude Cu from the leaf tissues by limiting translocation to the aerial parts, and to an efficient antioxidant system. A similar behavior was observed in *Solanum cheesmaniae* subjected to Cu stress [56]. This effect was further evidenced by the results shown in Figure 5A,B, as Cu accumulated particularly in root tissues independent of the concentration of the nutrient solution. In contrast, the stems were only slightly affected by the 100 μM Cu treatment, and the leaf tissues were totally unaffected.

According to Lange et al. [57], most Cu-tolerant species act as Cu excluders, with very low Cu translocation from root to shoots. Contrarily, Chen et al. [5] reported on 25 plant species identified as Cu hyperaccumulators and provided literature data concerning tolerant and accumulator species, with leaf Cu content ranging from 45 to 596 mg/kg DW and root content ranging from 33 to 3768 mg kg^{-1} DW. In plants, Cu uptake is generally dependent on the species, plant organ, concentration in the growing medium, and the time of exposure. For example, according to Adrees et al. [4], maize plants exposed for six days to 100 µM Cu in hydroponics accumulated 1070 and 56 mg kg^{-1} DW in roots and shoots, respectively; the same species was reported to accumulate 7790 mg kg^{-1} DW in the roots after 15 days treatment with 80 µM Cu. Chen et al. [5] reported that in hydroponically grown *Moso bamboo* with 100 µM Cu in the nutrient solution, the Cu content in leaf and root tissues were, respectively, 24 and 417 mg kg^{-1} DW after 15 days, and 91 and 809 mg kg^{-1} DW after 30 days exposure. Saleem et al. [13] reported that pot-grown flax accumulated Cu mainly in the root tissues after 35 days cultivation, while the contaminant was accumulated mainly in the shoots in mature plants (105–140 days). In our experiments, despite a dose-dependent Cu accumulation in the roots of up to 468.14 mg kg^{-1} DW, Cu content in the leaf tissues remained at the same level as the control both at 35 and 49 DAT, further characterizing *P. graveolens* as a Cu excluder species.

In a very recent paper, Tschinkel et al. [58] reported that the permitted concentration of impurities for drug substances and excipients set by the United States Pharmacopoeia Convention (USP) is 300 mg kg^{-1}. Additionally, according to a recent review from the European Food Safety Authority [59], the maximum residue level (MRL) for Cu compounds (Cu) in leaves and herbs for herbal infusions is 100 mg kg^{-1}. These limits are much higher than the Cu concentrations that were found in the leaf tissues of *P. graveolens*, which were below 50 mg kg^{-1} DW (Figure 5A,B). Therefore, the leaves of Cu-treated *P. graveolens* plants had higher contents of antioxidants and, at the same time, the same Cu content of the control plants, and may be safely used in the pharmaceutical/herbal industry for the extraction of phenolic compounds and other beneficial constituents such as essential oils. Although the latter were not examined in this work, some authors showed that heavy metals had minimal impact on the quality of *P. graveolens* essential oil, even when the contaminant was partly translocated to the aboveground organs [38,39]. In general, despite a high translocation factor from the root system to the aerial parts being indispensable for species with edible roots, the opposite is preferable for plant species that are used for leaf tissues, like the one examined in this study. Considering the excess of Cu application in agriculture and the consequent contamination of soils and water bodies, selection of MAPs according to their tolerance and potential accumulation in the organs of interest becomes a crucial issue in managing the problem of Cu pollution and preserving the quality of plant materials. It is noteworthy to mention that expanded and unexpanded perlite have some properties that can favor the adsorption of metal ions, including Cu [60]. However, the strong root Cu uptake shown in Figure 5A,B showed that this microelement was available to the plants in all Cu treatments.

Copper had a strong influence on Zn uptake (Figure 5C,D). In the root tissues of younger plants, Zn content was inversely related to that of Cu; this indicated a competitive absorption mechanism for the two micronutrients, in agreement with what reported by Kabata-Pendias and Szteke [7]. The amount of Zn in the aerial parts at 35 DAT increased significantly across the Cu treatments, suggesting that higher Cu levels promoted Zn translocation to the stems. Conversely, Cu and Zn uptake followed the same trend in older plants, despite a much lower Zn accumulation at 49 than 35 DAT. A decrease of Zn uptake in plants exposed to excess Cu has been observed in several species [4]. The content of Na decreased with increasing Cu in all plant organs, especially at 49 DAT, in agreement with what reported by Chrysargyris et al. [49] for the roots of *Mentha spicata*. The opposite trend was observed by other authors in the leaves of *Vicia faba* [61] and the shoots of pistachio seedlings [62]. The addition of Cu to the nutrient solution generally did not have a significant influence on the uptake of the macronutrients N, P and K, which was confirmed the scarce effects observed on the biomass production (Table 1 and Figure 1). The only exception was root P content at 49 DAT, which was higher in Cu-treated plants than in controls (Figure 4E,F), in contrast with the results reported by Chrysargyris et al. [49] and Eskandari and Mozaffari [62]. According to Adrees et al. [4], although Cu supply generally affects mineral nutrition, the effect of this microelement on the uptake of other mineral nutrients is strongly dose-, time- and species-dependent. In addition, in polluted environments, Cu could interact with other heavy metal contaminants [9,26,51,63]. In general, we observed that the Cu treatments did not impair mineral nutrition and, overall, *P. graveolens* showed a high capacity to grow in Cu-enriched mediums of up to 100 μM.

Further work is necessary to provide a deeper insight into the response of *P. graveolens* to Cu stress. For example, the effects of severe Cu exposure conditions could be investigated through an extension of the growing period beyond 49 DAT, or with Cu concentrations higher than 100 μM in the nutrient solution; additionally, a similar experiment could be carried out in open field, where Cu bioavailability is conditioned by soil properties; finally, the profiling of individual metabolites of interest under Cu stress could help in highlighting the effects of this element on the bioactive properties of *P. graveolens*.

4. Materials and Methods

4.1. Plant Material and Cultivation Conditions

Pelargonium graveolens L'Hér. plants were selected for the present study, which was implemented at the experimental greenhouse of Cyprus University of Technology, in Limassol, Cyprus. Cuttings of 10 cm length were collected from mother plants (National Agricultural Department, Nicosia, Cyprus) and were grown in peat:perlite (4:1 v/v) substrate, in plastic seedling trays for 25 days, till roots formation. Plants at the stage of four-to-five leaves were transplanted in pots (one plant per pot; 1.5 L capacity) filled with expanded perlite and placed on plastic trays to achieve proper drainage (see Chrysargyris et al. [64]). Perlite properties have been described previously [65]. Plants were grown in an open (free drainage) hydroponic system and the drainage nutrient solution was available to plants through capillary suction. Plants were sampled at two different growth stages.

Plants were initially grown with the application of a full-strength nutrient solution (electrical conductivity (EC) and pH of 2.1 mS cm^{-1} and 5.7, respectively) for 21 days. Nutrient solution composition was: NO_3^--N = 15.00, K = 9.50, PO_4^{-3}-P = 1.80, Ca = 4.20, Mg = 1.63, SO_4^{-2}-S = 1.55 and Na = 1.85 mmol L^{-1}, respectively; and B = 30.00, Fe = 35.05, Mn = 6.10, Cu = 4.00, Zn = 4.10, and Mo = 0.52 μmol L^{-1}, respectively. The described above concentrations were obtained using mineral salts and chelate for iron with ethylenediamine-N-N′bis(2-hydroxy-4-methylphenylacetic) acid (6.5% Fe EDDHMA). After that period, plants were subjected to different Cu levels (treatments) in the nutrient solution, namely (i) 4 μM Cu (control); (ii) 25 μM Cu; (iii) 50 μM Cu; and (iv) 100 μM Cu (in the form of $CuSO_4$). Plants were grown under Cu excess for additional 28 days (in total 49 days after transplanting, DAT). A total of 96 plants were used (4 Cu levels × 2 sampling periods × 12 replicates).

4.2. Plant Growth and Physiological Measurements

Plant growth and physiological parameters were measured at two sampling periods (35 DAT and 49 DAT) with six replicates per treatment and growth period. Plant height and leaf number were recorded. After harvest, upper fresh and dry biomass parts (leaves, petioles, leaf stem) and roots were measured. Different parts of the plants were separated to evaluate the uptake and translocation of Cu from the roots to upper plant parts and the relevant effects on nutrient accumulation. Individual samples were collected and put at 85 °C in a forced-air oven until constant weight was achieved to determine their dry weight.

Leaf stomatal conductance was measured with a ΔT-Porometer AP4 (Delta-T Devices Cambridge, Burwell, Cambridge, UK) [66]. Leaf chlorophyll fluorescence (chlorophyll fluorometer, opti-sciences OS-30p, Hertfordshire, UK) was measured on two fully developed, light-exposed leaves per plant. Following leaf incubation in the dark for 20 min, the Fv/Fm ratio was measured [66]. Leaf chlorophyll was extracted with dimethyl sulfoxide (DMSO) and chlorophyll a (Chl a), chlorophyll b (Chl b) and total chlorophylls (total Chl) were assayed and expressed as $\mu g\ g^{-1}$ FW [66].

4.3. Antioxidant Activity, Total Phenols and Total Flavonoids Content

The antioxidant activity of the methanolic leaf plant extracts was determined with four replicates per treatment and sampling date by the assays of 2,2-diphenyl-1-picrylhydrazyl (DPPH) and ferric reducing antioxidant power (FRAP), as previously described by Chrysargyris et al. [67], as well as the 2,2′-azino-bis(3-ethylbenzothiazoline-6-sulphonic acid) (ABTS) assay according to the methodology described by Woidjylo et al. [68]. The Folin–Ciocalteu method was used for determining the total phenols content, as previously described [69] and results were expressed as gallic acid equivalents (mg GAE per g FW). The total flavonoid content was determined according to aluminum chloride colorimetric method [70] and results were expressed as rutin equivalents (mg rutin per g FW).

4.4. Plant Stress Indicators

Cell damage index of lipid peroxidation in leaves was assessed in terms of malondialdehyde (MDA) content, which was determined by the thiobarbituric acid reaction [71]. Hydrogen peroxide (H_2O_2) content was measured according to the method of Loreto and Velikova [47]. The results were expressed as nmol MDA or µmol H_2O_2 per g FW. Four replicates were analyzed for each treatment and sampling date.

4.5. Nutrient Content

Dried tissue (0.5 g) from leaves, stems, leaf petioles and roots from each treatment (4 biological replications; each replication was a pool of 2 individual plants) at both sampling dates, was subjected to dry ashing at 450 °C and acid extraction (2N HCl). The extracts were used for the determination of sodium (Na) and potassium (K) by flame photometry (Lasany Model 1832, Lasany International, Panchkula, India), phosphorus (P) with the molybdate/vanadate method (yellow method) by spectrophotometry (Multiskan GO, Thermo Fischer Scientific, Waltham, MA, USA), zinc (Zn) and copper (Cu) by atomic absorption spectrometry (PG Instruments AA500FG, Leicestershire, UK). Nitrogen (N) was determined using the Kjeldahl method (BUCHI, Digest automat K-439 and Distillation Kjelflex K-360, Flawil, Switzerland) following Chrysargyris et al. [64]. In particular, the measured Cu content in this study refers to total dissolved Cu content, which was almost totally (≥98.21%) available as Cu^{2+} [49]. Plant nutrient content was expressed in $g\ kg^{-1}$ and $mg\ kg^{-1}$ DW, for macronutrients and micronutrients, respectively.

The Cu accumulation rate (AR), bioaccumulation coefficient (BAC), translocation factor (TF) and tolerance index (TI) of pelargonium were calculated by equations described by Benimeli et al. [72], Amin et al. [2] and Azooz et al. [73], as follows.

The accumulation rate (AR) was calculated as the sum up of Cu concentration in each plant tissue x plant DW divided by the number of days under Cu levels by the total plant DW [72].

$$\text{Accumulation rate mg per (kg DW x day)} = \frac{([\text{Cu}]\ \text{leave x DW leave} + ([\text{Cu}]\ \text{stem x DW stem} + ([\text{Cu}]\ \text{petiole x DW petiole} + ([\text{Cu}]\ \text{root x DW root})}{\text{Days x (DW leave} + \text{DW stem} + \text{DW petiole} + \text{DW root)}} \quad (1)$$

The bioaccumulation coefficient (BAC) was calculated as the ratio of Cu concentration in plant tissue to that of Cu concentration in nutrient solution, according to Amin et al. [2]:

$$\text{Bioaccumulation coefficient} = \frac{\text{Cu concentration in plant tissue (mg kg DW)}}{\text{Cu concentration in nutrient solution (mg per L)}} \quad (2)$$

The translocation factor (TF) was calculated as the ratio of Cu concentration in plant tissue to that of Cu concentration in plant roots according to Amin et al. [2]:

$$\text{Translocation factor} = \frac{\text{Cu concentration in plant tissue (mg kg DW)}}{\text{Cu concentration in plant root (mg per kg DW)}} \quad (3)$$

Copper tolerance index (TI) was calculated as the quotient of the dry weight of plants grown under copper treated and control conditions according to the following the equations described by Benimeli et al. [72] and Azooz et al. [73], with the following modifications:

$$\text{Tolerance index } (\%) = \frac{\text{Dry weight of Cu} - \text{treated plants } \times 100}{\text{Dry weight of Cu} - \text{untreated plants (control)}} \quad (4)$$

4.6. Statistical Analysis

For plant growth and physiological measurements, six samples were used per treatment, whereas chemical composition/antioxidants were recorded from four samples per treatment. The analysis of the data was accomplished with the use of SPSS v. 22.0 program (IBM Corp., Armonk, NY, USA) and the one-way analysis of variance (ANOVA) was carried out for the Cu concentration for each sampling date, while means were compared with the Duncan multiple range test (DMRT) at $p < 0.05$, when significant differences were detected. Results were expressed as mean values and standard error (SE). The two-way ANOVA was also performed, with both Cu concentration and sampling date as the sources of variation. Finally, a regression analysis was applied to the content of Cu in plant tissues and the biochemical parameters associated with antioxidant response and oxidative stress.

5. Conclusions

Hydroponically grown *P. graveolens* resulted in a species tolerant toward high Cu concentrations in the root zone and the initial symptoms of Cu toxicity. Namely, declines of photosynthesis-related parameters and increases in leaf H_2O_2 along with considerable Cu accumulation in root tissues were evidenced only at the 100 μM Cu concentration in the nutrient solution. However, the extent of the toxicity symptoms did not have an impact on biomass production; in addition, high Cu levels stimulated plant secondary metabolism, enhancing the production of bioactive antioxidant molecules. Due to low Cu translocation to the aerial organs during the whole growing cycle, this microelement did not reach the leaf tissues, which resulted in suitable plant material for the safe extraction of bioactive compounds. These results show that plant stress from excess Cu does not necessarily preclude the use of MAPs for medicinal purposes, depending on the target organ where the metal accumulates. The outcome of this study showed that the leaves of *P. graveolens* plants exposed to excess Cu could be safely employed for their medicinal properties in herbal or pharmaceutical preparations.

Author Contributions: Conceptualization, A.C., R.M. and N.T.; methodology, A.C. and R.M.; software, A.C.; validation, A.C., R.M., L.I., and A.P.; formal analysis, A.C., R.M. and L.I.; investigation, A.C. and R.M.; resources, A.C. and N.T.; data curation, A.C., R.M., L.I., and A.P.; writing—original draft preparation, R.M. and N.T.; writing—review and editing, R.M., L.I., A.P. and N.T.; visualization, R.M. and L.I.; supervision, A.C. and N.T.; project administration, N.T.; funding acquisition, L.I., A.P. and N.T. All authors have read and agreed to the published version of the manuscript.

Funding: This research has been co-financed by Cyprus University of Technology and University of Pisa Open Access Author Fund.

Acknowledgments: Authors would like to thank Filio Athinodorou and Panayiota Xylia for their technical assistance.

Conflicts of Interest: The authors declare no conflict of interest.

References

1. Gautam, S.; Anjani, K.; Srivastava, N. In vitro evaluation of excess copper affecting seedlings and their biochemical characteristics in *Carthamus tinctorius* L. (variety PBNS-12). *Physiol. Mol. Biol. Plants* **2016**, *22*, 121–129. [CrossRef]
2. Amin, H.; Arain, B.A.; Jahangir, T.M.; Abbasi, A.R.; Mangi, J.; Abbasi, M.S.; Amin, F. Copper (Cu) tolerance and accumulation potential in four native plant species: A comparative study for effective phytoextraction technique. *Geol. Ecol. Landsc.* **2019**, *5*, 1–12. [CrossRef]

3. MacKie, K.A.; Müller, T.; Kandeler, E. Remediation of copper in vineyards—A mini review. *Environ. Pollut.* **2012**, *167*, 16–26. [CrossRef]
4. Adrees, M.; Ali, S.; Rizwan, M.; Ibrahim, M.; Abbas, F.; Farid, M.; Zia-ur-Rehman, M.; Irshad, M.K.; Bharwana, S.A. The effect of excess copper on growth and physiology of important food crops: A review. *Environ. Sci. Pollut. Res.* **2015**, *22*, 8148–8162. [CrossRef]
5. Chen, J.; Shafi, M.; Li, S.; Wang, Y.; Wu, J.; Ye, Z.; Peng, D.; Yan, W.; Liu, D. Copper induced oxidative stresses, antioxidant responses and phytoremediation potential of Moso bamboo (*Phyllostachys pubescens*). *Sci. Rep.* **2015**, *5*, 13554. [CrossRef]
6. Council of the European Union. Directive 86/278/EEC of 12 June 1986 on the protection of the environment, and in particular of the soil, when sewage sludge is used in agriculture (2018). *Official Journal of the European Union*, 4 July 2018; L 181, consolidated version.
7. Kabata-Pendias, A.; Szteke, B. *Copper: Trace Elements in Abiotic and Biotic Environments*; CRC Press, Taylor & Francis Group: Abingdon, UK, 2015; p. 468.
8. Conry, R.R. *Copper: Inorganic and Coordination Chemistry. Encyclopedia of Inorganic and Bioinorganic Chemistry*; John Wiley and Sons: Hoboken, NJ, USA, 2011.
9. Wintz, H.; Fox, T.; Vulpe, C. Responses of plants to iron, zinc and copper deficiencies. *Biochem. Soc. Trans.* **2002**, *30*, 766–768. [CrossRef]
10. Mahmood, T.; Islam, K.R. Response of rice seedlings to copper toxicity and acidity. *J. Plant Nutr.* **2006**, *29*, 943–957. [CrossRef]
11. Li, S.; Zhang, G.; Gao, W.; Zhao, X.; Deng, C.; Lu, L. Plant growth, development and change in GSH level in Safflower (*Carthamus tinctorius* L.) exposed to copper and lead. *Arch. Biol. Sci.* **2015**, *67*, 385–396. [CrossRef]
12. Rizwan, M.; Meunier, J.D.; Davidian, J.C.; Pokrovsky, O.S.; Bovet, N.; Keller, C. Silicon alleviates Cd stress of wheat seedlings (*Triticum turgidum* L. cv. Claudio) grown in hydroponics. *Environ. Sci. Pollut. Res.* **2016**, *23*, 1414–1427. [CrossRef]
13. Saleem, M.H.; Kamran, M.; Zhou, Y.; Parveen, A.; Rehman, M.; Ahmar, S.; Malik, Z.; Mustafa, A.; Ahmad Anjum, R.M.; Wang, B.; et al. Appraising growth, oxidative stress and copper phytoextraction potential of flax (Linum usitatissimum L.) grown in soil differentially spiked with copper. *J. Environ. Manag.* **2020**, *257*. [CrossRef]
14. Kabata-Pendias, A. *Trace Elements in Soils and Plants*, 5th ed.; CRC Press: Boca Ratón, FL, USA, 2011.
15. Yruela, I. Copper in plants: Acquisition, transport and interactions. *Funct. Plant Biol.* **2009**, *36*, 409–430. [CrossRef] [PubMed]
16. Zhao, S.; Liu, Q.; Qi, Y.; Duo, L. Responses of root growth and protective enzymes to copper stress in turfgrass. *Acta Biol. Crac. Ser. Bot.* **2010**, *52*, 7–11. [CrossRef]
17. Liu, J.; Wang, J.; Lee, S.; Wen, R. Copper-caused oxidative stress triggers the activation of antioxidant enzymes via ZmMPK3 in maize leaves. *PLoS ONE* **2018**, *13*, e0203612. [CrossRef]
18. Ku, H.-M.; Tan, C.-W.; Su, Y.-S.; Chiu, C.-Y.; Chen, C.-T.; Jan, F.-J. The effect of water deficit and excess copper on proline metabolism in *Nicotiana benthamiana*. *Biol. Plant.* **2011**, *56*, 337–343. [CrossRef]
19. Monteoliva, M.I.; Rizzi, Y.S.; Cecchini, N.M.; Hajirezaei, M.R.; Alvarez, M.E. Context of action of Proline Dehydrogenase (ProDH) in the Hypersensitive Response of Arabidopsis. *BMC Plant Biol.* **2014**, *14*, 21. [CrossRef] [PubMed]
20. Hall, J.L. Cellular mechanisms for heavy metal detoxification and tolerance. *J. Exp. Bot.* **2002**, *53*, 1–11. [CrossRef]
21. Emamverdian, A.; Ding, Y.; Mokhberdoran, F.; Xie, Y. Heavy metal stress and some mechanisms of plant defense response. *Sci. World J.* **2015**, *2015*, 756120. [CrossRef]
22. Mani, D.; Kumar, C. Biotechnological advances in bioremediation of heavy metals contaminated ecosystems: An overview with special reference to phytoremediation. *Int. J. Environ. Sci. Technol.* **2014**, *11*, 843–872. [CrossRef]
23. Carolin, C.F.; Kumar, P.S.; Saravanan, A.; Joshiba, G.J.; Naushad, M. Efficient techniques for the removal of toxic heavy metals from aquatic environment: A review. *J. Environ. Chem. Eng.* **2017**, *5*, 2782–2799. [CrossRef]
24. Derwich, E.; Benziane, Z.; Boukir, A. Chemical composition of leaf essential oil of Juniperus phoenicea and evaluation of its antibacterial activity. *Int. J. Agric. Biol.* **2010**, *12*, 199–204.
25. Tripathy, V.; Basak, B.B.; Varghese, T.S.; Saha, A. Residues and contaminants in medicinal herbs—A review. *Phytochem. Lett.* **2015**, *14*, 67–78. [CrossRef]
26. Clemens, S. Toxic metal accumulation, responses to exposure and mechanisms of tolerance in plants. *Biochimie* **2006**, *88*, 1707–1719. [CrossRef]
27. Handique, G.K.; Handique, A.K. Proline accumulation in lemongrass (Cymbopogon flexuosus Stapf.) due to heavy metal stress. *J. Environ. Biol.* **2009**, *30*, 299–302.
28. Kováčik, J.; Klejdus, B.; Hedbavny, J.; Štork, F.; Bačkor, M. Comparison of cadmium and copper effect on phenolic metabolism, mineral nutrients and stress-related parameters in Matricaria chamomilla plants. *Plant Soil* **2009**, *320*, 231–242. [CrossRef]
29. Ali, I.B.E.; Tajini, F.; Boulila, A.; Jebri, M.A.; Boussaid, M.; Messaoud, C.; Sebaï, H. Bioactive compounds from Tunisian *Pelargonium graveolens* (L'Hér.) essential oils and extracts: α-amylase and acethylcholinesterase inhibitory and antioxidant, antibacterial and phytotoxic activities. *Ind. Crops Prod.* **2020**, *158*, 112951. [CrossRef]
30. Fiz, O.; Vargas, P.; Alarcón, M.; Aedo, C.; García, J.L.; Aldasoro, J.J. Phylogeny and historical biogeography of geraniaceae in relation to climate changes and pollination ecology. *Syst. Bot.* **2008**, *33*, 326–342. [CrossRef]
31. Ćavar, S.; Maksimović, M.; Vidic, D.; Parić, A. Chemical composition and antioxidant and antimicrobial activity of essential oil of *Artemisia annua* L. from Bosnia. *Ind. Crops Prod.* **2012**, *37*, 479–485. [CrossRef]

32. Boukhatem, M.N.; Kameli, A.; Saidi, F. Essential oil of Algerian rose-scented geranium (*Pelargonium graveolens*): Chemical composition and antimicrobial activity against food spoilage pathogens. *Food Control* **2013**, *34*, 208–213. [CrossRef]
33. Tahan, F.; Yaman, M. Can the Pelargonium sidoides root extract EPs®7630 prevent asthma attacks during viral infections of the upper respiratory tract in children? *Phytomedicine* **2013**, *20*, 148–150. [CrossRef] [PubMed]
34. Moyo, M.; Van Staden, J. Medicinal properties and conservation of Pelargonium sidoides DC. *J. Ethnopharmacol.* **2014**, *152*, 243–255. [CrossRef]
35. Colling, J.; Groenewald, J.H.; Makunga, N.P. Genetic alterations for increased coumarin production lead to metabolic changes in the medicinally important Pelargonium sidoides DC (Geraniaceae). *Metab. Eng.* **2010**, *12*, 561–572. [CrossRef] [PubMed]
36. Mehrarad, F.; Ziarati, P.; Mousavi, Z. Removing heavy metals from pharmaceutical effluent by Pelargonium grandiflorum. *Biomed. Pharmacol. J.* **2016**, *9*, 151–161. [CrossRef]
37. Lam, E.J.; Gálvez, M.E.; Cánovas, M.; Montofré, Í.L.; Keith, B.F. Assessment of the adaptive capacity of plant species in copper mine tailings in arid and semiarid environments. *J. Soils Sediments* **2018**, *18*, 2203–2216. [CrossRef]
38. Patel, A.; Patra, D.D. Phytoextraction capacity of *Pelargonium graveolens* L'Hér. grown on soilamended with tannery sludge—Its effect on the antioxidant activityand oil yield. *Ecol. Eng.* **2015**, *74*, 20–27. [CrossRef]
39. Chand, S.; Singh, G.; Rajkumari; Patra, D.D. Performance of rose scented geranium (*Pelargonium graveolens*) in heavy metal polluted soil vis-à-vis phytoaccumulation of metals. *Int. J. Phytoremediat.* **2016**, *18*, 754–760. [CrossRef]
40. Pandey, J.; Chand, S.; Chaurasiya, S.; Kumari, R.; Patra, D.D.; Verma, R.K.; Singh, S. Effect of tannery sludge amendments on the activity of soil enzymes and phytoremediation potential of two economically important cultivars of geranium (*Pelargonium graveolens*). *Soil Sediment Contam.* **2019**, *28*, 395–410. [CrossRef]
41. Dimitrova, M.; Mihaylova, D.; Popova, A.; Alexieva, J.; Tana Sapundzhieva, T.; Fidan, H. Phenolic profile, antibacterial and antioxidant activity of *Pelargonium graveolens* leaves' extracts. Scientific Bulletin. *Ser. F Biotechnol.* **2015**, *19*, 130–135. [CrossRef]
42. El Ouadi, Y.; Bendaif, H.; Mrabti, H.N.; Elmsellem, H.; Kadmi, Y.; Shariati, M.A.; Abdel-Rahman, I.; Hammouti, B.; Bouyanzer, A. Antioxidant activity of phenols and flavonoids contents of aqueous extract of *Pelargonium graveolens* orgin in the North-East Morocco. *J. Microbiol. Biotechnol. Food Sci.* **2017**, *6*, 1218–1220. [CrossRef]
43. Ennaifer, M.; Bouzaiene, T.; Chouaibi, M.; Hamdi, M. *Pelargonium graveolens* Aqueous Decoction: A New Water-Soluble Polysaccharide and Antioxidant-Rich Extract. *Biomed Res. Int.* **2018**, *2018*, 11. [CrossRef]
44. El Aanachi, S.; Gali, L.; Nacer, S.N.; Bensouici, C.; Dari, K.; Aassila, H. Phenolic contents and in vitro investigation of the antioxidant, enzyme inhibitory, photoprotective, and antimicrobial effects of the organic extracts of *Pelargonium graveolens* growing in Morocco. *Biocatal. Agric. Biotechnol.* **2020**, *29*, 101819. [CrossRef]
45. Savvas, D.; Ntatsi, G.; Passam, H.C. The European journal of plant science and biotechnology plant nutrition and physiological disorders in greenhouse grown tomato, pepper and eggplant. *Eur. J. Plant Sci. Biotechnol.* **2008**, *2*, 45–61.
46. Signore, A.; Serio, F.; Santamaria, P. A targeted management of the nutrient solution in a soilless tomato crop according to plant needs. *Front. Plant Sci.* **2016**, *7*, 391. [CrossRef]
47. Loreto, F.; Velikova, V. Isoprene produced by leaves protects the photosynthetic apparatus against ozone damage, quenches ozone products, and reduces lipid peroxidation of cellular membranes. *Plant Physiol.* **2001**, *127*, 1781–1787. [CrossRef] [PubMed]
48. Martins, L.L.; Mourato, M.P. Effect of excess copper on tomato plants: Growth parameters, enzyme activities, chlorophyll, and mineral content. *J. Plant Nutr.* **2006**, *29*, 2179–2198. [CrossRef]
49. Chrysargyris, A.; Papakyriakou, E.; Petropoulos, S.A.; Tzortzakis, N. The combined and single effect of salinity and copper stress on growth and quality of *Mentha spicata* plants. *J. Hazard. Mater.* **2019**, *368*, 584–593. [CrossRef]
50. Reichman, S.M.; Menzies, N.W.; Asher, C.J.; Mulligan, D.R. Responses of four Australian tree species to toxic concentrations of copper in solution culture. *J. Plant Nutr.* **2006**, *29*, 1127–1141. [CrossRef]
51. Öquist, G.; Chow, W.S.; Anderson, J.M. Photoinhibition of photosynthesis represents a mechanism for the long-term regulation of photosystem II. *Planta* **1992**, *186*, 450–460. [CrossRef]
52. Yruela, I. Copper in plants. *Braz. J. Plant Physiol.* **2005**, *17*, 145–156. [CrossRef]
53. Guidi, L.; Lo Piccolo, E.; Landi, M. Chlorophyll fluorescence, photoinhibition and abiotic stress: Does it make any difference the fact to be a C3 or C4 species? *Front. Plant Sci.* **2019**, *10*, 174. [CrossRef] [PubMed]
54. Zheng, Y.; Wang, L.; Dixon, M.A. Response to copper toxicity for three ornamental crops in solution culture. *HortScience* **2004**, *39*, 1116–1120. [CrossRef]
55. Osmond, C.B.; Grace, S.C. Perspectives of photoinhibition and photorespiration in the field: Quintessential inefficiencies of the light and dark reaction of photosynthesis? *J. Exp. Bot.* **1995**, *46*, 1415–1422. [CrossRef]
56. Branco-Neves, S.; Soares, C.; de Sousa, A.; Martins, V.; Azenha, M.; Gerós, H.; Fidalgo, F. An efficient antioxidant system and heavy metal exclusion from leaves make *Solanum cheesmaniae* more tolerant to Cu than its cultivated counterpart. *Food Energy Secur.* **2017**, *6*, 123–133. [CrossRef]
57. Lange, B.; van der Ent, A.; Baker, A.J.M.; Echevarria, G.; Mahy, G.; Malaisse, F.; Meerts, P.; Pourret, O.; Verbruggen, N.; Faucon, M.P. Copper and cobalt accumulation in plants: A critical assessment of the current state of knowledge. *New Phytol.* **2017**, *213*, 537–551. [CrossRef] [PubMed]
58. Tschinkel, P.F.S.; Melo, E.S.P.; Pereira, H.S.; Silva, K.R.N.; Arakaki, D.G.; Lima, N.V.; Fernandes, M.R.; Leite, L.C.S.; Melo, E.S.P.; Melnikov, P.; et al. The hazardous level of heavy metals in different medicinal plants and their decoctions in water: A public health problem in Brazil. *BioMed Res. Int.* **2020**, *2020*. [CrossRef]

59. European Food Safety Authority (EFSA). Review of the existing maximum residue levels for copper compounds according to Article 12 of Regulation (EC) No 396/2005. *EFSA J.* **2018**, *16*. [CrossRef]
60. Ghassabzadeh, H.; Mohadespour, A.; Torab-Mostaedi, M.; Zaheri, P.; Maragheh, M.G.; Taheri, H. Adsorption of Ag, Cu and Hg from aqueous solutions using expanded perlite. *J. Hazard. Mater.* **2010**, *177*, 950–955. [CrossRef]
61. Matijevic, L.; Romic, D.; Romic, M. Soil organic matter and salinity affect copper bioavailability in root zone and uptake by *Vicia faba* L. plants. *Environ. Geochem. Health* **2014**, *36*, 883–896. [CrossRef]
62. Eskandari, S.; Mozaffari, V. Interactive effect of soil salinity and copper application on growth and chemical composition of pistachio seedlings (cv. Badami). *Commun. Soil Sci. Plant Anal.* **2014**, *45*, 688–702. [CrossRef]
63. Li, L.; Long, M.; Islam, F.; Farooq, M.A.; Wang, J.; Mwamba, T.M.; Shou, J.; Zhou, W. Synergistic effects of chromium and copper on photosynthetic inhibition, subcellular distribution, and related gene expression in *Brassica napus* cultivars. *Environ. Sci. Pollut. Res.* **2019**, *26*, 11827–11845. [CrossRef] [PubMed]
64. Chrysargyris, A.; Antoniou, O.; Tzionis, A.; Prasad, M.; Tzortzakis, N. Alternative soilless media using olive-mill and paper waste for growing ornamental plants. *Environ. Sci. Pollut. Res.* **2018**, *25*, 35915–35927. [CrossRef] [PubMed]
65. Tzortzakis, N.G.; Economakis, C.D. Shredded maize stems as an alternative substrate medium: Effect on growth, flowering and yield of tomato in soilless culture. *J. Veg. Sci.* **2005**, *11*, 57–70. [CrossRef]
66. Chrysargyris, A.; Antoniou, O.; Athinodorou, F.; Vassiliou, R.; Papadaki, A.; Tzortzakis, N. Deployment of olive-stone waste as a substitute growing medium component for Brassica seedling production in nurseries. *Environ. Sci. Pollut. Res.* **2019**, *26*. [CrossRef] [PubMed]
67. Chrysargyris, A.; Laoutari, S.; Litskas, V.D.; Stavrinides, M.C.; Tzortzakis, N. Effects of water stress on lavender and sage biomass production, essential oil composition and biocidal properties against Tetranychus urticae (Koch). *Sci. Hortic.* **2016**, *213*, 96–103. [CrossRef]
68. Wojdyło, A.; Oszmiański, J.; Czemerys, R. Antioxidant activity and phenolic compounds in 32 selected herbs. *Food Chem.* **2007**, *105*, 940–949. [CrossRef]
69. Tzortzakis, N.G.; Tzanakaki, K.; Economakis, C.D. Effect of origanum oil and vinegar on the maintenance of postharvest quality of tomato. *Food Nutr. Sci.* **2011**, *2*, 974–982. [CrossRef]
70. Meyers, K.J.; Watkins, C.B.; Pritts, M.P.; Liu, R.H. Antioxidant and antiproliferative activities of strawberries. *J. Agric. Food Chem.* **2003**, *51*, 6887–6892. [CrossRef]
71. De Azevedo Neto, A.D.; Prisco, J.T.; Enéas-Filho, J.; De Abreu, C.E.B.; Gomes-Filho, E. Effect of salt stress on antioxidative enzymes and lipid peroxidation in leaves and roots of salt-tolerant and salt-sensitive maize genotypes. *Environ. Exp. Bot.* **2006**, *56*, 87–94. [CrossRef]
72. Benimeli, C.S.; Medina, A.; Navarro, C.M.; Medina, R.B.; Amoroso, M.J.; Gómez, M.I. Bioaccumulation of copper by Zea mays: Impact on root, shoot and leaf growth. *Water Air Soil Pollut.* **2010**, *210*, 365–370. [CrossRef]
73. Azooz, M.M.; Abou-Elhamd, M.F.; Al-Fredan, M.A. Biphasic effect of copper on growth, proline, lipid peroxidation and antioxidant enzyme activities of wheat (*Triticum aestivum* cv. Hasaawi) at early growing stage. *Aust. J. Crop Sci.* **2012**, *6*, 688–694.

plants

MDPI

Review

The Alleviation of Metal Stress Nuisance for Plants—A Review of Promising Solutions in the Face of Environmental Challenges

Mateusz Labudda [1], Kinga Dziurka [2], Justyna Fidler [1], Marta Gietler [1], Anna Rybarczyk-Płońska [1], Małgorzata Nykiel [1], Beata Prabucka [1], Iwona Morkunas [3] and Ewa Muszyńska [4,*]

1 Department of Biochemistry and Microbiology, Institute of Biology, Warsaw University of Life Sciences-SGGW, Nowoursynowska 159, 02-776 Warsaw, Poland
2 Department of Biotechnology, The Franciszek Górski Institute of Plant Physiology, Polish Academy of Sciences, Niezapominajek 21, 30-239, Kraków, Poland
3 Department of Plant Physiology, Poznań University of Life Sciences, Wołyńska 35, 60-637 Poznań, Poland
4 Department of Botany, Institute of Biology, Warsaw University of Life Sciences-SGGW, Nowoursynowska 159, 02-776 Warsaw, Poland
* Correspondence: ewa_muszynska@sggw.edu.pl; Tel.: +48-22-59326-61

Abstract: Environmental changes are inevitable with time, but their intensification and diversification, occurring in the last several decades due to the combination of both natural and human-made causes, are really a matter of great apprehension. As a consequence, plants are exposed to a variety of abiotic stressors that contribute to their morpho-physiological, biochemical, and molecular alterations, which affects plant growth and development as well as the quality and productivity of crops. Thus, novel strategies are still being developed to meet the challenges of the modern world related to climate changes and natural ecosystem degradation. Innovative methods that have recently received special attention include eco-friendly, easily available, inexpensive, and, very often, plant-based methods. However, such approaches require better cognition and understanding of plant adaptations and acclimation mechanisms in response to adverse conditions. In this succinct review, we have highlighted defense mechanisms against external stimuli (mainly exposure to elevated levels of metal elements) which can be activated through permanent microevolutionary changes in metal-tolerant species or through exogenously applied priming agents that may ensure plant acclimation and thereby elevated stress resistance.

Keywords: abiotic stress; adaptation; priming; defense mechanisms; metallophyte; oxidative stress; phytoremediation; tolerance

Citation: Labudda, M.; Dziurka, K.; Fidler, J.; Gietler, M.; Rybarczyk-Płońska, A.; Nykiel, M.; Prabucka, B.; Morkunas, I.; Muszyńska, E. The Alleviation of Metal Stress Nuisance for Plants—A Review of Promising Solutions in the Face of Environmental Challenges. *Plants* **2022**, *11*, 2544. https://doi.org/10.3390/plants11192544

Academic Editor: Juan Barceló

Received: 26 August 2022
Accepted: 25 September 2022
Published: 28 September 2022

Publisher's Note: MDPI stays neutral with regard to jurisdictional claims in published maps and institutional affiliations.

1. Introduction

Rapid industrialization and urbanization, chemicalization of agriculture, and the lack of a proper attitude to the surroundings in which we live are the main causes of unpredictable climate changes, as well as the deterioration of natural environments and ecosystems [1]. As a consequence of such imprudent human domination of the Earth, plants are constantly exposed to a wide array of adverse environmental events, including water deficits, salinity, imbalances in elements (resulting from their deficiency and/or pollution), extremes of temperature, ultraviolet radiation, etc. All the above-mentioned physical and chemical factors, collectively referred to as abiotic stress, may occur singly, sequentially, or simultaneously, and their effects may also act synergistically or additively on plant fitness [2]. Moreover, the effect of each stress factor depends on its intensity and the exposure time of the plants. Despite the impact of such a wide variety of stressors, plant exposure to any of them has one similar outcome, namely the overgeneration of reactive oxygen species (ROS) that are responsible for oxidative damage of cellular components such as proteins, lipids, nucleic acids, carbohydrates, and other metabolites [3–5]. Therefore, as shown in Figure 1, oxidative stress is a secondary but common reaction of plants subjected to various factors which, by interacting with each other, contribute to disturbances in the optimal growth and development of plants, finally leading to a considerable reduction in their productivity and yield [6,7].

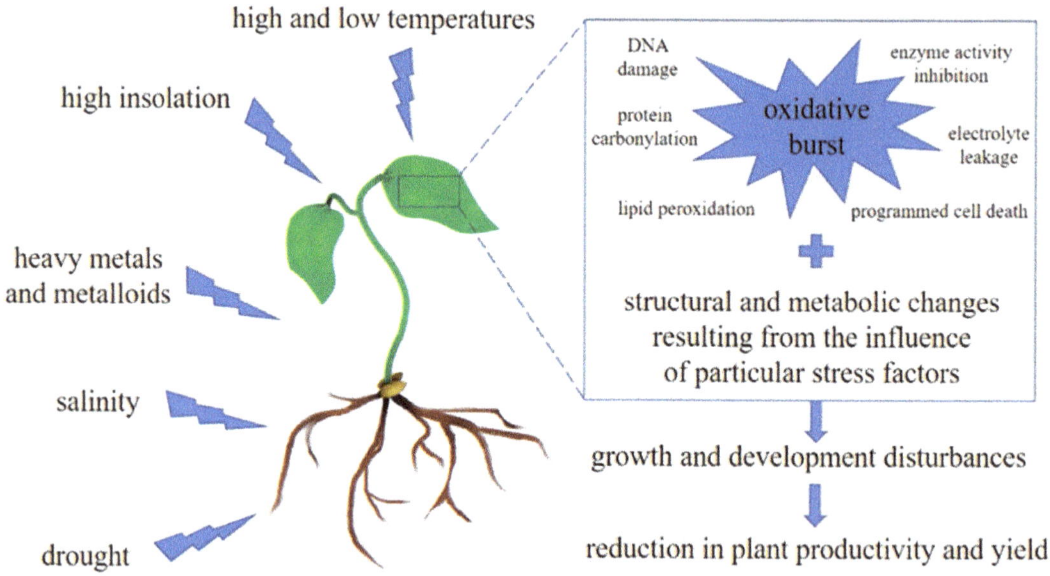

Figure 1. A simplified diagram of complex plant responses to stressful environmental stimuli: oxidative burst and its consequences as a universal reaction to different stressors are shown, as well as stressor-dependent reactions leading to plant growth retardation and a decline in productivity.

It should be emphasized that, among the various xenobiotics released to the environment due to anthropogenic activity, heavy metals and metalloids, classified also as metallic trace elements because of their presence at trace concentrations (parts per billion or less than 10 parts per million) in various matrices, have pulled ahead of other well-known contaminants such as plant protection products and carbon and sulfur dioxides [8]. Environmental pollution with metals is particularly prominent in point source terrains such as metalliferous mines, smelters and foundries, and other metal-based industrial operations. However, among the sources of these most common inorganic contaminants, fossil fuel burning and the use of fertilizers, pesticides, livestock manure, municipal wastes, and sewage should also be mentioned [9]. The problem of heavy metal accumulation may be aggravated by salinity stress, which causes disturbances in the homeostasis of macro- and micro-elements in the soil and facilitates metallic ion uptake by plants [10,11]. Such changes in soil composition are peculiarly important in the case of crops for which a sufficient supply of essential elements has to be ensured, whilst potentially toxic elements should be present only at very low levels. Since heavy metals are now ranked in second place when taking into account the degree of risk they pose to the human population all over the world, in recent years there has been increasing concern about environmental contamination from them [12].

Metallic trace elements can spread over long distances through interactions with wind, surface and ground waters, and herbivores. These metals pose a serious threat to our health due to food chain accumulation, dust inhalation, and skin contact which result in cardiovascular, respiratory, and neurodegenerative diseases [13]. They also have a negative impact on the majority of plant species and other living organisms. In the concentrations exceeding the maximum tolerable amount, they cause disturbances in the ultrastructure of cells and affect physiological and biochemical processes such as the biosynthesis of chlorophylls, photosynthetic capacity, transpiration, nutrient and water uptake, and the activity of enzymes involved in various metabolic pathways, as well as lead to increased ROS formation by direct involvement in redox reactions (in the case of highly reactive metals) or indirectly through depletion of antioxidant pools [5,9,14–17]. All these cellular effects result in morphological changes, such as shortening of shoots and roots, necrotic and chlorotic stains, decreases in leaf number and size, and premature aging [5,18,19]. As a consequence, it leads to limiting the productivity of agricultural crops [20]. The danger of metallic elements lies also in the fact that many of them are dispersed in the environment for a long time and, therefore, they are considered to be persistent [12]. As an example, half-life time varies from 75 to 380 years for cadmium (Cd) and from 1000 to 3000 years for copper (Cu), nickel (Ni), lead (Pb), zinc (Zn), and selenium (Se) in the soils of temperate climates [21].

Recently, research on the mechanisms by which plants recognize and cope with toxic metals and other stressful and dynamic circumstances has undergone a very exciting period leading to significant breakthroughs. The development of knowledge in this field is necessary to relieve the pressure of environmental changes and to ensure global food security for an increasing population, as well as to restore areas degraded by human activity. Since it is well known that plants have developed different adaptation strategies which can occur as a result of adaptation and acclimation, the purpose of this concise review is to indicate what can make the life of stressed plants a little easier, especially in respect to metallic trace elements. We have highlighted only two main possibilities, although many more issues are taken into consideration in the current research. The first one is based on natural defense mechanisms arising through evolutionary changes, the understanding of which enables the development of new strategies to alleviate metal danger for both plants and surroundings (and/or the improvement of those remediation techniques that already exist). The second one refers to the use of novel priming techniques that may provide plants with intracellular acclimation and thereby enhanced stress tolerance. Both of them represent the latest solutions for sustainable, cost-effective, and efficient approaches to environmental challenges.

2. Functional Traits of Plants Developed in Response to Severe External Pressures

Climate change has caused serious impacts on the ecosystem, including devastating its stability and affecting biodiversity. Plants, as an important component of terrestrial ecosystems, respond to climate change in an all-round way; therefore, changes in the functional traits of plants can be indicative of climate changes. The novel developmental direction of this research is to determine the interrelationships among various indicators based on physiological, biochemical, and ecological plant characteristics and to establish a network indicator system from individual plants and communities towards ecosystem functions.

Since plants are unable to avoid environmental stressors due to their sessile lifestyle, they have evolved effective mechanisms to combat stress which ensure their survival in uncomfortable conditions. Defense response can be attributed to phenotypic plasticity leading to changes within a single organism, that are reversible and result from subsequently occurring, occasional stress events ('priming'), or from chronic exposure to a new environment, to which plant metabolism adjusts ('acclimation') [22,23]. Both of these terms differ from 'adaptation', which describes permanent genotypic changes resulting in phenotypic traits that improve plant fitness or survival over multiple generations [24]. Morphological, anatomical, and physiological adaptations are characterized for metallophytes that have been gradually developed in habitats naturally or artificially enriched with metallic elements. Although in these first conditions metal tolerance may evolve over thousands or even millions of years, on human-influenced metalliferous soils it may be achieved in a relatively short time, i.e., less than 100–150 years [25]. Such genetically altered ecotypes of common species (i.e., pseudometallophytes or facultative metallophytes), as well as genera restricted only to metalliferous soils (i.e., obligate or absolute metallophytes), exhibit a higher toxicity threshold or even slightly beneficial metal effects compared to their counterparts from unpolluted areas due to a special tolerance mechanism which is not available to non-metalliferous genotypes [26].

2.1. Specific Characteristics of Metal-Tolerant Species and Their Application in Soil Remediation

Metallophytes utilize several adaptation mechanisms to control the uptake, mobility, and activity of potentially toxic ions in the cell. Firstly, modifications to cell wall components and structure favor the retention of metals and provide a mechanical and chemical barrier against their free penetration into the protoplast [27,28]. Similarly, various membrane transporters belonging to the following families: HMAs (heavy metal ATPases, also known as P-type ATPase), NRAMP (natural resistance-associated macrophage protein), CDF (cation diffusion facilitators), YSL (yellow stripe-like), ABC (ATP-binding cassette), COPT (copper transporter), and ZIP (zinc-regulated transporter, iron-regulated transporter-like protein), play an important role in the regulation of toxic ion influx into the protoplast and organelles, and therefore have been extensively discussed in recent research and numerous review articles [7,29–31]. Subsequently, in the cytosol, harmful ions are effectively detoxified and stored in places safe for metabolism in order to prevent deleterious physiological damage. The important cytoplasmic ligands responsible for the chelation and neutralization of metallic elements include phytochelatins, glutathione (GSH), amino acids, and organic acids [14,26,32]. In turn, metal sequestration may take place in the vacuole, dictiosomal vesicles, or the endoplasmic reticulum [33,34]. From an organismic point of view, ions can be withdrawn into aging leaves and trichomes or drawn outside by secretary glands, which has been observed in *Arabidopsis thaliana* and *A. halleri* [35], *Alyssum montanum* [36], and *Biscutella laevigata* [37].

Other cellular features that make the life of metallophytes easier are related to efficient antioxidant defense systems that confront oxidative stress. Cell redox homeostasis is kept by a synchronous action of various enzymes, such as superoxide dismutase (SOD), catalase (CAT), peroxidases (POD; such as guaiacol peroxidase, GPOX, glutathione peroxidase, GPX, and ascorbate peroxidase, APX), glutathione *S*-transferase (GST), glutathione reductase (GR), and nonenzymatic antioxidants, such as ascorbate (AsA), glutathione (GSH), carotenoids (CAR), α-tocopherols, phenolics, and amino acids such as proline [4]. Although oxidative stress as a reaction to metals is one of the most studied issues recently [3,5,38,39], ROS transformation pathways, as a basis of adaptation to their excess amounts, have not been frequently compared between the representatives of different species sharing the same ecological niches or described for ecotypes of the same species representing different habitats. In this regard, our previous studies on serpentine and calamine ecotypes of *Silene vulgaris* (Caryophyllaceae) and the calamine ecotype of *Alyssum montanum* (Brassicaceae) have shown both species- and ecotype-dependent features [14,18,19,36,40,41]. The response of *S. vulgaris* to metallic elements was mainly related to the activity of antioxidant enzymes during in vitro cultivation on media enriched with Zn, Pb, and Cd at the same concentration as in the post-industrial habitat of the calamine ecotype [41]. In turn, the response of *A. montanum* was associated with the transformations of phenolic compounds, which, in the metallicolous ecotype, led to the synthesis of phenolic acids with a high ability to ROS scavenging and, in the non-metallicolous ecotype, to the synthesis of other compounds not involved in alleviating oxidative stress [19]. Interestingly, the common reaction of the metallicolous *S. vulgaris* and *A. montanum* individuals was the activity of GPX. Nevertheless, the importance of this enzyme in particular ecotypes differentiated both calamine specimens from the serpentine ones. In the former, the increased activity of this enzyme correlated with the increased accumulation of phenylpropanoids which, acting together, contribute to the formation of a lignified cell wall preventing the easy penetration of ions into the protoplast; whereas GPX activity in the serpentine ecotype provided only ROS neutralization [36,41]. Curiously, we have also found that the exposure of the calamine ecotype of *S. vulgaris* to the concentration of metallic elements reflecting their level in the zinc–lead substrate resulted in a significant increase in the efficiency of all analyzed components of the antioxidant apparatus. As a consequence, the studied ions stimulated the growth of calamine specimens, which was manifested in accelerated growth and biomass accretion [40,41]. It is therefore likely that the trace element ions, at the doses which the calamine ecotype has adapted to in the selection process, play a pivotal physiological role, perhaps even as micro- or ultra-elements.

The above-mentioned mechanisms guarantee a high propensity of metallophytes to take up metallic trace elements; however, tolerant species, ecotypes, or particular populations differ in their degree of accumulation and the element distributions in their organs, even if they grow on the same soil. Furthermore, the enhanced ability of metal-tolerant species to accumulate one metal does not mean that other ions will be stored with the same intensity and distributed over the organs in a similar way [32]. As proposed by Baker [42], plants appearing in metal-enriched environments can be divided, on the basis of the relationship between ion content in tissue and soil, into:

(1) 'excluders' that detoxify most of the toxic ions in roots and minimize their translocation to shoots; for these plants the accumulation coefficient, i.e., the ratio of metal concentration in the shoot to the soil, is always lower than one;
(2) 'indicators', whose shoots contain a similar concentration of metals as the soil (the accumulation coefficient is close to one);
(3) 'accumulators', which are characterized by effective metal uptake, transport, and storage in shoots (the accumulation coefficient is higher than one); among them, approximately 720 species are considered to be hyperaccumulators that are able to accumulate extraordinary amounts of metallic ions without suffering any phytotoxic effects [43].

The amazing biology and behavior of metallophytes, in respect to metal accumulation and detoxification, make them to useful in various phytoremediation techniques. It is plant-based, environmentally friendly, non-invasive, and low-cost technology which is applied to remediate contaminated soils by accumulation, immobilization, or degradation of these pollutants [44]. In phytoextraction, which constitutes the most popular method of phytoremediation relying on the total removal of contaminants from the environment, hyperaccumulating plants may work the best due to their ability (about 100–1000-fold higher than in other plants) regarding effective uptake and translocation of metallic elements [45]. Many studies have demonstrated the phytoextraction potential of metallophytes from various genera, such as *Alyssum murale* [46], *Arabidopsis halleri* [35], *Biscutella laevigata* [37], and *Stackhousia tryonii* among others [47]. Nowadays, phytoextraction achieves two goals at once. It is not only exploited to clean up soil, but also to mine metal (so-called phytomining), mainly in places where the use of conventional methods for ore exploitation is economically unprofitable. As an example, the cultivation of Ni-hyperaccumulators *Alyssum corsicum* and *A. murale* allows the extraction of about 400 kg of Ni per hectare [48]. In turn, metal excluders are excellent candidates for phytostabilization, which is aimed at reducing metal mobility in the soils in order to prevent them leaching deeper into the ground water and to also prevent the dust blowing into the atmosphere [44]. Despite metal stabilization, this technique involves the permanent establishment of a vegetative cover, which performs anti-erosive and soil-forming functions. Currently, some studies have indicated that the recovery of vegetation on heavy metal polluted terrains should be performed by native metalliferous species, which spontaneously occur on degraded areas and are thus better adapted to local ecological conditions than introduced ones [49]. Such an approach was first used in the 1960s, when Zn-Pb tolerant populations of *Agrostis tenuis, A. stolonifera, Anthoxanthum odoratum, Festuca rubra*, and *F. ovina* were investigated [50]. Recently, the potential of native metal-tolerant species for revegetation has been successfully verified for *Agropyron smithii* and *Artemisia tridentate* [51], *Lygeum spartum* [52], *Achillea wilhelmsii* [53], *and Matthiola dagestanica* and *Draba stylaris* [46]. The usefulness of metallophytes for revegetation and the phytostabilization of Zn-Pb rich soils in the Olkusz Ore-bearing Region, one of the biggest industrial areas in Poland, has been also proven in our earlier studies for *Biscutella laevigata* [49,54], *Dianthus carthusianorum* [49], *Gypsophila fastigiata* [55], and *Silene vulgaris* [56]. Undoubtedly, phytoremediation combined with the biological reclamation of destroyed or degraded ecosystems may constitute a new and safe opportunity for humans to positively interact with the environment.

2.2. Relationship between Chosen Metal Tolerance Traits and Other Stresses

Besides evolving metal tolerance, metallicolous species or their ecotypes were co-selected for tolerance to other adverse site conditions because soils contaminated with heavy metals are often salinized and dry [49]. Therefore, metallophytes share tolerance mechanisms with other specialized groups of plants, which makes their biology even more interesting.

Apart from activation of the antioxidant defense system constituting the basic response to various types of stress, metal-tolerant species exhibit specific adaptations that ensure a high degree of resistance to salinity and drought as well, both of which may cause a lack or deficiency of water for plants. The increased resistance to water deficit in metallophytes may result from their ability to accumulate toxic ions in large amounts. Since metallic ions can be preferentially accumulated within epidermal leaf cells, reduced cuticular transpiration can be achieved [47]. Furthermore, one hypothesis justifying reasons for metal (hyper)accumulation postulates that elements stored within cells overcome the effect of water constraints by acting as an osmolyte [45]. Confirmation of the osmoregulatory role of metals can be found in a study conducted by Bhatia et al. [47], who proved that Ni content in the shoots of *Stackhousia tryonii*, a Ni hyperaccumulator, increased significantly as the soil moisture levels decreased. Another excellent osmolyte that also accrues during metal stress is proline. This important amino acid contributes to maintaining water balance and cell turgor through osmotic regulation, and also contributes to the stability of cell membranes by preventing electrolyte leakage, which in turn can help during water deficit [57,58]. Besides osmoregulation function, proline also acts as a metal chelator and an antioxidative defense molecule which prevents oxidative burst due to ROS scavenging, thus mitigating a wide array of adverse effects from toxic ions [4,41,59]. Subsequent features of metallophytes which may provide simultaneous protection against water losses or their better adjustment to drought are related to morpho-anatomical structure. Leaves of metal-tolerant species often have reduced transpiration surfaces, are less numerous, narrower, thicker, and waxy, and possess a limited number of stomata and increased mesophyll cell size [17,60,61]. Furthermore, plants from metalliferous areas, probably in response to dry substrate, may produce deeper roots covered with dense root hairs; however, root architecture and size do not form a rule enabling metal-tolerant individuals to be distinguished from non-tolerant ones [32].

Metallophytes also show some similarities with halophytes, natives of saline soils mostly rich in sodium (Na^+) and chloride (Cl^-) ions. Both these specialized groups of plants possess specific and more common functional mechanisms of tolerance towards numerous stresses, which refer not only to strong antioxidant defense systems and the synthesis of compatible solutes, but also to ion sequestration and detoxification pathways [62]. The vacuolar compartmentalization of salt and heavy metals through the enhanced activity of membrane transporters is one of them [63]. Nevertheless, it has not been fully explained if the same proton pumps are involved in this process, although the role of vacuolar H^+-ATPase was proven to protect against salt and Cd stress in the halophyte *Tamarix hispida* [64]. Moreover, halophytes, similarly to metal-tolerant species, are able to excrete excess deleterious ions from photosynthetically active tissues on leaf surfaces by different structures, such as salt glands, bladders, and trichomes; however, they are not specific to salt alone, as other toxic ions can be also removed in this way [62]. As an example, *Armeria maritima* ssp. *halleri*, an obligate metallophyte, can remove Cu ions via salt glands [65], while *Limoniastrum monopetalum*, a halophytic plant, uses these structures to excrete salt, Cd, and Pb as a detoxification mechanism [66]. On the other hand, all the above-mentioned mechanisms indicate clearly that the adaptation of halophytes for survival in the presence of high salt concentrations may also confer their tolerance to metallic elements. For this reason, halophytes can be good candidates for the phytoremediation of heavy metal polluted soils. These specimens with exclusion ability, rapid growth, and deep root systems can form dense vegetation cover and therefore be utilized for the purposes of phytostabilization. *Atriplex halimus* [67], *Cochearia anglica*, *C. x hollandica*, *C. danica*, *C. py-renaica* [63], and others are good examples (Table 1). Among halophytes, species that are able to accumulate both heavy metals and salt in extraordinary amounts in the shoots without suffering phytotoxic effects can be also found. One of the most effective in removing toxic ions seems to be an annual halophyte, *Chemopodium botrys*, which accumulates several times more Cd than *Noccaea caerulescens*, a well-known hyperaccumulator of Cd and Zn [68]. The study of Mazharia and Hoameed [68] indicated that the total amount of Cd removed by shoots of *Ch. botrys* was 120 g/ha; whereas the average Cd extraction ability of *N. caerulescens* may stay at a level of about 35 g/ha. Such salt/metal-accumulating species are extremely important for the decontamination of metal polluted saline soils, although recent findings also encourage their use for reclamation of purely saline soils, mainly in arid and semiarid regions [69]. Some more examples of halophytic species and their potential usefulness in particular soil phytoremediation methods are shown in Table 1, whilst the HALOPH database, which is available at http://www.sussex.ac.uk/affiliates/halophytes/ (accessed on 3 August 2022), presents probably the largest collection of halophyte examples for various applications. In turn, more aspects of halophyte responses to metallic elements (including common and specific mechanisms of metal and salt tolerance in this group of plants), their potential utilization for the phytoremediation of metal-contaminated soils, and their relevance to the phytodesalination of saline lands have been broadly discussed in some recently published reviews and books [62,70–72].

Table 1. Examples of the usefulness of halophytes in particular phytoremediation techniques for the removal of various metallic trace elements.

Technique	Halophyte Species	Accumulated Metal (s)	References
phytostabilization	*Atriplex atacamensis*	As	[62]
	Atriplex halimus	Cd, Pb	[67]
	Cochearia species	Zn, Pb	[63]
	Halimione portulacoides	Zn, Cu, Ni, Co	[73]
	Tamarix hispida	Zn, Pb	[64]
phytoextraction	*Chenopodium botrys*	Cd	[68]
	Halogeton glomeratus	Cr, Ni, Cu, Zn, As, Cd, Hg	[74]
	Limoniastrum monopetalum	Cd, Pb	[66]
	Sesuvium portulacastrum	Cr, Cd, Cu, Zn	[75]
	Tamarix gallica	As	[76]

3. Chemical and Physical Agents for Enhancing Plant Resistance to Abiotic Stress

Until now, many different techniques developed by humans have been applied to improve plant tolerance to abiotic stress factors. Some of them are based on conventional breeding; however, they have many limitations, such as being time-consuming, possessing the possibility of transferring numerous undesirable genes along with desirable ones, and having no guarantee of obtaining a particular gene combination responsible for better resistance [77]. Other techniques are related to plant biotechnology and genetic engineering, but these last options are unacceptable in many countries and remain in the laboratory experiments phase [78,79]. As an alternative, increasing attention is being paid to the priming process, i.e., short-lasting pre-exposure of plants to a variety of exogenously applied agents in order to induce a rapid and/or effective defense response to subsequently occurring stress [23,80]. There are many different types of priming methods, which are generally classified into chemical, physical, and biological methods, depending on the source of priming agents. Thus, plants can be primed by natural or synthetic chemical compounds (e.g., phytohormones), by physical factors such as (non-)ionizing radiations, and by colonization with beneficial microorganisms such as bacteria and mycorrhizal fungi [81]. Moreover, priming can be applied to various organs and at various stages of the plant life cycle. The most frequently used is seed priming, which provides faster and more uniform seed germination, ensures efficient nutrient and water uptake, releases photo- and thermo- dormancy, as well as improves seedling vigor in relation to their further growth and yield under both optimal and adverse conditions [82–84]. Less often, priming concerns seedlings, young plants, or their parts although they show significantly greater tolerance to different abiotic stresses than untreated ones [85,86]. Importantly, priming acts on the phenotypic level without any permanent DNA modification, and therefore its effects can be reversed [22]. Moreover, its performance can vary in respect to plant species, temperature, priming duration, priming agents, and their concentration [81].

Currently, priming seems to be the most promising approach for the mitigation of abiotic stress due to various possibilities regarding application. In the present review, we have briefly summarized the latest achievements in the techniques which have attracted the greatest interest recently. Their types, and the general mode of action discussed in this text, are shown in Figure 2.

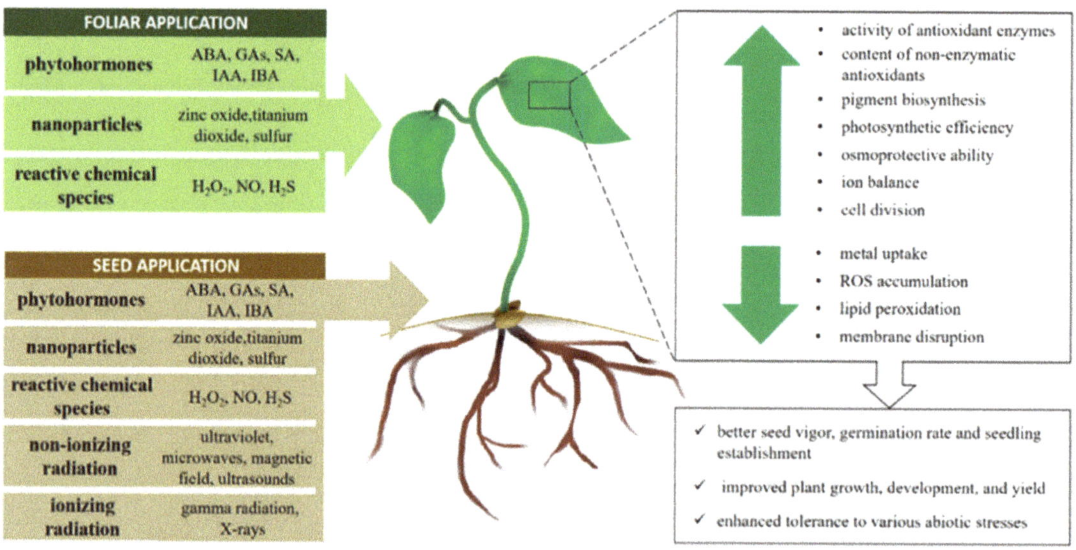

Figure 2. Types of exogenously applied priming agents discussed in this review, as well as their possible applications and the beneficial consequences of treatments that protect plant cells and provide improved defense potential.

3.1. Chemical Priming Agents

Chemical priming is one of the most popular strategies and one which has some advantages. One of them is versatility, since chemical compounds work in a broad number of species and improve tolerance to multiple stress types. Furthermore, chemical agents might be applied directly to selected plant tissues/organs, or during specified developmental stages, in order to minimize growth inhibition [87]. It is a good technique, especially for producing tolerant plants when more conventional methods are difficult to perform [88]. On the other hand, little is still known about the impact of priming agents on ecosystems and their persistence in the environment, although it is anticipated that their application may be a widespread tool in agriculture in the near future. The mode of action for three main groups of chemical priming agents, the ones that are most frequently used, is presented below on the basis of the latest scientific achievements.

3.1.1. Phytohormonal Priming

As has been shown in recent years, exogenous application of phytohormones may increase the metabolic status of plants in response to various abiotic and biotic stresses. In this respect, **abscisic acid** (ABA) is one of the more promising priming agents. The effectiveness of ABA lies in both reducing the ROS pool and activating non-enzymatic and enzymatic ROS scavenging. In research conducted by Saha et al. [89], seedlings of two rice genotypes were pre-treated with 10 µM ABA for 24 h and then exposed to arsenite (As (III)). In contrast to the untreated control, seedlings of both ABA-primed genotypes had reduced accumulation of superoxide anion ($O_2^{\bullet -}$) and hydrogen peroxide (H_2O_2) under arsenite toxicity. Furthermore, lipid oxidative damage, measured by 2-tribarbituric acid reactive substances (TBARS), was reduced by 25% and 48% under metal stress for ABA-treated individual genotypes compared to non-pre-treated ones. Mitigation of oxidative stress in primed seedlings was associated with higher concentrations of total glutathione, non-protein thiols, cysteine, and phytochelatins, as well as the increased activity of glutathione reductase [89]. Similarly, previous studies by Rehman et al. [90] and Leng et al. [16] demonstrated that Cd inhibited plant growth parameters; whereas the application of ABA (10 µM ABA) on seedlings considerably counteracted the Cd-caused negative effect and improved the root length, plant height, and biomass of shoots and roots of mung bean. Also in this case, the enhanced growth of Cd-stressed individuals sprayed with ABA was due to modification of the antioxidant defense systems. Interestingly, it is supposed that leaves-applied ABA can be then transferred to roots in order to regulate the response of the whole plant to metal stress [90–92]. ABA may also act positively on growth and physiological parameters under alkaline stress via effective control of ROS homeostasis, as found for alfalfa seedlings in which the enhanced activity SOD and POD was observed [93]. Furthermore, a significant increase in Ca^{2+} and Mg^{2+} content, as well as higher Ca^{2+}/Na^{+} and Mg^{2+}/Na^{+} ratios, was noticed in primed seedlings under alkaline conditions. In addition, genes encoding some important proteins involved in the sequestration of Na^{+} in vacuoles, i.e., vacuolar Na^{+}/H^{+} exchanger (NHX) and vacuolar H^{+}-PPase (AVP), which might help in neutralization of its excess amount, were overexpressed in primed seedlings [93].

Recent studies showed that, as well as ABA, priming with **gibberellins** (GAs) has a positive effect on plant growth under stress conditions. A study by Ahmad et al. [39] showed that foliar application of GA_3 (1 µM) on chickpea seedlings resulted in the increased activity of antioxidant enzymes (SOD, CAT, GST), which provided effective ROS scavenging and reduced membrane disruption, thus ensuring tolerance to Cd stress. The better response of plants treated with GA_3 to Cd presence can be also attributed to the reduced uptake and translocation of toxic ions, as well as increased accumulation of nutrient minerals (Ca, Na, Mg, K, Cu, P, Fe) [39]. This could possibly be achieved through the regulation of H^{+}-ATPase activity, as shown for soybean [94]. The advantageous impact of GA_3 application on morpho-physiological parameters and stress mitigation was also determined in *Lolium perenne* under Ni and Cd exposure and in *Lepidium sativum* under As treatment [95,96].

The latest articles also indicate the beneficial role that priming with **salicylic acid** (SA) has regarding plant tolerance to abiotic stresses; however, in the case of this phytohormone, seed priming seems to be the most effective technique. It has been recently found that the soaking of wheat seeds in SA at a concentration of 100 µM for 24 h results in significant improvements in germination rate and growth parameters in the presence of chromium (Cr) and Zn due to the prevention of ROS imbalance associated with the increase in the concentration of non-enzymatic antioxidants, mainly AsA and GSH [97]. In turn, SA-primed sunflower seeds exposed to Zn showed better germination properties because the exogenous SA application modulated the endogenous profile of the phytohormones [98]. It was noted that concentrations of SA and GA were increased, while ABA accumulation was inhibited as a result of the overexpression of genes related to SA and GA biosynthesis and the decrease in the expression of ABA-related genes that occurred in combination with a simultaneous increase in the expression of genes engaged in the catabolism of this phytohormone. Additionally, the role of SA in metal stress mitigation may also result from the upregulation of genes encoding proteins related to ion transport, such as heavy metal ATPases and metal tolerance protein (MTP), the overexpression of which provided a reduced accumulation of Zn in sunflowers [98] and Cr in tomato [99]. In these latter examples, seed soaking or foliar spraying with SA at a concentration of 0.5 mM ameliorated growth and the physiological reaction to Cr, in respect to chlorophyll biosynthesis and photosynthetic efficiency, through modulation of the ascorbate-glutathione (AsA-GSH) cycle that contributes to a decline in ROS accumulation and lipid peroxidation [99].

It is well known that the exogenous application of auxin, especially **indole-3-acetic acid** (IAA), or its precursors improves growth and development of plants; however, its role in the mitigation of metal stress is not fully understood, and the physiology of these tolerance mechanisms remains largely unknown. Nevertheless, a study by Mir et al. [85] revealed that foliar application of IAA (at a dose of 10 nM) on *Brassica juncea* plants under Cu stress significantly mitigated adverse responses due to the activation of cell division and elongation, as well as lateral root formation in which this phytohormone is involved. Furthermore, in *B. juncea* plants sprayed with IAA, effective ROS scavenging was observed which, together with improved photosynthesis and chlorophyll fluorescence parameters, sugar metabolism, and N, P, and K content, led to biomass accretion [85]. In turn, priming with indole-3-butyric acid (IBA), an IAA precursor, provided antioxidant protection through the stimulation of glutathione peroxidase activity and greater accumulation of nitric oxide (NO) that effectively reduced the elevated level of superoxides and organic peroxides in the root cells of barley seedlings under Cd stress [100].

3.1.2. Nanoparticle Priming

Nanotechnology is an emerging field with potentially wide-ranging applications in agriculture. The use of nanoparticles (NPs) in plant production, as well as in enhancing plant growth under stressful conditions, including those related to environmental pollution with heavy metals, has increased significantly in recent years [101,102]. Several studies on the seed priming of various plant species with **zinc oxide NPs** (ZnO NPs) have been published, with results indicating the beneficial effects of this NP on germination and growth. Wheat seeds primed with ZnO NPs (at a concentration of 10 mg/L) exhibited better germination rates and vigor index values compared to untreated seeds. In seeds primed with ZnO NPs, increased α-amylase activity was observed that could facilitate the efficient mobilization of starch reserves. Moreover, in plants 30 days after seed priming, increased photosynthetic pigment content (chlorophyll *a*, chlorophyll *b*, and total chlorophylls) and improved photosynthetic efficiency compared to untreated plants were determined. Additionally, the use of nanopriming had a positive effect on redox homeostasis in wheat plants [103]. ZnO nanoparticles, sodium selenite (Na-selenite), sodium selenate (Na-selenate), and their combinations as priming agents for direct-seeded rice seeds were also investigated [84]. It was observed that all tested combinations of the priming agents (10 μmol ZnO-NPs; 50 μmol Na-selenite; 50 μmol Na-selenate; and the following combinations at the mentioned concentrations: Na-selenite + Na-selenate; ZnO-NPs + Na-selenite; ZnO-NPs + Na-selenate; ZnO-NPs + Na-selenite + Na-selenate) resulted in the early emergence of seedlings with increased vigor compared to the control. Furthermore, in the field experiment, all tested combinations improved the plant growth parameters and yield, which was the result of increased photosynthetic pigments, increased phenol and protein content, and the increased uptake of nutrients such as N, P, and K [84]. Salam et al. [80] showed that priming maize seeds with ZnO NPs nanoparticles (500 mg/L for 24 h) significantly improved plant growth, biomass, and photosynthesis efficiency under cobalt (Co) stress. In this case, priming also caused a reduction in ROS accumulation and lipid peroxidation due to increased antioxidant activity in maize shoots. Additionally, priming with ZnO NPs reduced the toxic effect of Co by reducing its absorption. More importantly, the ultrastructures of cell organelles, guard cells, and stomatal aperture were stabilized and able to reduce the adverse effects of Co stress. In turn, the study by Zafar et al. [86] showed the effect of seed priming and the foliar application of Zn NPs (0.1–0.3%) on spinach salinity tolerance. It was found that external use of ZnNPs enhanced the growth of spinach plants, as well as improved biochemical parameters under stress conditions compared to untreated plants. Seed soaking and foliar application of ZnNPs provided a decline in H_2O_2 content accompanied by the activation of enzymatic and non-enzymatic antioxidant defense systems, as well as simultaneous accumulation of osmolytes.

The positive effect of priming was also demonstrated in the case of **titanium dioxide NPs** (TiO_2 NPs). Shah et al. [104] investigated the effect of seed priming with TiO_2 NPs on the germination and growth of maize seedlings under salinity conditions. Priming with TiO_2 NPs (60 ppm) resulted in improved germination percentage and energy, improved seedling vigor index values, increased root and shoot length, and improved fresh and dry weights of seedlings. Moreover, priming increased the activity of antioxidant enzymes and ROS scavenging capacity. This experiment showed that priming with TiO_2 NPs reduced the adverse effects of salinity stress in maize seedlings, as evidenced by a reduction in membrane lipid peroxidation and the relative electrolyte leakage level.

Recently, the effect of priming sunflower seeds with **sulfur NPs** (S NPs) on the cellular defense of seedlings against manganese (Mn) toxicity was also investigated. Priming with S NPs (50 and 100 μM) had a significant impact on reducing oxidative damage caused by excess H_2O_2, which was reflected in decreased lipid peroxidation. In primed seedlings, the values of these parameters under Mn stress were similar to those observed in seedlings growing under the control conditions [105].

3.1.3. Priming by Reactive Chemical Species

A significant amount of research has confirmed that the pre-treatment of plants or seeds with low concentrations of reactive oxygen, nitrogen, and sulphur species (such as H_2O_2, sodium nitroprusside (SNP), one of the donors for NO, or sodium hydrosulfide (NaHS), a donor for hydrogen sulphide (H_2S)) strengthens their resilience to later stress events [106–109]. Improved resistance to abiotic stress may be due to the fact that, at low concentration, these compounds can act as a stress signal transduction which induces stress acclimation and alleviates abiotic stress injury [87]. They play a significant protective role, mainly due to the induction of tolerance to oxidative stress caused by drought, salinity, temperature, or metal toxicity [107,110]. On the other hand, too high a concentration of these reactive chemical species results in oxidative burst and damage to cellular compounds [87].

The exogenously sourced H_2O_2 (at a concentration ranging from 100 to 500 μM) has the potential to counteract the toxicity of metallic trace elements in a number of plants, and its mode of action was briefly summarized in some review articles, such as those written by Hossain et al. [111] and Cuypers et al. [112] which discussed H_2O_2 interaction with signaling components (e.g., transcription factors, phytohormones, mitogen-activated protein kinases) as well as its involvement in the regulation of ROS homeostasis and gene expression during metal stress. Based on various studies, it can be assumed that the positive effects of H_2O_2 priming prior to metal exposure include the reduced accumulation of ROS accompanied by an enhanced activity of antioxidant enzymes, such as SOD, CAT, GPX, APX, and GST, as well as elevated levels of reduced forms of non-enzymatic antioxidants such as GSH and AsA [111]. This may be related to proactive protection of the thiol groups present in proteins that are particularly exposed to oxidation under stressful conditions [113]. Besides suppressing oxidative damage, the accumulation of GSH plays a role in metal detoxification in the cytosol through direct ion binding to thiol groups of its cysteine residues and acts as a precursor of metal-chelating phytochelatins [112]. Indeed, the reduced translocation of Cd ions from root to shoot was demonstrated in *Oryza sativa* cultivars pre-treated with H_2O_2 [114]; whereas an opposite result was obtained for Cr in *Brassica napus* seedlings in which foliar application of H_2O_2 increased metal movement from roots to aerial organs [115].

Interestingly, more and more recent studies concern the simultaneous application of H_2O_2 and other compounds in order to explore their cumulative role in metal stress resilience. As an example, the combination of H_2O_2 with 24-epibrassnolide (EBL), an effective by-product from brassinolide biosynthesis, provided tolerance and helped *Solanum lycopersicum* plants to cope well with Cu stress [58]. The positive morpho-physiological response of tomato to Cu treatment was related to a decreased accumulation of these metallic ions in the roots and shoots. Such an effect resulted from the complementary action of both applied molecules, since H_2O_2 may affect the absorption and transport of excess Cu ions to above-ground organs due to Cu precipitation at the root surface and preferentially affect the uptake of Ca; whereas EBL improves the accumulation of K, Ca, Fe, and Mg, which are translocated to younger leaves to minimize oxidative damage in photosynthetic machinery [58]. Although, in the study conducted by Nazir et al. [58], H_2O_2 and EBL were implemented through distinct modes, i.e., root dipping and foliar spraying, respectively, they both minimized ROS content (H_2O_2 and $O_2^{\bullet-}$) and electrolyte leakage in Cu-stressed plants by modulating the activities of antioxidant enzymes (CAT, POD, SOD) and providing osmotic adjustment through increased storage of proline. In turn, Verna and Prasad [116] investigated the involvement of H_2O_2 and NO when applied jointly in the regulation of Cd toxicity in cyanobacteria (from genera of Nostoc and Anabena). Their findings demonstrated the synergistic action of both molecules towards the improved growth and enhanced tolerance of cyanobacteria to Cd. In this case, H_2O_2 and NO reduced the intracellular content of Cd through an increased secretion of exopolysaccharides, which make a slimy physical barrier against ion penetration into the protoplast. Furthermore, tested cells were characterized by a well-operating antioxidant defense system, and ROS homeostasis was provided by the enhanced activity of antioxidant enzymes and the endogenous content of reactive nitrogen species that indirectly responded to the balancing of antioxidants in order to cope up with Cd stress [116].

Taking into account other abiotic stresses, research by dos Santos Araújo et al. [117] showed that H_2O_2 promoted salt tolerance in maize by protecting chloroplast ultrastructures, as reflected in more efficient photosynthetic performance. Furthermore, plants treated with 15 mM H_2O_2 and then exposed to salinity showed increased accumulation of metabolites, such as arabitol, glucose, asparagine, and tyrosine, which may contribute to the maintenance of osmotic stability and reductions in oxidative stress [117]. The role of H_2O_2 in salt stress prevention can also be attributed to ion homeostasis. After priming, a decline in Na^+ and Cl^- content in the leaves of sunflowers was observed during salinity stress, as well as positive control of K^+ and NO_3^- uptake [109]. In turn, the beneficial activity of H_2O_2 and NO towards drought stress was noticed by Habib et al. [107]. Despite stress conditions, pre-treated wheat plants exhibited increased growth and grain yield as a result of osmolyte storage and the effective functioning of an antioxidant defense mechanism, leading to a reduced accumulation of H_2O_2 and membrane lipid peroxidation [107]. Drought stress effects on agronomic features of plants were also minimized in the case of *Oryza sativa* after both seed soaking and foliar spraying with H_2O_2 [118]. Regardless of the application form, rice plants pre-treated with this molecule showed improved yield components such as tiller numbers, number of panicles, number of filled grains, filled grain weight, and harvest index [118].

The beneficial role of **exogenously applied NO**, used in the form of donor compounds (mainly SNP) due to its gaseous nature, has also been well-documented. In experiments that involve increasing stress tolerance, NO is applied the most frequently via foliar spraying [119] or seed soaking [120] at a concentration of 50 μM to 200 μM. Similar to other reactive chemical species, NO can prevent the spread of oxidative stress in cells. As an example, SNP enhanced the activity of enzymatic antioxidants and the AsA-GSH cycle in soybean cultivars under Cu stress [121]. The alleviation of Co stress by foliar-applied SNP in *Lactuca sativa* var. *capitata* resulted in a notable reduction in H_2O_2 and malonyldialdehyde (MDA) content, enhanced accumulation of photosynthetic pigments, and biomass accretion that was accompanied by the better nutritional status of plants [122]. In turn,

Basit et al. [120] studied the impact of SNP under Cr stress on rice seedlings. It has been shown that seed priming improved carbon assimilation and minimized oxidative damage, since NO-treated plants were characterized by lower accumulation of oxidative markers (such as H_2O_2, $O_2^{\bullet-}$, and MDA) and electrolyte leakage as compared to control plants. Consequently, their morphological traits were also improved [120]. It was also proven that NO stimulated seed germination and counteracted the inhibitory effect of Cd and Pb (and salinity as well) on the root growth of *Lupinus luteus*. Additionally, in this case, the increased activity of antioxidant enzymes, mainly SOD which is responsible for the neutralization of $O_2^{\bullet-}$, was correlated with a decreased level of ROS [123]. Although it would appear most likely that NO modulation of antioxidant enzyme activities and phenol and flavonoid production provides stress amelioration, Hassanein et al. [124] observed the opposite tendency in *Lupinus albus* subsp. *termis* in response to SNP and Ni treatment. Therefore, it was postulated that NO may act as an antioxidant molecule, interacting directly with ROS and giving rise to a number of reactive nitrogen species and their derivatives, which are rapidly degraded to nitrite and nitrate [124]. This is in accordance with the study by He et al. [125] which found that, regardless of aluminum presence, SNP significantly suppressed the generation of $O_2^{\bullet-}$ and H_2O_2 by mitochondria in peanut root tips.

The addition of NO (pre-sowing and foliar) can also minimize the adversaries of salinity stress, not through the activation of antioxidant machinery, but mainly due to osmotic adjustment and Na ion homeostasis. As an example, NO-increased tolerance in broccoli was associated with higher amounts of proline and glycine betaine keeping water potential in cells below the external solution under stress conditions [57]; whereas, in wheat, the antagonistic uptake of toxic Na^+ with key mineral elements, such as N, K, and Ca, reduced the deleterious effects of salt [119]. Importantly, Alnusairi et al. [119] showed that the application of NO may dismiss salt stress-mediated ravaging by the overexpression of genes encoding both antiporters that are responsible for excluding Na ions from the cytosol to outside the plasma membrane or inside the vacuole (SOS1/NHX1), and aquaporin (AQP) as well as osmotin (OSM-34) which are involved in the maintenance of proper plant–water relations. In turn, the protective effect of exogenous NO under drought stress may be dose-dependent. Majeed et al. [108] found that a foliar spray of 100 μM of SNP markedly improved water status and chlorophyll content and alleviated drought-induced oxidative damages through increased antioxidant enzyme activities (CAT, APX, SOD) in maize hybrids. Moreover, an exogenous supply of SNP increased nitrite and nitrate reductase activities and upregulated GR, GST, and GPX compared to plants not supplied with SNP [108]. In contrast, higher SNP doses (150 and 200 μM) intensified the toxic effects of oxidative stress through increased MDA, H_2O_2, and NO content and inhibited the enzymatic activities of antioxidants.

Many studies have indicated the significant role of **H_2S** priming in the response of plants to various abiotic factors [126]. It has been reported that pre-treating seedlings or a mature plant with NaHS as a H_2S donor may increase the tolerance of the plant upon following exposure to heavy metals such as Pb, Ni, and As [96,127]. Although many studies have assessed the positive effect of the pre-treatment of seedlings or mature plants with H_2S in relation to enhancing plant tolerance, few studies have employed H_2S for seed priming. Valivand et al. [128] reported that seed priming with Ca^{2+} and NaHS influenced the induction of cross-adaptation in seedlings under Ni stress. The authors reported that seed priming with H_2S and Ca^{2+} triggered signaling pathways, which resulted in the systemic accumulation of dormant stress memory in embryo cells in seeds. Upon subsequent exposure to Ni ions, stress memory was activated and primed plants showed enhanced tolerance-related responses, e.g., enhanced AsA-GSH cycle activity, redox homeostasis, and expression of phytochelatin genes [128]. In turn, Zanganeh et al. [127] reported that pre-treatment with NaHS, applied separately and together with SA, reduced Pb toxicity and improved Fe homeostasis in maize plants. The mechanism of their action was related to modulation of the glyoxalase system consisting of enzymes detoxifying methylglyoxal, which is a potent reactive cytotoxin capable of a complete disturbance of cellular roles, including oxidation of lipids and proteins [127,129].

Christou et al. [130] studied the effect of NaHS (100 μM for 48 h) on the tolerance of strawberry plants to subsequent exposure to salinity. Pre-treatment of roots resulted in increased leaf chlorophyll fluorescence, stomatal conductance, and leaf relative water content, as well as lower lipid peroxidation levels. Additionally, synthesis of NO and H_2O_2 in leaves was reduced and high ascorbate and glutathione redox states were maintained. The observed positive changes correlated with the stimulated gene expression of antioxidant enzymes (cytosolic APX, CAT, MnSOD, GR), enzymes involved in ascorbate and glutathione biosynthesis (glutamylcysteine synthetase; L-galactose dehydrogenase; glutathione synthetase), a transcription factor (DREB), and salt overly sensitive (SOS) pathways (SOS2-like, SOS3-like, SOS4) [130]. Hydrogen sulfide pre-treatment (500 μM NaHS for 72 h) also mitigated growth inhibition and regulated root architecture under salt stress in *Malus hupehensis* seedlings, not only through the activation of antioxidant defense (mainly CAT and POD activities), but also through maintaining the balance of water (by proline accumulation) and Na^+/K^+ (by higher uptake of K than Na ions) as well [131]. Under drought conditions, H_2S may improve tolerance by regulating stomatal closure and reducing water loss thanks to ABA synthesis and signaling, which was noticed in *Oryza sativa* seedlings together with an increase in endogenous H_2S production and antioxidant capacity [132]. However, an innovative approach in the use of H_2S as a priming agent is its application in combination with NO. In this respect, NOSH is a novel hybrid synthetic compound that simultaneously releases NO and H_2S. Antoniou et al. [133] demonstrated that NOSH synthetic compounds provide significant protection in *Medicago sativa* plants against drought stress. This protection appears to be achieved through a coordinated modification of improved physiological performance, reactive oxygen/nitrogen species homeostasis, and transcriptional regulation of defense-related pathways [133].

3.2. Physical Priming

Priming with physical factors includes a number of methods, especially those related to radiation. Among them, both non-ionizing radiations, such as UV radiation, microwaves, magnetic field radiation, and sonication, and ionizing radiations, i.e., X-ray radiation and γ-radiation, can be distinguished [134]. Physical priming is considered to be an accessible, affordable, and eco-friendly technique which brings beneficial effects on seed parameters, the metabolic activities of plants, and plant development and growth [135]. It has the advantage, over chemical priming, that it does not pollute the environment, which is an important aspect in agriculture, especially if the contemporary injudicious application of chemical compounds during food production is taken into account [136]. Therefore, until recently, physical treatment was successfully employed in crops, mainly for stimulating seed germination and seedling establishment since these stages are considered to be the most critical stages in the life cycle and ultimately determine field production. This aspect of physical priming application has been widely discussed over the past few years [6,134–137].

Despite increasing understanding of the effects of physical priming performance on plants under optimal conditions, data in the literature on its application to alleviating stress nuisance are still limited, especially in respect to metal toxicity. Thus, in the present review we have focused on the latest achievements that are related mostly to drought and salinity, during which physical priming strengthens antioxidant response. It is therefore likely that physical treatment of plants subsequently exposed to excess amounts of metallic trace elements will bring comparable responses. Nevertheless, the mode of the physical agent's actions in plants under metal stress is also mentioned whenever the most recent studies were available.

3.2.1. Priming with Non-Ionizing Radiation

Ultraviolet (UV) radiation is a type of electromagnetic radiation with a vibration frequency between 30 PHz and 750 THz, photon energy between 3 and 124 eV, and a wavelength between 10 and 400 nm, which is shorter than visible light but longer than X-ray radiation. UV radiation is divided into UV-A, UV-B, and UV-C, with UV-A radiation being the least harmful to living organisms and UV-C the most [134]. Both seed and seedling UV-B priming, applied for 45 min at 4 kJ/m^2 intensity, was shown to effectively alleviate oxidative stress and its resulting damage by significant reductions in superoxide, H_2O_2, and MDA content in stress-sensitive rice variety (*Oryza sativa* cv. Aiswarya) seedlings under stressful conditions caused by NaCl, PEG, and UV-B treatments [138]. The study also demonstrated that UV-B priming led to significant increases in glutathione and ascorbate contents, SOD, CAT, and APX activity, gene expression levels, photosystem activities, foliar gas exchange parameters, and, finally, in mitochondrial activity. The increases were the most pronounced in seedlings subjected to NaCl stress. Similar results were reported for UV-B primed seeds for two varieties of rice: Neeraja and Vaisakh. Additionally, reductions in leaf osmolarity level, increases in proline, total sugar, and free amino acids content, and induced expression levels of stress-related proteins (Hsp90 and Group 3 late embryogenesis abundant proteins) under NaCl and PEG stress were observed [139]. The observed differences were significantly higher in the tolerant variety (Kanchana) than in the sensitive one (Aiswarya) and were also reported by Thomas et al. [138], who found that UV-B priming at low doses (4 and 6 kJ/m^2) led to increased levels of flavonoids and anthocyanins, the increased activity of phenylalanine ammonia lyase, and increased levels of cuticular wax in rice seedlings under UV-B, NaCl, and PEG stress. The UV-C seedling priming of a cumulative dose of 10.2 kJ/m^2 was also found to significantly reduce leaf spot disease severity in strawberry plants due to induced accumulation of pathogenesis-related proteins, terpenes, phenolic compounds with triggered ROS, and antioxidant enzymes, while also inducing plant hormone synthesis [140]. Moreover, the transgenerational effect of UV-B priming was shown by the rice seedlings of the drought-tolerant Vaisakh variety being characterized by the increased expression of genes encoding antioxidant enzymes and stress-related proteins in F0 generation, with even more of an increase in the F1 generation after re-priming. This resulted in better protection against PEG stress [141]. The UV-B priming protection against UV-B stress was proven to be related to the UV RESISTANCE LOCUS (UVR8) pathway in *Arabidopsis thaliana*, since 14-day old seedlings without UVR8, primed for 10 min with UV-B at 35 μW/cm^2, did not acquire UV-B resistance [142].

Microwave radiation is a form of electromagnetic radiation with a frequency ranging between 300 MHz and 300 GHz [143]. Physical seed priming with microwave radiation at 2.45 GHz for a short time had stimulatory effects on seed germination, seedling growth, and biomass accumulation in different cereals, such as barley, rice, and wheat [144]. A study by Bian et al. [145] proved that treatment of *Fagopyrum tataricum* with microwaves with a power of 300 W and a frequency of 2.45 GHz for 75 s optimally increased the activity of antioxidant enzymes (SOD, CAT, POD, and APX), leading to the increase in the total reduction potential of plants and the ability of the seedlings to neutralize radicals such as 2,2-diphenyl-1-picrylhydrazyl (DPPH), 2,2′-azino-bis(3-ethylbenzothiazoline-6-sulfonic acid (ABTS), $O_2^{\bullet-}$, and $^{\bullet}OH$ [145]. Microwaves have been used as factors for improving the resistance of crops to a number of stress factors. Maswada et al. [146] applied microwave priming prior to sowing two *Triticum aestivum* genotypes, Giza 168 and Gharbiya. The results of the conducted experiments proved that microwave priming (with 700 W of power, a variable frequency of 2.45 GHz, and a wavelength of 125 nm with a power intensity of 126 mW/cm^2) improved wheat resistance to drought, increasing both the yield and growth parameters through improvement of tissue water content and a reduction in membrane permeability. Furthermore, osmotic adjustment and decreased H_2O_2 accumulation through increasing proline content and ROS scavenging activity were also observed [146]. In turn, Farid et al. [147] showed that microwaves can also help to alleviate heavy metal stress in *Brassica napus*. Pre-saw treatment of the genotype Faisal Canola (RBN-03060), with microwaves with a frequency of 2.45 GHz for 30 s, led to a greater ability to grow and biomass accretion for plants treated with Ni. Heavy metal-stressed plants also showed a higher concentration of photosynthetic pigments, including chlorophyll *a* and *b* and carotenoids, and higher antioxidant enzyme activity (SOD, POD, APX, CAT), which was associated with a reduction in ROS (H_2O_2) content and the oxidative damage caused by them (MDA, electrolyte leakage). Moreover, it was shown that microwave priming resulted in greater accumulation of Ni from the soil, especially in roots, stems, and leaves [147].

Magnetic fields can be used for priming in several variants: as alternative magnetic field (AMF), electromagnetic field (EMF), pulsed magnetic field (PMF), static magnetic field (SMF), and sinusoidal magnetic field (SSMF) priming. All of those techniques were used in research on crops, and their application improved the germination and vigor of plants, as well as the response to unfavorable environmental factors, although SMF is the most common one [148]. Mohammadi and Roshandel [149] applied SMFs of 90 mT, 200 mT, and 250mT on *Hyssopus officinalis* plants for 5 min. The best effect was obtained at 200 mT. In response to drought stress, the plants subjected to magnetopriming showed higher dry matter content, total chlorophyll and phenol content, and a higher reduction capacity (DPPH, $O_2^{\bullet-}$ scavenging) resulting from, among other factors, higher CAT, APX, and GPX activity. At the same time, magnetopriming led to a reduction in oxidative damage to biological membranes, which reduced electrolyte leakage [149]. Kataria et al. [150] applied an SMF of 200mT on soybean for 1 h, which resulted in increased resistance to salinity. Plants subjected to magnetopriming showed greater leaf area, leaf mass, photosynthetic activity, and nitrogenase activity than plants subjected to salinity stress only. However, the content of H_2O_2 and AsA, and the activity of antioxidant enzymes, was reduced due to magnetopriming. These changes resulted in higher biomass accumulation, yield, and harvest index values for soybean under both the saline and non-saline conditions [150]. Baghel et al. [151] showed that the use of a 200 mT SMF for 1 h on *Zea mays* plants reduced their susceptibility to salinity. Magnetopriming increased the content of photosynthetic pigments, as well as increasing photosynthesis parameters such as the quantum yield of PSII photochemistry (F_v/F_m), electron transport per leaf CS (ETo/CSm), the density of reaction centers (RC/CSm), and the performance index (PI). Moreover, the maize leaves showed lower H_2O_2 accumulation, which proves the reduction in oxidative stress. These changes resulted in better plant growth and increased maize yield under salinity conditions [151].

Ultrasound priming involves treating plants or seeds with the energy of acoustic waves with a frequency greater than 20 kHz [134]. Xia et al. [137] applied high-intensity ultrasound (HIU) with a frequency of 28 kHz and a power of 17.83 W/cm^2 for 5 to 30 min on brown rice seeds. Ultrasound priming led to both an increase in starch content and a simultaneous reduction in the size of grains and in the content of reducing sugars. Moreover, the accumulation of free amino acids, γ-aminobutyric acid, antioxidants, and proline (as stress-responsive secondary metabolites) may also have potentially positive effects on plant response to adverse environmental factors [137]. Dashab and Omidi [152] primed *Brassica napus* with ultrasound at 40 kHz and 59 kHz with a power of 60, 80, and 100 W for 2, 4, 6, 8, and 20 min. Depending on the combination of those parameters, different physiological effects were achieved. The greatest increase in seed germination was observed with 40 and 59 kHz at 100 W for 2 min of exposure, while an increase in vigor and seedling weight was observed with 59 kHz at 100 W. At 40 kHz with 80 W and an exposure time of 8 min, an increase in the content of photosynthetic pigments was determined [152]. In turn, Rao et al. [135] treated canola cultivars Youyanzao18 and Zaoshu104 for 1 min with ultrasound at a frequency of 20 kHz in order to reduce susceptibility to Cd stress. It has been shown that ultrasound, depending on the Cd dose, can improve such parameters as germination, shoot and root length, and fresh mass. Moreover, in the Youyanzao18 cultivar, ultrasound priming increased the activity of SOD, POD, CAT, and APX, as well as the increased content of proline, GSH, and soluble protein. This translated to a reduction in MDA content, which indicates less oxidative damage to biological membranes in response to Cd. In both cultivars, ultrasound increased pods per plant, seeds per pod, and rapeseed yield. Importantly, the accumulation of Cd in all parts of the plant decreased [135]. Similarly, Chen et al. [153] demonstrated that ultrasonic vibration can help wheat seedlings eliminate an excess amount of ROS resulting from Cd and Pb treatment, as well as improve the biosynthesis of molecules and division of cells, leading to biomass accretion despite the metal stress.

3.2.2. Priming with Ionizing Radiation

Gamma (γ) radiation is a high-energy type of ionizing radiation capable of penetrating and interacting with living tissues, whose absorbed dose is expressed in units of Gray (Gy). Usually, Cobalt-60 is used for this type of priming [144]. When Hussein [83] used 5, 10, and 20 Gy gamma radiation on barley plants, it was shown that both lower doses improved plant growth and yield, while the highest one (20 Gy) increased shoot growth and tiller number; however, only at the lowest radiation dose (5 Gy) was an increase in the content of photosynthetic pigments observed. Gamma radiation enhanced the accumulation of phenols, flavonoids, free amino acids, and antioxidant enzymes (APX, POD, CAT), but it also elevated H_2O_2 content. Moreover, it led to a reduction in the content of sugars and proline [83]. Researchers have proven that low doses of γ-rays not only modify redox homeostasis, but they also change the protein pattern and the metabolic profile in plants, leading to improved growth and yielding. As an example, a study by Hanafy and Akladious [154] showed that a dose of 100 Gy improved growth and yield for *Trigonella foenum-graecum* plants, as well as the content of soluble proteins in leaves and the content of phenols and flavonoids. Moreover, a significant rise in the content of AsA, α-tocopherol, retinol, and proline was observed. In turn, the highest dose of radiation (400 Gy) caused a decrease in the content of all tested parameters and induced changes in the DNA profile that consisted of the appearance and disappearance of polymorphic bands [154]. Pradhan et al. [155] used gamma radiation at a dose of 10 Gy on the microalgae *Chlamydomonas reinhardtii* (which is considered to be a model organism for studying the effects of heavy metals on photosynthetic organisms) and exposed it to Cd stress. As a result of Cd treatment, redox homeostasis was disturbed due to a decline in antioxidant enzyme activity and in the content of photosynthetic pigments. Consequently, cell death was induced and growth was minimized. On the contrary, the application of γ-radiation had positive effects on the mentioned parameters, and cell growth and biochemical synthesis were not injured. As a consequence, an increased resistance to toxic Cd ions was achieved [155].

X-rays are characterized by a wavelength ranging from 0.01 to 10 nm of the electromagnetic spectrum, which corresponds to frequencies ranging from 30 to 30,000 PHz and energies oscillating from 120 eV to 120 keV [144]. Currently, there are very few new studies concerning the effects of this type of priming on plants; however, in 2019, Rezk et al. [82] used X-rays in doses from 0 to 100 Gy on two genotypes of okra (*Hibiscus esculentus*), genotypes of Hassawi and Clemson. It was shown that radiation doses up to 5 Gy improved plant morphological parameters, the content of photosynthetic pigments, the activity of antioxidant enzymes (CAT, SOD, APX), and the content of low-molecular weight antioxidants (AsA, GSH, anthocyanins). In contrast, higher doses of radiation (at levels above 5 Gy) had the opposite effect, and plants treated with this type of priming showed greater lipid peroxidation caused by the increased concentration of ROS (mainly H_2O_2 and $O_2^{\bullet -}$) [82]. This confirmed the previous discoveries of Al-Enezi et al. [156] regarding the influence of X-rays on date palm (*Phoenix dactylifera* cv. Khalas). In this case, the inhibitory impact of radiation on seed germination was noticed even at a dose of 0.25 Gy, and a graduated increase in X-ray dose up to 15 Gy contributed to further reductions in germination; however, at the same time, an increase in root length was observed. A similar stimulatory effect was found for the leaf length of the date palm plants, but it concerned only X-ray doses between 0.05 and 0.25 Gy [156]. It can be summarized that only low doses of this type of radiation may improve plant growth parameters, but little is still known about its ameliorative actions under various stress conditions and further research is therefore required.

4. Concluding Remarks

In the present review, we have briefly discussed the adaptative traits of metallophytes, whose application may be an antidote to environmental pollution with heavy metals, perceived as one of the most dangerous factors for all living organisms. The amazing biology of metallophytes, especially in respect to metal detoxification and accumulation, as well as tolerance to drought and salinity, make them applicable for the phytoremediation and reclamation of chemically degraded areas which, after returning to their original state before contamination, can be reused for different goals. Furthermore, deeper insight into plants with evolutionarily developed tolerance mechanisms, may help to obtain specimens with ideal survival levels and fertility under stressful conditions. It seems to be particularly important to take into account that the majority of plants do not exhibit tolerance to abiotic stresses developing as a result of severe selection pressure due to the complexity associated with the inheritance of adaptive traits. Therefore, to combat the most important global problems, including metal pollution, drought, and salinity, through biological methods and to provide sustainable agriculture and food security for continuing global population growth, increasing attention is being given to priming strategies which make plants capable of responding more effectively and more rapidly to stress. Since priming offers a large variety of priming factors, doses, and application forms, the diversified morphological and biochemical responses of plants can be observed. Thus, the chemical and physical treatment for stress amelioration requires extensive future research for the elaboration of specific protocols in respect to optimal dosage and duration of exposure, which certainly vary between genotypes and environmental conditions. Furthermore, further understanding of both the mode of actions of particular priming agents and the mechanisms underlying the better performance of primed plants can lead to combined usage of various priming methods, preferably with synergistic effects that would allow a reduction in the dose of each agent compared to the dose used individually. Undoubtedly, the joint knowledge gathered here clearly indicates that all priming agents contribute to the scavenging of excess amounts of ROS via efficiently operating antioxidant machinery and thus put oxidative mitigation at the core of enhanced tolerance to various stressors.

Author Contributions: Conceptualization, E.M.; formal analysis, E.M.; writing—original draft preparation, M.L., K.D., J.F., M.G., A.R.-P., M.N., B.P., I.M. and E.M; visualization, M.G. and E.M.; supervision, E.M.; funding acquisition, M.L., K.D. and E.M. All authors have read and agreed to the published version of the manuscript.

Funding: This research received no external funding.

Institutional Review Board Statement: Not applicable.

Informed Consent Statement: Not applicable.

Data Availability Statement: Not applicable.

Conflicts of Interest: The authors declare no conflict of interest.

References

1. Chakraborti, S.; Bera, K.; Sadhukhan, S.; Dutta, P. Bio-priming of seeds: Plant stress management and its underlying cellular, biochemical and molecular mechanisms. *Plant Stress* **2022**, *3*, 100052. [CrossRef]
2. Savvides, A.; Ali, S.; Tester, M.; Fotopoulos, V. Chemical priming of plants against multiple abiotic stresses: Mission possible? *Trends Plant Sci.* **2016**, *21*, 329–340. [CrossRef]
3. AbdElgawad, H.; Zinta, G.; Hamed, B.A.; Selim, S.; Beemster, G.; Hozzein, W.N.; Wadaan, M.A.M.; Asard, H.; Abuelsoud, W. Maize roots and shoots show distinct profiles of oxidative stress and antioxidant defense under heavy metal toxicity. *Environ. Pollut.* **2020**, *258*, 113705. [CrossRef]
4. García-Caparrós, P.; De Filippis, L.; Gul, A.; Hasanuzzaman, M.; Ozturk, M.; Altay, V.; Lao, M.T. Oxidative stress and antioxidant metabolism under adverse environmental conditions: A review. *Bot. Rev.* **2021**, *87*, 421–466. [CrossRef]
5. Liu, D.; Gao, Z.; Li, J.; Yao, Q.; Tan, W.; Xing, W.; Lu, Z. Effects of cadmium stress on the morphology, physiology, cellular ultrastructure, and *BvHIPP24* gene expression of sugar beet (*Beta vulgaris* L.). *Int. J. Phytoremediation* **2022**, *30*, 1–11. [CrossRef]
6. Farooq, M.; Usman, M.; Nadeem, F.; Rehman, H.; Wahid, A.; Basra, S.M.A.; Siddique, K.H.M. Seed priming in field crops: Potential benefits, adoption and challenges. *Crop Pasture Sci.* **2019**, *70*, 731–771. [CrossRef]
7. He, G.; Tian, W.; Qin, L.; Meng, L.; Wu, D.; Huang, Y.; Li, D.; Zhao, D.; He, T. Identification of novel heavy metal detoxification proteins in *Solanum tuberosum*: Insights to improve food security protection from metal ion stress. *Sci. Total Environ.* **2021**, *779*, 146197. [CrossRef]

8. Sytar, O.; Kumari, P.; Yadav, S.; Brestic, M.; Rastogi, A. Phytohormone priming: Regulator for heavy metal stress in plants. *J. Plant Growth Regul.* **2019**, *38*, 739–752. [CrossRef]
9. Chandra, R.; Kumar, V. Phytoextraction of heavy metals by potential native plants and their microscopic observation of root growing on stabilised distillery sludge as a prospective tool for in situ phytoremediation of industrial waste. *Environ. Sci. Pollut. Res.* **2017**, *24*, 2605–2619. [CrossRef] [PubMed]
10. Zeiner, M.; Juranović Cindrić, I.; Nemet, I.; Franjković, K.; Salopek Sondi, B. Influence of soil salinity on selected element contents in different *Brassica* species. *Molecules* **2022**, *27*, 1878. [CrossRef] [PubMed]
11. Šamec, D.; Linić, I.; Salopek-Sondi, B. Salinity stress as an elicitor for phytochemicals and minerals accumulation in selected leafy vegetables of Brassicaceae. *Agronomy* **2021**, *11*, 361. [CrossRef]
12. Tan, B.; Wang, H.; Wang, X.; Ma, C.; Zhou, J.; Dai, X. Health risks and source analysis of heavy metal pollution from dust in Tianshui, China. *Minerals* **2021**, *11*, 502. [CrossRef]
13. Bi, C.; Zhou, Y.; Chen, Z.; Jia, J.; Bao, X. Heavy metals and lead isotopes in soil, road dust and leafy vegetables and health risks via vegetable consumption in the industrial areas of Shanghai, China. *Sci. Total Environ.* **2018**, *619–620*, 1349–1357. [CrossRef] [PubMed]
14. Muszyńska, E.; Labudda, M.; Kamińska, I.; Górecka, M.; Bederska-Błaszczyk, M. Evaluation of heavy metal-induced responses in *Silene vulgaris* ecotypes. *Protoplasma* **2019**, *256*, 1279–1297. [CrossRef]
15. Giannakoula, A.; Therios, I.; Chatzissavvidis, C. Effect of lead and copper on photosynthetic apparatus in citrus (*Citrus aurantium* L.) plants. The role of antioxidants in oxidative damage as a response to heavy metal stress. *Plants* **2021**, *10*, 155. [CrossRef] [PubMed]
16. Leng, Y.; Li, Y.; Ma, Y.-H.; He, L.-F.; Li, S.-W. Abscisic acid modulates differential physiological and biochemical responses of roots, stems, and leaves in mung bean seedlings to cadmium stress. *Environ. Sci. Pollut. Res.* **2021**, *28*, 6030–6043. [CrossRef]
17. Sitko, K.; Opała-Owczarek, M.; Jemioła, G.; Gieroń, Z.; Szopiński, M.; Owczarek, P.; Rudnicka, M.; Małkowski, E. Effect of drought and heavy metal contamination on growth and photosynthesis of silver birch trees growing on post-industrial heaps. *Cells* **2022**, *11*, 53. [CrossRef]
18. Muszyńska, E.; Labudda, M.; Hanus-Fajerska, E. Changes in proteolytic activity and protein carbonylation in shoots of *Alyssum montanum* ecotypes under multi-metal stress. *J. Plant Physiol.* **2019**, *232*, 61–64. [CrossRef]
19. Muszyńska, E.; Tokarz, K.; Dziurka, M.; Labudda, M.; Dziurka, K.; Piwowarczyk, B. Photosynthetic apparatus efficiency, phenolic acid profiling and pattern of chosen phytohormones in metal-tolerant and intolerant *Alyssum montanum* ecotypes. *Sci. Rep.* **2021**, *11*, 4135. [CrossRef]
20. Shahid, M.; Khalid, S.; Abbas, G.; Shahid, N.; Nadeem, M.; Sabir, M.; Aslam, M.; Dumat, C. Heavy Metal Stress and Crop Productivity. In *Crop Production and Global Environmental Issues*; Hakeem, K., Ed.; Springer: Cham, Switzerland, 2015; pp. 1–25.
21. Muszyńska, E.; Labudda, M. Dual role of metallic trace elements in stress biology—From negative to beneficial impact on plants. *Int. J. Mol. Sci.* **2019**, *20*, 3117. [CrossRef]
22. Hilker, M.; Schwachtje, J.; Baier, M.; Balazadeh, S.; Bäurle, I.; Geiselhardt, S.; Hincha, D.K.; Kunze, R.; Mueller-Roeber, B.; Rillig, M.C.; et al. Priming and memory of stress responses in organisms lacking a nervous system. *Biol. Rev.* **2016**, *91*, 1118–1133. [CrossRef] [PubMed]
23. Leuendorf, J.E.; Frank, M.; Schmülling, T. Acclimation, priming and memory in the response of *Arabidopsis thaliana* seedlings to cold stress. *Sci. Rep.* **2020**, *10*, 1–11. [CrossRef] [PubMed]
24. Demmig-Adams, B.; Dumlao, M.R.; Herzenach, M.K.; Adams, W.W. Acclimation. In *Encyclopedia of Ecology*; Jørgensen, S.E., Fath, B.D., Eds.; Elsevier: Amsterdam, The Netherlands, 2008; pp. 15–23.
25. Ernst, W.H.O. Evolution of metal tolerance in higher plants. *For. Snow Landsc. Res.* **2006**, *80*, 251–274.
26. Sharma, S.S.; Dietz, K.-J. The significance of amino acids and amino acid-derived molecules in plant responses and adaptation to heavy metal stress. *J. Exp. Bot.* **2006**, *57*, 711–726. [CrossRef] [PubMed]
27. Le Gall, H.; Philippe, F.; Domon, J.M.; Gillet, F.; Pelloux, J.; Rayon, C. Cell wall metabolism in response to abiotic stress. *Plants* **2015**, *4*, 112–166. [CrossRef]
28. Sujkowska-Rybkowska, M.; Muszyńska, E.; Labudda, M. Structural adaptation and physiological mechanisms in the leaves of *Anthyllis vulneraria* L. from metallicolous and non-metallicolous populations. *Plants* **2020**, *9*, 662. [CrossRef]
29. Guerinot, M.L. The ZIP family of metal transporters. *Biochim. Biophys. Acta (BBA)-Biomembr.* **2000**, *1465*, 190–198. [CrossRef]
30. Krämer, U. Metal hyperaccumulation in plants. *Annu. Rev. Plant Biol.* **2010**, *61*, 517–534. [CrossRef]
31. Kajala, K.; Walker, K.L.; Mitchell, G.S.; Krämer, U.; Cherry, S.R.; Brady, S.M. Real-time whole-plant dynamics of heavy metal transport in *Arabidopsis halleri* and *Arabidopsis thaliana* by gamma-ray imaging. *Plant Direct* **2019**, *3*, e00131. [CrossRef]
32. Wójcik, M.; Gonnelli, C.; Selvi, F.; Dresler, S.; Rostański, A.; Vangronsveld, J. Metallophytes of serpentine and calamine soils—Their unique ecophysiology and potential for phytoremediation. *Adv. Bot. Res.* **2017**, *83*, 1–42.
33. Wierzbicka, M.; Potocka, A. Lead tolerance in plants growing on dry and moist soils. *Acta Biol. Crac. Ser. Bot.* **2002**, *44*, 21–28.
34. Saraswat, S.; Rai, J.P.N. Complexation and detoxification of Zn and Cd in metal accumulating plants. *Rev. Environm. Sci. Biotechnol.* **2011**, *10*, 327–339. [CrossRef]
35. Ricachenevsky, F.K.; Punshon, T.; Salt, D.E.; Fett, J.P.; Guerinot, M.L. *Arabidopsis thaliana* zinc accumulation in leaf trichomes is correlated with zinc concentration in leaves. *Sci. Rep.* **2021**, *11*, 5278. [CrossRef] [PubMed]

36. Muszyńska, E.; Labudda, M.; Różańska, E.; Hanus-Fajerska, E.; Znojek, E. Heavy metal tolerance in contrasting ecotypes of *Alyssum montanum*. *Ecotoxicol. Environ. Saf.* **2018**, *161*, 305–317. [CrossRef] [PubMed]
37. Wierzbicka, M.; Pielichowska, M.; Kalabun, O.B.; Wąsowicz, P. Microevolution on anthropogenically changed areas on the example of *Biscutella laevigata* plants from calamine waste heap in Poland. *J. Environ. Anal. Toxicol.* **2017**, *7*, 1–10. [CrossRef]
38. Woźniak, A.; Bednarski, W.; Dancewicz, K.; Gabryś, B.; Borowiak-Sobkowiak, B.; Bocianowski, J.; Samardakiewicz, S.; Rucińska-Sobkowiak, R.; Morkunas, I. Oxidative stress links response to lead and *Acyrthosiphon pisum* in *Pisum sativum* L. *J. Plant Physiol.* **2019**, *240*, 152996. [CrossRef]
39. Ahmad, P.; Raja, V.; Ashraf, M.; Wijaya, L.; Bajguz, A.; Alyemeni, M.N. Jasmonic acid (JA) and gibberellic acid (GA3) mitigated Cd-toxicity in chickpea plants through restricted Cd uptake and oxidative stress management. *Sci. Rep.* **2021**, *11*, 19768. [CrossRef]
40. Muszyńska, E.; Labudda, M. Effects of lead, cadmium and zinc on protein changes in *Silene vulgaris* shoots cultured in vitro. *Ecotoxicol. Environ. Saf.* **2020**, *204*, 111086. [CrossRef]
41. Muszyńska, E.; Labudda, M.; Kral, A. Ecotype-specific pathways of reactive oxygen species deactivation in facultative metallophyte *Silene vulgaris* (Moench) Garcke treated with heavy metals. *Antioxidants* **2020**, *9*, 102. [CrossRef]
42. Baker, A.J.M. Accumulators and excluders strategies in the response of plants to heavy metals. *J. Plant Nutr.* **1981**, *3*, 643–654. [CrossRef]
43. Reeves, R.D.; Baker, A.J.M.; Jaffré, T.; Erskine, P.D.; Echevarria, G.; van der Ent, A. A global database for plants that hyperaccumulate metal and metalloid trace elements. *New Phytol.* **2018**, *218*, 407–411. [CrossRef]
44. Muszyńska, E.; Hanus-Fajerska, E.; Ciarkowska, K. Phytoremediation as an antidote to environmental pollution. In *Buckler Mustard (Biscutella laevigata L.) an Extraordinary Plant on Ordinary Mine Heaps Near Olkusz*; Szarek-Łukaszewska, G., Ed.; W. Szafer Institute of Botany, Polish Academy of Sciences: Kraków, Poland, 2020; pp. 231–259.
45. Rascio, N.; Navari-Izzo, F. Heavy metal hyperaccumulating plants: How and why do they do it? And what makes them so interesting? *Plant Sci.* **2011**, *180*, 169–181. [CrossRef] [PubMed]
46. Drozdova, I.; Alekseeva-Popova, N.; Kalimova, I.; Bech, J.; Roca, N. Research of reclamation of polluted mine soils by native metallophytes: Some cases. *Geochem. Explor. Environ. Anal.* **2019**, *19*, 164–170. [CrossRef]
47. Bhatia, N.P.; Baker, A.J.M.; Walsh, K.B.; Midmore, D.J. A role for nickel in osmotic adjustment in drought-stressed plants of the nickel hyperaccumulator *Stackhousia tryonii* Bailey. *Planta* **2005**, *223*, 134–139. [CrossRef]
48. Chaney, R.L.; Angle, J.S.; Broadhurst, C.L.; Peters, C.A.; Tappero, R.V.; Sparks, D.L. Improved understanding of hyperaccumulation yields commercial phytoextraction and phytomining technologies. *J. Environ. Qual.* **2007**, *36*, 1429–1443. [CrossRef]
49. Ciarkowska, K.; Hanus-Fajerska, E.; Gambuś, F.; Muszyńska, E.; Czech, T. Phytostabilization of Zn-Pb ore flotation tailings with *Dianthus carthusianorum* and *Biscutella laevigata* after amending with mineral fertilizers or sewage sludge. *J. Environ. Manag.* **2017**, *189*, 75–83. [CrossRef]
50. Gadgil, R.L. Tolerance of heavy metals and the reclamation of industrial waste. *J. Appl. Ecol.* **1969**, *6*, 247–259. [CrossRef]
51. Mummey, D.L.; Stahl, P.D.; Buyer, J.S. Soil microbiological properties 20 years after surface mine reclamation: Spatial analysis of reclaimed and undisturbed sites. *Soil Biol. Biochem.* **2002**, *34*, 1717–1725. [CrossRef]
52. Conesa, H.M.; Robinson, B.H.; Schulin, R.; Nowack, B. Growth of *Lygeum spartum* in acid mine tailings: Response of plants developed from seedlings, rhizomes and at field conditions. *Environ. Pollut.* **2007**, *145*, 700–707. [CrossRef] [PubMed]
53. Jamali Hajiani, N.; Ghaderian, S.M.; Karimi, N.; Schat, H. A comparison of antimony accumulation and tolerance among *Achillea wilhelmsii, Silene vulgaris* and *Thlaspi arvense*. *Plant Soil* **2017**, *412*, 267–281. [CrossRef]
54. Muszyńska, E.; Hanus-Fajerska, E.; Piwowarczyk, B.; Augustynowicz, J.; Ciarkowska, K.; Czech, T. From laboratory to field studies—The assessment of *Biscutella laevigata* suitability to biological reclamation of areas contaminated with lead and cadmium. *Ecotoxicol. Environ. Saf.* **2017**, *142*, 266–273. [CrossRef] [PubMed]
55. Muszyńska, E.; Hanus-Fajerska, E.; Koźmińska, A. Differential tolerance to lead and cadmium of micropropagated *Gypsophila fastigiata* ecotype. *Water Air Soil Pollut.* **2018**, *229*, 42. [CrossRef]
56. Hanus-Fajerska, E.; Ciarkowska, K.; Muszyńska, E. Long-term field study on stabilization of contaminated wastes by growing clonally reproduced *Silene vulgaris* calamine ecotype. *Plant Soil* **2019**, *439*, 431–445. [CrossRef]
57. Akram, N.A.; Hafeez, N.; Farid-ul-Haq, M.; Ahmad, A.; Sadiq, M.; Ashraf, M. Foliage application and seed priming with nitric oxide causes mitigation of salinity-induced metabolic adversaries in broccoli (*Brassica oleracea* L.) plants. *Acta Physiol. Plant* **2020**, *42*, 155. [CrossRef]
58. Nazir, F.; Fariduddin, Q.; Hussain, A.; Khan, T.A. Brassinosteroid and hydrogen peroxide improve photosynthetic machinery, stomatal movement, root morphology and cell viability and reduce Cu-triggered oxidative burst in tomato. *Ecotoxicol. Environ. Saf.* **2021**, *207*, 111081. [CrossRef] [PubMed]
59. Hayat, S.; Hayat, Q.; Alyemeni, M.N.; Wani, A.S.; Pichtel, J.; Ahmad, A. Role of proline under changing environments: A review. *Plant Signal. Behav.* **2012**, *7*, 1456–1466. [CrossRef]
60. Pereira, M.P.; de Almeida Rodrigues, L.C.; Corrêa, F.F.; de Castro, E.M.; Ribeiro, V.E.; Pereira, F.J. Cadmium tolerance in *Schinus molle* trees is modulated by enhanced leaf anatomy and photosynthesis. *Trees* **2016**, *30*, 807–814. [CrossRef]
61. Muszyńska, E.; Labudda, M.; Różańska, E.; Hanus-Fajerska, E.; Koszelnik-Leszek, A. Structural, physiological and genetic diversification of *Silene vulgaris* ecotypes from heavy metal-contaminated areas and their synchronous in vitro cultivation. *Planta* **2019**, *249*, 1761–1778. [CrossRef] [PubMed]

62. Nikalje, G.C.; Saini, N.; Suprasanna, P. Halophytes and heavy metals: Interesting partnerships. In *Plant-Metal Interactions*; Srivastava, S., Srivastava, A., Suprasanna, P., Eds.; Springer: Cham, Switzerland, 2019; pp. 99–118.
63. Nawaz, I.; Iqbal, M.; Bliek, M.; Schat, H. Salt and heavy metal tolerance and expression levels of candidate tolerance genes among four extremophile Cochlearia species with contrasting habitat preferences. *Sci. Total Environ.* **2017**, *584*, 731–741. [CrossRef]
64. Gao, C.; Wang, Y.; Jiang, B.; Liu, G.; Yu, L.; Wei, Z.; Yang, C. A novel vacuolar membrane H^+-ATPase c subunit gene (ThVHAc1) from *Tamarix hispida* confers tolerance to several abiotic stresses in *Saccharomyces cerevisiae*. *Mol. Biol. Rep.* **2011**, *38*, 957–963. [CrossRef]
65. Neumann, D.; Nieden, U.Z.; Lichtenberger, O.; Leopold, I. How Does *Armeria maritima* tolerate high heavy metal concentrations? *J. Plant Physiol.* **1995**, *146*, 704–717. [CrossRef]
66. Manousaki, E.; Kosmoula, G.; Lamprini, P.; Kalogerakis, N. Metal phytoremediation by the halophyte *Limoniastrum monopetalum* (L.) Boiss: Two contrasting ecotypes. *Int. J. Phytorem.* **2014**, *16*, 755–769. [CrossRef] [PubMed]
67. Manousaki, E.; Kalogerakis, N. Phytoextraction of Pb and Cd by the Mediterranean saltbush (*Atriplex halimus* L.): Metal uptake in relation to salinity. *Eviron. Pollut. Res.* **2009**, *16*, 844–854. [CrossRef] [PubMed]
68. Mazharia, M.; Homaeed, M. Annual halophyte *Chenopodium botrys* can phytoextract cadmium from contaminated soils. *J. Basic Appl. Sci. Res.* **2012**, *2*, 1415–1422.
69. Wang, L.; Wang, X.; Jiang, L.; Zhang, K.; Tanveer, M.; Tian, C.; Zhao, Z. Reclamation of saline soil by planting annual euhalophyte *Suaeda salsa* with drip irrigation: A three-year field experiment in arid northwestern China. *Ecol. Eng.* **2021**, *159*, 106090. [CrossRef]
70. Van Oosten, M.J.; Maggio, A. Functional biology of halophytes in the phytoremediation of heavy metal contaminated soils. *Environ. Exp. Bot.* **2015**, *111*, 135–146. [CrossRef]
71. Nikalje, G.C.; Suprasanna, P. Coping with metal toxicity—Cues from halophytes. *Front. Plant Sci.* **2018**, *9*, 777. [CrossRef] [PubMed]
72. Hasanuzzaman, M.; Nahar, K.; Öztürk, M. *Ecophysiology, Abiotic Stress Responses and Utilization of Halophytes*; Springer Nature: Singapore, 2019.
73. Milić, D.; Luković, J.; Ninkov, J.; Zeremski-Škorić, T.; Zorić, L.; Vasin, J.; Milić, S. Heavy metal content in halophytic plants from inland and maritime saline areas. *Cent. Eur. J. Biol.* **2012**, *7*, 307–317. [CrossRef]
74. Li, B.; Wang, J.; Yao, L.; Meng, Y.; Ma, X.; Si, E.; Ren, P.; Yang, K.; Shang, X.; Wang, H. Halophyte *Halogeton glomeratus*, a promising candidate for phytoremediation of heavy metal-contaminated saline soils. *Plant Soil* **2019**, *442*, 323–331. [CrossRef]
75. Ayyappan, D.; Sathiyaraj, G.; Ravindran, K.G. Phytoextraction of heavy metals by *Sesuvium portulacastrum* L. A salt marsh halophyte from tannery effluent. *Int. J. Phytorem.* **2016**, *18*, 453–459. [CrossRef] [PubMed]
76. Sghaier, D.B.; Pedro, S.; Diniz, M.S.; Duarte, B.; Caçador, I.; Sleimi, N. Tissue localization and distribution of As and Al in the halophyte *Tamarix gallica* under controlled conditions. *Front. Mar. Sci.* **2016**, *3*, 274. [CrossRef]
77. Anwar, A.; Kim, J.-K. Transgenic breeding approaches for improving abiotic stress tolerance: Recent progress and future perspectives. *Int. J. Mol. Sci.* **2020**, *21*, 2695. [CrossRef] [PubMed]
78. Martínez, M.; Bernal, P.; Almela, C.; Vélez, D.; García-Agustín, P.; Serrano, R.; Navarro-Aviñó, J. An engineered plant that accumulates higher levels of heavy metals than *Thlaspi caerulescens*, with yields of 100 times more biomass in mine soils. *Chemosphere* **2006**, *64*, 478–485. [CrossRef] [PubMed]
79. Xu, W.; Li, Y.; Cheng, Z.; Cheng, Z.; Xia, G.; Wang, M. A wheat histone variant gene *TaH2A.7* enhances drought tolerance and promotes stomatal closure in Arabidopsis. *Plant Cell Rep.* **2016**, *35*, 1853–1862. [CrossRef] [PubMed]
80. Salam, A.; Khan, A.R.; Liu, L.; Yang, S.; Azhar, W.; Ulhassan, Z.; Zeeshan, M.; Wu, J.; Fan, X.; Gan, Y. Seed priming with zinc oxide nanoparticles downplayed ultrastructural damage and improved photosynthetic apparatus in maize under cobalt stress. *J. Hazard. Mater.* **2022**, *423*, 127021. [CrossRef]
81. Waqas, M.; Korres, N.E.; Khan, M.D.; Nizami, A.-S.; Deeba, F.; Ali, I.; Hussain, H. Advances in the concept and methods of seed priming. In *Priming and Pretreatment of Seeds and Seedlings*; Hasanuzzaman, M., Fotopoulos, V., Eds.; Springer: Singapore, 2019; pp. 11–41.
82. Rezk, A.A.; Al-Khayri, J.M.; Al-Bahrany, A.M.; El-Beltagi, H.S.; Mohamed, H.I. X-ray irradiation changes germination and biochemical analysis of two genotypes of okra (*Hibiscus esculentus* L.). *J. Radiat. Res. Appl. Sci.* **2019**, *12*, 393–402. [CrossRef]
83. Hussein, H.-A.A. Influence of radio-grain priming on growth, antioxidant capacity, and yield of barley plants. *Biotechnol. Rep.* **2022**, *34*, e00724. [CrossRef] [PubMed]
84. Adhikary, S.; Biswas, B.; Chakraborty, D.; Timsina, J.; Pal, S.; Tarafdar, J.C.; Banerjee, S.; Hossain, A.; Roy, S. Seed priming with selenium and zinc nanoparticles modifies germination, growth, and yield of direct-seeded rice (*Oryza sativa* L.). *Sci. Rep.* **2022**, *12*, 7103. [CrossRef]
85. Mir, A.R.; Alam, P.; Hayat, S. Auxin regulates growth, photosynthetic efficiency and mitigates copper induced toxicity via modulation of nutrient status, sugar metabolism and antioxidant potential in *Brassica juncea*. *Plant Physiol. Biochem.* **2022**, *185*, 244–259. [CrossRef]
86. Zafar, S.; Perveen, S.; Khan, M.K.; Shaheen, M.R.; Hussain, R.; Sarwar, N.; Rashid, S.; Nafees, M.; Farid, G.; Alamri, S.; et al. Effect of zinc nanoparticles seed priming and foliar application on the growth and physio-biochemical indices of spinach (*Spinacia oleracea* L.) under salt stress. *PLoS ONE* **2022**, *17*, e0263194. [CrossRef]
87. Sako, K.; Nguyen, H.M.; Seki, M. Advances in chemical priming to enhance abiotic stress tolerance in plants. *Plant Cell Physiol.* **2021**, *61*, 1995–2003. [CrossRef] [PubMed]

88. Bechtold, U.; Field, B. Molecular mechanisms controlling plant growth during abiotic stress. *J. Exp. Bot.* **2018**, *69*, 2753–2758. [CrossRef] [PubMed]
89. Saha, I.; Hasanuzzaman, M.; Adak, M.K. Abscisic acid priming regulates arsenite toxicity in two contrasting rice (*Oryza sativa* L.) genotypes through differential functioning of Sub1A quantitative trait loci. *Environ. Pollut.* **2021**, *287*, 117586. [CrossRef]
90. Rehman, S.; Abbas, G.; Shahid, M.; Saqib, M.; Umer Farooq, A.B.; Hussain, M.; Murtaza, B.; Amjad, M.; Naeem, M.A.; Farooq, A. Effect of salinity on cadmium tolerance, ionic homeostasis and oxidative stress responses in conocarpus exposed to cadmium stress: Implications for phytoremediation. *Ecotoxicol. Environ. Saf.* **2019**, *171*, 146–153. [CrossRef] [PubMed]
91. Fidler, J.; Graska, J.; Gietler, M.; Nykiel, M.; Prabucka, B.; Rybarczyk-Płońska, A.; Muszyńska, E.; Morkunas, I.; Labudda, M. PYR/PYL/RCAR receptors play a vital role in the abscisic-acid-dependent responses of plants to external or internal stimuli. *Cells* **2022**, *11*, 1352. [CrossRef] [PubMed]
92. Gietler, M.; Fidler, J.; Labudda, M.; Nykiel, M. Abscisic acid—Enemy or savior in the response of cereals to abiotic and biotic stresses? *Int. J. Mol. Sci.* **2020**, *21*, 4607. [CrossRef]
93. Wei, T.-J.; Wang, M.-M.; Jin, Y.-Y.; Zhang, G.-H.; Liu, M.; Yang, H.-Y.; Jiang, C.-J.; Liang, Z.-W. Abscisic acid priming creates alkaline tolerance in alfalfa seedlings (*Medicago sativa* L.). *Agriculture* **2021**, *11*, 608. [CrossRef]
94. Ghassemi-Golezani, K.; Farhangi-Abriz, S. Foliar sprays of salicylic acid and jasmonic acid stimulate H^+-ATPase activity of tonoplast, nutrient uptake and salt tolerance of soybean. *Ecotoxicol. Environ. Saf.* **2018**, *166*, 18–25. [CrossRef] [PubMed]
95. Akar, M.; Atis, I. The effects of priming pretreatments on germination and seedling growth in perennial ryegrass exposed to heavy metal stress. *Fresenius Environ. Bull.* **2018**, *27*, 6677–6685.
96. Nouri, M.; Haddioui, A. Improving seed germination and seedling growth of *Lepidium sativum* with different priming methods under arsenic stress. *Acta Ecol. Sin.* **2021**, *41*, 64–71. [CrossRef]
97. Mazumder, M.K.; Sharma, P.; Moulick, D.; Tata, S.K.; Choudhury, S. Salicylic acid ameliorates zinc and chromium-induced stress responses in wheat seedlings: A biochemical and computational analysis. *Cereal Res. Commun.* **2021**, *50*, 407–418. [CrossRef]
98. Huang, Y.T.; Cai, S.Y.; Ruan, X.L.; Chen, S.Y.; Mei, G.F.; Ruan, G.H.; Cao, D.D. Salicylic acid enhances sunflower seed germination under Zn^{2+} stress via involvement in Zn^{2+} metabolic balance and phytohormone interactions. *Sci. Hortic.* **2021**, *275*, 109702. [CrossRef]
99. Gupta, S.; Seth, C.S. Salicylic acid alleviates chromium (VI) toxicity by restricting its uptake, improving photosynthesis and augmenting antioxidant defense in *Solanum lycopersicum* L. *Physiol. Mol. Biol. Plants* **2021**, *27*, 2651–2664. [CrossRef] [PubMed]
100. Demecsová, L.; Zelinová, V.; Liptáková, L'.; Valentovičová, K.; Tamás, L. Indole-3-butyric acid priming reduced cadmium toxicity in barley root tip via NO generation and enhanced glutathione peroxidase activity. *Planta* **2020**, *252*, 46. [CrossRef] [PubMed]
101. Santo Pereira, A.E.; Caixeta Oliveira, H.; Fernandes Fraceto, L.; Santaella, C. Nanotechnology potential in seed priming for sustainable agriculture. *Nanomaterials* **2021**, *11*, 267. [CrossRef]
102. Sridharan, K.; Puthur, J.T.; Dhankher, O.P. Priming with nanoscale materials for boosting abiotic stress tolerance in crop plants. *J. Agric. Food Chem.* **2021**, *69*, 10017–10035.
103. Rai-Kalal, P.; Jajoo, A. Priming with zinc oxide nanoparticles improve germination and photosynthetic performance in wheat. *Plant Physiol. Biochem.* **2021**, *160*, 341–351. [CrossRef]
104. Shah, T.; Latif, S.; Saeed, F.; Ali, I.; Ullah, S.; Abdullah Alsahli, A.; Jan, S.; Ahmad, P. Seed priming with titanium dioxide nanoparticles enhances seed vigor, leaf water status, and antioxidant enzyme activities in maize (*Zea mays* L.) under salinity stress. *J. King Saud Univ. Sci.* **2021**, *33*, 101207. [CrossRef]
105. Ragab, G.; Saad-Allah, K. Seed priming with greenly synthesized sulfur nanoparticles enhances antioxidative defense machinery and restricts oxidative injury under manganese stress in *Helianthus annuus* (L.) seedlings. *J. Plant Growth Regul.* **2021**, *40*, 1894–1902. [CrossRef]
106. Ellouzi, H.; Sghayar, S.; Abdelly, C. H_2O_2 seed priming improves tolerance to salinity; drought and their combined effect more than mannitol in *Cakile maritima* when compared to *Eutrema salsugineum*. *J. Plant Physiol.* **2017**, *210*, 38–50. [CrossRef]
107. Habib, N.; Ali, Q.; Ali, S.; Javed, M.T.; Zulqurnain Haider, M.; Perveen, R.; Shahid, M.R.; Rizwan, M.; Abdel-Daim, M.M.; Elkelish, A.; et al. Use of nitric oxide and hydrogen peroxide for better yield of wheat (*Triticum aestivum* L.) under water deficit conditions: Growth, osmoregulation, and antioxidative defense mechanism. *Plants* **2020**, *9*, 285. [CrossRef] [PubMed]
108. Majeed, S.; Nawaz, F.; Naeem, M.; Ashraf, M.Y. Effect of exogenous nitric oxide on sulfur and nitrate assimilation pathway enzymes in maize (*Zea mays* L.) under drought stress. *Acta Physiol. Plant.* **2018**, *40*, 206. [CrossRef]
109. Silva, P.C.C.; de Azevedo Neto, A.D.; Gheyi, H.R.; Ribas, R.F.; dos Reis Silva, C.R.; Cova, A.M.W. Salt tolerance induced by hydrogen peroxide priming on seed is related to improvement of ion homeostasis and antioxidative defense in sunflower plants. *J. Plant Nutr.* **2021**, *44*, 1207–1221. [CrossRef]
110. Ellouzi, H.; Oueslati, S.; Hessini, K.; Rabhi, M.; Abdelly, C. Seed-priming with H_2O_2 alleviates subsequent salt stress by preventing ROS production and amplifying antioxidant defense in cauliflower seeds and seedlings. *Sci. Hortic.* **2021**, *288*, 110360. [CrossRef]
111. Hossain, M.A.; Bhattacharjee, S.; Armin, S.-M.; Qian, P.; Xin, W.; Li, H.-Y.; Burritt, D.J.; Fujita, M.; Tran, L.-S.P. Hydrogen peroxide priming modulates abiotic oxidative stress tolerance: Insights from ROS detoxification and scavenging. *Front. Plant Sci.* **2015**, *6*, 420. [CrossRef]
112. Cuypers, A.; Hendrix, S.; dos Reis, R.A.; De Smet, S.; Deckers, J.; Gielen, H.; Jozefczak, M.; Loix, C.; Vercampt, H.; Vangronsveld, J.; et al. Hydrogen peroxide, signaling in disguise during metal phytotoxicity. *Front. Plant Sci.* **2016**, *7*, 470. [CrossRef]

113. Singh, S.; Husain, T.; Kushwaha, B.K.; Suhel, M.; Fatima, A.; Mishra, V.; Singh, S.K.; Tripathi, D.K.; Rai, M.; Prasad, S.M.; et al. Regulation of ascorbate-glutathione cycle by exogenous nitric oxide and hydrogen peroxide in soybean roots under arsenate stress. *J. Hazard. Mater.* **2020**, *409*, 123686. [CrossRef]
114. Bai, X.-J.; Liu, L.-J.; Zhang, C.-H.; Ge, Y.; Cheng, W.-D. Effect of H_2O_2 pretreatment on Cd tolerance of different rice cultivars. *Rice Sci.* **2011**, *18*, 29–35. [CrossRef]
115. Yıldız, M.; Terzi, H.; Bingül, N. Protective role of hydrogen peroxide pretreatment on defense systems and *BnMP1* gene expression in Cr(VI)-stressed canola seedlings. *Ecotoxicology* **2013**, *22*, 1303–1312. [CrossRef]
116. Verna, N.; Prasad, S.M. Regulation of redox homeostasis in cadmium stressed rice field cyanobacteria by exogenous hydrogen peroxide and nitric oxide. *Sci. Rep.* **2021**, *11*, 2893. [CrossRef]
117. dos Santos Araújo, G.; de Oliveira Paula-Marinho, S.; de Paiva Pinheiro, S.K.; de Castro Miguel, E.; de Sousa Lopes, L.; Camelo Marques, E.; de Carvalho, H.H.; Gomes-Filho, E. H_2O_2 priming promotes salt tolerance in maize by protecting chloroplasts ultrastructure and primary metabolites modulation. *Plant Sci.* **2021**, *303*, 110774. [CrossRef] [PubMed]
118. Jira-anunkul, W.; Pattanagul, W. Effects of hydrogen peroxide application on agronomic traits of rice (*Oryza sativa* L.) under drought stress. *Plant Soil Environ.* **2021**, *67*, 221–229. [CrossRef]
119. Alnusairi, G.S.H.; Mazrou, Y.S.A.; Qari, S.H.; Elkelish, A.A.; Soliman, M.H.; Eweis, M.; Abdelaal, K.; El-Samad, G.A.; Ibrahim, M.F.M.; ElNahhas, N. Exogenous nitric oxide reinforces photosynthetic efficiency, osmolyte, mineral uptake, antioxidant, expression of stress-responsive genes and ameliorates the effects of salinity stress in wheat. *Plants* **2021**, *10*, 1693. [CrossRef]
120. Basit, F.; Ulhassan, Z.; Mou, Q.; Nazir, M.M.; Hu, J.; Hu, W.; Song, W.; Sheteiwy, M.S.; Zhou, W.; Bhat, J.A.; et al. Seed priming with nitric oxide and/or spermine mitigate the chromium toxicity in rice (*Oryza sativa*) seedlings by improving the carbon-assimilation and minimising the oxidative damages. *Funct. Plant Biol.* **2022**. [CrossRef] [PubMed]
121. Ahmad, P.; Alyemeni, M.N.; Wijaya, L.; Ahanger, M.A.; Ashraf, M.; Alam, P.; Paray, B.A.; Rinklebe, J. Nitric oxide donor, sodium nitroprusside, mitigates mercury toxicity in different cultivars of soybean. *J. Hazard. Mater.* **2021**, *408*, 124852. [CrossRef] [PubMed]
122. Samet, H. Alleviation of cobalt stress by exogenous sodium nitroprusside in iceberg lettuce. *Chil. J. Agric. Res.* **2020**, *80*, 161–170. [CrossRef]
123. Kopyra, M.; Gwóźdź, E.A. Nitric oxide stimulates seed germination and counteracts the inhibitory effect of heavy metals and salinity on root growth of *Lupinus luteus*. *Plant Physiol. Biochem.* **2003**, *41*, 1011–1017. [CrossRef]
124. Hassanein, A.; Esmail, N.; Hashem, H. Sodium nitroprusside mitigates the inhibitory effect of salt and heavy metal stress on lupine yield and downregulates antioxidant enzyme activities. *Acta Agrobot.* **2020**, *73*, 7336. [CrossRef]
125. He, H.; Oo, T.L.; Huang, W.; He, L.F.; Gu, M. Nitric oxide acts as an antioxidant and inhibits programmed cell death induced by aluminum in the root tips of peanut (*Arachis hypogaea* L.). *Sci. Rep.* **2019**, *9*, 9516. [CrossRef]
126. Li, Z.-G.; Min, X.; Zhou, Z.-H. Hydrogen sulfide: A signal molecule in plant cross-adaptation. *Front. Plant Sci.* **2016**, *7*, 1621. [CrossRef]
127. Zanganeh, R.; Jamei, R.; Rahmani, F. Pre-sowing seed treatment with salicylic acid and sodium hydrosulfide confers Pb toxicity tolerance in maize (*Zea mays* L.). *Ecotoxicol. Environ. Saf.* **2020**, *206*, 111392. [CrossRef] [PubMed]
128. Valivand, M.; Amooaghaie, R.; Ahadi, A. Seed priming with H_2S and Ca^{2+} trigger signal memory that induces cross-adaptation against nickel stress in zucchini seedlings. *Plant Physiol. Biochem.* **2019**, *143*, 286–298. [CrossRef] [PubMed]
129. Hasanuzzaman, M.; Nahar, K.; Hossain, M.S.; Mahmud, J.; Rahman, A.; Inafuku, M.; Oku, H.; Fujita, M. Coordinated actions of glyoxalase and antioxidant defense systems in conferring abiotic stress tolerance in plants. *Int. J. Mol. Sci.* **2017**, *18*, 200. [CrossRef] [PubMed]
130. Christou, A.; Manganaris, G.A.; Papadopoulos, I.; Fotopoulos, V. Hydrogen sulfide induces systemic tolerance to salinity and non-ionic osmotic stress in strawberry plants through modification of reactive species biosynthesis and transcriptional regulation of multiple defence pathways. *J. Exp. Bot.* **2013**, *64*, 1953–1966. [CrossRef] [PubMed]
131. Li, H.; Shi, J.; Wang, Z.; Zhang, W.; Yang, H. H_2S pretreatment mitigates the alkaline salt stress on *Malus hupehensis* roots by regulating Na^+/K^+ homeostasis and oxidative stress. *Plant Physiol. Biochem.* **2020**, *156*, 233–241. [CrossRef] [PubMed]
132. Zhou, H.; Chen, Y.; Zhai, F.; Zhang, J.; Zhang, F.; Yuan, X.; Xie, Y. Hydrogen sulfide promotes rice drought tolerance via reestablishing redox homeostasis and activation of ABA biosynthesis and signaling. *Plant Physiol. Biochem.* **2020**, *155*, 213–220. [CrossRef]
133. Antoniou, C.; Xenofontos, R.; Chatzimichail, G.; Christou, A.; Kashfi, K.; Fotopoulos, V. Exploring the potential of nitric oxide and hydrogen sulfide (NOSH)-releasing synthetic compounds as novel priming agents against drought stress in *Medicago sativa* plants. *Biomolecules* **2020**, *10*, 120. [CrossRef]
134. Bera, K.; Dutta, P.; Sadhukhan, S. Seed priming with non-ionizing physical agents: Plant responses and underlying physiological mechanisms. *Plant Cell Rep.* **2022**, *41*, 53–73. [CrossRef]
135. Rao, G.; Huang, S.; Ashraf, U.; Mo, Z.; Duan, M.; Pan, S.; Tang, X. Ultrasonic seed treatment improved cadmium (Cd) tolerance in *Brassica Napus* L. *Ecotoxicol. Environ. Saf.* **2019**, *185*, 109659. [CrossRef]
136. Dutta, P. Seed priming: New vistas and contemporary perspectives. In *Advances in Seed Priming*; Rakshit, A., Singh, H., Eds.; Springer: Singapore, 2018; pp. 3–22.

137. Xia, Q.; Tao, H.; Li, Y.; Pan, D.; Cao, J.; Liu, L.; Zhou, X.; Barba, F.J. Characterizing physicochemical, nutritional and quality attributes of wholegrain *Oryza sativa* L. subjected to high intensity ultrasound-stimulated pre-germination. *Food Control* **2020**, *108*, 106827. [CrossRef]
138. Thomas, T.T.D.; Dinakar, C.; Puthur, J.T. Effect of UV-B priming on the abiotic stress tolerance of stress-sensitive rice seedlings: Priming imprints and cross-tolerance. *Plant Physiol. Biochem.* **2020**, *147*, 21–30. [CrossRef]
139. Sen, A.; Challabathula, D.; Puthur, J.T. UV-B priming of *Oryza sativa* seeds augments the innate tolerance potential in a tolerant variety more effectively toward NaCl and PEG stressors. *J. Plant Growth Regul.* **2021**, *40*, 1166–1180. [CrossRef]
140. Xu, Y.; Charles, M.T.; Luo, Z.; Mimee, B.; Tong, Z.; Véronneau, P.-Y.; Roussel, D.; Rolland, D. Ultraviolet-C priming of strawberry leaves against subsequent *Mycosphaerella fragariae* infection involves the action of reactive oxygen species, plant hormones, and terpenes. *Plant Cell Environ.* **2019**, *42*, 815–831. [CrossRef]
141. Sen, A.; Puthur, J.T.; Challabathula, D.; Brestič, M. Transgenerational effect of UV-B priming on photochemistry and associated metabolism in rice seedlings subjected to PEG-induced osmotic stress. *Photosynthetica* **2022**, *60*, 219–229. [CrossRef]
142. Xiong, Y.; Xing, Q.; Müller-Xing, R. A novel UV-B priming system reveals an UVR8-depedent memory, which provides resistance against UV-B stress in Arabidopsis leaves. *Plant Signal. Behav.* **2021**, *16*, 1879533. [CrossRef]
143. Kanwal, S.; Tariq, M.; Dawar, S. Effect of microwave radiation on plants infected with root rot pathogens. *Pak. J. Bot.* **2018**, *50*, 2389–2393.
144. Araújo, S.S.; Paparella, S.; Dondi, D.; Bentivoglio, A.; Carbonera, D.; Balestrazzi, A. Physical methods for seed invigoration: Advantages and challenges in seed technology. *Front. Plant Sci.* **2016**, *7*, 646. [CrossRef]
145. Bian, Z.-X.; Wang, J.-F.; Ma, H.; Wang, S.-M.; Luo, L.; Wang, S.-M. Effect of microwave radiation on antioxidant capacities of tartary buckwheat sprouts. *J. Food Sci. Technol.* **2020**, *57*, 3913–3919. [CrossRef]
146. Maswada, H.F.; Sunoj, V.S.J.; Prasad, P.V.V. A comparative study on the effect of seed pre-sowing treatments with microwave radiation and salicylic acid in alleviating the drought-induced damage in wheat. *J. Plant Growth Regul.* **2021**, *40*, 48–66. [CrossRef]
147. Farid, M.; Ali, S.; Rizwan, M.; Saeed, R.; Tauqeer, H.M.; Sallah-Ud-Din, R.; Azam, A.; Raza, N. Microwave irradiation and citric acid assisted seed germination and phytoextraction of nickel (Ni) by *Brassica napus* L.: Morpho-physiological and biochemical alterations under Ni stress. *Environ. Sci. Pollut. Res.* **2017**, *24*, 21050–21064. [CrossRef]
148. Radhakrishnan, R. Magnetic field regulates plant functions, growth and enhances tolerance against environmental stresses. *Physiol. Mol. Biol. Plants Int. J. Funct. Plant Biol.* **2019**, *25*, 1107–1119. [CrossRef] [PubMed]
149. Mohammadi, R.; Roshandel, P. Ameliorative effects of a static magnetic field on hyssop (*Hyssopus officinalis* L.) growth and phytochemical traits under water stress. *Bioelectromagnetics* **2020**, *41*, 403–412. [CrossRef] [PubMed]
150. Kataria, S.; Baghel, L.; Jain, M.; Guruprasad, K.N. Magnetopriming regulates antioxidant defense system in soybean against salt stress. *Biocatal. Agric. Biotechnol.* **2019**, *18*, 101090. [CrossRef]
151. Baghel, L.; Kataria, S.; Jain, M. Mitigation of adverse effects of salt stress on germination, growth, photosynthetic efficiency and yield in maize (*Zea mays* L.) through magnetopriming. *Acta Agrobot.* **2019**, *72*, 1757. [CrossRef]
152. Dashab, S.; Omidi, H. Effect of intensity, duration and power of ultrasonic waves on germination indices and photosynthetic pigments of canola seedling. *Agroecol. J.* **2020**, *15*, 13–24.
153. Chen, Y.P.; Liu, Q.; Yue, X.Z.; Meng, Z.W.; Liang, J. Ultrasonic vibration seeds showed improved resistance to cadmium and lead in wheat seedling. *Environ. Sci. Pollut. Res.* **2013**, *20*, 4807–4816. [CrossRef]
154. Hanafy, R.S.; Akladious, S.A. Physiological and molecular studies on the effect of gamma radiation in fenugreek (*Trigonella foenum-graecum* L.) plants. *J. Genet. Eng. Biotechnol.* **2018**, *16*, 683–692. [CrossRef]
155. Pradhan, B.; Patra, S.; Nayak, R.; Swain, S.S.; Jit, B.P.; Behera, C.; Ragusa, A.; Ki, J.-S.; Jena, M. Low-dose priming of gamma radiation enhanced cadmium tolerance in *Chlamydomonas reinhardtii* by modulating physio-biochemical pathways. *Environ. Sci. Pollut. Res.* **2022**. [CrossRef]
156. Al-Enezi, N.A.; Al-Bahrany, A.M.; Al-Khayri, J.M. Effect of X-irradiation on date palm seed germination and seedling growth. *Emir. J. Food Agric.* **2012**, *24*, 415–424.

plants

MDPI

Article

Salinity Tolerance, Ion Accumulation Potential and Osmotic Adjustment In Vitro and In Planta of Different *Armeria maritima* Accessions from a Dry Coastal Meadow

Līva Purmale, Astra Jēkabsone, Una Andersone-Ozola and Gederts Ievinsh *

Department of Plant Physiology, Faculty of Biology, University of Latvia, 1 Jelgavas Str., LV-1004 Rīga, Latvia
* Correspondence: gederts.ievins@lu.lv

Abstract: The aim of the present study was to compare tolerance to salinity and ion accumulation potential of *Armeria maritima* subsp. *elongata*. Three accessions (AM1 and AM2, both from Latvia, and AM3 from Sweden) from relatively dry sandy soil habitats in the Baltic Sea region were selected and compared using both in vitro cultivated shoot explants and long-term soil-cultivated plants at flowering stage. Growth of root non-forming explants treated with increasing concentrations of NaCl was significantly inhibited starting from 110 mmol L^{-1}, and the rate of shoot formation was even more sensitive. Significant differences in morphology and responses to salinity were found between different accessions. For soil-grown plants, biomass accumulation in above-ground parts was relatively little affected by salinity in AM1 and AM2 in comparison to that in AM3. Differences in ion accumulation were evident between the accessions as well as in respect to cultivation system used. Maximum accumulation capacity for Na^+ was up to 2.5 mol kg^{-1} both in shoot explant tissues and in old leaves of soil-grown plants treated with NaCl, but that for K^+ reached 4.0 mol kg^{-1} in old leaves of soil-grown plants treated with KCl. Non-ionic component of osmotic value was relatively high in old leaves and significantly increased under NaCl treatment, especially for AM2 and AM3 plants at moderate salinity, but in AM1 only at high salinity. In contrast, it significantly decreased in old leaves of AM2 plants treated with increasing concentration of KCl. It can be concluded that a wide salinity tolerance exists within *A. maritima* accessions from dry sandy soil habitats, associated with the ability to accumulate surplus ions both in salt glands and old leaves.

Keywords: *Armeria maritima*; electrical conductivity; functional differences; ion accumulation; non-ionic osmolytes; osmotic adjustment; potassium; salinity; sodium; tissue culture

Citation: Purmale, L.; Jēkabsone, A.; Andersone-Ozola, U.; Ievinsh, G. Salinity Tolerance, Ion Accumulation Potential and Osmotic Adjustment In Vitro and In Planta of Different *Armeria maritima* Accessions from a Dry Coastal Meadow. *Plants* **2022**, *11*, 2570. https://doi.org/10.3390/plants11192570

Academic Editors: Kinga Dziurka, Mateusz Labudda and Ewa Muszyńska

Received: 9 September 2022
Accepted: 27 September 2022
Published: 29 September 2022

1. Introduction

Increasing soil salinity is one of the major threats to agricultural production, especially, in a light of global climate change-dependent increase in environmental heterogeneity [1–3]. Therefore, studies aiming to understand the mechanisms of salt tolerance are gaining special importance. In this respect, plant species native to salt-affected habitats are especially valuable models for understanding physiological characteristics of adaptive value important for life in saline environments [4].

In contrast to salt exclusion approach used by salt tolerant glycophytes, tight control of compartmentation of salinity-related ions is one of the adaptive mechanisms used by plants native to saline habitats [5]. As ionic species are both electrolytically and osmotically active, appropriate internal adjustment of osmotic balance is an important strategy for salinity tolerance. A major nonessential substance in soils of sea-water affected habitats is NaCl, but other types of soil salinity may be common in other situations [3]. Usually, when characterizing plant responses and resistance to salinity, emphasis is placed on Na^+ toxicity in the form of NaCl. In the context of salinity, K^+ has been analyzed mostly in a view of necessity to maintain high cytoplasmic K^+/Na^+ concentration ratio [6,7]. It was shown that coastal plant species from salt-affected habitats represent either Na^+ excluders, regulating tissue electrical conductivity (EC) by changes in K^+ concentration, K^+ excluders, regulating EC by changes in Na^+ concentration, or tight EC regulators [8]. There is a reason to suggest that for typical high salt-adapted species native to saline habitats (halophytes) high level of K^+ in substrate will have the same negative effects than Na^+ [9]. It was also shown that for halophytes, NaCl toxicity is related mainly to the effect of Cl^- [10].

Halophyte species, native to salt-affected habitats, represent a valuable resource for studies of physiological mechanisms related to salinity tolerance [11,12]. Usually the most common approach has been to use single halophytic model species or one halophyte and one glycophyte species [13], and to assess changes in different biochemical and physiological parameters, preferably over a gradient of salinity. Comparative studies of salinity tolerance have been performed recently involving a number of halophyte species from the same or similar habitats [14,15]. However, studies comparing different accessions or genotypes of a single halophyte species have been seldomly performed [16–18]. Recently, we performed a study involving a number of accessions of crop wild relative legume species *Trifolium fragiferum* and concluded that high intraspecies variability in morphological and physiological responses to salinity exist between geographically isolated populations [19].

Tissue culture has been often used as a tool for screening crop plant genotypes for their salinity tolerance [20–22]. In addition, it offers an opportunity to study salt tolerance-related responses at tissue level excluding whole plant-level responses, which can be important aspect of salinity tolerance [23]. However, in vitro studies of salinity tolerance with halophytic plant species have been relatively seldom performed. Thus, shoot explant culture was used for multiplication and selection of salt-tolerant genotypes of *Atriplex halimus* [24]. Axillary shoot culture was used to study antioxidative defense and osmotic adjustment of *Sesuvium portulacastrum* during high salinity [25]. Shoot cultures of *Salicornia europaea* [26], *Salicornia brachiata* [27] and *Limoniastrum monopetalum* [28] were used to reveal essentiality of NaCl for efficient in vitro propagation of these halophyte species. However, NaCl gradually reduced shoot proliferation in explants of *Crithmum maritimum* [29]. Several studies used seeds as explants to assess salinity tolerance of seedlings in tissue culture [30,31]. Comparative salinity tolerance studies using both tissue culture and whole plants grown in the substrate are rarely carried out [32].

Rosette-forming evergreen perennial species *Armeria maritima* (Mill.) Willd. (Plumbaginaceae) is characteristically found in temperate open habitats with dry, saline, sandy or metal-rich soil [33]. In a study with *A. maritima* plants from eight populations representing five different habitats in Britain (mountain, sea-cliff, salt marsh, shingle, and pasture) it was found that each population showed different ecological responses viewed as a result of local adaptation [34]. However, some traits of *A. maritima* exhibit phenotypic plasticity [33]. Presence of *A. maritima* in salt-affected habitats raises the question of salt tolerance of the species and its mechanisms. Previously, all ecotypes of *A. maritima* from different habitats have shown relatively high salinity tolerance, similar to that of species from brackish conditions of upper salt marsh, and even shoot growth stimulation at low salinity [35]. However, it is still an open question whether salinity tolerance of *A. maritima* plants from different populations is related to phenotypic plasticity or local genetic adaptation. In this respect, it is reasonable to ask whether there are differences between *A. maritima* accessions growing in habitats with relatively similar conditions but being geographically isolated, in respect to their responses to salinity. Recently we performed a comparative study of heavy metal tolerance and metal accumulation potential of three geographically isolated *A. maritima* accessions from a dry coastal meadow, and showed significant species-wide metal tolerance and extremely high accumulation potential of *A. maritima* but with some accession-specific differences [36]. Therefore, the aim of the present study was to analyze salinity tolerance of the three *A. maritima* subsp. *elongata* accessions from a dry coastal meadow in controlled conditions. A special attention was paid to possible different growth responses and ion accumulation potential to variable treatment with NaCl or KCl. It was hypothesized that similar effect of Na^+ and K^+ on growth of *A. maritima* plants will be found.

2. Materials and Methods

2.1. Plant Material

Seeds of *A. maritima* subsp. *elongata* from three geographically isolated micropopulations growing on sandy soils in water reservoir-associated meadows were used as propagation material (Table 1). Two small micropopulations in Latvia were located on a shore meadow of River Vecdaugava (AM1) and a shore meadow of Buļļupe (AM2). Micropopulation from Sweden (AM3) was located on a coastal meadow of the Baltic Sea. Seeds of AM1 and AM2 were used for initiation of tissue culture as described further. Multiplied shoot explants were used for tissue culture experiment or were rooted and acclimatized for soil culture experiment. Seeds of AM3 were used for establishment of plants for soil culture experiment. Salt treatments for soil-grown plants were performed at the beginning of appearance of reproductive structures.

Table 1. Accessions of *Armeria maritima* used in the present study, characterization of their habitats and experiments performed.

Code	Associated Water Reservoir	Habitat	Electrical Conductivity (mS m^{-1})	Location	Coordinates	Performed Experiments (Treatments)
AM1	River Vecdaugava	Dry shore meadow	97 ± 5 c	City of Riga, Ziemeļu District, Vecdaugava, Latvia	57°03′29″ N 24°05′47″ E	In vitro (NaCl 44, 110, 174, 217 mol L^{-1}).In planta (NaCl 22, 44, 87, 217 mol L^{-1}; KCl 22, 44, 87, 217 mol L^{-1})
AM2	River Buļļupe	Dry shore meadow	127 ± 4 b	City of Riga, Kurzeme District, Island of Buļļu Sala, Vakarbuļļi, Latvia	56°59′54″ N 23°57′31″ E	In vitro (NaCl 44, 110, 174, 217 mol L^{-1}).In planta (NaCl 22, 44, 87, 217 mol L^{-1}; KCl 22, 44, 87, 217 mol L^{-1})
AM3	The Baltic Sea	Dry coastal meadow	223 ± 11 a	Nybrostrand, Ystad Municipality, Skåne County, Sweden	55°25′40″ N 13°57′27″ E	In planta (NaCl 22, 44, 87, 217 mol L^{-1}; KCl 22, 44, 87, 217 mol L^{-1}); NaCl + KCl 44, 87, 217 mol L^{-1})

Different letters for soil electrical conductivity indicate statistically significant differences ($p < 0.05$).

Soil electrical conductivity (EC) was measured in natural habitats using HH2 m equipped with WET-2 sensor (Delta-T Devices, Burwell, UK) at five spots at least 5 m apart. For all micropopulations, soil was non-saline or slightly saline (Table 1).

2.2. In Vitro Experiment

Seeds of AM1 and AM2 were used as explants for establishment of tissue culture. All details on culture initiation and multiplication were as described previously [36]. Briefly, Murashige and Skoog medium supplemented with 30 g L^{-1} sucrose and 6 g L^{-1} agar was used for initial culture, followed by multiplication on Murashige and Skoog medium containing sucrose and agar supplemented with 0.1 mg L^{-1} 1-naphthaleneacetic acid and 1 mg L^{-1} 6-benzylaminopurine. Rooting was performed on the same medium with 0.2 mg L^{-1} 1-naphthaleneacetic acid. Rooted explants were acclimatized ex vitro in a peat substrate for two weeks and further used for in planta experiment.

Shoot explants at multiplication stage were used for treatment with NaCl. Necessary concentration of NaCl (Table 1) was added to multiplication medium before autoclaving. Medium was poured in 200 mL jars (five per treatment) and five shoot explants were placed in each jar. Cultures were placed in a growth cabinet under 16 h photoperiod provided by a fluorescent light with photon flux density 50 μmol m^{-2} s^{-1} of photosynthetically active radiation at 25 °C. After 4 weeks, the experiment was terminated. Multiplication rate was evaluated by counting number of shoots per explant. Fresh and dry mass (after drying at

60 °C for 72 h) of tissues were measured. Tissue water content was expressed as g of H_2O per g dry mass.

2.3. *In Planta Experiments*

For accessions AM1 and AM2, rooted and acclimatized explants were used for establishment of experimental material as described previously [36]. Plants were individually planted in 1.3 L plastic containers filled with 1 L of a mixture of quartz sand (Saulkalne S, Saulkalne, Latvia) and heat-treated (60 °C, 24 h) garden soil (Biolan, Eura, Finland) 1:3 (*v*/*v*), and placed in an experimental automated greenhouse (HortiMaX, Maasdijk, Netherlands) with supplemented light from Master SON-TPIA Green Power CG T 400 W (Philips, Amsterdam, Netherlands) and Powerstar HQI-BT 400 W/D PRO (Osram, Munich, Germany) lamps (photon flux density of photosynthetically active radiation 380 μmol m^{-2} s^{-1} at the plant level), 16 h photoperiod, day/night temperature 24/16 °C, relative air humidity 60 to 70%. Salinity treatment (Table 1) was started after a week-long period of additional acclimatization in greenhouse. Salt treatment was performed gradually, by 44 mmol L^{-1} increments during 5 weeks, using NaCl and KCl solution. Necessary amount of salt was dissolved in deionized water and 0.1 L per container was applied to soil. During treatments, plants started to develop generative structures. Plants were cultivated for 8 more weeks after reaching full treatment.

Plant material for accession AM3 was established by seeds as described previously [36]. Seeds were surface sterilized with a half-diluted commercial bleach (ACE, Procter & Gamble, Warszawa, Poland) and sown in 1 L plastic plant tissue culture containers filled with autoclaved (1 atm, 20 min) garden soil (Biolan, Eura, Finland), closed with lids and further cultivated for two weeks in a growth cabinet (light/dark period of 16/8 h, photosynthetically active radiation with a photon flux density 100 μmol m^{-2} s^{-1}, day/night temperature 5/15 °C). After change of temperature regime to day/night temperature 15/20 °C for two additional weeks, seedlings were transplanted to 0.25 L plastic containers with a mixture of quartz sand (Saulkalne S, Saulkalne, Latvia) and garden soil (Biolan, Eura, Finland) 1:3 (*v*/*v*) and acclimatized to greenhouse conditions. Final transplantation to 0.5 L plastic containers with the same substrate was performed after two weeks. Fully developed 2-month-old plants in a vegetative stage were assigned to one of 12 treatments (Table 1), five individual plants per treatment. Salt treatment was performed gradually, by not more than 44 mmol L^{-1} increments during 5 weeks, using NaCl and KCl solution. During treatments, plants started to develop generative structures. Plants were cultivated for 7 more weeks after reaching full treatment.

During cultivation, individual containers were randomly redistributed weekly on a greenhouse bench. Substrate water content was monitored with HH2 moisture meter equipped with WET-2 sensor (Delta-T Devices, Burwell, UK) and kept at 50 to 60%. Every third week plants were fertilized with Yara Tera Kristalon Red and Yara Tera Calcinit fertilizers (Yara International, Oslo, Norway). A stock solution was prepared for each fertilizer (100 g L^{-1}) and working solution contained 25 mL of each per 10 L deionized water, used with a rate 100 mL per container.

At termination of each experiment, plants were individually separated in different parts (roots, flower stalks, inflorescences (flowers), leaves of different age). In experiment with AM1 and AM2, older leaves did not fully decay, as only part of the particular leaf was becoming brown and dry. Therefore, instead of designating these leaves as "decayed", all leaves were sorted either as "old" or "young" according to their position and general morphological appearance. In experiment with AM3, individual older leaves decayed fully, therefore, all leaves were sorted either as "decayed" or "living" but for sake of comparison were indicated as "old" or "young", respectively. Inflorescences were counted, and the length of flower stalks was measured. Plant material was weighed separately before and after drying in an oven at 60 °C for 72 h. Water content was calculated as g H_2O per g dry mass.

2.4. Measurements

All analyses were performed in triplicate, using representative tissue samples from individual biological replicates. Plant tissues were homogenized by crushing and a sample (0.2 g) was taken for analysis of EC, Na^+ concentration and K^+ concentration in water extract by respective compact meters and analysis of osmotic activity by a freezing point osmometer. Tissues were ground with mortar and pestle to a fine powder and 10 mL of deionized water was added. The homogenate was stirred with pestle for 1 min. After filtration through nylon mesh cloth (No. 80) homogenate was used for measurement of ion concentration by LAQUAtwin compact meters B-722 (Na^+) and B-731 (K^+), and electrical conductivity by LAQUAtwin conductivity meter B-771 (Horiba, Kyoto, Japan) and measurement of osmotic value. For osmotic value analysis, 50 µL of extract were transferred in a 1.5 mL Eppendorf tube and placed in a freezing point osmometer Osmomat 3000 Basic (Gonotec Meβ- und Regeltechnik, Berlin, Germany) and operated according to the manufacturer's instructions. Using a standard curve for different concentrations of NaCl and KCl, the osmotic value caused by the total concentration of Na^+ and K^+ was calculated according to the actual Na^+ and K^+ concentration of each sample extract. For each sample, the difference between the total osmotic value and the osmotic value due to Na^+, K^+ and Cl^- ions was calculated and designated as "non-ionic osmotic value", which showed the osmotic effect of other osmotically active ions (besides Na^+, K^+ and Cl^-) or non-ionic compounds. At least three analytical replicates were performed for each sample and the average value was calculated.

2.5. Data Analysis

Results were analyzed by KaleidaGraph (v. 5.0, Synergy Software, Reading, PA, USA). Statistical significance of differences was evaluated by one-way ANOVA using post-hoc analysis with minimum significant difference. Significant differences were indicated at $p < 0.05$.

3. Results

3.1. In Vitro Experiment

Growth and proliferation of root non-forming explants of *A. maritima* cultivated on multiplication medium was negatively affected by increasing concentration of NaCl in the medium (Figure 1). Biomass of explants from accession AM1 tended to be higher at all NaCl concentrations in comparison to AM2, but significant difference was evident only at 44 mmol L^{-1} (Figure 1A). At the highest NaCl concentration (217 mol L^{-1}) biomass accumulation was inhibited by 46 and 49%, and proliferation by 56 and 54%, for AM1 and AM2, respectively. Water content in explants showed a tendency to increase at 44 mmol L^{-1} NaCl, followed by a significant decrease at 110 mmol L^{-1} (Figure 1C). Further increase in NaCl concentration did not result in changes of water content.

Concentration of Na^+ increased in explant tissues cultivated in presence of increasing medium NaCl, but the accumulation response seemed to be saturable at 110 mmol L^{-1} (Figure 2A). However, at the highest medium NaCl concentration, explants of AM1 showed significant further increase in tissue Na^+ concentration. In contrast, explant K^+ concentration decreased with increasing medium NaCl concentration up to 174 mmol L^{-1} (Figure 2B). As a result, summed $Na^+ + K^+$ concentration was relatively stable over a range of medium NaCl concentration (Figure 2C). However, increase in tissue electrical conductivity with increasing salinity was relatively more pronounced (Figure 2D).

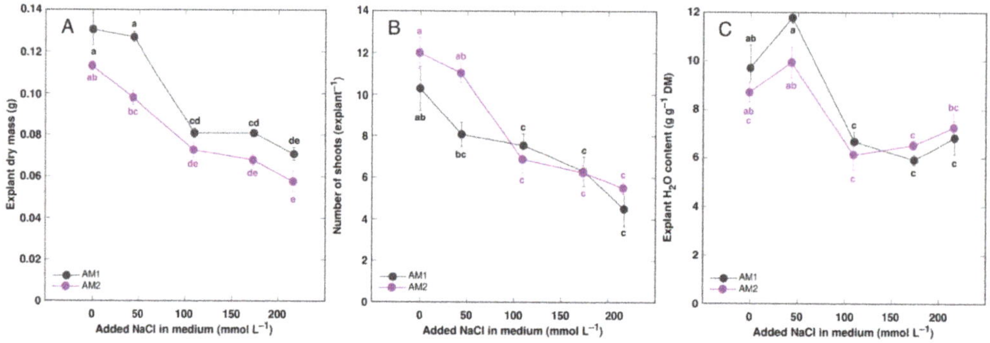

Figure 1. Effect of added Na concentration in tissue culture medium on dry mass (**A**), number of shoots (**B**) and tissue water content (**C**) of explants of *Armeria maritima* accessions AM1 and AM2 after 4 weeks of cultivation. Data are means ± SE from 5 replicates, with five explants each. Different letters of respective color between accessions and treatments indicate statistically significant differences ($p < 0.05$).

3.2. In Planta Experiments: Effect of Salinity on Growth

Effect of Na^+ and K^+ in a form of chloride on growth of *A. maritima* plants cultivated in substrate was identical. Total biomass of substrate-cultivated plants for accessions AM1 and AM2 was relatively stable over increasing NaCl and KCl concentration range, and no significant decrease was evident even at 217 mol L^{-1} salinity (Figure 3). However, as biomass tended to increase for AM1 plants cultivated at 22 mol L^{-1}, there was a significant difference between plants at 22 and 217 mL L^{-1} (Figure 3A).

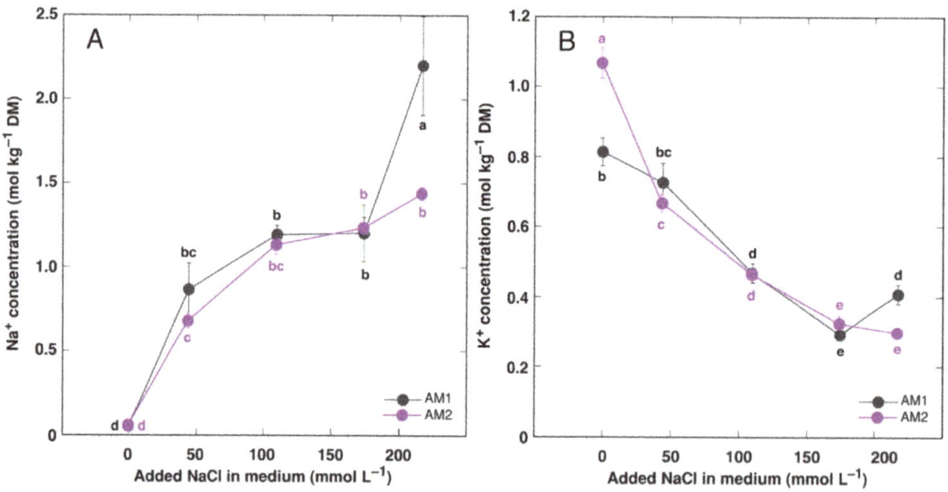

Figure 2. *Cont.*

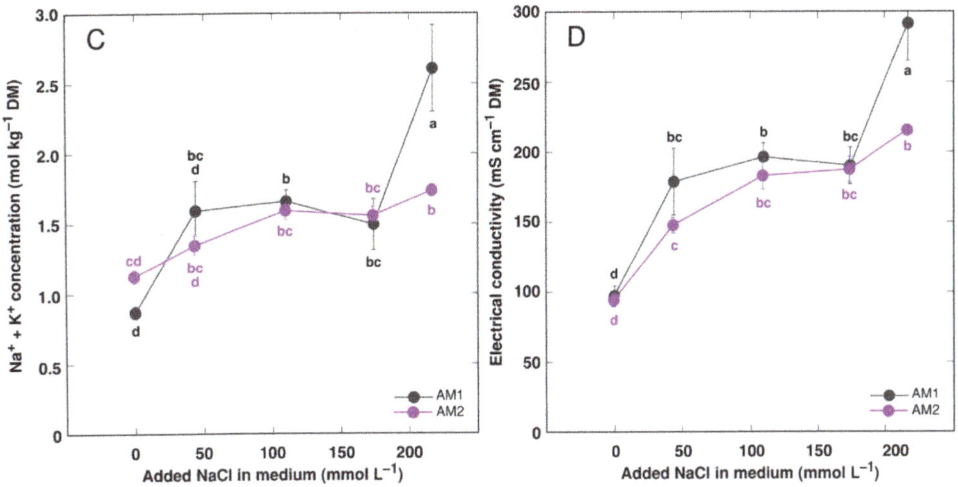

Figure 2. Effect of added Na^+ concentration in tissue culture medium on Na^+ concentration (**A**), K^+ concentration (**B**), summed $Na^+ + K^+$ concentration (**C**) and electrical conductivity (**D**) in explant tissues of *Armeria maritima* accessions AM1 and AM2 after 4 weeks of cultivation. DM, dry mass. Data are means ± SE from 4–5 replicates. Different letters of respective color between accessions and treatments indicate statistically significant differences ($p < 0.05$).

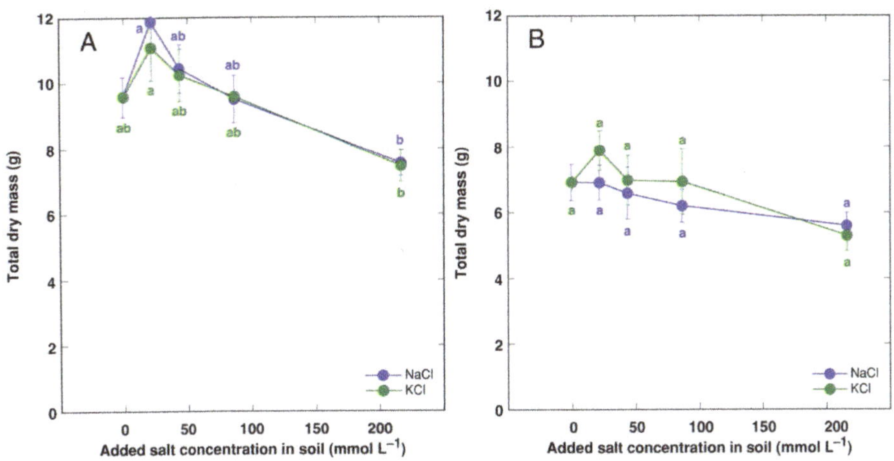

Figure 3. Effect of added NaCl and KCl concentration in soil on total dry mass of *Armeria maritima* plants from accessions AM1 (**A**) and AM2 (**B**) after 8 weeks of cultivation. Data are means ± SE from 5 replicates. Different letters of respective color between accessions and treatments indicate statistically significant differences ($p < 0.05$).

Number of flower stalks and dry mass of leaves (Table 2), as well as dry mass of both flower stalks and flowers was not significantly affected by increasing salinity for both AM1 and AM2 (Table 3). However, total length of flowers talks was significantly decreased for AM2 at 217 mol L^{-1} KCl (Table 2). Biomass of roots was significantly decreased for AM1 at 87 and 217 mmol L^{-1} salinity for both NaCl- and KCl-treated plants, but was not significantly affected for AM2 plants (Table 3).

Table 2. Effect of salinity treatment on morphological parameters of *Armeria maritima* accessions AM1 and AM2 cultivated for 8 weeks in soil.

Salt	Concentration (mmol L^{-1})	Flower Stalks (*n*)		Total Length of Flower Stalks (m $Plant^{-1}$)		Dry Mass of Leaves (g)	
		AM1	AM2	AM1	AM2	AM1	AM2
Control	0	8.0 ± 0.7 abc	8.4 ± 1.0 ab	1.77 ± 0.14 abc	2.00 ± 0.25 ab	6.26 ± 0.47 ab	3.77 ± 0.52 a
NaCl	22	13.2 ± 1.2 a	9.8 ± 1.3 a	2.52 ± 0.30 a	2.13 ± 0.25 ab	7.64 ± 0.56 a	3.70 ± 0.35 a
	44	10.4 ± 0.5 ab	7.6 ± 0.6 ab	1.91 ± 0.09 ab	1.71 ± 0.17 abc	6.88 ± 0.55 ab	3.58 ± 0.40 a
	87	9.8 ± 0.9 abc	7.8 ± 0.5 ab	1.76 ± 0.12 abc	1.61 ± 0.13 abc	6.65 ± 0.61 ab	3.36 ± 0.27 a
	217	6.2 ± 0.6 c	6.6 ± 0.6 ab	1.02 ± 0.06 c	1.18 ± 0.16 bc	5.59 ± 0.30 ab	3.75 ± 0.37 a
KCl	22	10.8 ± 0.6 ab	10.0 ± 0.7 a	2.07 ± 0.15 a	2.26 ± 0.15 a	7.61 ± 0.75 a	3.66 ± 0.23 a
	44	9.8 ± 1.2 abc	9.6 ± 0.7 ab	2.07 ± 0.17 a	2.16 ± 0.20 a	6.73 ± 0.62 ab	3.66 ± 1.02 a
	87	9.4 ± 1.3 abc	8.4 ± 1.3 ab	1.86 ± 0.26 ab	1.89 ± 0.32 abc	6.48 ± 0.20 ab	4.13 ± 0.69 a
	217	7.6 ± 0.7 bc	5.4 ± 1.0 b	1.16 ± 0.13 bc	1.00 ± 0.21 c	5.06 ± 0.28 b	3.61 ± 0.26 a

Data are means ± SE from 5 replicates. Different letters between treatments for a particular parameter for each accession separately indicate statistically significant differences ($p < 0.05$).

Table 3. Effect of salinity treatment on dry mass of generative parts and roots of *Armeria maritima* accessions AM1 and AM2 cultivated for 8 weeks in soil.

Salt	Concentration (mmol L^{-1})	Dry Mass of Flower Stalks (g)		Dry Mass of Flowers (g)		Dry Mass of Roots (g)	
		AM1	AM2	AM1	AM2	AM1	AM2
Control	0	0.70 ± 0.03 abc	0.97 ± 0.06 abc	0.83 ± 0.08 ab	1.01 ± 0.05 ab	1.79 ± 0.20 a	1.18 ± 0.14 ab
NaCl	22	1.01 ± 0.13 a	1.13 ± 0.11 a	1.30 ± 0.17 a	1.08 ± 0.12 ab	1.93 ± 0.10 a	1.01 ± 0.19 ab
	44	0.80 ± 0.05 ab	0.89 ± 0.13 abc	1.03 ± 0.06 ab	0.95 ± 0.16 ab	1.73 ± 0.15 a	1.18 ± 0.14 ab
	87	0.68 ± 0.05 abc	0.89 ± 0.09 abc	0.99 ± 0.08 ab	1.04 ± 0.11 ab	1.19 ± 0.09 bc	0.93 ± 0.11 ab
	217	0.43 ± 0.01 c	0.57 ± 0.09 bc	0.79 ± 0.06 b	0.69 ± 0.05 b	0.75 ± 0.07 c	0.60 ± 0.03 b
KCl	22	0.91 ± 0.06 ab	1.33 ± 0.10 a	1.11 ± 0.08 ab	1.46 ± 0.11 a	1.44 ± 0.18 ab	1.45 ± 0.22 a
	44	0.87 ± 0.08 ab	1.07 ± 0.04 ab	1.13 ± 0.08 ab	1.14 ± 0.04 ab	1.51 ± 0.11 ab	1.12 ± 0.20 ab
	87	0.82 ± 0.12 ab	0.96 ± 0.20 abc	1.13 ± 0.16 ab	0.98 ± 0.14 ab	1.16 ± 0.07 bc	0.88 ± 0.03 ab
	217	0.58 ± 0.07 bc	0.49 ± 0.10 c	1.06 ± 0.08 ab	0.63 ± 0.13 b	0.77 ± 0.06 c	0.56 ± 0.05 b

Data are means ± SE from 5 replicates. Different letters between treatments for a particular parameter for each accession separately indicate statistically significant differences ($p < 0.05$).

Significant decrease in water content of old leaves of accession AM1 at 217 mmol L^{-1} salinity for both NaCl- and KCl-treated plants indicated similar degree of stimulation of leaf decay (Figure 4A). However, for plants from accession AM2, the degree of decay was significantly more pronounced for plants treated with 87 and 217 mol L^{-1}, in comparison to NaCl-treated plants (Figure 4B).

A. maritima plants of accession AM3 was treated with identical concentrations of NaCl, KCl or combination of NaCl with KCl, and all treatments had identical negative effect on total biomass starting from 44 mol L^{-1} (Figure 5A). Biomass of decayed leaves had a tendency to increase at 217 mol L^{-1} salinity, especially, for KCl-treated plants, but the effect was not statistically significant (Figure 5B). Biomass of living leaves tended to increase at low salinity and decreased with increasing salt concentration, but the effect was not statistically significant due to high variability between individual plants (Figure 5C). As a result, total dry mass of leaves did not change significantly in salt-treated AM3 plants (Table 4). However, growth of reproductive structures and roots was more negatively affected by increasing salinity (Table 4). Thus, number of flower stalks significantly decreased from 87 mml L^{-1} salinity, total length and dry mass of flower stalks significantly decreased from 44 mol L^{-1} salinity, but dry mass of flowers significantly decreased from 217 mmol L^{-1} for NaCl, 44 mmol L^{-1} for KCl and NaCl + KCl treatments. Root biomass tended to decrease from 44 mmol L^{-1}, and the effect was not statistically significant for NaCl, but was significant starting from 87 mmol L^{-1} for plants treated with KCl and NaCl + KCl.

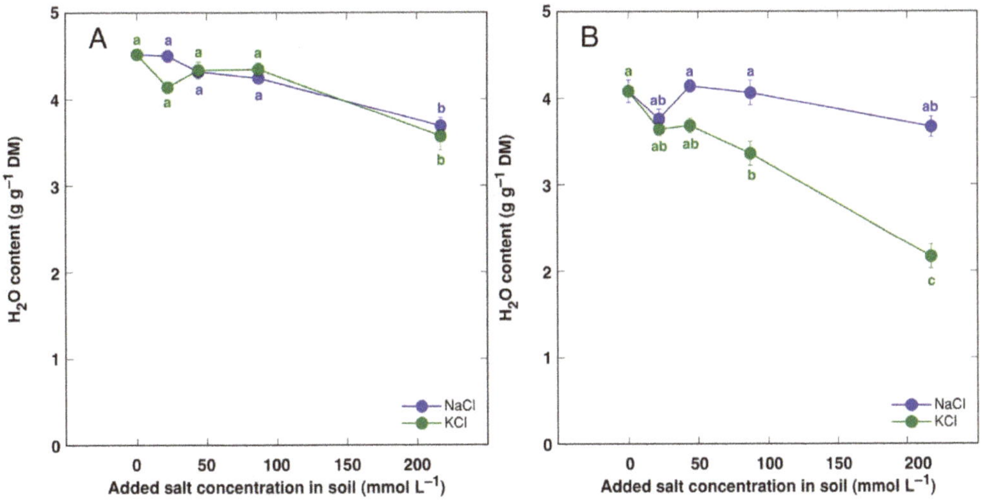

Figure 4. Effect of added NaCl and KCl concentration in soil on water content in old leaves of *Armeria maritima* plants from accessions AM1 (**A**) and AM2 (**B**) after 8 weeks of cultivation. DM, dry mass. Data are means ± SE from 5 replicates. Different letters of respective color between accessions and treatments indicate statistically significant differences ($p < 0.05$).

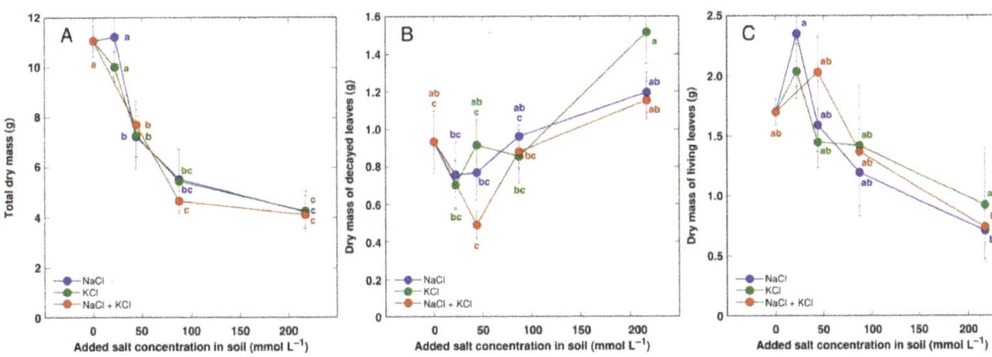

Figure 5. Effect of added NaCl, KCl and NaCl + KCl concentration in soil on total dry mass (**A**), dry mass of decayed leaves (**B**) and dry mass of living leaves (**C**) of *Armeria maritima* plants from accession AM3 after 7 weeks of cultivation. Data are means ± SE from 5 replicates. Different letters of respective color between accessions and treatments indicate statistically significant differences ($p < 0.05$).

Table 4. Effect of salinity treatment on morphological parameters of *Armeria maritima* accession AM3 cultivated for 7 weeks in soil.

Salt	Concentration (mol L^{-1})	Flower Stalks (*n*)	Total Length of Flower Stalks (m)	Dry Mass of Leaves (g)	Dry Mass of Flower Stalks (g)	Dry Mass of Flowers (g)	Dry Mass of Roots (g)
Control	0	10.6 ± 0.7 a	4.34 ± 0.26 a	2.63 ± 0.16 a	3.50 ± 0.14 a	2.64 ± 0.11 a	2.28 ± 0.34 ab
NaCl	22	8.2 ± 0.9 ab	3.80 ± 0.33 ab	3.11 ± 0.60 a	2.80 ± 0.31 ab	2.39 ± 0.18 abc	2.93 ± 0.49 a
	44	7.2 ± 0.6 abc	2.49 ± 0.24 bc	2.36 ± 0.33 a	1.77 ± 0.17 bcd	1.74 ± 0.18 abcde	1.38 ± 0.34 bcd
	87	5.4 ± 0.5 bc	1.81 ± 0.27 cd	2.16 ± 0.38 a	1.07 ± 0.24 cde	1.29 ± 0.40 abcde	1.02 ± 0.22 bcd
	217	4.5 ± 1.2 bc	1.19 ± 0.29 d	1.91 ± 0.20 a	0.65 ± 0.19 de	0.83 ± 0.26 de	1.13 ± 0.21 bcd

Table 4. *Cont.*

Salt	Concentration (mol L^{-1})	Flower Stalks (*n*)	Total Length of Flower Stalks (m)	Dry Mass of Leaves (g)	Dry Mass of Flower Stalks (g)	Dry Mass of Flowers (g)	Dry Mass of Roots (g)
KCl	22	10.6 ± 1.1 a	4.03 ± 0.44 ab	2.74 ± 0.11 a	2.83 ± 0.23 ab	2.59 ± 0.35 ab	1.87 ± 0.39 abc
	44	7.4 ± 1.8 ab	2.51 ± 0.68 bc	2.36 ± 0.20 a	1.69 ± 0.44 cd	1.97 ± 0.60 abcde	1.23 ± 0.20 bcd
	87	5.4 ± 0.9 bc	1.69 ± 0.40 cd	2.27 ± 0.61 a	1.05 ± 0.27 cde	1.23 ± 0.29 bcde	0.90 ± 0.20 cd
	217	2.8 ± 0.7 c	0.64 ± 0.22 d	2.44 ± 0.64 a	0.39 ± 0.12 e	0.60 ± 0.18 e	0.82 ± 0.17 cd
NaCl + KCl	44	6.6 ± 0.8 abc	2.44 ± 0.17 bc	2.52 ± 0.26 a	1.97 ± 0.12 bc	2.18 ± 0.23 abcd	1.05 ± 0.12 bcd
	87	4.4 ± 0.3 bc	1.44 ± 0.37 cd	2.25 ± 0.26 a	0.92 ± 0.03 cde	1.00 ± 0.16 cde	0.61 ± 0.11 cd
	217	4.4 ± 0.7 bc	1.26 ± 0.22 d	1.89 ± 0.15 a	0.74 ± 0.14 de	0.86 ± 0.14 de	0.48 ± 0.03 d

Data are means ± SE from 5 replicates. Different letters between treatments for a particular parameter indicate statistically significant differences ($p < 0.05$).

3.3. *In Planta Experiments: Effect of Salinity on Ion Accumulation and Osmotic Adjustment*

Possible differences in accumulation of electrolytically and osmotically active ions Na^+ and K^+ in *A. maritima* plants from three different accessions treated with NaCl or KCl were evaluated after long-term cultivation in substrate. Plants treated with 217 mmol L^{-1} NaCl as well as plants treated with 87 and 217 mmol L^{-1} KCl actively deposited salt crystals on surface of leaves and flower stalks (Figure S1), but intensity of deposition was not estimated. Nevertheless, growth at increased substrate salinity caused either by NaCl or KCl resulted in increased accumulation of respective cations in all plant parts, but with clear organ specific pattern of accumulation (Figures 6, 7D–F, S2 and S3). The highest Na^+ concentration reached was 2.48 mol kg^{-1} (57 g kg^{-1}) in old leaves of plants from accession AM3 treated with 217 mmol L^{-1} NaCl (Figure 6A), and that for K^+ was 4.03 mol kg^{-1} (157 g kg^{-1}) in old leaves of plants from accession AM2 treated with 217 mmol L^{-1} KCl (Figure 7D). Accumulation capacity of either Na^+ or K^+ in plants treated with NaCl and KCl, respectively, increased in an order roots < flowers < flower stalks < new leaves < old leaves. There was a stable tendency that AM3 plants accumulated more Na^+ in old leaves of NaCl-treated plants in comparison to AM1 and AM2 plants (Figure 6A), and AM2 plants accumulated more K^+ in old leaves of KCl-treated plants in comparison to AM1 and AM3 plants (Figure 7D). It is also important that accumulation capacity for K^+ in all plant parts was higher than that for Na^+ at identical salinity caused either by KCl or NaCl treatment, respectively.

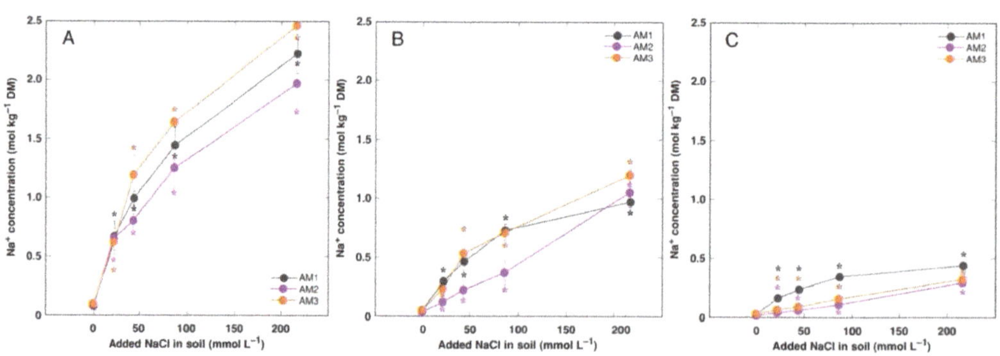

Figure 6. Effect of added NaCl concentration in soil on accumulation of Na^+ in old leaves (**A**), new leaves (**B**) and flowers (**C**) of *Armeria maritima* plants from different accessions after 7–8 weeks of cultivation. DM, dry mass. Data are means ± SE from 3 replicates. Asterisks of respective color indicate statistically significant differences ($p < 0.05$) from control.

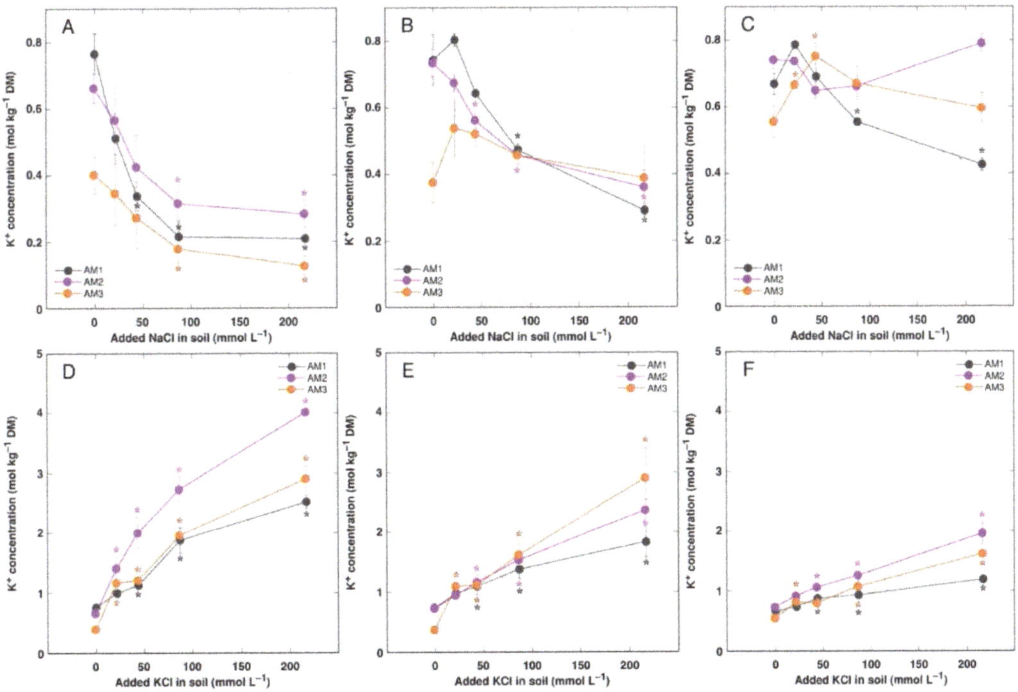

Figure 7. Effect of added NaCl (**A–C**) and KCl (**D–F**) concentration in soil on accumulation of K^+ in old leaves (**A**,**D**), new leaves (**B**,**E**) and flowers (**C**,**F**) of *Armeria maritima* plants from different accessions after 7–8 weeks of cultivation. DM, dry mass. Data are means ± SE from 3 replicates. Asterisks of respective color indicate statistically significant differences ($p < 0.05$) from control.

For NaCl-treated plants, increased salinity resulted in a significant reduction in K^+ concentration in old leaves for all accessions (Figure 7A) and new leaves for AM1 and AM2 (Figure 7B). Effect of NaCl treatment on K^+ concentration in generative parts and roots differed between accessions. Thus, it significantly increased in flowers (Figure 7C) and flower stalks (Figure S3B) of plants from accession AM3, but decreased in flowers of AM1 (Figure 7C) and flower stalks of AM1 and AM2 (Figure S3B). Moreover, K^+ concentration tended to decrease in roots of NaCl-treated AM1 and AM2 plants (Figure S3A).

Level of intensity of electrical conductivity was partitioned in *A. maritima* in a same way as Na^+ and K^+, and it tended to be higher in KCl-treated plants in comparison to the NaCl-treated plants (Figure 8 and Figure S4). Distribution and levels of summed $Na^+ + K^+$ concentration showed the similar character (Figure S5). Control plants of *A. maritima* showed pronounced gradient in K^+: Na^+ concentration ratio between plant parts, decreasing in an order flowers > new leaves > flower stalks > old leaves > roots, and plants from accession AM2 had higher values of the ratio in generative parts and new leaves (Figure S6).

Total osmotic value in *A. maritima* tissues under rising salinity increased in all plant parts in a concentration-dependent manner, but the increase was relatively less pronounced than that for electrical conductivity (Figures 9 and S7). The highest values were reached in old leaves followed by new leaves and generative parts, and was the lowest in roots. There were no pronounced differences in osmotic value between various accessions, but it tended to be higher in some combinations of KCl-treated plants in comparison to NaCl-treated plants. However, there were pronounced differences in non-ionic osmotic values between plant parts, accession and treatments (Figures 10 and S8). In old leaves of AM2 and AM3 plants treated with NaCl, non-ionic osmotic value significantly increased at low to moderate salinity, but decreased at high salinity (Figure 10A). However, for AM1, there was initial decrease of non-ionic osmotic value at low salinity, followed by increase at high salinity. In contrast, for KCl-treated plants of AM2, non-ionic osmotic value decreased with increasing salinity and completely disappeared at moderate and high salinity, while it did not significantly change for AM3 and increased for AM1 at high salinity (Figure 10D). For new leaves, non-ionic osmotic value tended to increase in AM2 plants at moderate salinity, and significantly increased in AM1 and AM3 plants at high salinity, when NaCl treatment was taken into account (Figure 10B). In contrast, in new leaves of KCl-treated plants, non-ionic osmotic value decreased in AM2 and in part in AM3, followed by recovery in high salinity, and increased in high salinity for AM1 (Figure 10E). Contrasting pattern in respect to non-ionic osmotic value was evident also for flowers between NaCl-treated (Figure 10C) and KCl-treated (Figure 10F) plants, as well as flower stalks (Figure S8B,D). In roots, non-ionic osmotic value increased at high salinity of AM1 plants at both treatments, but there was no non-ionic osmotic activity present in roots of AM3 (Figure S8A,C).

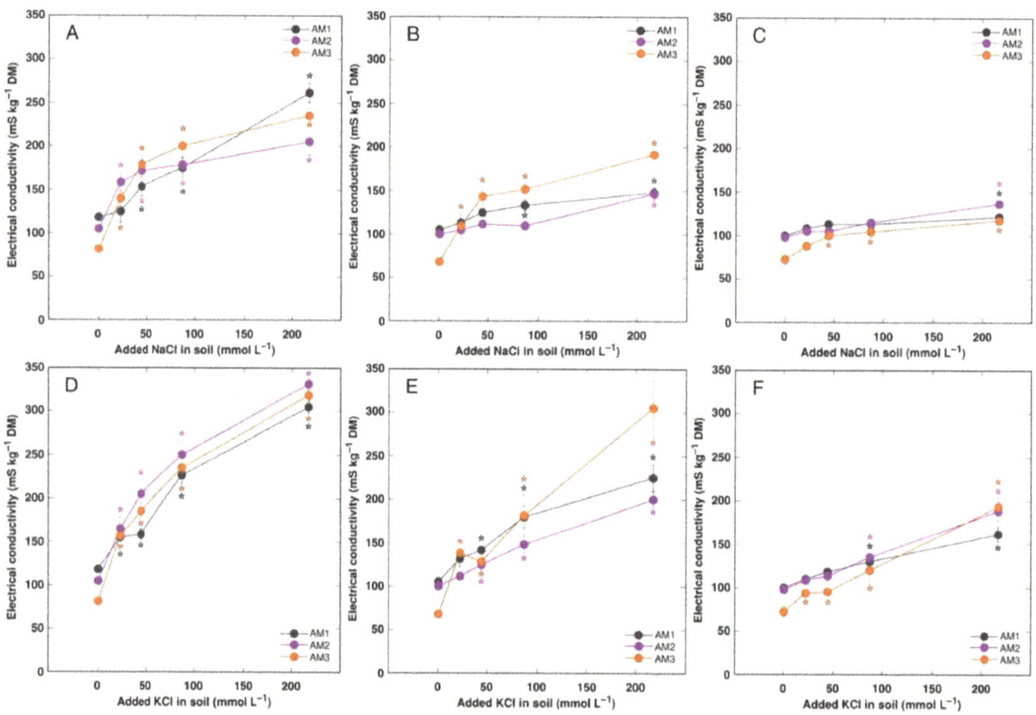

Figure 8. Effect of added NaCl (**A–C**) and KCl (**D–F**) concentration in soil on electrical conductivity in old leaves (**A**,**D**), new leaves (**B**,**E**) and flowers (**C**,**F**) of *Armeria maritima* plants from different accessions after 7–8 weeks of cultivation. DM, dry mass. Data are means ± SE from 3 replicates. Asterisks of respective color indicate statistically significant differences ($p < 0.05$) from control.

Relationships between summed $Na^+ + K^+$ concentration and osmotic value in tissues confirmed differences in osmotic regulation between NaCl and KCl treatments in plants from accession AM2 and, partially, for AM3 (Figure 11).

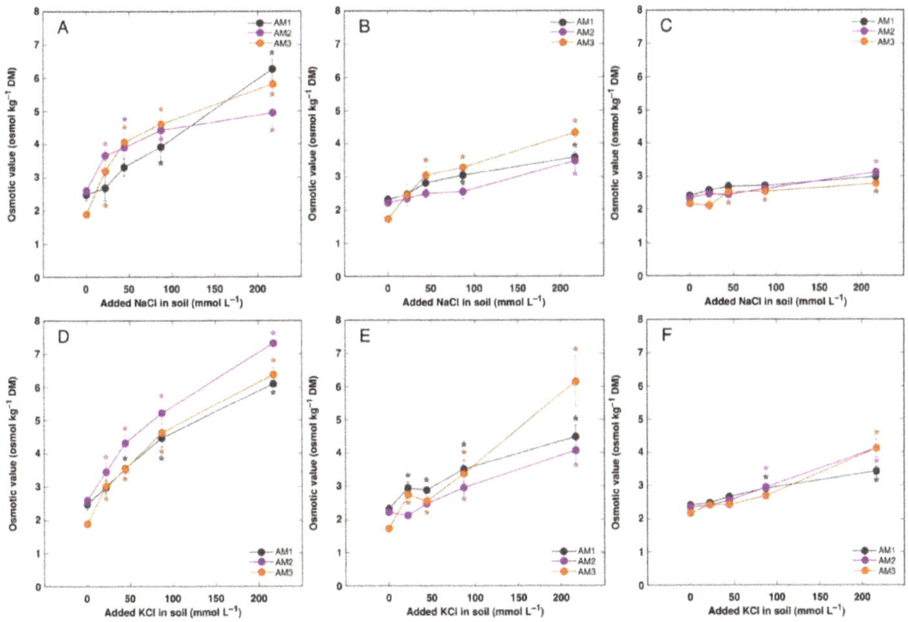

Figure 9. Effect of added NaCl (**A–C**) and KCl (**D–F**) concentration in soil on osmotic value in old leaves (**A**,**D**), new leaves (**B**,**E**) and flowers (**C**,**F**) of *Armeria maritima* plants from different accessions after 7–8 weeks of cultivation. DM, dry mass. Data are means ± SE from 3 replicates. Asterisks of respective color indicate statistically significant differences ($p < 0.05$) from control.

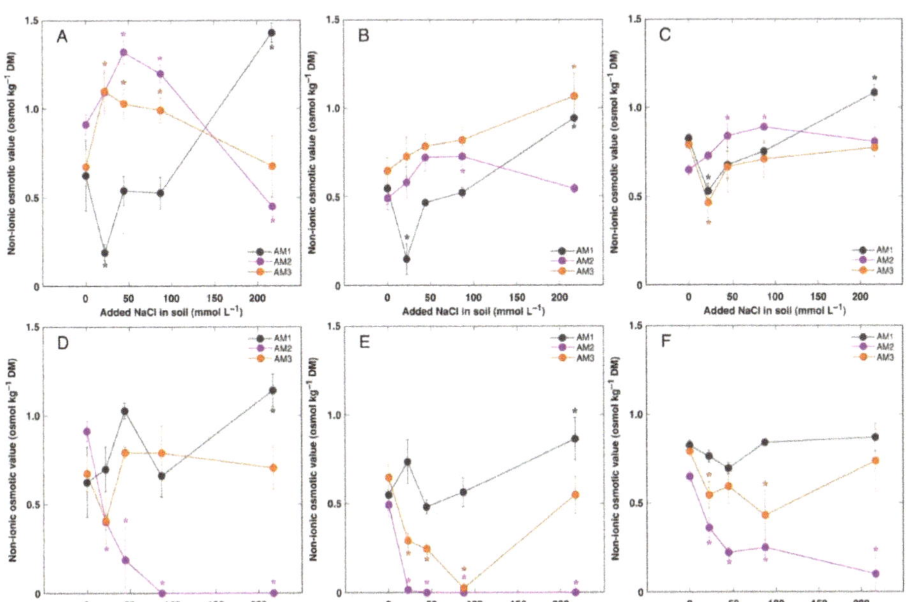

Figure 10. Effect of added NaCl (**A–C**) and KCl (**D–F**) concentration in soil on non-ionic osmotic value in old leaves (**A**,**D**), new leaves (**B**,**E**) and flowers (**C**,**F**) of *Armeria maritima* plants from different accessions after 7–8 weeks of cultivation. DM, dry mass. Data are means ± SE from 3 replicates. Asterisks of respective color indicate statistically significant differences ($p < 0.05$) from control.

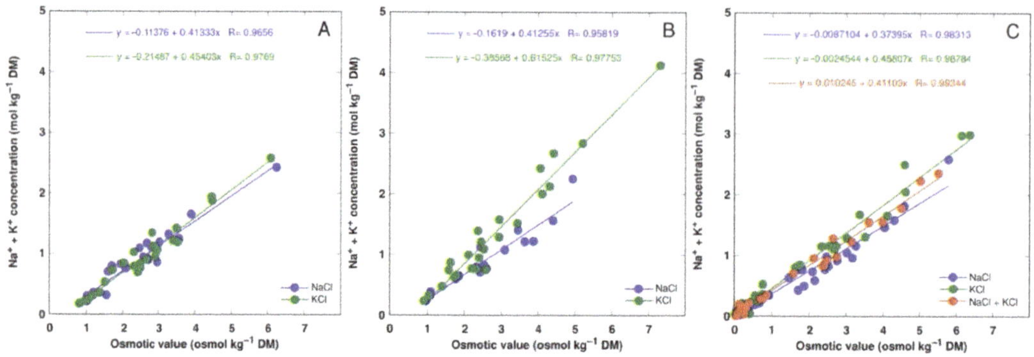

Figure 11. Relationship between Na^+ + K^+ concentration and osmotic value in all parts of *Armeria maritima* accessions AM1 (**A**), AM2 (**B**) and AM3 (**C**) treated with different salts. DM, dry mass.

4. Discussion

4.1. Salinity Tolerance of A. maritima

Given high ecological variability of *A. maritima* accessions, two earlier studies addressed question on possible differences in adaptations of *A. maritima* to high soil salinity between different geographically and ecologically isolated micropopulations [35,37]. In general, both salt marsh and inland populations of *A. maritima* showed relatively high salinity tolerance, possibly related to the ability of plants to allocate Na^+ to leaves and to accumulate organic osmolytes. However, root growth of inland plants was extremely sensitive to salinity and leaf biomass significantly decreased by increasing salinity in contrast to salt marsh plants. Another difference was that only plants from a salt marsh habitat accumulated more Na^+ in leaves in comparison to that in roots. Still, the question on whole-species abiotic stress tolerance vs. local genetic adaptation of *A. maritima* has not been definitively solved. Recently we have shown that the same accessions of *A. maritima* used also in the present study native to uncontaminated soil are highly tolerant to heavy metals and have an extreme metal accumulation ability, exceeding threshold concentration values for hyperaccumulation of Cd, Cu, Mn, Pb, and Zn [36]. As a next step, within the present study, we concentrated on characterization of ion distribution and osmotic protection of *A. maritima* plants under the effect of Na^+ and K^+ salinity.

According to the typology used for coastal plant species in respect to control of total concentration of electrolytically-active ions [8], *A. maritima* can be designated as EC controlling species, with both Na^+ and K^+ internal concentration concomitantly changing to keep relatively stable tissue concentration of soluble ions. Based on characteristic accumulation of ions in leaves and exclusion of them from roots and partially from reproductive structures, *A. maritima* can be characterized as salt-accumulating halophyte. Increased succulence is usually regarded as one of the mechanisms of salinity tolerance in Na^+-accumulating halophyte species allowing to dilute salts and efficiently decrease salt concentration in tissues on a tissue water basis [38]. Thus, both NaCl and KCl treatments increased water content in leaves of halophyte *Atriplex halimus* at moderate salinity [39]. However, this was not the case with *A. maritima*. This difference is most probably related to the fact that *A. maritima* is a salt-secreting species, but it only secretes about 4% of the absorbed Na^+ [40], thus, it can be characterized as salt-accumulating recretohalophyte.

Perennial plants with rosette type of growth are able for indeterminate production of new leaves from multiple apical meristems, allowing for induction of senescence and replacement of older leaves [41]. This mechanism seems to be especially important for evergreen species *A. maritima*, allowing deposition of harmful compounds in older leaves together with physiological exclusion of them from actively photosynthesizing leaves and reproductive structures. This phenomenon has been reflected in preferential accumulation of both heavy metals [36,42] and Na^+ [35] in decaying leaves of *A. maritima*, and confirmed also by the results of the present study.

4.2. Differences between Accessions

In the earlier study, an ability to tolerate salinity was compared for sandy soil, heavy metal and salt marsh ecotypes of *A. maritima* cultivated in artificial soil in controlled conditions [35]. Low salinity (40 and 100 mM NaCl in soil solution) stimulated biomass accumulation in leaves for all ecotypes. However, root growth of sandy soil and heavy metal ecotypes was inhibited even at low salinity. High salinity (150 and 200 mM NaCl) resulted in growth inhibition of plants from both sandy soil and heavy metal populations, but growth of plants (both shoots and roots) from salt marsh population was not affected. In the present study, growth of plants from accessions AM1 and AM2 was not significantly affected by increasing salinity (Figure 3), but that of plants from accession AM3 was significantly inhibited already at 50 mmol L^{-1} (Figure 5A). Another difference between salt-adapted and inland populations was that only salt marsh plants accumulated more Na^+ in green leaves in comparison to roots, and only at moderate to high salinity, but Na^+ accumulation ability in leaves was similar for plants from all ecotypes, reaching 0.6–0.9 mol kg^{-1} at 200 mM NaCl [35]. In the present study, all accessions of *A. maritima* preferentially accumulated Na^+ in above-ground parts, especially, leaves (Figures 6 and S2), and Na^+ concentration in young leaves reached 1.0–1.3 mol kg^{-1}, with no significant differences between the accessions. In comparison, the potential for Na^+ accumulation in proliferating explants in tissue culture reached 2.0 mol kg^{-1} (Figure 2A) showing very high tissue tolerance to Na^+. Interestingly, Na^+ concentration in decayed leaves and inflorescences was not ecotype-dependent and was 0.7–1.0 and 0.1–0.4 mol kg^{-1}, respectively [35]. Consequently, according to the morphological responses and ion accumulation characteristics in conditions of increasing salinity, *A. maritima* accessions from sandy soil salt-unaffected habitats used in the present study have large similarities with these of salt-marsh specific ecotype of *A. maritima* studied earlier [35].

Leaves of plants from salt-marsh population had higher K^+ concentration in control conditions, and it decreased with increasing salinity [35]. Similar decrease was evident for plants from sandy-soil population, but K^+ concentration in leaves was stimulated by low to moderate salinity for plants from heavy-metal population. Similarly, K^+ concentration significantly decreased in explant tissues (Figure 2B) as well as in leaves (Figure 7A,B) and flower stalks (Figure S3B) of soil-grown plants with increasing salinity.

4.3. Effect of Na^+ vs. K^+

Traditionally, Na^+ is opposed to K^+ in the manner of the classical battle of "evil" and "good" [6]. Of course, it is not debatable that K^+ is an essential element for all plant species, but Na^+ is necessary only for certain C_4 species. However, "toxicity" of Na^+, while widely advertised, has not been much experimentally proven, especially, in comparison to that of K^+, and in salt-adapted species. As it has been correctly pointed out by Kronzucker et al. (2013) [43], "... several leading paradigms in the field, such as on the roles of Na^+ influx and tissue accumulation or the cytosolic K^+/Na^+ ratio in the development of toxicity, are currently insufficiently substantiated and require a new, critical approach".

One of the most important finding of the present study was that effect of Na^+ and K^+ in a form of chloride salt on growth of *A. maritima* plants was nearly identical. However, there were differences in water content in older leaves as well as in osmotic adjustment between the two cations. First, leaves of accession AM2 showed decreasing water content with increasing KCl concentration, reflecting increase of partial dieback of these leaves (Figure 4B). Second, treatment with KCl did not induce increase of non-ionic osmotic values in old leaves (Figure 10D) in comparison to that with NaCl (Figure 10A).

Evidence for equal effects of Na^+ and K^+ on plant growth contradicts the generally accepted view of Na^+ as a particularly toxic chemical element in contrast to K^+. While both elements have similar chemical properties and remain in a soluble form in plant cells, K^+ is an essential plant nutrient with important physiological functions. For typical glycophyte species, effect of Na^+ is characteristically more negative in comparison to that of surplus K^+ [44]. However, for several halophyte species, KCl has the same of even more pronounced negative effect on plant growth in comparison to that of NaCl, as for *Atriplex nummularia* [9], *Sesuvium portulacastrum* [45], *Atriplex halimus* [39].

It appears that there are no Na^+-specific effects, but the adverse effect of Na^+ and K^+ salts on growth is related to the general effect of surplus salinity, possibly, due to high electrolyte activity. It has been shown, at least, for glycophyte species, that the negative effect of salinity is related to osmotic stress [46], but several studies have shown that treatment with isoosmotic concentration of organic osmolytes (as sorbitol or polyethylene glycol) result in more negative effect on halophyte growth in comparison to Na^+ or K^+ salts [47]. In fact, presence of ions can alleviate negative consequence of osmotic stress on plants [48]. However, it is reasonable to suggest that particularly anionic component of salts plays an important role in determining the nature of salinity responses. For example, both carbonate and hydrogen carbonate of Na^+ have been shown to be more harmful in comparison to chloride form, which has been associated mainly with the alkalinity of these salts [10,49]. However, the role of chloride itself in salinity tolerance of halophytes has been recently questioned, and it was concluded that in the case of halophytes, Cl^- represents an essential osmoticum [50]. On the other hand, in halophyte species *Sesuvium portulacastrum* NaCl toxicity has been fully attributed to negative effects of Cl^- [45].

4.4. Osmotic Adjustment

According to the generally accepted mechanism of salinity tolerance in halophytes, Na^+ and Cl^- are compartmented in vacuoles while osmotic balance is maintained by accumulation of organic solutes in cytoplasm [51]. Role of organic osmolytes in environmental stress tolerance has been shown also for species of genus *Armeria*. In high mountain specialist species, including *Armeria caespitosa*, both osmotically active carbohydrates and proline simultaneously act as osmoregulators in drought conditions [52]. Early studies with *A. maritima* have shown that proline accumulates in plant tissues as a result of salinity treatment [53]. There is no doubt that tissue proline concentration increases with increasing salinity or under the effect of other adverse environmental conditions in many species, but its contribution to osmotic control could be relatively small. Instead, a role of proline as a regulator of plant defense responses has been considered recently [54]. In addition, later it has been shown that *A. maritima* contain betaines [55] and even respond to increasing salinity by accumulation of betaine [35]. However, the role of "compatible solutes" in osmotic adjustment has been seriously questioned [56]. According to this analysis, it appears that inorganic osmolytes (K^+, Na^+, Cl^-) are responsible for most of osmotic adjustment both in halophytes as well as glycophytes. In *A. maritima*, degree of adjustment by inorganic ions vs. organic osmolytes varied between NaCl and KCl treatments and different accessions, as well as between different plant parts, (Figure 10). Non-ionic component of osmotic value was relatively high in old leaves (about 40% of the total value) and significantly increased under NaCl treatment, especially for AM2 and AM3 plants at moderate salinity, but in AM1 only at high salinity (Figure 10A). In contrast, it significantly decreased in old leaves of AM2 plants treated with increasing concentration of KCl (Figure 10D). It has been also shown that for succulent halophytes, as *Salicornia bigelovii*, contribution of K^+ to osmotic regulation is low in comparison to that of Na^+ [57].

In Na-accumulating halophyte species, Na^+ is stored in vacuoles of mesophyll cells [51], and there is an experimental evidence that surplus K^+ is also stored in vacuoles, both regulating osmotic potential and acting as the main cellular reserve of K^+ [58]. Earlier it was shown that *A. maritima* plants from salt marsh redistributed K^+ from shoots to roots with increasing salinity [59], but no such response was found in the present study in spite of significant decrease of K^+ concentration in leaves (Figure 7).

4.5. In Vitro vs. In Planta Effects

It has been a matter of long scientific debate if plant salinity tolerance assessed in conditions of tissue culture fully reflects that in whole plants cultivated in soil-like substrate [23]. For six halophytic species, higher tolerance to NaCl was evident in in vivo conditions, in comparison to in vitro culture [32]. In vitro cultivated explants of two out of six species could not propagate and survive at 400 mM NaCl, but the other species showed drastic inhibition of growth and development. In contrast, all plants survived 600 mM NaCl treatment when cultivated in substrate. This phenomenon was confirmed also in the present study, as multiplication intensity of *A. maritima* accessions AM1 and AM2 was negatively affected by increasing NaCl concentration (Figure 1B), and also biomass of AM2 decreased (Figure 1A) in conditions of tissue culture, but total biomass of both accessions was not significantly negatively affected by increasing salinity when cultivated in substrate (Figure 3). One of the reasons for this difference could possibly be related to root system acting as a barrier for Na^+ transfer to shoots. Stimulated development of Casparian strips in roots under salinity can act as a barrier for passive apoplastic flux of Na^+, resulting in a large concentration gradient between roots and shoots [60,61], similar to the limitation of heavy metal uptake [62]. However, this clearly was not the case in the present study, as roots of plant grown in soil accumulated much lower concentration of Na^+ (Figure S2) in comparison to that in leaves (Figure 6), and Na^+ concentration in cultivated shoot explants ($\leq$1 mol kg^{-1}, Figure 2A) at the highest salinity was similar to that in new leaves of soil-grown plants (Figure 6B).

5. Conclusions

It can be concluded that a species-wide salinity tolerance exists within dry sandy soil accessions of *A. maritima*, associated with the ability to accumulate surplus ions both in salt glands and old leaves. Both Na^+ and K^+ in a form of chloride salts had similar effect on plant growth, but with significant differences in osmotic adjustment by inorganic ions vs. organic osmolytes. Further studies aiming at dissecting physiological and biochemical mechanisms related to salinity tolerance of *A. maritima* and differences in osmotic and antioxidative protection of internal environment in different accessions due to various salinity types are necessary. Most importantly, functional diversity needs to be related to possible genetic diversity of different ecologically and geographically isolated micropopulations of *A. maritima* at the Northern range of distribution, to get further insight into abiotic stress adaptation mechanisms of the species.

Supplementary Materials: The following supporting information can be downloaded at: https://www.mdpi.com/article/10.3390/plants11192570/s1, Figure S1: Typical *Armeria maritima* (AM2) individuals 6 weeks after full treatment with 217 mol L^{-1} NaCl (A) and 217 mol L^{-1} KCl (B) showing formation of salt crystals on surface of leaves and flower stems; Figure S2: Effect of added NaCl concentration in soil on accumulation of Na^+ in roots (A) and flower stalks (B) of *Armeria maritima* plants from different accessions after 7–8 weeks of cultivation; Figure S3: Effect of added NaCl (A,B) and KCl (C,D) concentration in soil on accumulation of K^+ in roots (A,C) and flower stalks (B,D) of *Armeria maritima* plants from different accessions after 7–8 weeks of cultivation; Figure S4: Effect of added NaCl (A,B) and KCl (C,D) concentration in soil on electrical conductivity in roots (A,C) and flower stalks (B,D) of *Armeria maritima* plants from different accessions after 7–8 weeks of cultivation; Figure S5: Effect of added NaCl (A–E) and KCl (F–J) concentration on summed concentration of $Na^+ + K^+$ in old leaves (A,F), new leaves (B,G), flowers (C,H), roots (D,I) and flower stalks (E,J) of *Armeria maritima* plants from different accessions after 7–8 weeks of cultivation; Figure S6: Effect of added NaCl (A–E) and KCl (F–J) concentration in K^+: Na^+ concentration ratio in old leaves (A,F), new leaves (B,G), flowers (C,H), roots (D,I) and flower stalks (E,J) of *Armeria maritima* plants from different accessions after 7–8 weeks of cultivation; Figure S7: Effect of added NaCl (A,B) and KCl (C,D) concentration in soil on osmotic value in roots (A,C) and flower stalks (B,D) of *Armeria maritima* plants from different accessions after 7–8 weeks of cultivation; Figure S8: Effect of added NaCl (A,B) and KCl (C,D) concentration in soil on non-ionic osmotic value in roots (A,C) and flower stalks (B,D) of *Armeria maritima* plants from different accessions after 7–8 weeks of cultivation.

Author Contributions: L.P. and G.I. proposed the research. L.P., A.J., U.A.-O. and G.I. performed the experiments and analyzed the data. G.I. drafted the manuscript. U.A.-O. and L.P. revised the manuscript. All authors have read and agreed to the published version of the manuscript.

Funding: This study was supported by the University of Latvia project "Functional diversity of ecosystems and their contribution to ecosystem services II". The funding source had no involvement in any phase of the study.

Institutional Review Board Statement: Not applicable.

Informed Consent Statement: Not applicable.

Data Availability Statement: All data reported here is available from the authors upon request.

Acknowledgments: Technical support of Māris Romanovs is kindly acknowledged.

Conflicts of Interest: The authors declare no conflict of interest. The funders had no role in the design of the study; in the collection, analyses, or interpretation of data; in the writing of the manuscript, or in the decision to publish the results.

References

1. Daliakopoulos, I.N.; Tsanis, I.K.; Koutroulis, A.; Kourgialas, N.N.; Varouchakis, A.E.; Karatzas, G.P.; Ritsema, C.J. The threat of soil salinity: A European scale review. *Sci. Total Envrion.* **2016**, *573*, 727–739. [CrossRef] [PubMed]
2. Qadir, M.; Quillérou, E.; Nangia, V.; Murtaza, G.; Singh, M.; Thomas, R.J.; Drechsel, P.; Noble, A.D. Economics of salt-induced land degradation and restoration. *Nat. Resour. Forum* **2018**, *38*, 282–295. [CrossRef]
3. Ondrasek, G.; Rengel, Z. Environmental salinization processes: Detection, implications & solutions. *Sci. Total Environ.* **2021**, *754*, 142432. [PubMed]
4. Meng, X.; Zhou, J.; Sui, N. Mechanisms of salt tolerance in halophytes: Current understanding and recent advances. *Open Life Sci.* **2018**, *13*, 149–154. [CrossRef]
5. Zhao, C.; Zhang, H.; Song, C.; Zhu, J.-K.; Shabala, S. Mechanisms of plant responses and adaptation to soil salinity. *Innovation* **2020**, *1*, 100017. [CrossRef]
6. Maathuis, F.J.M.; Amtmann, A. K^+ nutrition and Na^+ toxicity: The basis of cellular K^+/Na^+ ratios. *Ann. Bot.* **1999**, *84*, 123–133. [CrossRef]
7. Rubio, F.; Nieves-Cordones, M.; Horie, T.; Shabala, S. Doing 'business as usual' comes with a cost: Evaluating energy cost of maintaining plant intracellular K^+ homeostasis under saline conditions. *New Phytol.* **2019**, *225*, 1097–1104. [CrossRef]
8. Ievinsh, G.; Ieviņa, S.; Andersone-Ozola, U.; Samsone, I. Leaf sodium, potassium and electrolyte accumulation capacity of plant species from salt-affected coastal habitats of the Baltic Sea: Towards a definition of Na hyperaccumulation. *Flora* **2021**, *274*, 151748. [CrossRef]
9. Ramos, J.; López, M.J.; Benlloch, M. Effect of NaCl and KCl salts on the growth and solute accumulation of the halophyte *Atriplex nummularia*. *Plant Soil* **2004**, *259*, 163–168. [CrossRef]
10. Wang, J.; Zhang, Y.; Yan, X.; Guo, J. Physiological and transcriptomic analyses of yellow horn (*Xanthoceras sorbifiolia*) provide important insights into salt and saline-alkali stress tolerance. *PLoS ONE* **2012**, *15*, e0244365.
11. Rahman, M.M.; Mostofa, M.G.; Keya, S.S.; Siddiqui, M.N.; Ansary, M.M.U.; Das, A.K.; Rahman, M.A.; Tran, L.S.-P. Adaptive mechanisms of halophytes and their potential in improving salinity tolerance in plants. *Int. J. Mol. Sci.* **2021**, *22*, 10733. [CrossRef]
12. Lombardi, T.; Bertacchi, A.; Pistelli, L.; Pardossi, A.; Pecchia, S.; Toffanin, A.; Sanmartin, C. Biological and agronomic traits of the main halophytes widespread in the Mediterranean region as potential new vegetable crops. *Horticulturae* **2022**, *8*, 195. [CrossRef]
13. Ellouzi, H.; Ben Hamed, K.; Cela, J.; Munné-Bosch, S.; Abdelly, C. Early effects of salt stress on the physiological and oxidative status of *Cakile maritima* (halophyte) and *Arabidopsis thaliana* (glycophyte). *Physiol. Plant.* **2011**, *142*, 128–143. [CrossRef]
14. Bueno, M.; Lendinez, M.L.; Aparicio, C.; Cordovilla, M.P. Germination and growth of *Atriplex prostrata* and *Plantago coronopus*: Two strategies to survive in saline habitats. *Flora* **2017**, *227*, 56–63. [CrossRef]
15. Tabot, P.T.; Adams, J.B. Ecophysiology of salt marsh plants and predicted responses to climate change in South Africa. *Ocean. Coast. Manag.* **2013**, *80*, 89–99. [CrossRef]
16. Rilke, S.; Reimans, C. Morphological and ecophysiological differences between the subspecies of *Salsola kali* L. in Europe: Results of culture experiments. *Flora* **1996**, *191*, 363–376. [CrossRef]
17. Ghoulam, C.; Foursy, A.; Fares, K. Effects of salt stress om growth, inorganic ions and proline accumulation in relation to osmotic adjustment in five sugar beet cultivars. *Envrion. Exp. Bot.* **2002**, *47*, 39–50. [CrossRef]
18. Xu, X.; Zhou, Y.; Mi, P.; Wang, B.; Yuan, F. Salt-tolerance screening in *Limonium sinuatum* varieties with different flower colors. *Sci. Rep.* **2021**, *11*, 14562. [CrossRef]
19. Jēkabsone, A.; Andersone-Ozola, U.; Karlsons, A.; Romanovs, M.; Ievinsh, G. Effect of salinity on growth, ion accumulation and mineral nutrition of different accessions of a crop wild relative legume species, *Trifolium fragiferum*. *Plants* **2022**, *11*, 797. [CrossRef]
20. Winicov, I. Characterization of rice (*Oryza sativa* L.) plants regenerated from salt-tolerant cell lines. *Plant Sci.* **1996**, *113*, 105–111. [CrossRef]

21. Hossain, Z.; Mandal, A.K.A.; Datta, S.K.; Biswas, A.K. Development of NaCl-tolerant line in *Chrysanthemum morifolium* Ramat. through shoot organogenesis of selected callus line. *J. Biotechnol.* **2007**, *129*, 658–667. [CrossRef]
22. Rattana, K.; Bunnag, S. Differential salinity tolerance in calli and shoots of four rice cultivars. *Asian J. Crop Sci.* **2015**, *7*, 48–60. [CrossRef]
23. Winicov, I.; Bastola, D.R. Salt tolerance in crop plants: New approaches through tissue culture and gene regulation. *Acta Physiol. Plant.* **1997**, *19*, 435–449. [CrossRef]
24. Aldahhak, O.; Zaid, S.; Teixeira da Silva, J.A.; Abdul-Kader, A.M. In vitro approach to the multiplication of a halophyte species forage shrub *Atriplex halimus* L. and in vitro selection for salt tolerance. *Int. J. Plant Dev. Biol.* **2010**, *4*, 8–14.
25. Lokhande, V.H.; Nikam, T.D.; Patade, V.Y.; Ahire, M.L.; Suprasanna, P. Effects of optimal and supra-optimal salinity stress on antioxidative defence, osmolytes and in vitro growth response on *Sesuvium portulacastrum* L. *Plant Cell Tissue Organ Cult.* **2011**, *104*, 41–49. [CrossRef]
26. Shi, X.L.; Han, H.P.; Shi, W.L.; Li, Y.X. NaCl and TDZ are two key factors for the improvement of in vitro regeneration rate of *Salicornia europaea* L. *J. Integr. Plant Biol.* **2006**, *48*, 1185–1189. [CrossRef]
27. Joshi, M.; Mishra, A.; Jha, B. NaCl plays a key role for in vitro micropropagation of *Salicornia brachiata*, an extreme halophyte. *Industr. Crops Prod.* **2012**, *35*, 313–316. [CrossRef]
28. Martini, A.N.; Papafotiou, M. In vitro propagation and NaCl tolerance of the multipurpose medicinal halophyte *Limoniastrum monopetalum*. *HortScience* **2020**, *55*, 436–443. [CrossRef]
29. Grigoriadou, K.; Maloupa, E. Microporpagation and salt tolerance of in vitro grown *Crithmum maritimum* L. *Plant Cell Tissue Organ Cult.* **2008**, *94*, 209–217. [CrossRef]
30. Aghaleh, M.; Niknam, V.; Ebrahimzadeh, H.; Razavi, K. Salt stress effects on growth, pigments, proteins and lipid peroxidation in *Salicorrnia persica* and *S. europaea*. *Biol. Plant.* **2008**, *53*, 243–248. [CrossRef]
31. Samiei, L.; Pahnehkilayi, M.D.; Karimian, Z.; Nabati, J. Morpho-physiological responses of halophyte *Climacoptera crassa* to salinity and heavy metal stresses in in vitro condition. *S. Afr. J. Bot.* **2021**, *131*, 468–474. [CrossRef]
32. Xiong, Y.; Liang, H.; Yan, H.; Guo, B.; Niu, M.; Chen, S.; Jian, S.; Ren, H.; Zhang, X.; Li, Y.; et al. NaCl-induced stress: Physiological responses of six halophytic species in in vitro and in vivo culture. *Plant Cell Tissue Organ Cult.* **2019**, *139*, 531–546. [CrossRef]
33. Woodell, S.R.J.; Dale, A. *Armeria maritima* (Mill.) Willd. (*Statice armeria* L.; *S. maritima* Mill.). *J. Ecol.* **1993**, *81*, 573–588. [CrossRef]
34. Goldsmith, F.B. Interaction (competition) studies as a step towards the synthesis of sea-cliff vegetation. *J. Ecol.* **1978**, *66*, 921–931. [CrossRef]
35. Köhl, K.I. The effect of NaCl on growth, dry matter allocation and ion uptake in salt marsh and inland populations of *Armeria maritima*. *New Phytol.* **1997**, *135*, 213–225. [CrossRef]
36. Purmale, l.; Jēkabsone, A.; Andersone-Ozola, U.; Karlsons, A.; Osvalde, A.; Ievinsh, G. Comparison of in vitro and in planta heavy metal tolerance and accumulation potential of different *Armeria maritima* accessions from a dry coastal meadow. *Plants* **2022**, *11*, 2104. [CrossRef] [PubMed]
37. Köhl, K.I. NaCl homeostasis as a factor for the survival of the evergreen halophyte *Armeria maritima* (Mill.) Willd. under salt stress in winter. *Plant Cell Environ.* **1997**, *20*, 1253–1263. [CrossRef]
38. Jennings, D.H. Halophytes, succulence and sodium in plants—A unified theory. *New Phytol.* **1968**, *67*, 899–911. [CrossRef]
39. Belkheiri, O.; Mulas, M. The effects of salt stress on growth, water relations and ion accumulation in two halophyte *Atriplex* species. *Environ. Exp. Bot.* **2013**, *86*, 17–28. [CrossRef]
40. Rozema, J.; Gude, H.; Pollak, G. An ecophysiological study of the salt secretion of four halophytes. *New Phytol.* **1981**, *89*, 201–217. [CrossRef]
41. Wingler, A.; Stangberg, E.J.; Saxena, T.; Mistry, R. Interactions between temperature and sugars in the regulation of leaf senescence in the perennial herb *Arabis alpina* L. *J. Integr. Plant Biol.* **2012**, *54*, 595–605. [CrossRef]
42. Dahmani-Muller, H.; van Ort, F.; Gélie, B.; Balabane, M. Strategies of heavy metal uptake by three plant species growing near a metal smelter. *Envrion. Pollut.* **2000**, *109*, 231–238. [CrossRef]
43. Kronzucker, H.J.; Coskun, D.; Schulze, L.M.; Wong, J.R.; Britto, D.T. Sodium as nutrient and toxicant. *Plant Soil* **2013**, *369*, 1–23. [CrossRef]
44. Martínez-Ballesta, M.C.; Martínez, V.; Carvajal, M. Osmotic adjustment, water relations and gas exchange in pepper plants grown under NaCl and KCl. *Environ. Exp. Bot.* **2004**, *52*, 161–174. [CrossRef]
45. Wang, D.; Wang, H.; Han, B.; Wang, B.; Guo, A.; Zheng, D.; Liu, C.; Chang, L.; Peng, M.; Wang, X. Sodium instead of potassium and chloride is an important macronutrient to improve leaf succulence and shoot development for halophyte *Sesuvium portulacastrum*. *Plant Physiol. Biochem.* **2012**, *51*, 53–62. [CrossRef]
46. Ben-Gal, A.; Borochov-Neori, H.; Yermiyahu, U.; Shani, U. Is osmotic potential a more appropriate property than electrical conductivity for evaluating whole plant response to salinity? *Environ. Exp. Bot.* **2009**, *65*, 232–237. [CrossRef]
47. Nada, R.M.; Abogadallah, G.M. Developmental acquisition of salt tolerance in the halophyte *Atriplex halimus* L. is related to differential regulation of salt inducible genes. *Plant Growth Regul.* **2015**, *75*, 165–178. [CrossRef]
48. Song, J.; Feng, G.; Tian, C.; Zhang, F. Strategies for adaptation of *Suaeda physophora, Haloxylon ammodendron* and *Haloxylon persicum* to a saline environment during seed-germination stage. *Ann. Bot.* **2005**, *96*, 399–405. [CrossRef]
49. Li, J.; Hussain, T.; Feng, X.; Guo, K.; Chen, H.; Yang, C.; Liu, X. Comparative study on the resistance of *Suaeda glauca* and *Suaeda salsa* to drought, salt, and alkali stresses. *Ecol. Eng.* **2019**, *140*, 105593. [CrossRef]

50. Bazihizina, N.; Colmer, T.D.; Cuin, T.A.; Mancuso, S.; Shabala, S. Friend or foe? Chloride patterning in halophytes. *Trends Plant Sci.* **2019**, *24*, 142–151. [CrossRef]
51. Flowers, T.J.; Munns, R.; Colmer, T.D. Sodium chloride toxicity and the cellular basis of salt tolerance in halophytes. *Ann. Bot.* **2015**, *115*, 419–431. [CrossRef] [PubMed]
52. Ugarte, R.M.; Escudero, A.; Gavilán, R.G. Assessing the role of selected osmolytes in Mediterranean high-mountain specialists. *Front. Ecol. Evol.* **2021**, *9*, 576122. [CrossRef]
53. Stewart, C.R.; Lee, J.A. The role of proline accumulation in halophytes. *Planta* **1974**, *120*, 279–289. [CrossRef] [PubMed]
54. Ghosh, U.K.; Islam, M.N.; Siddiqui, M.N.; Cao, X.; Khan, M.A.R. Proline, a multifaceted signalling molecule in plant responses to abiotic stress: Understanding the physiological mechanisms. *Plant Biol.* **2022**, *24*, 227–239. [CrossRef] [PubMed]
55. Adrian-Romero, M.; Wilson, S.J.; Blunden, G.; Yang, M.-H.; Carabot-Cuervvo, A.; Bashir, A.K. Betaines in coastal plants. *Biochem. Syst. Ecol.* **1998**, *26*, 535–543. [CrossRef]
56. Shabala, S.; Shabala, L. Ion transport and osmotic adjustment in plants and bacteria. *BioMol. Concepts* **2011**, *2*, 407–419. [CrossRef]
57. Yamada, M.; Kuroda, C.; Fujiyama, H. Function of sodium and potassium in growth of sodium-lowing Amaranthaceae species. *Soil Sci. Plant Nutr.* **2016**, *62*, 20–26. [CrossRef]
58. Ragel, P.; Raddatz, N.; Leidi, E.O.; Quintero, F.J.; Pardo, J.M. Regulation of K^+ nutrition in plants. *Front. Plant Sci.* **2019**, *10*, 281. [CrossRef]
59. Skeffington, M.J.; Jeffrey, D.W. Response of *Armeria maritima* (MIll.) Willd. and *Plantago maritima* L. from an Irish salt marsh to nitrogen and salinity. *New Phytol.* **1988**, *110*, 399–408. [CrossRef]
60. Chen, M.; Yang, Z.; Liu, J.; Zhu, T.; Wei, X.; Fan, H.; Wang, B. Adaptation mechanisms of salt excluders under saline conditions and its applications. *Int. J. Mol. Sci.* **2018**, *19*, 3668. [CrossRef]
61. Chen, T.; Cai, X.; Wu, X.; Karahara, I.; Schreiber, L.; Lin, J. Casparian strip development and its potential function in salt tolerance. *Plant Signal. Behav.* **2011**, *6*, 1499–1502. [CrossRef]
62. Singh, S.; Parihar, P.; Singh, R.; Singh, V.P.; Prasad, S.M. Heavy metal tolerance in plants: Role of transcriptomics, proteomics, metabolomics, and ionomics. *Front. Plant Sci.* **2016**, *6*, 1143. [CrossRef]

plants

Article

Genome-Wide Analysis and Characterization of the Proline-Rich Extensin-like Receptor Kinases (PERKs) Gene Family Reveals Their Role in Different Developmental Stages and Stress Conditions in Wheat (*Triticum aestivum* L.)

Mahipal Singh Kesawat [1,2], Bhagwat Singh Kherawat [3], Anupama Singh [1], Prajjal Dey [1], Snehasish Routray [4], Chinmayee Mohapatra [4], Debanjana Saha [5], Chet Ram [6], Kadambot H. M. Siddique [7], Ajay Kumar [8], Ravi Gupta [9], Sang-Min Chung [10] and Manu Kumar [10,*]

1 Department of Genetics and Plant Breeding, Faculty of Agriculture, Sri Sri University, Cuttack 754006, Odisha, India; mahipal.s@srisriuniversity.edu.in (M.S.K.); anupama.s@srisriuniversity.edu.in (A.S.); prajjal.d@srisriuniversity.edu.in (P.D.)
2 School of Biological Sciences and Institute for Molecular Biology and Genetics, Seoul National University, Seoul 08826, Korea
3 Krishi Vigyan Kendra, Bikaner II, Swami Keshwanand Rajasthan Agricultural University, Bikaner 334603, Rajasthan, India; skherawat@gmail.com
4 Department of Entomology and Plant Pathology, Faculty of Agriculture, Sri Sri University, Cuttack 754006, Odisha, India; snehasish.r@srisriuniversity.edu.in (S.R.); chinmayee.m@srisriuniversity.edu.in (C.M.)
5 Department of Biotechnology, Centurion University of Technology and Management, Bhubaneshwar 752050, Odisha, India; debanjana.saha@cutm.ac.in
6 ICAR-Central Institute for Arid Horticulture, Bikaner 334006, Rajasthan, India; chetram.nbpgr@gmail.com
7 The UWA Institute of Agriculture, The University of Western Australia, Perth, WA 6009, Australia; kadambot.siddique@uwa.edu.au
8 Agriculture Research Organization, Volcani Center, Department of Postharvest Science, Rishon Lezzion 50250, Israel; ajaykumar_bhu@yahoo.com
9 College of General Education, Kookmin University, Seoul 02707, Korea; ravigupta07@ymail.com
10 Department of Life Science, Dongguk University, Dong-gu, Ilsan, Seoul 10326, Korea; smchung@dongguk.edu
* Correspondence: manukumar007@gmail.com

Citation: Kesawat, M.S.; Kherawat, B.S.; Singh, A.; Dey, P.; Routray, S.; Mohapatra, C.; Saha, D.; Ram, C.; Siddique, K.H.M.; Kumar, A.; et al. Genome-Wide Analysis and Characterization of the Proline-Rich Extensin-like Receptor Kinases (PERKs) Gene Family Reveals Their Role in Different Developmental Stages and Stress Conditions in Wheat (*Triticum aestivum* L.). *Plants* **2022**, *11*, 496. https://doi.org/10.3390/plants11040496

Academic Editors: Kinga Dziurka, Mateusz Labudda and Ewa Muszyńska

Received: 19 January 2022
Accepted: 9 February 2022
Published: 11 February 2022

Abstract: Proline-rich extensin-like receptor kinases (PERKs) are a class of receptor kinases implicated in multiple cellular processes in plants. However, there is a lack of information on the PERK gene family in wheat. Therefore, we identified 37 PERK genes in wheat to understand their role in various developmental processes and stress conditions. Phylogenetic analysis of PERK genes from *Arabidopsis thaliana*, *Oryza sativa*, *Glycine max*, and *T. aestivum* grouped them into eight well-defined classes. Furthermore, synteny analysis revealed 275 orthologous gene pairs in *B. distachyon*, *Ae. tauschii*, *T. dicoccoides*, *O. sativa* and *A. thaliana*. Ka/Ks values showed that most TaPERK genes, except TaPERK1, TaPERK2, TaPERK17, and TaPERK26, underwent strong purifying selection during evolutionary processes. Several cis-acting regulatory elements, essential for plant growth and development and the response to light, phytohormones, and diverse biotic and abiotic stresses, were predicted in the promoter regions of TaPERK genes. In addition, the expression profile of the TaPERK gene family revealed differential expression of TaPERK genes in various tissues and developmental stages. Furthermore, TaPERK gene expression was induced by various biotic and abiotic stresses. The RT-qPCR analysis also revealed similar results with slight variation. Therefore, this study's outcome provides valuable information for elucidating the precise functions of TaPERK in developmental processes and diverse stress conditions in wheat.

Keywords: PERK; kinase; RT-qPCR; promoter; drought; heat stress

1. Introduction

Protein phosphorylation is a key post-translational modification that regulates cell signaling networks and cellular processes in response to internal and external environmental stimulation through the reversible regulation of protein function by activation or deactivation, formation of protein complexes, and determination of the subcellular location of proteins [1–3]. Phosphorylation is the most common event in which the phosphoryl group transfers from adenosine triphosphate to hydroxyl residue of the protein substrate [4]. Although plants deploy receptor kinases at the cell surface to perceive, the signal, generated in the sudden changing environment, activates the various signaling pathways and regulates the growth, reproduction, and response against diverse stresses [2,5]. Receptor kinases are the most prominent gene family in crop species, including Arabidopsis, rice, maize, soybean, cotton, and sorghum [6–8]. For example, the Arabidopsis receptor kinase gene family encompasses ~610 members, and their homologs have been identified and characterized in many plant species [7,9–11]. The biological functions of these predicted receptor kinase genes remain to be elucidated. However, receptor kinases play a crucial roles in cell differentiation, pollen tube growth, pollen development, symbiosis, pathogen recognition, phytohormone response, signal transduction, self-incompatibility and response towards internal and external stimuli [8,9,12–16]. A few studies have identified and characterized the ligands that activate specific receptor kinases, as well as some signaling components [8,13]. Receptor kinases bind to different kinds of biomolecules such as polypeptides, steroids, carbohydrates, and cell wall components. These receptor kinases perceive and transduce the signals across the plasma membrane via diverse signaling complexes, which have been developed during the long course evolution of complex multicellular organisms [2,9].

Receptor kinases were divided into different groups based on the motifs structure in their extracellular domains [9,17]. For instance, the leucine-rich repeat receptor kinase family comprises the main class of receptor kinases in plants, including BAK1 and BRI1, which have been implicated in brassinosteroid signaling [18–20]. Proline-rich extensin-like receptor kinases (PERKs) are one of the main classes of receptor kinases. Fifteen PERK genes have been found in the Arabidopsis genome. However, their functions are poorly understood [6,12]. Most of the *AtPERK* gene family members are ubiquitously expressed, while few genes are specifically expressed [6]. For instance, *AtPERK1* is broadly expressed, whereas the expression of *AtPERK2* is mainly observed in rosette leaf veins, stems, and pollen [21]. In addition, the expression of *AtPERK8* and *AtPERK13* were detected in root hairs [6,22,23]. The expression of *AtPERK5*, *AtPERK6*, *AtPERK7*, *AtPERK11*, and *AtPERK12* were highly elevated in pollen. However, unnoticeable expressions were observed in the sporophytic tissues [23–25]. Furthermore, *AtPERK4* regulates the root growth function at an early stage of ABA signaling by perturbing calcium homeostasis in Arabidopsis [14]. A few studies have demonstrated that increased concentrations of calcium in the cells also enhances antioxidant enzyme activities and eventually regulates the lipid peroxidation of cell membranes and stomatal apertures [26–28]. The PERKs suppress the accumulation of reactive oxygen species (ROS) in the root, which is necessary for root hair growth [29,30]. MAPK cascade is an essential regulator of high light-induced Cu/Zn SODs and anti-PERK antibodies from animals, used to detect the presence of homologous proteins such as MPK3 and MPK6 in plants [30,31]. *AtPERK5* and *AtPERK12* are essential for the pollen tube growth in Arabidopsis [16]. Furthermore, *AtPERK8*, *AtPERK9*, and At*PERK10* negatively regulate root growth in Arabidopsis [23]. *PERK1* rapidly induces early perception and response to a wound in Chinese cabbage [12]. Antisense suppression of *BnPERK1* has exhibited various growth defects, such as amplified secondary branching, loss of apical dominance, and defects in floral organ formation. At the same time, the overexpression line showed increased lateral shoot production, seed set, and unusual deposition of callose and cellulose in *Brassica napus* [21]. A PERK-like receptor kinase specifically interacts with the nuclear shuttle protein (NSP), led viral infection, and positively regulates the NSP function in cabbage leaf curl virus and geminivirus [32].

Plants are sessile organisms, and constantly face fluctuating environmental conditions and various biotic and abiotic stresses during growth and development [33–35]. For example, wheat is an important cereal crop cultivated worldwide [36,37], and its quality and productivity are largely influenced by different biotic and abiotic factors [38,39]. With the recent advent of sequencing technology, a rapid increase in sequenced plant genomes has been accessed in the past few years [40]. However, identifying the genes in plant species' genomes is now a great challenge, particularly in terms of their structure to functionally characterization [41,42]. For example, the wheat genome sequenced, completed and identified 124,201 genes [43]. Thus, this project's completion has made it possible to complete genome-wide analysis and identification of the PERK gene family in wheat. We performed a comprehensive analysis of 37 PERK genes using several computational approaches in this work.

Furthermore, phylogenetic analysis, physical and biochemical properties, exon/intron, conserved motifs, chromosomal distribution, subcellular localization, gene duplication, Ka/Ks values, synteny analysis, and three-dimensional (3D) structure were also determined. In addition, tissue-specific expression profiles and responses to diverse stress conditions were also examined for the TaPERK genes. The outcome of the present study will be helpful in the detailed understanding of the TaPERK gene's role in plant growth, development, and survivability under different stress conditions.

2. Results

2.1. Identification of TaPERK in Wheat

In this study, we identified 37 PERK genes in the wheat genome using various computational approaches (Table 1).

Table 1. Nomenclature and characteristics of the putative proline-rich extensin-like receptor kinases (PERKs) proteins in wheat were predicted using various computational tools.

Proposed Gene Name	Gene ID	Genomic Location	Orientation	CDS Length (bp)	Intron Number	Protein Length (aa)	Molecular Weight (KDa)	Isoelectric Point (pI)	GRAVY	Predicted Subcellular Localization
TaPERK1	TraesCS1A02G127900	1A:155693812–155696618	Forward	1977	7	658	69.44	7.53	−0.531	Nucleus
TaPERK2	TraesCS1B02G1470000	1B:209130189–209130266	Reverse	1431	8	476	52.14	6.17	−0.5	Nucleus
TaPERK3	TraesCS1D02G00430	1D:2110107–2112027	Forward	1971	7	656	68.93	9.04	−0.393	Chloroplast outer membrane
TaPERK4	TraesCS1D02G126300	1D:137437684–137440387	Reverse	1962	7	653	69.07	7.21	−0.52	Nucleus
TaPERK5	TraesCS2A02G418200	2A:674030843–674031911	Forward	3048	23	1015	110.33	6.33	−0.193	Plasma membrane
TaPERK6	TraesCS2A02G418300	2A:674050369–674051442	Forward	3042	23	1013	110.39	7.06	−0.135	Plasma membrane
TaPERK7	TraesCS2A02G418400	2A:674061248–674062244	Forward	3159	23	1052	113.73	5.96	−0.127	Plasma membrane
TaPERK8	TraesCS2B02G437200	2B:629023953–629025021	Forward	3045	23	1014	110.38	6.3	−0.178	Plasma membrane
TaPERK9	TraesCS2B02G437300	2B:629106216–629107285	Forward	3048	23	1015	110.62	6.61	−0.147	Plasma membrane
TaPERK10	TraesCS2D02G415600	2D:529537635–529538701	Forward	3048	22	1015	110.34	6.6	−0.167	Plasma membrane
TaPERK11	TraesCS2D02G415700	2D:529548057–529548998	Forward	2775	23	924	101.07	7.29	−0.182	Plasma membrane
TaPERK12	TraesCS2D02G415800	2D:529558487–529559547	Forward	3156	23	1051	113.45	6.11	−0.102	Plasma membrane
TaPERK13	TraesCS3A02G003900	3A:1925607–1927275	Reverse	2064	7	687	72.42	5.96	−0.429	Plasma membrane
TaPERK14	TraesCS3A02G152200	3A:142891955–142894634	Forward	1893	7	630	67.43	6.28	−0.569	Endomembrane system

Table 1. *Cont.*

Proposed Gene Name	Gene ID	Genomic Location	Orientation	CDS Length (bp)	Intron Number	Protein Length (aa)	Molecular Weight (KDa)	Isoelectric Point (pI)	GRAVY	Predicted Subcellular Localization
TaPERK15	TraesCS3A02G229800	3A:429615911–429617422	Reverse	2163	6	720	74.97	7.93	−0.401	Chloroplast thylakoid lumen
TaPERK16	TraesCS3A02G278100	3A:507637093–507638935	Reverse	2028	7	675	72.42	7.31	−0.481	Plasma membrane
TaPERK17	TraesCS3A02G290300	3A:519244808–519246110	Reverse	2184	7	727	75.8	6.11	−0.535	Endomembrane system
TaPERK18	TraesCS3B02G008600	3B:4324660–4326408	Forward	2061	7	686	71.88	5.97	−0.437	Plasma membrane
TaPERK19	TraesCS3B02G179300	3B:187347873–187350697	Forward	1896	7	631	67.46	6.35	−0.569	Endomembrane system
TaPERK20	TraesCS3B02G259100	3B:416806224–416809608	Reverse	2097	6	698	72.98	7.63	−0.448	Plasma membrane
TaPERK21	TraesCS3B02G312300	3B:501498044–501499926	Reverse	2034	7	677	72.6	7.31	−0.493	Endomembrane system
TaPERK22	TraesCS3B02G325100	3B:525990462–525991846	Reverse	2436	7	811	84.77	6.09	−0.51	Plasma membrane
TaPERK23	TraesCS3D02G005400	3D:2141185–2143272	Forward	1206	6	401	44.37	5.55	−0.403	Nucleus
TaPERK24	TraesCS3D02G160000	3D:130928685–130931461	Forward	1899	7	632	67.49	6.36	−0.555	Endomembrane system
TaPERK25	TraesCS3D02G278400	3D:385473929–385474240	Reverse	2031	8	676	72.56	7.1	−0.471	Endomembrane system
TaPERK26	TraesCS3D02G290100	3D:400311470–400312883	Reverse	1317	6	438	47.14	5.97	−0.447	Nucleus
TaPERK27	TraesCS4A02G077500	4A:76627667–76628358	Forward	1866	5	621	64.49	5.58	−0.457	Endomembrane system
TaPERK28	TraesCS4A02G449700	4A:715718345–715719349	Forward	2604	19	867	94.75	7.03	−0.145	Plasma membrane
TaPERK29	TraesCS4B02G233600	4B:486206279–486206961	Reverse	1857	5	618	64.45	5.63	−0.487	Plasma membrane
TaPERK30	TraesCS5A02G411300	5A:599978835–599979642	Reverse	1722	4	573	60.33	7.96	−0.387	Chloroplast outer membrane
TaPERK31	TraesCS5B02G415000	5B:589228532–589228944	Reverse	1842	3	613	64.68	7.86	−0.368	Chloroplast outer membrane
TaPERK32	TraesCS7A02G038600	7A:17358644–17359648	Reverse	3030	23	1009	109.7	6.22	−0.108	Plasma membrane
TaPERK33	TraesCS7A02G231900	7A:202852283–202853761	Reverse	2187	6	728	76.24	5.32	−0.484	Nucleus
TaPERK34	TraesCS7B02G130400	7B:156752944–156754400	Reverse	2208	6	735	76.89	5.22	−0.49	Nucleus
TaPERK35	TraesCS7D02G034800	7D:17864178–17865182	Reverse	3021	23	1006	109.26	6.1	−0.09	Plasma membrane
TaPERK36	TraesCS7D02G232700	7D:194224547–194225929	Forward	2256	6	751	78.2	6.16	−0.645	Endomembrane system
TaPERK37	TraesCSU02G104700	Un:92294980–92296477	Reverse	2205	6	734	76.55	5.32	−0.483	Nucleus

ID: identity; bp: base pair; aa: amino acids; pI: isoelectric point; MW: molecular weight; KDa: Kilo dalton.

This number is relatively higher than the earlier reported PERK genes in Arabidopsis, soybean, rice, sorghum, maize, and cotton (Table 2).

This might be due to the higher chromosome number and big size of the wheat genome, which indicates that the PERK genes underwent a substantial expansion in wheat. In addition, wheat is derived from the hybridization of three progenitor genomes: A, B, and D. The TaPERK family had protein lengths ranging from 401–1052, and amino acid with molecular weight (MW) 44.37–113.73 kDa for TaPERK23 and TaPERK7, respectively. The isoelectric point (pI) ranged from 5.22 and 9.04 for TaPERK34 and TaPERK3, respectively. We also plotted the MW of TaPERK with their pI to understand the MW distribution of different TaPERK proteins (Figure S1). The plots showed that most of the TaPERKs had similar MW and pI. Hence, pI values ranged from acidic to basic, and the heaviest TaPERK was over twice the weight of the lightest. Furthermore, the grand average of hydropathy index values ranged from −0.09 to −0.645, indicating that TaPERK proteins are hydrophilic in nature. Moreover, the subcellular localization prediction of TaPERK proteins indicated that most of the TaPERKs were situated on the plasma membrane (Table 1).

Table 2. Number of PERK proteins in different plant species.

Plant Species	Genome Size (Approx.)	Coding Genes	PERK Genes
Triticum aestivum (6n)	17 Gb	107,891	37
Arabidopsis thaliana (2n)	135 Mb	27,655	15
Oryza sativa	500 Mb	37,960	8
Zea mays (2n)	2.4 Gb	39,591	23
Glycine max (2n)	1.15 Gb	55,897	16
Sorghum bicolor (2n)	730 Mb	28,120	15
Gossypium arboretum (2n)	1746 Mb	41,330	15
Gossypium raimondii (2n)	885 Mb	40,976	16
Gossypium hirsutum (4n)	2.43 Gb	75,376	33

To understand the origins and evolutionary dynamics between plant species PERKs, the phylogenetic tree was produced with TaPERKs, AtPERKs OsPERKs, and GmPERK proteins (Table S2). The phylogenetic analysis revealed that TaPERK proteins were classified into eight groups (Figure 1).

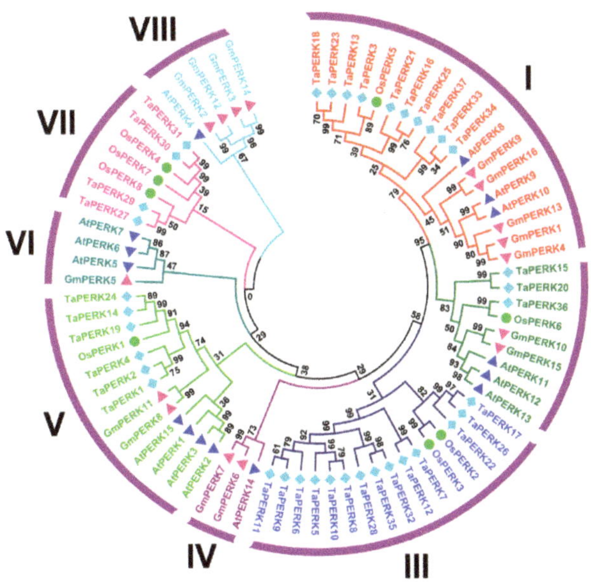

Figure 1. Phylogenetic analysis of TaPERK proteins with Arabidopsis (15), rice (8), and soybean (16). The phylogenetic analysis was executed using the ClustalW program as well as MEGAX software by the neighbor-joining method and bootstrap values of 1000 replicates. The numbers on the nodes indicate the bootstrap values. Distinct groups are represented by the different colors.

Group III was the biggest with 14 members, while Group I, II, IV, V, VI, VII, and VIII contained 10, 3, 0, 6, 0, 4, and 0 members, respectively (Figure S2).

2.2. Chromosomal Distribution, Gene Duplication, and Synteny Analysis

To map the chromosomal distribution of the identified TaPERK genes in wheat, corresponding to chromosomal locations of PERK genes were determined using the PhenGram online server. The TaPERK genes were found on 17 wheat chromosomes (Figure 2A and Table 1). TaPERK genes showed a higher presence on A sub-genomes (Figure 2B). Maximum TaPERK genes (Fourteen) were located on the chromosomes of the A sub-genome.

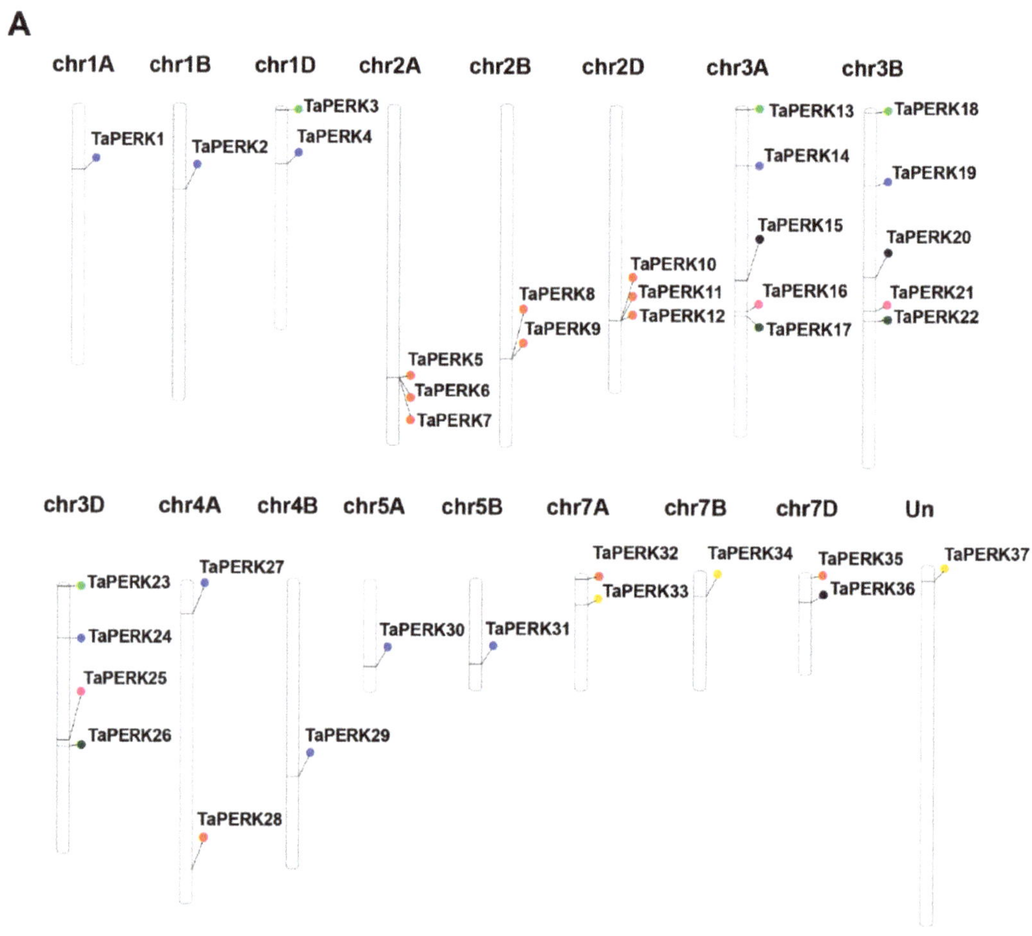

Figure 2. *Cont.*

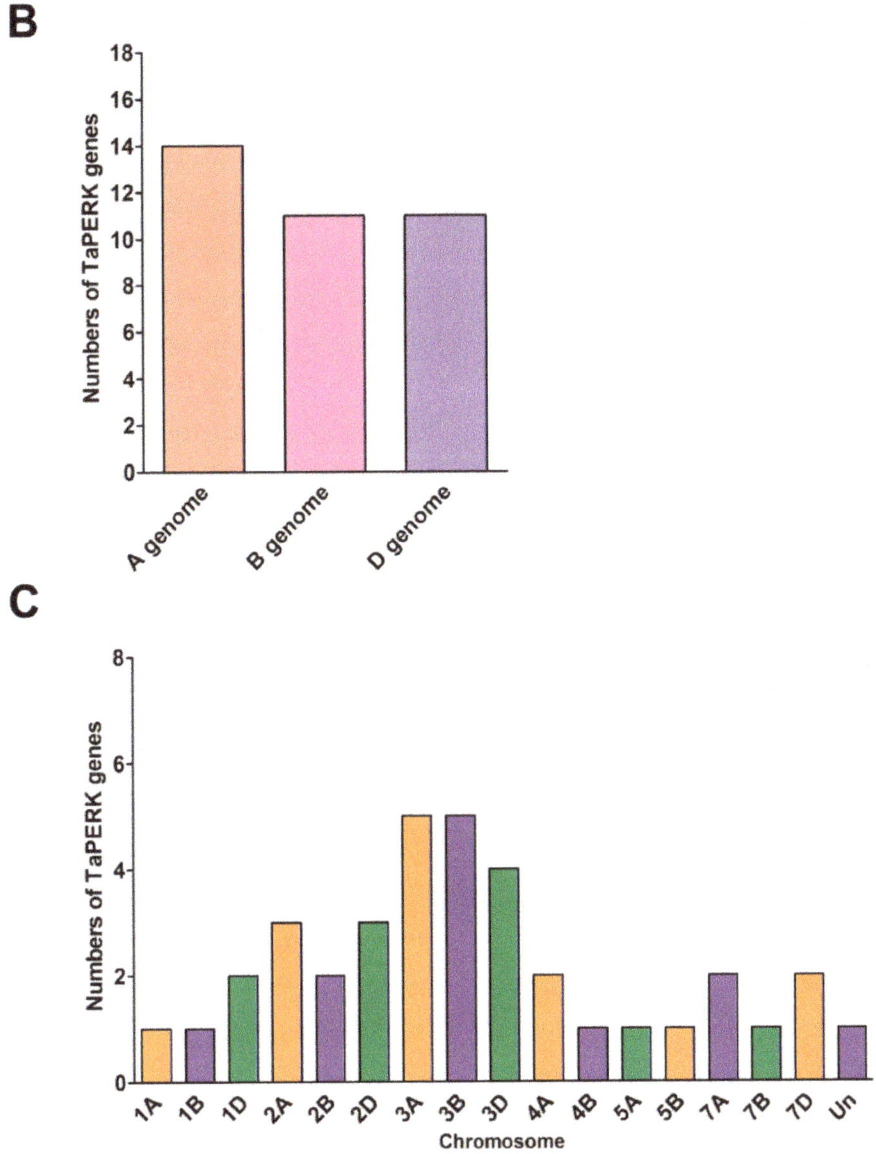

Figure 2. Genomic distribution of identified PERK genes on the 21 chromosomes of wheat and within the three sub-genomes. (**A**) Schematic representations of the chromosomal distribution of PERK genes on the 21 chromosomes of wheat and the name of the gene on the right. The colored circles on the chromosomes indicate the position of the PERK genes. The chromosome numbers of the three sub-genomes are indicated at the top of each bar. (**B**) Distribution of PERK genes in the three sub-genomes. (**C**) Distribution of PERK genes across 21 chromosomes, Un: unaligned contig.

The B and D sub-genome had a minimum number of TaPERK genes (Eleven). Five TaPERKs were mapped on chromosomes 3A and 3B (Figure 2C). The lowest number of TaPERKs was detected on the chromosomes 1A, 1B, 4B, 5A, 5B, and 7B (single gene, respectively). On the contrary, none of the TaPERK genes were located on the chromosomes

4D, 5D, and 6D. In addition, one TaPERK was located on an unaligned contig. Thus, all the PERK family members were uniformly distributed on the wheat's A, B, and D sub-genome.

To explore why the wheat was polyploidy with the largest genome, we further investigated the duplication events in the TaPERK gene family. The phylogenetic analysis of the TaPERK genes also revealed many duplication events (Figure S3). We observed that 26 PERK genes in wheat involved duplication events (Figure S4 and Table S3), indicating expanding the PERK gene family in wheat. Furthermore, to examine the selective pressure on the duplicated TaPERK genes, we analyzed the synonymous substitution (Ks), non-synonymous (Ka), and the Ka/Ks ratios for the 13 TaPERK genes pairs (Table S3). The value of Ka/Ks = 1 indicates that genes underwent a neutral selection; <1 denotes negative selection or purifying, and >1 suggests a positive selection [44]. The Ka/Ks values for all 11 gene pairs were <1, which indicates that TaPERK genes experienced a robust purifying selection pressure with slight alteration after duplication. However, 2 gene pairs, *TaPERK1/TaPERK2* and *TaPERK17/TaPERK26,* had more than 1, which suggests that two pairs of TaPERK genes experienced a positive selection (Table S3). These findings showed the conserved evolution of TaPERKs.

To further elucidate the synteny relationships of TaPERK genes with wheat relatives and other model plants, including *B. distachyon*, *Ae. tauschii*, *T. dicoccoides*, *O. sativa,* and *A. thaliana*, multiple collinearity scan tools were run to identify the orthologous genes between genomes of these plant species (Figure 3 and Table S4).

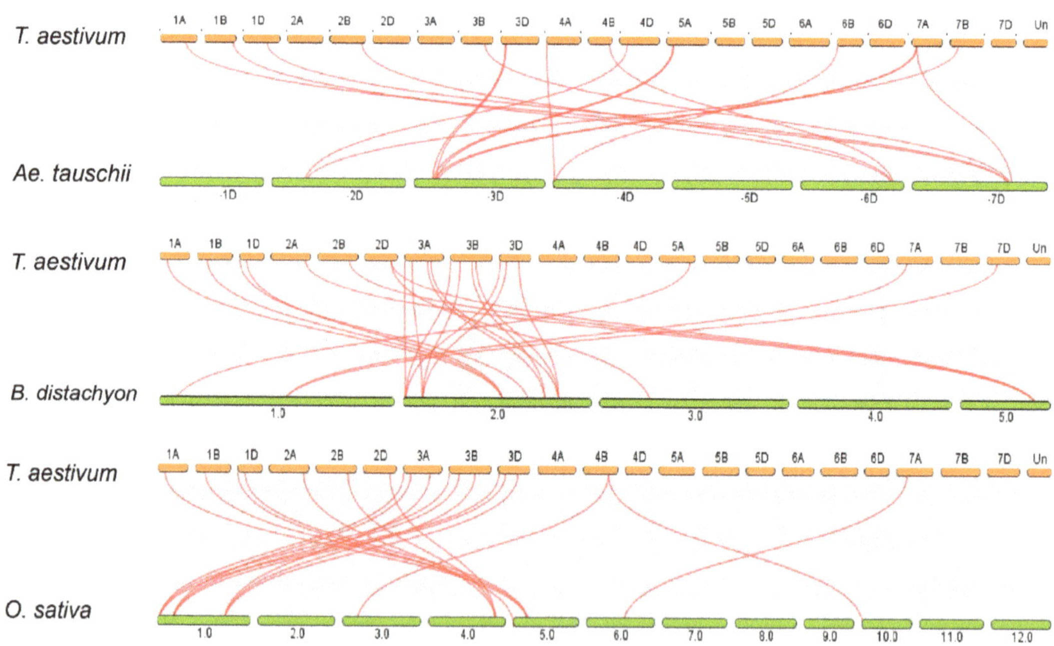

Figure 3. Syntenic relationships of TaPERK genes between *Aegilops tauschii*, *Brachypodium distachyon*, and *Oryza sativa*. The gray lines in the background represent the collinear blocks within *Triticum aestivum* and other plant genomes, while the red lines highlight the syntenic PERK gene pairs.

We found 35, 30, 61, 66, and 83 orthologous gene pairs between TaPERKs with other PERK genes in *B. distachyon*, *Ae. tauschii*, *T. dicoccoides*, *O. sativa*, and *A. thaliana*, respectively. The results showed that 26, 24, 37, 42, and 66 TaPERK genes were collinear with PERK genes in *B. distachyon*, *Ae. tauschii*, *T. dicoccoides*, *O. sativa*, and *A. thaliana*, respectively. Some of the TaPERK genes had five pairs of orthologous genes, for example; *TaPERK5*, *TaPERK6*, *TaPERK8*, *TaPERK9*, and *TaPERK11*, while few of the TaPERK genes had four pairs of orthologous genes,

TaPERK7, *TaPERK10*, *TaPERK12*, *TaPERK15*, *TaPERK20*, *TaPERK28*, *TaPERK32* and *TaPERK35* that might have played an essential role in the evolution of PERK genes. Thus, these results indicated that PERK genes in wheat-derived from a common ancestor.

2.3. Exon/Intron Structure and Motif analysis of TaPERK Genes

To elucidate the structural character of the TaPERK genes, the exon/intron organization and conserved motifs (Figure 4) of TaPERK genes were examined.

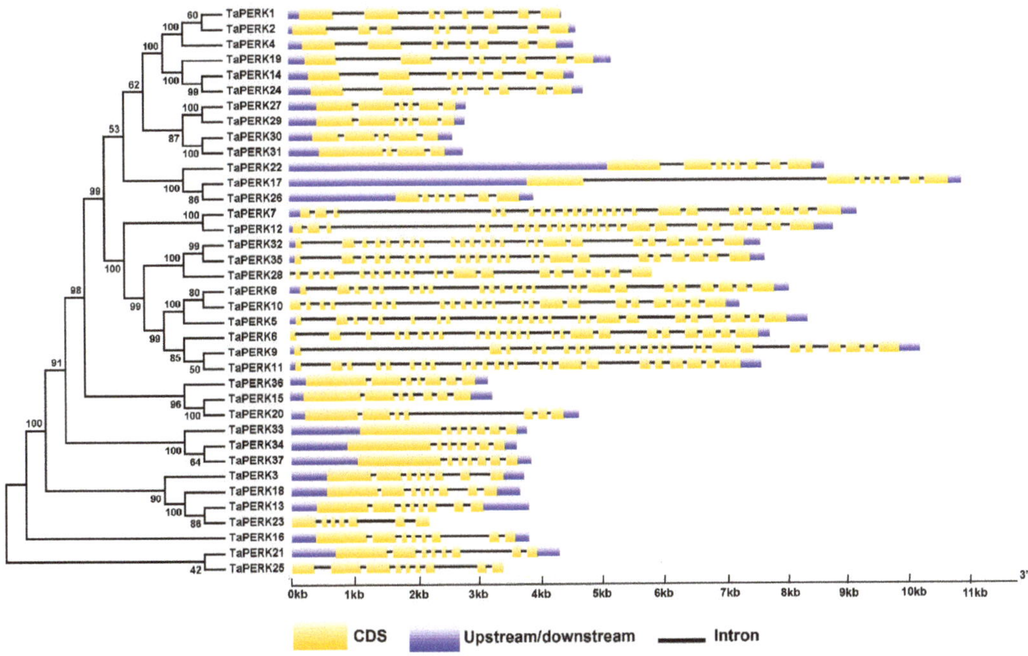

Figure 4. Diagrammatic representation of the exon–intron organization of the TaPERK genes. Yellow boxes represent exons, untranslated regions (UTRs) are indicated by blue boxes, and black lines represent introns. The lengths of the boxes and lines are scaled based on gene length. The exon and intron sizes can be estimated using the scale at the bottom.

Exon–intron analysis showed that the TaPERK gene family greatly varied in terms of gene structure. For instance, most TaPERK genes contain 3–23 introns. Maximum twenty-three introns were detected in the *TaPERK5*, *TaPERK6*, *TaPERK7*, *TaPERK8*, *TaPERK9*, *TaPERK11*, *TaPERK12*, *TaPERK32*, and *TaPERK35*, while *TaPERK31* had three introns (Figure S5). Furthermore, we also analyzed the conserved motif of TaPERK genes using the Multiple Em for Motif Elicitation (MEME) webserver. Eventually, ten well-preserved motifs were found in 37 TaPERK genes (Figure 5A,B).

Furthermore, the TaPERK gene family was detected by the presence of the tyrosine kinase domain (Pfam PF07714), and all TaPERKs consist of at least one tyrosine kinase domain (Table S5) involved in signal transduction. Furthermore, to understand the biological function of TaPERK genes in wheat, 3D protein models of all TaPERKs were produced using a phyre2 webserver. TaPERKs 3D protein structure had two distinct subdomains, a smaller N-terminal lobe and a bigger C-terminal lobe connected by a small hinge loop (Figure S6B). In addition, protein sequence alignment also showed that all TaPERK proteins consisted of a conserved tyrosine kinase domain (Figure 6 and Figure S6A).

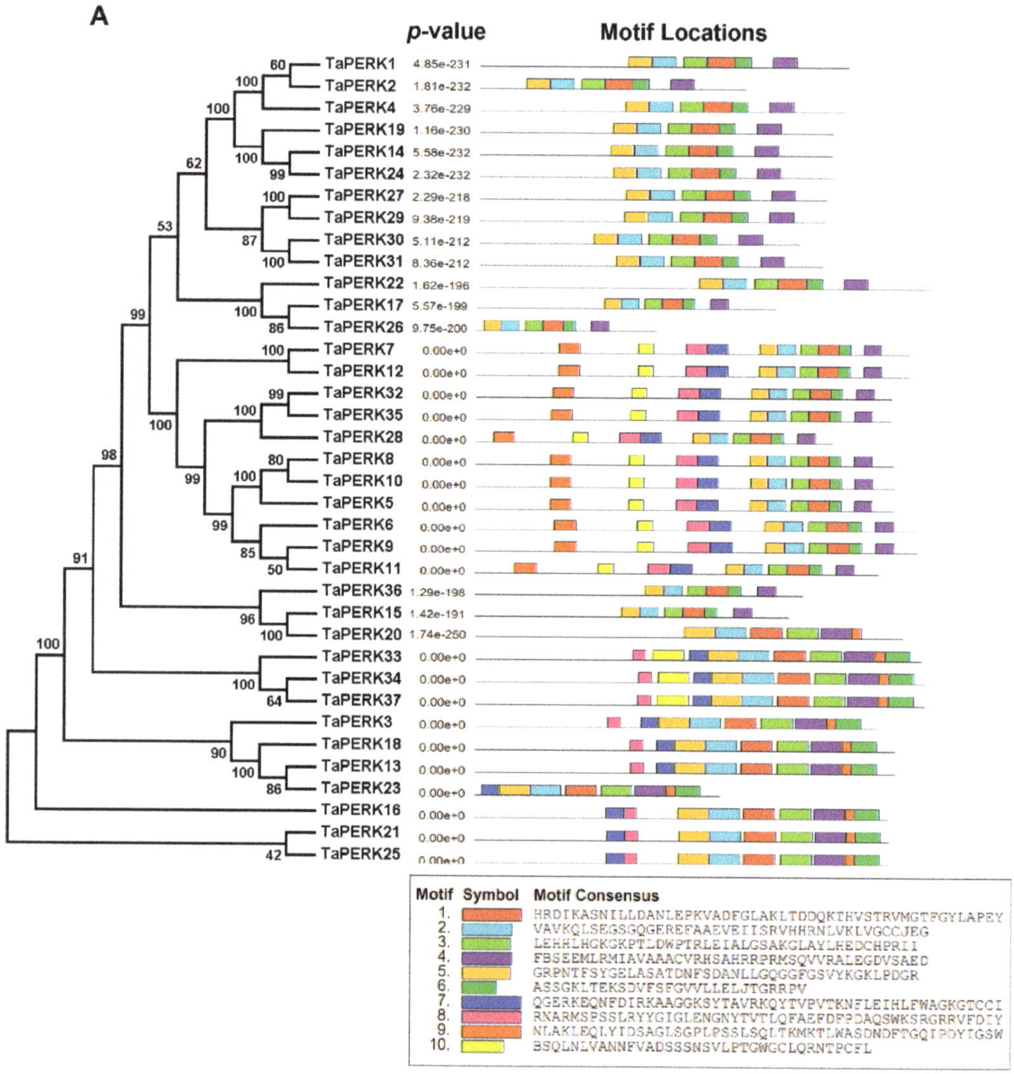

Figure 5. *Cont.*

B

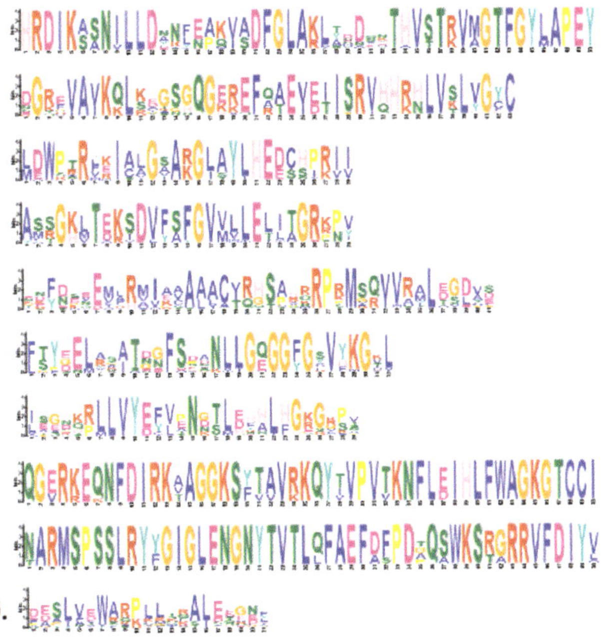

	Logo	E-value	Site	Width
1.		8.7e-1545	37	50
2.		1.4e-1304	37	43
3.		9.1e-821	37	29
4.		5.8e-754	37	29
5.		7.6e-882	37	41
6.		2.2e-751	37	32
7.		2.4e-534	34	29
8.		1.4e-447	11	50
9.		8.3e-404	11	50
10.		6.6e-287	37	21

Figure 5. Conserved motifs of TaPERK genes elucidated by MEME. Up to 10 motifs were shown in different colors. (**A**) Colored boxes representing different conserved motifs with different sequences and sizes. (**B**) Sequence logo conserved motif of the wheat PERK proteins. The overall height of each stack represents the degree of conservation at this position, while the height of individual letters within each stack indicates the relative frequency of the corresponding amino acids. The sequence of each motif, combined *p*-value, and length are shown on the left side of the figure. MEME Parameters: number of repetitions, any; maximum number of motifs, 10; optimum motif width, between 6 and 50.

This result will help understand and explain the substrate specificity and molecular function of TaPERK genes in activating the PERK signal transduction pathway.

2.4. Cis-Acting Regulatory Elements (CAREs) Analysis of TaPERK Genes

To further understand the function of TaPERK genes, upstream 2000 bp sequences from the transcription start site of TaPERKs were analyzed using the PlantCARE web server. This analysis revealed that the promoter region of TaPERKs gene families contained the multiple *cis*-elements related to phytohormones, developmental processes, and different stresses (Figure 7A and Table S6).

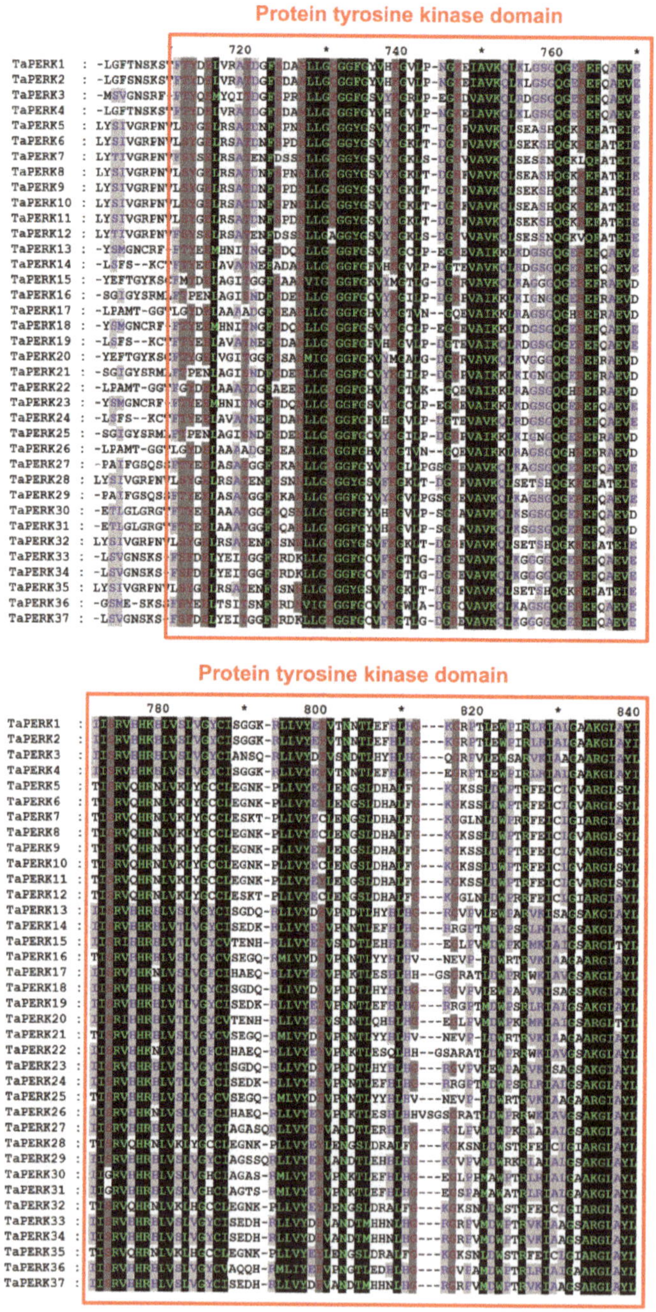

Figure 6. *Cont.*

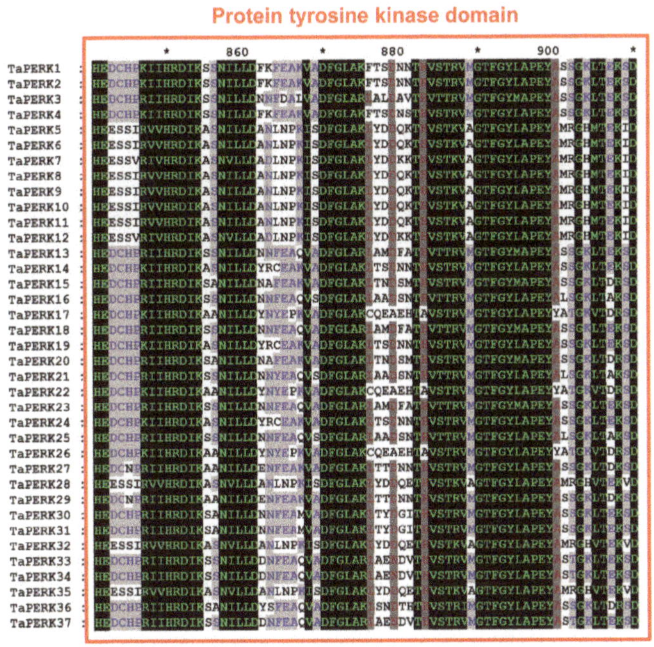

Figure 6. Multiple sequence alignment of the TaPERK protein sequences. The conserved protein tyrosine kinase domain is boxed in red. Colored and shaded amino acids are chemically similar residues. Dashes indicate gaps introduced to maximize the alignment of the homologous region. * indicates positions which have a single, fully conserved residue.

The TaPERKs gene family consists of five hormone response elements, including the auxin response element (AuxRE), gibberellin response element (GARE), methyl jasmonate response element (MeJARE), abscisic acid response element (ABRE), and salicylic acid response element (SARE). The response elements belong to light responses, MeJARE, defense, and stress response and ABRE were the most abundant CAREs in the TaPERK gene family (Figure 7B). This result indicates that TaPERKs play a crucial role in plant growth and development.

Furthermore, TaPERKs contain *cis*-elements related to zein metabolism, endosperm expression, circadian control, meristem expression, seed-specific, and cell cycle regulation. Thus, the CAREs found in the TaPERK gene family indicate that TaPERKs might be participating in a wide range of biological processes. Furthermore, various types of CAREs in the TaPERK genes suggest that these genes might be involved in diverse developmental processes. Therefore, these results provide valuable insights to understand the regulatory mechanism of the TaPERK gene family in response to phytohormone, defense, stress, and various developmental processes.

2.5. Gene Ontology (GO) Enrichment of TaPERK Genes

Gene ontology (GO) assists in understanding the biological function of any genes by comparing their sequence similarity with the known function of genes and gene products with other species [40,42]. All TaPERKs were successfully annotated and allotted GO terms using AgriGO, and further verified using eggNOG-Mapper (Figure S7; Table S7 and Table S8), giving almost the same results as AgriGO. In the biological process category, TaPERK genes were enriched in cell communication (GO:0007154), signaling (GO:0023052), cellular process (GO:0009987), and regulation of biological process (GO:0050789) categories (Figure S7A). In the cellular component category, TaPERK displayed enrichment in the cell (GO:0005623), cell junction (GO:0030054), and membrane (GO:0016020) (Figure S7B). Furthermore, subcellular localization prediction (Table 1) also provided indistinguishable results. In the molecular function category, molecular transducer activity (GO:0060089) and catalytic activity (GO:0003824) were the most prevalent category which was primarily involved in signal transduction (Figure S7C). Apart from cell communication and signaling, the GO term analysis also suggested a variety of roles of TaPERK genes, such as maintenance of dormancy, tissue development, organ formation, post-embryonic organ development, gametophyte development, seedling development, and regulation of developmental process and metabolism. Thus, these results demonstrate that TaPERK genes play a critical role in plant growth and development.

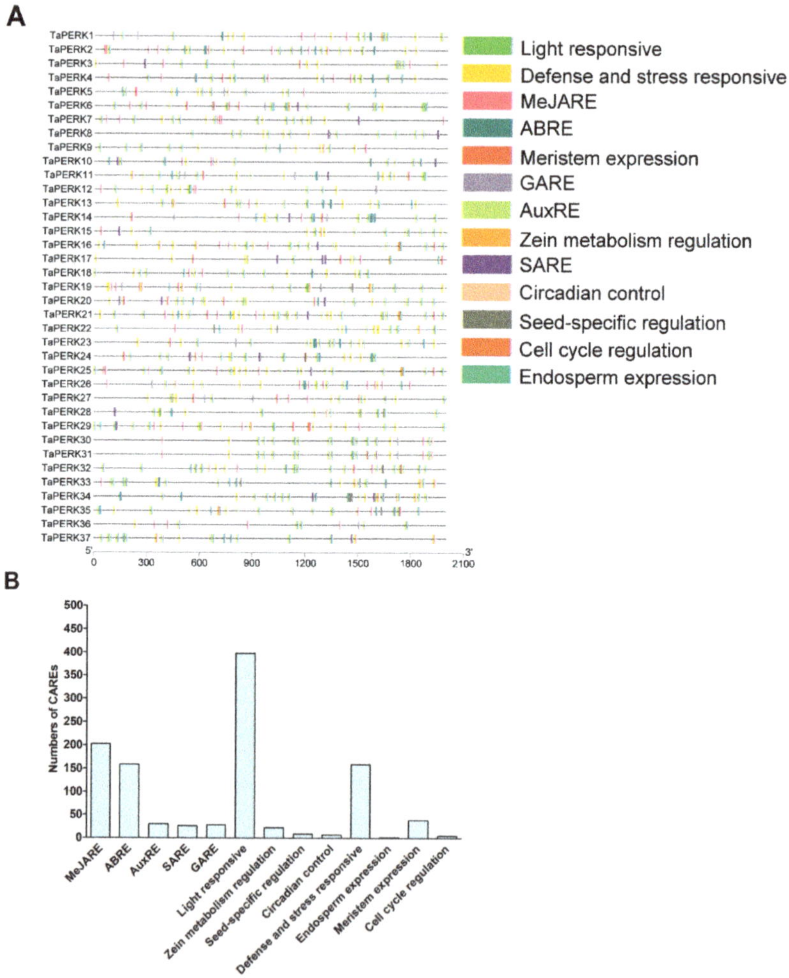

Figure 7. Cis-acting regulatory elements (CAREs) in the promoter region of the TaPERK genes family. The CAREs analysis was performed with a 2kb upstream region using PlantCARE online server. The different numbers of *cis*-regulatory elements represent different colors. (**A**) Hormone-responsive elements, stress-responsive elements, growth and development-related elements, light-responsive elements, and other elements with unknown functions are differentiated by color. (**B**) Most commonly occurring CAREs in TaPERKs.

2.6. Expression Profiling of TaPERK Genes in Various Developmental Stages and under Diverse Stress Conditions

To investigate the precise function of TaPERK genes, the expression pattern of TaPERK genes was examined during different developmental and in diverse stress conditions. The TPM values of all TaPERKs were retrieved from the wheat gene expression database. These TPM values were directly used to generate the PCA and heatmaps (Figure S8A,B, Figures 8 and 9).

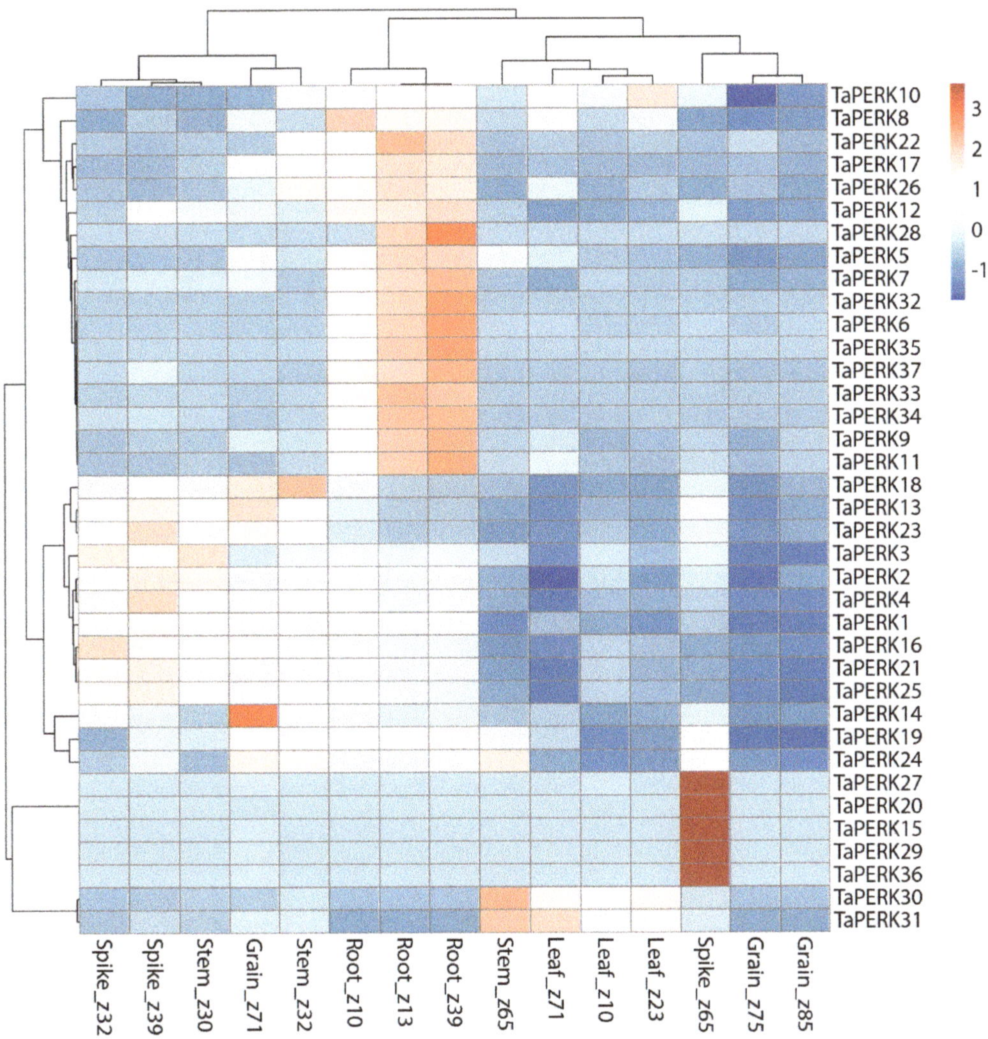

Figure 8. Heatmap representing expression profile of the TaPERK genes at various developmental stages. Columns represent genes, and rows represent different developmental stages. TPM values were used directly to create the heatmaps. The "z" nomenclature refers to Zadok's growth stage.

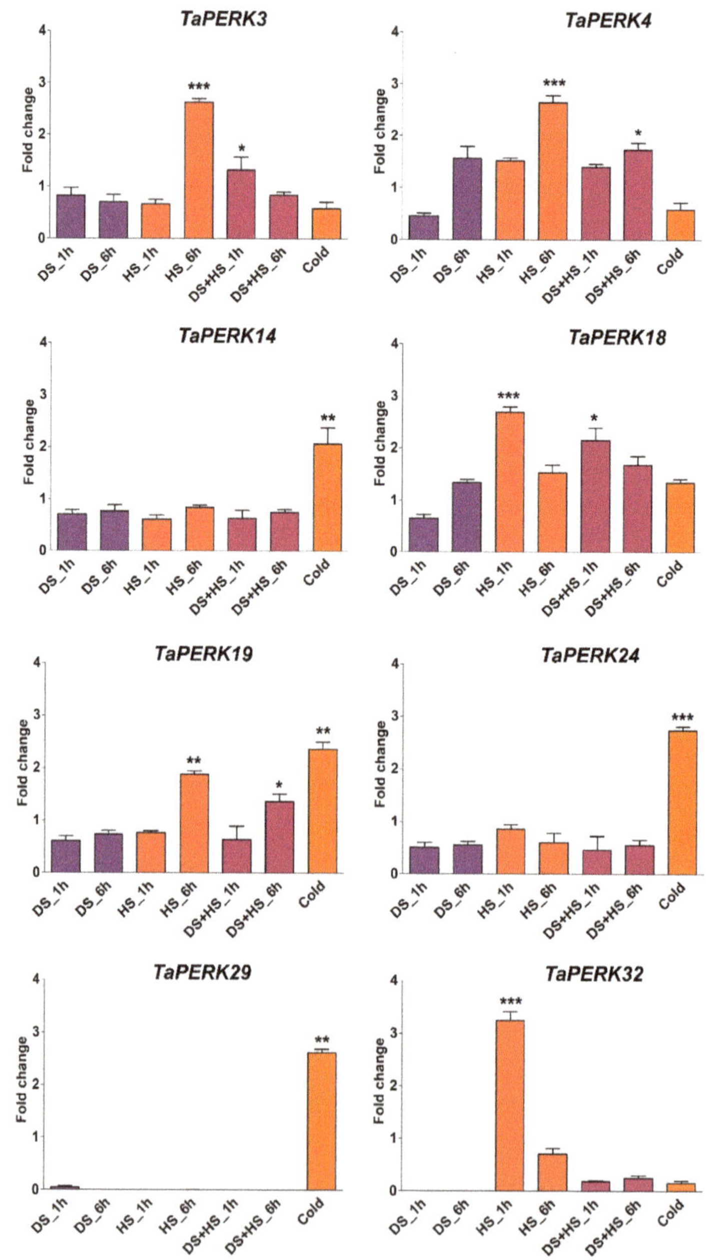

Figure 9. Quantitative real-time PCR analysis of selected TaPERK genes in response to drought stress (DS), heat stress (HS), and cold stress to verify RNA seq data. The wheat actin gene was used as the internal control to standardize the RNA samples for each reaction. Asterisks indicate significant differences compared with control. Bars represent results of Tukey's HSD test at the <0.05 and <0.001 level (* $p < 0.05$, ** p lies in between the values of 0.05 and 0.001, and *** $p < 0.001$). Error bars show standard deviation. Data are mean ± SD ($n = 3$).

To examine the expression pattern of TaPERKs, five tissues from three different developmental stages were taken in this work. The TaPERK genes displayed differential induction among the different tissues; for example, *TaPERK2, TaPERK4, TaPERK13, TaPERK21, TaPERK23,* and *TaPERK25* exhibited induction at the spike z39 stage, while *TaPERK15, TaPERK20, TaPERK27, TaPERK29,* and *TaPERK36* exhibited induction at spike z65 stage (Figure 8).

The expression of TaPERK5, TaPERK6, TaPERK7, TaPERK8, TaPERK9, TaPERK11, TaPERK12, TaPERK17, TaPERK22, TaPERK26, TaPERK28, TaPERK32, TaPERK33, TaPERK34, TaPERK35, and TaPERK37 were elevated in roots at z13 and z39 stage, respectively. TaPERK10, TaPERK30, and TaPERK31 were also up-regulated in the leaf at the z23 and z71 stages, respectively. In addition, TaPERK13, TaPERK14, TaPERK23, TaPERK24, and TaPERK31 showed induction at the grain z71 stage. TaPERK18 and TaPERK24 showed higher expression at the stem z30 stage, whereas TaPERK24, TaPERK30, and TaPERK31 expression was raised at the stem z65 stage (Figure 8). These results showed that the TaPERK gene family members might be involved in developing different tissues and stages.

Expression patterns of TaPERKs were also investigated under the different stress conditions, including septoria tritici blotch (STB), stripe rust, powdery mildew, drought, and heat stress. The expression of several members of the TaPERK gene family was elevated in biotic and abiotic stress (Figure S9). The expression of *TaPERK9, TaPERK13, TaPERK15, TaPERK17, TaPERK20, TaPERK22, TaPERK23, TaPERK26, TaPERK28, TaPERK33, TaPERK35,* and *TaPERK36* were induced during the septoria tritici blotch, while the expression of *TaPERK2, TaPERK6, TaPERK7, TaPERK8, TaPERK10, TaPERK11, TaPERK12 TaPERK27, TaPERK35,* and *TaPERK37* were significantly raised upon the powdery mildew infection. *TaPERK1, TaPERK8, TaPERK16, TaPERK21, TaPERK24, TaPERK25, TaPERK29, TaPERK30, TaPERK31,* and *TaPERK34* were highly elevated during the stripe rust infection. In the case of abiotic stress, the expression profile indicates the expression of a few members of the TaPERK family, for instance, *TaPERK3, TaPERK4, TaPERK18,* and *TaPERK32* were raised during the initial hours of heat stress. It seems that the TaPERK family does not participate in drought stress. However, only *TaPERK4* and *TaPERK18* genes were elevated during the combined drought and heat stress (Figure S9). The expression level of *TaPERK14, TaPERK19, TaPERK24,* and *TaPERK29* was significantly raised during cold stress. Furthermore, the expression patterns of a few selected TaPERK genes were validated through RT-qPCR, and the results displayed nearly similar expression patterns (Figure 9). Overall, these results demonstrated that different TaPERK genes respond to diverse stress conditions.

2.7. *Protein–Protein Network Analysis of the TaPERK Family Genes*

A protein network was produced using the STRING online webserver to examine the interactions between TaPERKs and other *T. aestivum* proteins (Figure 10 and Table S9).

We found eighteen TaPERKs interacting with 10 different wheat proteins according to the STRING results. TaPERK29 can interact with seven other wheat proteins (Traes_3AL_394642923.1, Traes_3AL_5E8DEE3E8.1, Traes_3B_514AAB5F3.1, Traes_3B_9EBD47B52.1, Traes_4AS_E28B34320.1, Traes_6BL_DFDCD5B11.1 and Traes_4DL_05BF7F181.1), which were cGMP-dependent protein kinase/PKG II, protein of unknown function (DUF1645) and BRASSINOSTEROID INSENSITIVE 1, and play critical roles in the Brassinosteroids signaling. TaPERK2 and TaPERK4 can interact with four other wheat proteins (Traes_7AL_5E0DD589E.1, Traes_4DL_E447FD9FD.1, Traes_6BL_DFDCD5B11.1 and Traes_7DL_909EA97B3.1) which were cGMP-dependent protein kinase/PKG II and non-specific serine/threonine-protein kinase. cGMP-dependent protein-kinase is a phosphorylated diverse biologically important pathway [45–47]. PKG is activated by cGMP and has been implicated in the regulation of cell division, nucleic acid synthesis response to biotic stress, stomata closure during osmotic stress, and development of adventitious roots [45–49]. These results provide important insight for further elucidating the complex biological functions of TaPERK genes.

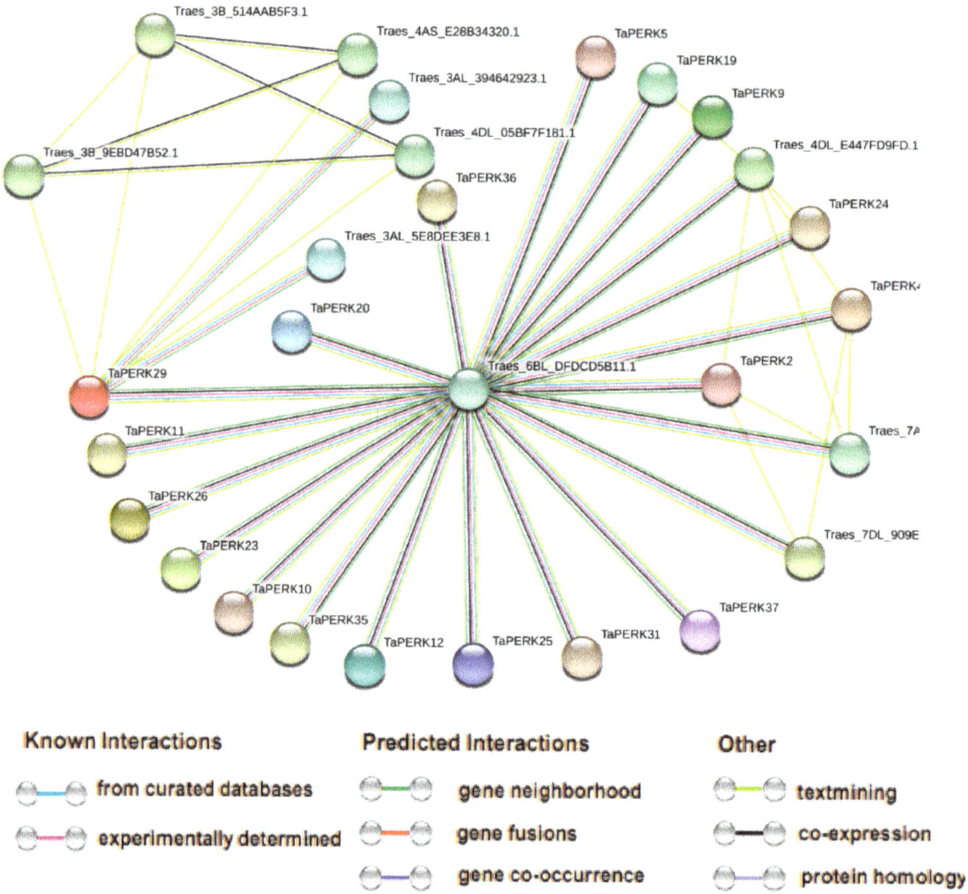

Figure 10. Protein–protein interaction analysis of TaPERKs proteins. Protein–protein interaction network produced by STRINGV9.1, each node represents a protein, and each edge represents an interaction, colored by evidence type. The figure highlights the connections between differentially represented proteins.

3. Discussion

PERKs are a class of receptor kinases that have been implicated during various stages of growth and developments in plants, including cell differentiation, pollen tube growth, pollen development, symbiosis, pathogen recognition, phytohormone response, signal transduction, self-incompatibility, and response to internal and external stimuli [8,9,12–16]. PERKS gene family members have also been identified in other plant species, such as 15 genes in Arabidopsis, 8 in *O. sativa*, 23 in *Z. mays*, 16 in *G. max*, 15 in *S. bicolor*, 15 in *G. arboreum*, 16 in *G. raimondii*, and 33 from *G. hirsutum* [6,7,12]. However, this is the first time we have identified the PERK gene family in the wheat genome. Many studies have reported PERKS genes in ancient land plants, which have expanded during evolutionary processes [6,7,15]. However, in this study, we identified 37 TaPERK genes in the wheat genome (Table 1), which, upon phylogenetic analysis, classified the TaPERK gene family into eight subfamilies or groups (Figure 1). Phylogenetic analysis revealed that groups III and VII were monocot-specific TaPERKs, while groups IV, VI, and VIII contained dicot-specific TaPERKs (Figure 1). The evolution of this type of gene indicates the monocot's specific functions that might play an essential role in establishing physiological and morphological development [40,42,50]. Although, TaPERK genes were distributed into the well-known rice, Arabidopsis, and the soybean cluster, indicating that TaPERKs might be derived from a common ancestor. In addition, most of the TaPERKs showed orthologous relationships with rice, Arabidopsis, and soybean PERKs.

Furthermore, the phylogenetic tree also displayed that all subfamilies have an expanded number of members (Figure 1 and Figure S2), suggesting that the duplication of TaPERKs results from a long course of evolution. Similar results were reported in Arabidopsis, *B. rapa,* and cotton [6,7,15]. Collectively, these results demonstrated a lineage-specific expansion of TaPERKs via the partial alteration of the genome to adapt to internal and external environments during evolution [40,42,50,51].

The wheat PERKS gene family was widely expanded and had comparatively more PERKs than the previously reported PERKs in *A. thaliana, O. sativa, G. max, S. bicolor, Z. mays, G. max, G. arboreum, G. raimondii,* and *G. hirsutum* [6,7,12]. Many previous studies have demonstrated that polyploidy enabled numerous plant species to adapt to adverse environmental conditions [44,52,53]. Mostly, polyploidy is linked with gene duplication and, in our study, we also found that tandem, segmental and whole-genome duplication was the critical driving force responsible for the duplication of TaPERK genes. Segmental duplication is the fundamental drive factor, and occurs in numerous plant genomes during evolution consisting of several duplicated chromosomal blocks [54]. For instance, several Arabidopsis gene families experienced coherent evolutionary dynamics directed to expanding the gene family [55,56]. Moreover, several gene families, including cotton GRAS, RH2FE3, MADS-Box, MIKC-Type, YABBY, WOX, sesame heat shock proteins, and soybean WRKY, experienced segmental expansion and whole-genome duplication events [7,57–62]. The chromosomal map of TaPERK genes revealed that the 37 TaPERKs were unequally distributed throughout chromosomes, excluding chromosome 6 (Figure 2). The gene number on each chromosome varied from one to five: chromosomes 3A and 3B had five genes; chromosome 3D had four genes; chromosome 2A and 2D contained three genes; chromosomes 1D, 2B, 4A,7A and 7D had two genes; and 1A, 1B, 4B, 5A, 5B and 7B consisted of a single gene. Hence, uneven distribution of the TaPERK genes on the 17 chromosomes of wheat indicates probable gene addition or loss via whole genome or segmental duplication events and errors during genome sequencing and assembly. Gene duplication analysis showed 13 pairs of duplicated genes, which shared high sequence similarity at the nucleotide level. The duplicated pairs were *TaPERK1:TaPERK2, TaPERK14:TaPERK24, TaPERK27:TaPERK29, TaPERK30:TaPERK31, TaPERK17:TaPERK26, TaPERK7:TaPERK12, TaPERK32:TaPERK35, TaPERK8:TaPERK10, TaPERK9:TaPERK11, TaPERK15:TaPERK20, TaPERK34:TaPERK37, TaPERK13:TaPERK23,* and *TaPERK16:TaPERK21*. Furthermore, the Ka/Ks value of 11 gene pairs was <1, indicating that TaPERK genes experienced a robust purifying selection pressure (Figure S3 and Table S3). However, two gene pairs, *TaPERK1:TaPERK2* and *TaPERK17:TaPERK26,* had more than 1, suggesting that two pairs of TaPERK genes underwent a positive selection. Therefore, these results indicate that TaPERK genes were not changed much in function after duplication and exhibited the conserved evolution of TaPERK genes. Qanmber and colleagues (2019) also reported similar results in cotton. Furthermore, ten gene pairs were the results of segmental duplications in cotton [7]. Furthermore, 146 out of 149 duplicated gene pairs had a Ka/Ks ratio of <1.0, and only three duplicated gene pairs displayed more than 1, which indicates the positive selection pressure. Similar type gene duplication events were also described in the BrPERKs genes [15]. Our gene duplication analysis also demonstrated that the TaPERK gene duplication events were similar, as previously reported in the cotton and *Brassica rapa* [7,15]. Thus, these results showed that segmental and whole-genome duplications might play a critical role in the evolution and expansion of the PERK genes in wheat.

To further elucidate the synteny relationships of TaPERK genes with wheat relatives and other model plants, we identified 35, 30, 61, 66, and 83 orthologous gene pairs between TaPERKs with other PERK genes in *B. distachyon, Ae. tauschii, T. dicoccoides, O. sativa,* and *A. thaliana*, respectively (Figure 3 and Table S4). Additionally, *Ae. speltoides* (BB, diploid) and *Ae. tauschii* (DD, diploid) were the foundation of B and D subgenomes of wheat. The synteny relationship displayed that nine orthologous gene pairs between *Ae. tauschii* with a wheat D subgenome were found on the same chromosomes with two on 1D, one on 2D, four on 3D, and two on 7D (Figure 3 and Table S4). Furthermore, twenty-two orthologous gene pairs between *T. dicoccoides* with a wheat AABB subgenome were detected on the same chromosomes with one on 1A, two on 2A, five on 3A, two on 4A, one on 5A, two on 7A, one on 1B, one on 2B, four on 3B, one on 4B, one on 5B, and one on 7B (Figure 3 and Table S4). These findings suggest that PERK genes might have come from *Ae. tauschii* and *T. dicoccoides* during natural hybridization events. Furthermore, more orthologous gene pairs were found in *T. aestivum* with *A. thaliana* and *O. sativa*, which exhibited that TaPERK and other PERKS genes might be derived from these orthologous genes during evolution.

The gene structure analysis of TaPERKs revealed that TaPERKs greatly varied in gene structure. The majority of the TaPERK genes contained more than five exons, except for TaPERK31 with four exons, while TaPERK30 had five exons (Figure 4). A maximum of twenty-four exons were detected in *TaPERK5, TaPERK6, TaPERK7, TaPERK8, TaPERK9, TaPERK11, TaPERK12, TaPERK32*, and *TaPERK35*. Furthermore, a maximum of twenty-three introns were found in *TaPERK5, TaPERK6, TaPERK7, TaPERK8, TaPERK9, TaPERK11, TaPERK12, TaPERK32*, and *TaPERK35*, while *TaPERK31* had a minimum of three introns (Figure S5). The size of an intron is a critical player that affects the gene size; for example, a notable difference in gene size was found between the biggest gene *TaPERK17* (4 kb) and the smallest gene *TaPERK23* (2.1 kb), and this was mainly caused by the total intron length (4 kb vs. 1.1 kb). Many studies have shown the significance of introns in the evolution of numerous plant genes [63,64]. Several gene families had less, lack, or more introns in their gene families [7,59,65,66]. The exon and intron differences might be due to deletion/insertion events, which would predict the evolutionary processes [67]. All PERK gene family members in cotton had no introns, indicating that GhPERK genes might have evolved comparatively quickly [7].

Furthermore, it has been established that gene families containing larger or more introns can acquire new functions during evolution processes. There were more intron gains than losses in the plant lineages and chordates, while in arthropods and fungi, losses prevailed over gains [63,64,67]. In our study, almost all TaPERK genes had more and larger introns. Hence, we can speculate that PERK genes gained new functions during evolution in wheat. Furthermore, conserved motif analysis showed ten different types of motif compositions amidst the TaPERK proteins. We observed that five motifs were found in all the TaTERK proteins (Motif 1, 2, 3, 4 and 6), and proteins of the same subfamilies usually shared the same motifs and were more conservative. Thus, we hypothesize that proteins of the same subfamilies may have the same function.

Additionally, amino acid sequence alignment of TaPERK with other plant species PERK proteins also showed that all TaPERK proteins consisted of a conserved tyrosine kinase domain (Figure 6 and Figure S6). The amino acid residues of PERK were highly conserved in rice, Arabidopsis, soybean, and wheat, which might be helpful to find the pattern of PERK protein sequence conservation in different plant species. Yang and colleagues also found that YABBY and WOX gene families were evolutionarily conserved in cotton [57,58]. Furthermore, 3D protein structure analysis revealed that TaPERKs had two distinct subdomains, a smaller N-terminal lobe, and a more prominent C-terminal lobe connected by a small hinge loop (Figure S6A,B). These findings will be helpful to understand and explain the substrate specificity and molecular function of TaPERK genes in activating the PERK signal transduction pathway.

The *cis*-acting regulatory element in the promoter plays an important role in regulating and functioning genes [68]. The promoter region of TaPERKs gene families contains the multiple cis-acting elements related to plant hormones, growth, development, defense, and stress-related functions (Figure 7A and Table S6). We predicted more than eight CAREs in the promoter region of each TaPERK (Table S6). A total of 15 CAREs related to light response were detected, including AE-box, Box 4 and ATCT motif, chs-Unit 1 m1, TCT-motif, I-box, chs-CMA1a, chs-CMA2a, GA-motif, GATA-motif, LAMP-element and TCCC-motif, ACE and GT1-motif, Sp1 and 3-AF1 binding site [69,70]. We also detected the six CAREs related to growth and development, such as MSA-like (cell cycle regulation), GCN4-motif (endosperm expression), O2-site (zein metabolism regulation), CAT-box (meristem expression), RY-element (seed-specific regulation) and CAAAGATATC-motif (circadian control) [71,72]. In addition, we also found the CARE related to phytohormone response, for instance, CGTCA-motif (MeJA-responsive element), ABRE (abscisic acid-responsive element), TCA-element (salicylic acid responsiveness), TGA-motif (auxin-responsive element), P-box, and GARE-motif (gibberellin-responsive element). The MeJA-responsive element was predicted in most TaPERK genes except *TaPERK8*, *TaPERK11*, *TaPERK23*, and *TaPERK28*. Moreover, we also predicted that other *cis*-elements had been involved in different stress conditions, such as LTR (low-temperature responsiveness), MBS (drought inducibility), and TC-rich repeats (defense and stress responsiveness) in the TaPERK promoters. Several studies have demonstrated that light plays a crucial role in plant growth and development processes [73]. Several CAREs related to low temperature, fungal elicitors, stress and defense, auxins, MeJA, gibberellin, ethylene abscisic acid, and the salicylic acid-responsive element were also predicted in *GhPERK* and *BrPERK* gene promoter regions [7,15]. In this study, almost all TaPERK genes contained the multiple CAREs involved in plant growth and the responses to diverse stress. *GhPERK8*, *GhPERK 9*, *GhPERK12*, *GhPERK23*, *GhPERK27*, and *GhPERK29* expression levels were elevated upon exposure to plant hormones such as indole-3-acetic acid, gibberellin, salicylic acid, and MeJA; however, the expression level of *GhPERK5* declined [7]. *PERK4* regulates the root growth function at an early stage of ABA signaling by perturbing calcium homeostasis in Arabidopsis [14]. *PERK1* rapidly induced early perception and response to a wound stimulus in Chinese cabbage [12]. Antisense suppression of *BnPERK1* exhibited various growth defects such as amplified secondary branching, loss of apical dominance, and defects in floral organ formation. At the same time, the overexpression line showed increased lateral shoot production, seed set, and unusual deposition of callose and cellulose in *Brassica napus* [21]. Collectively, these results showed that PERKS gene family members might regulate diverse biological processes, responses to phytohormones, and work against different biotic and abiotic stress. Of course, this needs to be established by experimental studies in the near future. Therefore, these data provide the valuable information to understand TaPERKs' function in plant growth and development, response to phytohormones, and different stresses.

Receptor kinases play a critical role in different biological processes and responses to internal and external stimuli [8,9,12–16,21,25]. Different TaPERK genes displayed differential expressions in various tissues. For example, *TaPERK2*, *TaPERK4*, *TaPERK13*, *TaPERK21*, *TaPERK23*, and *TaPERK25* exhibited induction at the spike z39 stage, while *TaPERK15*, *TaPERK20*, *TaPERK27*, *TaPERK29*, and *TaPERK36* exhibited induction at spike z65 stage (Figure 8). The expression of *TaPERK5*, *TaPERK6*, *TaPERK7*, *TaPERK8*, *TaPERK9*, *TaPERK11*, *TaPERK12*, *TaPERK17*, *TaPERK22*, *TaPERK26*, *TaPERK28*, *TaPERK32*, *TaPERK33*, *TaPERK34*, *TaPERK35*, and *TaPERK37* were elevated in the roots at the z13 and z39 stages, respectively. *TaPERK10*, *TaPERK30*, and *TaPERK31* were also up-regulated in the leaf at the z23 and z71 stages, respectively. In addition, *TaPERK13*, *TaPERK14*, *TaPERK23*, *TaPERK24*, and *TaPERK31* showed induction at the grain z71 stage. *TaPERK18* and *TaPERK24* showed higher expression at the stem z30 stage, whereas *TaPERK24*, *TaPERK30*, and *TaPERK31* expressions were raised at the stem z65 stage. PERKs proteins have been involved in various developmental processes, including cell differentiation, pollen tube growth, pollen development, symbiosis, pathogen recognition, phytohormone response, signal transduction, and self-incompatibility [9,12,14,16,17,21]. AtPERK gene family members are ubiquitously expressed, while few genes are specifically expressed [6]. For instance, *AtPERK1* is broadly expressed, whereas *AtPERK2* is mainly expressed in rosette leaf veins, stems, and pollen [21,32]. *AtPERK8* and *AtPERK13* expression were found in the root hairs [6,22,23]. In addition, *AtPERK5*, *AtPERK6*, *AtPERK7*, *AtPERK11*, and *AtPERK12* expression were up-regulated in the pollens [23–25]. Furthermore, PERK4 modulates the root tip growth at an early stage of ABA signaling via the disruption of calcium homeostasis in Arabidopsis [14]. Some researchers have shown that increased calcium concentration in the cells also enhances antioxidant enzyme activities to regulate the lipid peroxidation of cell membranes and stomatal aperture [26–28]. *AtPERK5* and *AtPERK12* play an essential role in pollen tube growth in Arabidopsis [16]. Furthermore, *AtPERK8*, *AtPERK9*, and *AtPERK10* negatively regulate root growth in Arabidopsis [23]. The expression level of twelve GhPERK genes was significantly elevated in leaves and ovule development in cotton [7]. BrPERK genes were differentially expressed in various tissues of Chinese cabbage, but some BrPERK genes were specially expressed in reproductive organs [15]. Our GO analysis also indicated the critical roles of the TaPERK gene in the cell (Figure S7A–C). Thus, this spatial and temporal expression of the TaPERK genes suggests that these PERKs might have an essential function in different wheat tissue.

Receptor kinases are crucial in plant adaptations and responses to internal and external stimuli [8,9,12–15,74]. Our results also showed that several TaPERK gene family members' expressions were elevated in different stress conditions (Figure S9). *TaPERK9*, *TaPERK13*, *TaPERK15*, *TaPERK17*, *TaPERK20*, *TaPERK22*, *TaPERK23*, *TaPERK26*, *TaPERK28*, *TaPERK33*, *TaPERK35*, and *TaPERK36* were induced during the septoria tritici blotch, while the expressions of *TaPERK2*, *TaPERK6*, *TaPERK7*, *TaPERK8*, *TaPERK10*, *TaPERK11*, *TaPERK12* *TaPERK27*, *TaPERK35*, and *TaPERK37* were significantly raised after powdery mildew infection. *TaPERK1*, *TaPERK8*, *TaPERK16*, *TaPERK21*, *TaPERK24*, *TaPERK25*, *TaPERK29*, *TaPERK30*, *TaPERK31*, and *TaPERK34* were up-regulated during the stripe rust infection. Furthermore, *TaPERK3*, *TaPERK4*, *TaPERK18*, and *TaPERK32* were induced during the initial hours of heat stress. However, none of the genes were expressed in drought stress. It seems that the TaPERK family does not participate in drought stress, and only *TaPERK4* and *TaPERK18* genes were elevated during the combined drought and heat stress (Figure S9). Furthermore, *TaPERK14*, *TaPERK19*, *TaPERK24*, and *TaPERK29* expression levels were significantly elevated in cold stress. Most TaPERKs respond similarly to biotic and abiotic stress; hence, all stress-responsive genes cluster together (Figure S8B). Several PERKS genes in *A. thaliana*, *G. hirsutum*, and *B. rapa* were responsive to diverse abiotic stresses, including cold, salt, heat, and PEG, indicating that PERK genes play a critical role in other plant species to adapt to different stress conditions [6,7,15,23]. PERK1 rapidly induced early perception and response to a wound stimulus in Chinese cabbage [12]. A PERK-like receptor kinase specifically interacts with the nuclear shuttle protein (NSP), led viral infection, and positively regulates the NSP function in cabbage leaf curl virus and geminivirus [32]. The expression profile of TaPERK genes under different stresses indicated that they might participate in the diverse biotic and abiotic stress tolerance in wheat. Therefore, these findings demonstrated that TaPERK genes respond to various stresses, and this might be used for breeding wheat lines to develop stress-tolerant varieties in wheat.

Many studies have shown that receptor kinases at the cell surface perceive a sudden changing environment that activates the various signaling pathways, regulating growth, reproduction, and response to diverse stress conditions [2,5,75]. Our protein–protein network analysis revealed that eighteen TaPERKs interacted with 10 different wheat proteins (Figure 10 and Table S9). TaPERK29 can interact with seven other wheat proteins (Traes_3AL_394642923.1, Traes_3AL_5E8DEE3E8.1, Traes_3B_514AAB5F3.1, Traes_3B_9EBD47B52.1, Traes_4AS_E28B34320.1, Traes_6BL_DFDCD5B11.1 and Traes_4DL_05BF7F181.1), which were cGMP-dependent protein kinase/PKG II, protein of unknown function (DUF1645) and BRASSINOSTEROID INSENSITIVE 1, playing a critical role in Brassinosteroid signaling. Brassinosteroids play an essential role in various cellular processes, such as cell division, seed germination, vascular differentiation, flowering, xylem cell differentiation, stomata formation, photomorphogenesis, and pollen tube growth [76–79]. In addition, TaPERK2 and TaPERK4 can interact with four other wheat proteins (Traes_7AL_5E0DD589E.1, Traes_4DL_E447FD9FD.1, Traes_6BL_DFDCD5B11.1 and Traes_7DL_909EA97B3.1) which were cGMP-dependent protein kinase/PKG II and non-specific serine/threonine-protein kinase/Threonine-specific protein kinase. cGMP-dependent protein kinase is a phosphorylated diverse biologically important pathway [45–47]. PKG activated by cGMP has been implicated in cell division regulation, nucleic acid synthesis response to biotic stress, stomata closure during osmotic stress, and the development of adventitious roots [45–49].

TaPERK29 was highly elevated during the stripe rust infection, and cold stress might interact with Traes_6BL_DFDCD5B11.1, that is, cGMP-dependent protein kinase/PKG II specifically activates the signal transduction pathways implicated in different stress tolerance in wheat.

Moreover, TaPERK2 was co-expressed with TaPERK4 in spike development (Figure 8 and Figure S9), indicating that TaPERK2 and TaPERK4 might have an essential function in spike development through interacting with each other. These results provided valuable insight and the complex biological functions of TaPERK genes. In summary, this study provides valuable information about the TaPERK gene family, functions in plant growth, and response to phytohormones and different stress. Therefore, the outcome of this work is significant to dissect and understand the precise functions of TaPERK in developmental processes and various biotic and abiotic stress in wheat.

4. Materials and Methods

4.1. Identification of PERK Genes in Wheat

To carry out the genome-wide survey in bread wheat (*Triticum aestivum*) cv. *Chinese Spring*, genome data including genomic, CDS, and protein sequences of TaPERK genes were downloaded from the Ensembl plants biomart (http://plants.ensembl.org/biomart/martview, accessed on 27 September 2021). Two approaches were used to identify the PERK genes family in wheat. In the first method, we prepared a local database of the wheat protein sequences in BioEdit ver. 7.2.6 [80]. The sixty-two PERK genes from *A. thaliana*, *G. max*, *O. sativa*, and *Z. mays* were used for BLASTp against the local database. To find PERK genes, the e-value of 10^{-5} and >100-bit score were kept cut-off, and eventually, the BLASTp result was tabulated. In the second approach, the protein sequences of PERK from the above plant species were retrieved from the Ensembl Plants (http://plants.ensembl.org/index.html, accessed on 27 September 2021) and the BLASTp search was performed against the *T. aestivum* proteome with an e-value 10^{-5} and bit-score > 100. Based on the above method, putative PERK candidates were selected. Further putative PERK candidates were confirmed for the presence of protein tyrosine kinase domain using other online databases: InterPro (https://www.ebi.ac.uk/interpro, accessed on 25 September 2021), Simple Modular 132 Architecture Research Tool tool (SMART, http://smart.emblheidelberg.de/, accessed on 25 September 2021), HMMscan (https://www.ebi.ac.uk/Tools/hmmer/search/hmmscan, accessed on 25 September 2021) and NCBI CDD (https://www.ncbi.nlm.nih.gov/Structure/cdd/cdd.shtml, accessed on 25 September 2021). Finally, the protein sequences with protein tyrosine kinase domains were taken and renamed according to their chromosomal positions.

4.2. Genomic Localization, Gene Duplication, and Synteny Analysis

To map the chromosomal locations of TaPERK genes, genomic positions of PERK genes were downloaded from Ensembl plants biomart (http://plants.ensembl.org/biomart/martview, accessed on 26 September 2021). The PERK genes were named with a 'Ta' prefix and numbered according to their chromosomal positions. PhenoGram was used to map the TaPERK genes on the chromosomes (http://visualization.ritchielab.org/phenograms/plot, accessed on 26 September 2021). MCScanX tool kit was used to examine gene duplication events and synteny analysis within species and other plant species [81,82]. We used default parameters in MCScanx for synteny analysis. The non-synonymous (Ka) and synonymous substitution (Ks) ratio was calculated to estimate the selection pressure of duplicated TaPERK genes using the TBtools [82].

4.3. Biophysical Characteristics, Subcellular Localization, and 3D Structure

The biophysical characteristics of TaPERK proteins were predicted using ExPASy [83] and an isoelectric point calculator [84]. Subcellular localization was evaluated using CELLO [85], softberry (www.softberry.com, accessed on 27 September 2021), and BUSCA [86]. Finally, the three-dimensional structure of TaPERKs was generated using the Phyre2 web server [87].

4.4. Exon/intron Structure, Protein Motif, and Gene Ontology Analysis

The coding sequence, genomic and protein sequences of TaPERK genes were downloaded from the Ensembl plants biomart (http://plants.ensembl.org/biomart/martview, accessed on 27 September 2021). Exon, intron positions, and untranslated regions were elucidated using the Gene Structure Display Server 2.0 (http://gsds.gao-lab.org/, accessed on 27 September 2021). The protein motifs in the TaPERK were visualized using MEME (Multiple Em for Motif Elicitation ver.5.3.3; http://meme-suite.org/tools/meme, accessed on 27 September 2021) with default settings. TaPERK protein sequences were explored to detect GO terms enrichment using EggNOG (http://eggnogdb.embl.de/#/app/emapper, accessed on 27 September 2021) and agriGO [88].

4.5. Promoter Cis-Acting Regulatory Elements (CAREs) and Protein Interaction Network Analysis

To identify *cis*-elements, 2 kb upstream sequences of PERK genes were retrieved from Ensembl plants and examined using a PlantCARE online webserver (http://bioinformatics.psb.ugent.be/webtools/plantcare/html/, accessed on 28 September 2021). The number of occurrences for each *cis*-element motif was counted for TaPERK genes, and the most frequently occurring CAREs were used to generate Figure 7 using TBtools [82]. The TaPERK protein interaction network was predicted using the STRING webserver (https://string-db.org/cgi, accessed on 28 September 2021).

4.6. Expression Analysis of TaPERK Genes

Transcripts per million (TPM) values for five different tissues, including leaf, stem, root, spike, grain, and under various stress conditions, were downloaded from the Wheat Expression database (http://www.wheat-expression.com/, accessed on 30 September 2021). Heatmaps and principal component analysis (PCA) were performed using ClustVis [89] and TBtools software [82].

4.7. Plant Growth Conditions, Stress Treatment, and RT-qPCR Analysis

Wheat (*Triticum aestivum* L.) cv. HI 1612 was used for the experiments. Seeds of HI 1612 were sown on soil in plastic pots and reared in a greenhouse. Ten-day-old wheat seedlings were acclimatized for two days in growth chamber conditions. They were further subjected to drought and high-temperature stress (40 °C) for 1 h and 6 h [90], and cold stress for 3 days (4 °C). For the combined drought and high-temperature stress, first, wheat seedling was exposed for the drought stress, then given a heat shock for 1 h and 6 h at 40 °C in an incubator. Controls were kept at 25 °C. The cold, drought, and high-temperature stressed seedlings were collected for RNA extraction and stored at −80 °C. The RNA was isolated from control, drought, and heat-stressed seedlings, as described by [91,92]. cDNA was synthesized using the iScriptTM cDNA synthesis kit (Bio-Rad, Hercules, CA, USA). Quantitative real-time PCR (RT-qPCR) was performed using the Applied Biosystems 7500 Fast Real-Time PCR (Applied Biosystems) with the SYBR Premix (Toyobo, Osaka, Japan). Wheat actin (AB181991) was used as a control to normalize the gene expression data. Transcript abundance was analyzed using the RT-qPCR. Each qRT-PCR reaction was carried out with three biological samples with two technical replicates and repeated three times. The fold change was calculated based on mean $2^{-\Delta\Delta CT}$ values and, eventually, this fold value was used to plot the graph [93,94]. Furthermore, one-way ANOVA, followed by Tukey's HSD for multiple pairwise comparisons were applied. Means, standard errors and statistical significances for each sample were represented in figures (* $p < 0.05$, ** $p < 0.01$). All primers used in this study are mentioned in Table S10.

5. Conclusions

Wheat is the most important cereal crop and widely consumed staple food worldwide. However, global warming is becoming a severe threat to food security due to the constant climate changes, largely influencing plant development and productivity. This has raised a major challenge for plant biologists to increase yield and improve wheat's quality, biotic and abiotic stress tolerance. The PERKS gene family plays a critical role in plant development and responses to various stresses. We identified and characterized the PERK gene family in wheat in this work. Expression patterns also revealed the role of TaPERKs in different developmental stages and stress conditions. Thus, this study facilitates a detailed understanding of PERK genes' biological functions in wheat under different developmental processes and stress conditions.

Supplementary Materials: The following are available online at https://www.mdpi.com/article/10.3390/plants11040496/s1. Figure S1: Molecular weight (kDa) vs. isoelectric point plots of TaPERK genes. The distinct round shape colors represent the TaPERK gene family members. Figure S2: Distribution of TaPERKs in a different group of the phylogenetic tree. The *y*-axis indicates the number of TaPERK genes, and the *x*-axis indicates the phylogenetic groups. Figure S3: Phylogenetic analysis of TaPERK genes. A phylogenetic tree was constructed using MEGAX with the neighbor-joining (NJ) method and 1000 bootstrap replications. A black asterisk indicates the duplicated genes. Figure S4: Chromosomal distribution and duplicated PERK gene pairs in wheat. Duplicated PERK gene pairs are connected with lines with distinct colors. The figure was generated using TB tools. Figure S5: Distribution of exon and introns in TaPERKs gene family. The *y*-axis indicates the number of exons and introns, and the *x*-axis indicates the TaPERK genes. Exons and introns are represented by purple and orange boxes, respectively. Figure S6: Alignment and 3-dimensional structure of the TaPERK protein sequences. A. The conserved protein tyrosine kinase domain is boxed with red color. Colored and shaded amino acids are chemically similar residues. Dashes indicate gaps introduced to maximize the alignment of the homologous region. B. Predicted 3D structures TaPERK proteins. Figure S7: Gene ontology term distribution TaPERK gene family predicted using AgriGO A. Biological Process. B. Cellular component. C. Molecular function. Figure S8: PCA plots displaying grouping of different (A) developmental stages (B) biotic and abiotic stress conditions based on the TaPERK expression pattern. DS: drought stress, HS: heat stress, Zt: *Zymoseptoria tritici*, PM: powdery mildew; SR: stripe rust, h: hour and d: days. Figure S9: Heatmaps representing the expression pattern of TaPERK genes in different stress conditions. TPM values were directly used to construct the heatmaps. DS: drought stress, HS: heat stress, Zt: *Zymoseptoria tritici*, PM: powdery mildew; SR: stripe rust, h: hour and d: days, Table S1: TaPERK genomic, CDS protein and promoter sequence. Table S2: PERK proteins from Arabidopsis, rice, soybean, and wheat used to generate a phylogenetic tree. Table S3: Ratio of Ka/Ks and distribution of duplicate wheat PERK genes. Table S4: Orthologous relationships of TaPERK genes with other PERK genes in *B. distachyon*, *Ae. tauschii*, *T. dicoccoides*, *O. sativa* and *A. thaliana*. Table S5: Domain organization of *TaPERK* genes predicted using Pfam with default parameters. Table S6: *cis*-regulatory elements present in the *TaPERK* gene promoter region. Table S7: Significant Go term predicted in TaPERK gene family by AgriGo analysis. Table S8: *TaPERK* gene annotation using eggNOGmapper. Table S9: The protein-protein interaction network between TaPERK and other proteins in wheat. Table S10: qRT-PCR primers of TaPERK genes.

Author Contributions: M.K. and M.S.K. designed and wrote the manuscript; M.K. acquired funding; M.K. and M.S.K. supervised the study; B.S.K., A.S., P.D., S.R., C.M., D.S., C.R., K.H.M.S., A.K., R.G. and S.-M.C. provided valuable feedback to this study. All authors have read and agreed to the published version of the manuscript.

Funding: This work was supported by the Dongguk University Research Fund of 2021.

Institutional Review Board Statement: Not applicable.

Informed Consent Statement: Not applicable.

Data Availability Statement: Data is available in the manuscript and in the Supplementary Materials.

Acknowledgments: M.K. would like to acknowledge the Department of Life Science, Dongguk University, for providing infrastructure to carry out experiments.

Conflicts of Interest: The authors declare no conflict of interest.

References

1. Day, E.K.; Sosale, N.G.; Lazzara, M.J. Cell signaling regulation by protein phosphorylation: A multivariate, heterogeneous, and context-dependent process. *Curr. Opin. Biotechnol.* **2016**, *40*, 185–192. [CrossRef] [PubMed]
2. Hohmann, U.; Lau, K.; Hothorn, M. The structural basis of ligand perception and signal activation by receptor kinases. *Annu. Rev. Plant Biol.* **2017**, *68*, 109–137. [CrossRef] [PubMed]
3. Sim, J.-S.; Kesawat, M.S.; Kumar, M.; Kim, S.-Y.; Mani, V.; Subramanian, P.; Park, S.; Lee, C.-M.; Kim, S.-R.; Hahn, B.-S. Lack of the α1, 3-fucosyltransferase gene (OsFucT) affects anther development and pollen viability in rice. *Int. J. Mol. Sci.* **2018**, *19*, 1225. [CrossRef] [PubMed]
4. Hanks, S.K.; Quinn, A.M.; Hunter, T. The protein kinase family: Conserved features and deduced phylogeny of the catalytic domains. *Science* **1988**, *241*, 42–52. [CrossRef]
5. DeFalco, T.A.; Anne, P.; James, S.R.; Willoughby, A.; Johanndrees, O.; Genolet, Y.; Pullen, A.-M.; Zipfel, C.; Hardtke, C.S.; Nimchuk, Z.L. A conserved regulatory module regulates receptor kinase signaling in immunity and development. *bioRxiv* **2021**. [CrossRef]
6. Nakhamchik, A.; Zhao, Z.; Provart, N.J.; Shiu, S.-H.; Keatley, S.K.; Cameron, R.K.; Goring, D.R. A comprehensive expression analysis of the Arabidopsis proline-rich extensin-like receptor kinase gene family using bioinformatic and experimental approaches. *Plant Cell Physiol.* **2004**, *45*, 1875–1881. [CrossRef]
7. Qanmber, G.; Liu, J.; Yu, D.; Liu, Z.; Lu, L.; Mo, H.; Ma, S.; Wang, Z.; Yang, Z. Genome-wide identification and characterization of the PERK gene family in Gossypium hirsutum reveals gene duplication and functional divergence. *Int. J. Mol. Sci.* **2019**, *20*, 1750. [CrossRef]
8. Diévart, A.; Clark, S.E. LRR-containing receptors regulating plant development and defense. *Development* **2004**, *131*, 251–261. [CrossRef]
9. Shiu, S.-H.; Bleecker, A.B. Receptor-like kinases from Arabidopsis form a monophyletic gene family related to animal receptor kinases. *Proc. Natl. Acad. Sci. USA* **2001**, *98*, 10763–10768. [CrossRef]
10. Morris, E.R.; Walker, J.C. Receptor-like protein kinases: The keys to response. *Curr. Opin. Plant Biol.* **2003**, *6*, 339–342. [CrossRef]
11. Shiu, S.-H.; Karlowski, W.M.; Pan, R.; Tzeng, Y.-H.; Mayer, K.F.; Li, W.-H. Comparative analysis of the receptor-like kinase family in Arabidopsis and rice. *Plant Cell* **2004**, *16*, 1220–1234. [CrossRef]
12. Silva, N.F.; Goring, D.R. The proline-rich, extensin-like receptor kinase-1 (PERK1) gene is rapidly induced by wounding. *Plant Mol. Biol.* **2002**, *50*, 667–685. [CrossRef]
13. Haffani, Y.Z.; Silva, N.F.; Goring, D.R. Receptor kinase signalling in plants. *Can. J. Bot.* **2004**, *82*, 1–15. [CrossRef]
14. Bai, L.; Zhang, G.; Zhou, Y.; Zhang, Z.; Wang, W.; Du, Y.; Wu, Z.; Song, C.P. Plasma membrane-associated proline-rich extensin-like receptor kinase 4, a novel regulator of Ca^{2+} signalling, is required for abscisic acid responses in Arabidopsis thaliana. *Plant J.* **2009**, *60*, 314–327. [CrossRef]
15. Chen, G.; Wang, J.; Wang, H.; Wang, C.; Tang, X.; Li, J.; Zhang, L.; Song, J.; Hou, J.; Yuan, L. Genome-wide analysis of proline-rich extension-like receptor protein kinase (PERK) in Brassica rapa and its association with the pollen development. *BMC Genom.* **2020**, *21*, 401. [CrossRef]
16. Borassi, C.; Sede, A.; Mecchia, M.A.; Mangano, S.; Marzol, E.; Denita-Juarez, S.P.; Salter, J.D.S.; Velasquez, S.M.; Muschietti, J.P.; Estevez, J. Proline-rich Extensin-like Receptor Kinases PERK5 and PERK12 are involved in Pollen Tube Growth. *bioRxiv* **2021**. [CrossRef]
17. Shiu, S.-H.; Bleecker, A.B. Expansion of the receptor-like kinase/Pelle gene family and receptor-like proteins in Arabidopsis. *Plant Physiol.* **2003**, *132*, 530–543. [CrossRef]
18. Li, J.; Chory, J. A putative leucine-rich repeat receptor kinase involved in brassinosteroid signal transduction. *Cell* **1997**, *90*, 929–938. [CrossRef]
19. Li, J.; Wen, J.; Lease, K.A.; Doke, J.T.; Tax, F.E.; Walker, J.C. BAK1, an Arabidopsis LRR receptor-like protein kinase, interacts with BRI1 and modulates brassinosteroid signaling. *Cell* **2002**, *110*, 213–222. [CrossRef]
20. Nam, K.H.; Li, J. BRI1/BAK1, a receptor kinase pair mediating brassinosteroid signaling. *Cell* **2002**, *110*, 203–212. [CrossRef]
21. Haffani, Y.; Silva-Gagliardi, N.; Sewter, S.; Aldea, M.G.; Zhao, Z.; Nakhamchik, A.; Cameron, R.; Goring, D. Altered expression of PERK receptor kinases in Arabidopsis leads to changes in growth and floral organ formation. *Plant Signal. Behav.* **2006**, *1*, 251–260. [CrossRef]
22. Won, S.-K.; Lee, Y.-J.; Lee, H.-Y.; Heo, Y.-K.; Cho, M.; Cho, H.-T. Cis-element-and transcriptome-based screening of root hair-specific genes and their functional characterization in Arabidopsis. *Plant Physiol.* **2009**, *150*, 1459–1473. [CrossRef]
23. Humphrey, T.V.; Haasen, K.E.; Aldea-Brydges, M.G.; Sun, H.; Zayed, Y.; Indriolo, E.; Goring, D.R. PERK–KIPK–KCBP signalling negatively regulates root growth in Arabidopsis thaliana. *J. Exp. Bot.* **2015**, *66*, 71–83. [CrossRef]
24. Hwang, I.; Kim, S.Y.; Kim, C.S.; Park, Y.; Tripathi, G.R.; Kim, S.-K.; Cheong, H. Over-expression of the IGI1 leading to altered shoot-branching development related to MAX pathway in Arabidopsis. *Plant Mol. Biol.* **2010**, *73*, 629–641. [CrossRef]
25. Borassi, C.; Sede, A.R.; Mecchia, M.A.; Salgado Salter, J.D.; Marzol, E.; Muschietti, J.P.; Estevez, J.M. An update on cell surface proteins containing extensin-motifs. *J. Exp. Bot.* **2016**, *67*, 477–487. [CrossRef]
26. Mansfield, T.; Hetherington, A.; Atkinson, C. Some current aspects of stomatal physiology. *Annu. Rev. Plant Biol.* **1990**, *41*, 55–75. [CrossRef]

27. Webb, A.A.; McAinsh, M.R.; Taylor, J.E.; Hetherington, A.M. Calcium ions as intracellular second messengers in higher plants. *Adv. Bot. Res.* **1996**, *22*, 45–96. [CrossRef]
28. Gong, M.; Chen, S.-N.; Song, Y.-Q.; Li, Z.-G. Effect of calcium and calmodulin on intrinsic heat tolerance in relation to antioxidant systems in maize seedlings. *Funct. Plant Biol.* **1997**, *24*, 371–379. [CrossRef]
29. Hwang, Y.; Lee, H.; Lee, Y.S.; Cho, H.T. Cell wall-associated ROOT HAIR SPECIFIC 10, a proline-rich receptor-like kinase, is a negative modulator of Arabidopsis root hair growth. *J. Exp. Bot.* **2016**, *67*, 2007–2022. [CrossRef]
30. Samuel, M.A.; Ellis, B.E. Double jeopardy: Both overexpression and suppression of a redox-activated plant mitogen-activated protein kinase render tobacco plants ozone sensitive. *Plant Cell* **2002**, *14*, 2059–2069. [CrossRef]
31. Xing, Y.; Cao, Q.Q.; Zhang, Q.; Qin, L.; Jia, W.S.; Zhang, J.H. MKK5 Regulates High Light-Induced Gene Expression of Cu/Zn Superoxide Dismutase 1 and 2 in Arabidopsis. *Plant Cell Physiol.* **2013**, *54*, 1217–1227. [CrossRef] [PubMed]
32. Florentino, L.H.; Santos, A.A.; Fontenelle, M.R.; Pinheiro, G.L.; Zerbini, F.M.; Baracat-Pereira, M.C.; Fontes, E.P. A PERK-like receptor kinase interacts with the geminivirus nuclear shuttle protein and potentiates viral infection. *J. Virol.* **2006**, *80*, 6648–6656. [CrossRef] [PubMed]
33. Singh, K.M.; Kumar, D.B.; Kumar, D.S. Manorama. Assessment of genetic diversity among Indian Sesame (*Sesamum indicum* L.) accessions using RAPD, ISSR and SSR markers. *Res. J. Biotechnol.* **2015**, *10*, 35–47.
34. Kumar, M.; Kesawat, M.S.; Ali, A.; Lee, S.-C.; Gill, S.S.; Kim, H.U. Integration of abscisic acid signaling with other signaling pathways in plant stress responses and development. *Plants* **2019**, *8*, 592. [CrossRef] [PubMed]
35. Kesawat, M.S.; Shivaraj, S.; Kim, D.K.; Kumar, M.; Hahn, B.S.; Deshmukh, R. Metalloids and Their Role in the Biological System. *Met. Plants Adv. Future Prospect.* **2020**, 1–17. [CrossRef]
36. The International Wheat Genome Sequencing Consortium. A chromosome-based draft sequence of the hexaploid bread wheat (Triticum aestivum) genome. *Science* **2014**, *345*, 1251788. [CrossRef]
37. Gill, B.S.; Appels, R.; Botha-Oberholster, A.-M.; Buell, C.R.; Bennetzen, J.L.; Chalhoub, B.; Chumley, F.; Dvorák, J.; Iwanaga, M.; Keller, B. A workshop report on wheat genome sequencing: International Genome Research on Wheat Consortium. *Genetics* **2004**, *168*, 1087–1096. [CrossRef]
38. Afzal, F.; Chaudhari, S.K.; Gul, A.; Farooq, A.; Ali, H.; Nisar, S.; Sarfraz, B.; Shehzadi, K.J.; Mujeeb-Kazi, A. Bread wheat (*Triticum aestivum* L.) under biotic and abiotic stresses: An overview. *Crop Prod. Glob. Environ. Issues* **2015**, 293–317. [CrossRef]
39. Kumar, P.; Yadava, R.; Gollen, B.; Kumar, S.; Verma, R.K.; Yadav, S. Nutritional contents and medicinal properties of wheat: A review. *Life Sci. Med. Res.* **2011**, *22*, 1–10.
40. Kumar, M.; Kherawat, B.S.; Dey, P.; Saha, D.; Singh, A.; Bhatia, S.K.; Ghodake, G.S.; Kadam, A.A.; Kim, H.-U.; Chung, S.-M. Genome-Wide Identification and Characterization of PIN-FORMED (PIN) Gene Family Reveals Role in Developmental and Various Stress Conditions in *Triticum aestivum* L. *Int. J. Mol. Sci.* **2021**, *22*, 7396. [CrossRef]
41. Kesawat, M.S.; Das, B.K.; Bhaganagare, G.R. Genome-wide identification, evolutionary and expression analyses of putative Fe–S biogenesis genes in rice (*Oryza sativa*). *Genome* **2012**, *55*, 571–583. [CrossRef]
42. Kesawat, M.S.; Kherawat, B.S.; Singh, A.; Dey, P.; Kabi, M.; Debnath, D.; Saha, D.; Khandual, A.; Rout, S.; Ali, A. Genome-wide identification and characterization of the brassinazole-resistant (BZR) gene family and its expression in the various developmental stage and stress conditions in wheat (*Triticum aestivum* L.). *Int. J. Mol. Sci.* **2021**, *22*, 8743. [CrossRef]
43. Appels, R.; Eversole, K.; Stein, N.; Feuillet, C.; Keller, B.; Rogers, J.; Pozniak, C.J.; Choulet, F.; Distelfeld, A.; Poland, J. Shifting the limits in wheat research and breeding using a fully annotated reference genome. *Science* **2018**, *361*. [CrossRef]
44. Lawton-Rauh, A. Evolutionary dynamics of duplicated genes in plants. *Mol. Phylogenet. Evol.* **2003**, *29*, 396–409. [CrossRef]
45. Hou, Y.; Gupta, N.; Schoenlein, P.; Wong, E.; Martindale, R.; Ganapathy, V.; Browning, D. An anti-tumor role for cGMP-dependent protein kinase. *Cancer Lett.* **2006**, *240*, 60–68. [CrossRef]
46. Gross, I.; Durner, J. In search of enzymes with a role in 3′, 5′-cyclic guanosine monophosphate metabolism in plants. *Front. Plant Sci.* **2016**, *7*, 576. [CrossRef]
47. Shen, Q.; Zhan, X.; Yang, P.; Li, J.; Chen, J.; Tang, B.; Wang, X.; Hong, Y. Dual activities of plant cGMP-dependent protein kinase and its roles in gibberellin signaling and salt stress. *Plant Cell* **2019**, *31*, 3073–3091. [CrossRef]
48. Wheeler, J.I.; Wong, A.; Marondedze, C.; Groen, A.J.; Kwezi, L.; Freihat, L.; Vyas, J.; Raji, M.A.; Irving, H.R.; Gehring, C. The brassinosteroid receptor BRI 1 can generate cGMP enabling cGMP-dependent downstream signaling. *Plant J.* **2017**, *91*, 590–600. [CrossRef]
49. Isner, J.-C.; Maathuis, F.J. cGMP signalling in plants: From enigma to main stream. *Funct. Plant Biol.* **2016**, *45*, 93–101. [CrossRef]
50. Lespinet, O.; Wolf, Y.I.; Koonin, E.V.; Aravind, L. The role of lineage-specific gene family expansion in the evolution of eukaryotes. *Genome Res.* **2002**, *12*, 1048–1059. [CrossRef]
51. Jordan, I.K.; Makarova, K.S.; Spouge, J.L.; Wolf, Y.I.; Koonin, E.V. Lineage-specific gene expansions in bacterial and archaeal genomes. *Genome Res.* **2001**, *11*, 555–565. [CrossRef]
52. Ramsey, J.; Schemske, D.W. Pathways, mechanisms, and rates of polyploid formation in flowering plants. *Annu. Rev. Ecol. Syst.* **1998**, *29*, 467–501. [CrossRef]
53. Moore, R.C.; Purugganan, M.D. The early stages of duplicate gene evolution. *Proc. Natl. Acad. Sci. USA* **2003**, *100*, 15682–15687. [CrossRef]
54. Cannon, S.B.; Mitra, A.; Baumgarten, A.; Young, N.D.; May, G. The roles of segmental and tandem gene duplication in the evolution of large gene families in Arabidopsis thaliana. *BMC Plant Biol.* **2004**, *4*, 10. [CrossRef]

55. Baumberger, N.; Doesseger, B.; Guyot, R.; Diet, A.; Parsons, R.L.; Clark, M.A.; Simmons, M.; Bedinger, P.; Goff, S.A.; Ringli, C. Whole-genome comparison of leucine-rich repeat extensins in Arabidopsis and rice. A conserved family of cell wall proteins form a vegetative and a reproductive clade. *Plant Physiol.* **2003**, *131*, 1313–1326. [CrossRef]
56. Wang, D.; Guo, Y.; Wu, C.; Yang, G.; Li, Y.; Zheng, C. Genome-wide analysis of CCCH zinc finger family in Arabidopsis and rice. *BMC Genom.* **2008**, *9*, 44. [CrossRef]
57. Yang, Z.; Gong, Q.; Qin, W.; Yang, Z.; Cheng, Y.; Lu, L.; Ge, X.; Zhang, C.; Wu, Z.; Li, F. Genome-wide analysis of WOX genes in upland cotton and their expression pattern under different stresses. *BMC Plant Biol.* **2017**, *17*, 113. [CrossRef]
58. Yang, Z.; Gong, Q.; Wang, L.; Jin, Y.; Xi, J.; Li, Z.; Qin, W.; Yang, Z.; Lu, L.; Chen, Q. Genome-wide study of YABBY genes in upland cotton and their expression patterns under different stresses. *Front. Genet.* **2018**, *9*, 33. [CrossRef]
59. Zhang, B.; Liu, J.; Yang, Z.E.; Chen, E.Y.; Zhang, C.J.; Zhang, X.Y.; Li, F.G. Genome-wide analysis of GRAS transcription factor gene family in *Gossypium hirsutum* L. *BMC Genom.* **2018**, *19*, 348. [CrossRef]
60. Yin, G.; Xu, H.; Xiao, S.; Qin, Y.; Li, Y.; Yan, Y.; Hu, Y. The large soybean (*Glycine max*) WRKY TF family expanded by segmental duplication events and subsequent divergent selection among subgroups. *BMC Plant Biol.* **2013**, *13*, 148. [CrossRef]
61. Ren, Z.; Yu, D.; Yang, Z.; Li, C.; Qanmber, G.; Li, Y.; Li, J.; Liu, Z.; Lu, L.; Wang, L. Genome-wide identification of the MIKC-type MADS-box gene family in *Gossypium hirsutum* L. unravels their roles in flowering. *Front. Plant Sci.* **2017**, *8*, 384. [CrossRef] [PubMed]
62. Dossa, K.; Diouf, D.; Cissé, N. Genome-wide investigation of Hsf genes in sesame reveals their segmental duplication expansion and their active role in drought stress response. *Front. Plant Sci.* **2016**, *7*, 1522. [CrossRef] [PubMed]
63. Roy, S.W.; Gilbert, W. The evolution of spliceosomal introns: Patterns, puzzles and progress. *Nat. Rev. Genet.* **2006**, *7*, 211–221. [PubMed]
64. Roy, S.W.; Penny, D. Patterns of intron loss and gain in plants: Intron loss–dominated evolution and genome-wide comparison of *O. sativa* and *A. thaliana*. *Mol. Biol. Evol.* **2007**, *24*, 171–181. [CrossRef]
65. Serrano, M.; Parra, S.; Alcaraz, L.D.; Guzmán, P. The ATL gene family from Arabidopsis thaliana and Oryza sativa comprises a large number of putative ubiquitin ligases of the RING-H2 type. *J. Mol. Evol.* **2006**, *62*, 434–445. [CrossRef]
66. Dan, Y.; Niu, Y.; Wang, C.; Yan, M.; Liao, W. Genome-wide identification and expression analysis of the trehalose-6-phosphate synthase (TPS) gene family in cucumber (*Cucumis sativus* L.). *PeerJ* **2021**, *9*, e11398. [CrossRef]
67. Lecharny, A.; Boudet, N.; Gy, I.; Aubourg, S.; Kreis, M. Introns in, introns out in plant gene families: A genomic approach of the dynamics of gene structure. *J. Struct. Funct. Genom.* **2003**, *3*, 111–116. [CrossRef]
68. Hernandez-Garcia, C.M.; Finer, J.J. Identification and validation of promoters and cis-acting regulatory elements. *Plant Sci.* **2014**, *217*, 109–119. [CrossRef]
69. Roy, A.L.; Sen, R.; Roeder, R.G. Enhancer–promoter communication and transcriptional regulation of Igh. *Trends Immunol.* **2011**, *32*, 532–539. [CrossRef]
70. Roy, A.L.; Singer, D.S. Core promoters in transcription: Old problem, new insights. *Trends Biochem. Sci.* **2015**, *40*, 165–171. [CrossRef]
71. Zhang, Y.; Wong, C.-H.; Birnbaum, R.Y.; Li, G.; Favaro, R.; Ngan, C.Y.; Lim, J.; Tai, E.; Poh, H.M.; Wong, E. Chromatin connectivity maps reveal dynamic promoter–enhancer long-range associations. *Nature* **2013**, *504*, 306–310. [CrossRef]
72. Cheng, Y.; Tang, Q.; Li, Y.; Zhang, Y.; Zhao, C.; Yan, J.; You, H. Folding/unfolding kinetics of G-quadruplexes upstream of the P1 promoter of the human BCL-2 oncogene. *J. Biol. Chem.* **2019**, *294*, 5890–5895. [CrossRef]
73. Fankhauser, C.; Chory, J. Light control of plant development. *Annu. Rev. Cell Dev. Biol.* **1997**, *13*, 203–229. [CrossRef]
74. Dievart, A.; Gottin, C.; Périn, C.; Ranwez, V.; Chantret, N. Origin and diversity of plant receptor-like kinases. *Annu. Rev. Plant Biol.* **2020**, *71*, 131–156. [CrossRef]
75. Li, S.; Liu, Z.; Chen, G.; Qanmber, G.; Lu, L.; Zhang, J.; Ma, S.; Yang, Z.; Li, F. Identification and analysis of GhEXO gene family indicated that GhEXO7_At promotes plant growth and development through brassinosteroid signaling in cotton (*Gossypium hirsutum* L.). *Front. Plant Sci.* **2021**, *12*, 719889. [CrossRef]
76. Šimura, J.; Antoniadi, I.; Široká, J.; Tarkowská, D.e.; Strnad, M.; Ljung, K.; Novák, O. Plant hormonomics: Multiple phytohormone profiling by targeted metabolomics. *Plant Physiol.* **2018**, *177*, 476–489. [CrossRef]
77. Clouse, S.D. Brassinosteroid signal transduction: From receptor kinase activation to transcriptional networks regulating plant development. *Plant Cell* **2011**, *23*, 1219–1230. [CrossRef]
78. Clouse, S.D.; Langford, M.; McMorris, T.C. A brassinosteroid-insensitive mutant in Arabidopsis thaliana exhibits multiple defects in growth and development. *Plant Physiol.* **1996**, *111*, 671–678. [CrossRef]
79. Kim, E.-J.; Russinova, E. Brassinosteroid signalling. *Curr. Biol.* **2020**, *30*, R294–R298. [CrossRef]
80. Hall, T.A. BioEdit: A User-Friendly Biological Sequence Alignment Editor and Analysis Program for Windows 95/98/NT. *Nucleic Acids Symp. Ser.* **1999**, *41*, 95–98.
81. Wang, Y.; Tang, H.; DeBarry, J.D.; Tan, X.; Li, J.; Wang, X.; Lee, T.-h.; Jin, H.; Marler, B.; Guo, H. MCScanX: A toolkit for detection and evolutionary analysis of gene synteny and collinearity. *Nucleic Acids Res.* **2012**, *40*, e49. [CrossRef]
82. Chen, C.; Chen, H.; Zhang, Y.; Thomas, H.R.; Frank, M.H.; He, Y.; Xia, R. TBtools: An integrative toolkit developed for interactive analyses of big biological data. *Mol. Plant* **2020**, *13*, 1194–1202. [CrossRef] [PubMed]
83. Gasteiger, E.; Gattiker, A.; Hoogland, C.; Ivanyi, I.; Appel, R.D.; Bairoch, A. ExPASy: The proteomics server for in-depth protein knowledge and analysis. *Nucleic Acids Res.* **2003**, *31*, 3784–3788. [CrossRef] [PubMed]

84. Kozlowski, L.P. IPC–isoelectric point calculator. *Biol. Direct* **2016**, *11*, 1–16. [CrossRef] [PubMed]
85. Yu, C.S.; Chen, Y.C.; Lu, C.H.; Hwang, J.K. Prediction of protein subcellular localization. *Proteins Struct. Funct. Bioinform.* **2006**, *64*, 643–651. [CrossRef] [PubMed]
86. Savojardo, C.; Martelli, P.L.; Fariselli, P.; Profiti, G.; Casadio, R. BUSCA: An integrative web server to predict subcellular localization of proteins. *Nucleic Acids Res.* **2018**, *46*, W459–W466. [CrossRef] [PubMed]
87. Kelley, L.A.; Mezulis, S.; Yates, C.M.; Wass, M.N.; Sternberg, M.J. The Phyre2 web portal for protein modeling, prediction and analysis. *Nat. Protoc.* **2015**, *10*, 845–858. [CrossRef]
88. Tian, T.; Liu, Y.; Yan, H.; You, Q.; Yi, X.; Du, Z.; Xu, W.; Su, Z. agriGO v2. 0: A GO analysis toolkit for the agricultural community, 2017 update. *Nucleic Acids Res.* **2017**, *45*, W122–W129. [CrossRef]
89. Metsalu, T.; Vilo, J. ClustVis: A web tool for visualizing clustering of multivariate data using Principal Component Analysis and heatmap. *Nucleic Acids Res.* **2015**, *43*, W566–W570. [CrossRef]
90. Grigorova, B.; Vaseva, I.; Demirevska, K.; Feller, U. Combined Drought and Heat Stress in Wheat: Changes in Some Heat Shock Proteins. *Biol. Plant.* **2011**, *55*, 105–111. [CrossRef]
91. Kim, D.K.; Kesawat, M.S.; Hong, C.B. One gene member of the ADP-ribosylation factor family is heat-inducible and enhances seed germination in Nicotiana tabacum. *Genes Genom.* **2017**, *39*, 1353–1365. [CrossRef]
92. Kesawat, M.S.; Kim, D.K.; Zeba, N.; Suh, M.C.; Xia, X.; Hong, C.B. Ectopic RING zinc finger gene from hot pepper induces totally different genes in lettuce and tobacco. *Mol. Breed.* **2018**, *38*, 1–24. [CrossRef]
93. Livak, K.J.; Schmittgen, T.D. Analysis of relative gene expression data using real-time quantitative PCR and the $2^{-\Delta\Delta CT}$ method. *Methods* **2001**, *25*, 402–408. [CrossRef]
94. Narancio, R.; John, U.; Mason, J.; Spangenberg, G. Selection of optimal reference genes for quantitative RT-PCR transcript abundance analysis in white clover (*Trifolium repens* L.). *Funct. Plant Biol.* **2018**, *45*, 737–744. [CrossRef]

plants

MDPI

Article

Unique N-Terminal Interactions Connect F-BOX STRESS INDUCED (FBS) Proteins to a WD40 Repeat-like Protein Pathway in Arabidopsis

Edgar Sepulveda-Garcia [1,2], Elena C. Fulton [3], Emily V. Parlan [3], Lily E. O'Connor [3], Anneke A. Fleming [3], Amy J. Replogle [3], Mario Rocha-Sosa [2], Joshua M. Gendron [4] and Bryan Thines [3,*]

1 Instituto de Biotecnología, Universidad del Papaloapan, Tuxtepec 68301, Mexico; esepulveda@unpa.edu.mx
2 Departamento de Biología Molecular de Plantas, Instituto de Biotecnología, Universidad Nacional Autónoma de México, Cuernavaca 62250, Mexico; rocha@ibt.unam.mx
3 Biology Department, University of Puget Sound, Tacoma, WA 98416, USA; efulton@alumni.pugetsound.edu (E.C.F.); eparlan@alumni.pugetsound.edu (E.V.P.); l.e.oconnor@wustl.edu (L.E.O.); afleming@alumni.pugetsound.edu (A.A.F.); areplogle@pugetsound.edu (A.J.R.)
4 Department of Molecular, Cellular and Developmental Biology, Yale University, New Haven, CT 06511, USA; joshua.gendron@yale.edu
* Correspondence: bthines@pugetsound.edu

Citation: Sepulveda-Garcia, E.; Fulton, E.C.; Parlan, E.V.; O'Connor, L.E.; Fleming, A.A.; Replogle, A.J.; Rocha-Sosa, M.; Gendron, J.M.; Thines, B. Unique N-Terminal Interactions Connect F-BOX STRESS INDUCED (FBS) Proteins to a WD40 Repeat-like Protein Pathway in Arabidopsis. *Plants* **2021**, *10*, 2228. https://doi.org/10.3390/plants10102228

Academic Editors: Ewa Muszyńska, Kinga Dziurka and Mateusz Labudda

Received: 16 September 2021
Accepted: 11 October 2021
Published: 19 October 2021

Abstract: SCF-type E3 ubiquitin ligases provide specificity to numerous selective protein degradation events in plants, including those that enable survival under environmental stress. SCF complexes use F-box (FBX) proteins as interchangeable substrate adaptors to recruit protein targets for ubiquitylation. FBX proteins almost universally have structure with two domains: A conserved N-terminal F-box domain interacts with a SKP protein and connects the FBX protein to the core SCF complex, while a C-terminal domain interacts with the protein target and facilitates recruitment. The F-BOX STRESS INDUCED (FBS) subfamily of plant FBX proteins has an atypical structure, however, with a centrally located F-box domain and additional conserved regions at both the N- and C-termini. FBS proteins have been linked to environmental stress networks, but no ubiquitylation target(s) or biological function has been established for this subfamily. We have identified two WD40 repeat-like proteins in Arabidopsis that are highly conserved in plants and interact with FBS proteins, which we have named FBS INTERACTING PROTEINs (FBIPs). FBIPs interact exclusively with the N-terminus of FBS proteins, and this interaction occurs in the nucleus. FBS1 destabilizes FBIP1, consistent with FBIPs being ubiquitylation targets SCF^{FBS1} complexes. This work indicates that FBS proteins may function in stress-responsive nuclear events, and it identifies two WD40 repeat-like proteins as new tools with which to probe how an atypical SCF complex, SCF^{FBS}, functions via FBX protein N-terminal interaction events.

Keywords: F-box protein; SCF complex; stress response; WD40 repeat-like protein

1. Introduction

At the onset of environmental stress, the ubiquitin 26S proteasome system (UPS) selectively degrades key cellular proteins to initiate plant responses that promote resilience and survival. Protein targets destined for removal are ubiquitylation substrates for E3 ubiquitin ligases, where one prevalent E3 ligase subtype is the SKP1-CUL1-F-box (SCF) complex [1]. SCF complexes use an interchangeable F-box (FBX) protein subunit as a substrate adaptor to specifically interact with unique protein targets [2–5]. FBX proteins almost universally have a structure with two domains: An N-terminal F-box domain facilitates interaction with a SKP protein and the core SCF complex, and a C-terminal domain interacts specifically with the target(s) [2]. This two-domain structure directly bridges core UPS components to precise protein targets under specific conditions, and it places FBX proteins at a dynamic interface that regulates diverse cellular pathways critical for plant life.

A very small number of FBX proteins, however, deviate from this typical two-domain protein structure. Many of these atypical FBX proteins have a centrally located F-box domain, a C-terminal target interaction domain, and an additional protein interaction domain at the N-terminus [6–8]. In humans, N-terminal domains can control subcellular localization [9], bind to an accessory protein that assists with C-terminal targeting events [10] or mediate regulatory interactions with other proteins [6,11,12]. The only plant FBX proteins with established N-terminal interaction dynamics belong to the ZEITLUPE (ZTL), FLAVIN-BINDING KELCH REPEAT F-BOX1 (FKF1), and LOV KELCH PROTEIN2 (LKP2) subfamily, which regulate the circadian clock and flowering time [8,13–16]. In addition to a central F-box domain, the ZTL/FKF1/LKP2 subfamily has a N-terminal blue-light sensing LOV domain and C-terminal kelch repeats [16], which are both used to recruit distinct ubiquitylation substrates [8,15,17,18]. The N-terminal LOV domain has additional roles that regulate the FBX function through an interaction with GIGANTEA (GI), which controls subcellular localization and protein stability [13,14]. Therefore, across kingdoms, a few atypical FBX proteins with a N-terminal protein interaction domain, in addition to a C-terminal targeting domain, achieve an expanded function by having further regulatory capacity and/or coordinating multiple cellular outputs through dual targeting.

F-BOX STRESS INDUCED (FBS) proteins constitute a far less understood subfamily of plant FBX proteins with an atypical structure [19–21]. Arabidopsis FBS1 is the founding member of this subfamily and is noteworthy for its broad biotic and abiotic stress-inducible gene expression profiles [19,21]. In FBS1, a centrally located F-box domain is flanked by two conserved regions present at the N- and C-termini, which do not match any known protein interaction domains or motifs [19]. FBS1 interacts with Arabidopsis SKP1 (ASK1) and can auto-ubiquitylate [19,20], suggesting that it forms a functional SCF-type E3 ligase in vivo. At least five of 13 Arabidopsis 14-3-3 regulatory proteins bind to FBS1 [20]. However, since this interaction requires both the N-terminal region and the F-box domain of FBS1 [20], and ubiquitylation presumably requires an unhindered F-box domain to interact with the SKP subunit of the SCF complex [1], the 14-3-3 proteins are unlikely ubiquitylation targets. Furthermore, an inducible *FBS1* gene construct had no discernable effect on FBS1 interactor 14-3-3λ protein abundance [20]. Importantly though, FBS1-interacting 14-3-3 proteins are negative regulators of Arabidopsis responses to cold and salt stress [22–26], which demonstrates another important link between FBS1 and environmental stress response networks in plant cells.

A more complete understanding of the FBS family protein function in plants has been stymied by two primary limitations. First, not knowing selective targeting relationships between SCF^{FBS} complexes and their putative substrates has left FBS action on cellular output pathways completely enigmatic. Second, functional redundancy within this family has likely thwarted past efforts seeking to establish a biological function. Arabidopsis *fbs1* plants have no obvious phenotype [19,21], however, three additional FBS family members that may be functionally redundant are encoded in the genome. Here, we identify two highly conserved WD40 repeat-like proteins that interact with multiple FBS family members in Arabidopsis, which we have named FBS INTERACTING PROTEINs (FBIPs). Interactions between all four FBS subfamily members and FBIP proteins occur in the nucleus, and interactions occur exclusively via the N-terminal domain of FBS proteins. These findings connect a stress network involving FBS proteins to nuclear processes, and they provide new tools with which to probe unique N-terminal interactions in FBX proteins in the context of plant stress responses.

2. Results

2.1. FBS Protein Interaction with ASK1

FBS1 is the founding member of a four-member FBX protein subfamily (FBS1–FBS4) in Arabidopsis. FBS2–FBS4, similar to FBS1, share a non-canonical structure with a centrally located F-box domain and conserved regions at their N- and C-termini (Figure 1A). The conserved region at the N-termini of FBS proteins spans approximately 20 residues, while the conserved region at the C-terminus encompasses about 35 residues (Figure 1A). FBS1 interacts with ASK1 and auto-ubiquitylates, indicating that FBS1 likely participates in functional SCF complexes [19,20]. However, the ability of other FBS family members to interact with ASK proteins remains unknown, as does the possibility of functional redundancy among family members. To interrogate this possibility, all four FBS family members were tested as bait constructs (DBD, GAL4 DNA-binding domain) for the interaction with ASK1 as prey (AD, GAL4 activation domain) under less stringent (-TLH) and more stringent (-TLHA) nutritional selection. Interactions were apparent between all four FBS family members on -TLH, although only very minimal growth was observed for FBS2 (Figure 1B). Only the interactions between FBS1 and FBS4 with ASK1 were apparent under the most stringent selection (-TLHA) (Figure 1B). Since Arabidopsis has 21 ASK proteins, it is possible that the FBS proteins showing minimal partnering with ASK1 interact more strongly with the other untested ASKs [27]. These interactions show, however, that FBS2–FBS4 are viable candidates for functional SCF complex substrate adapters, similar to FBS1.

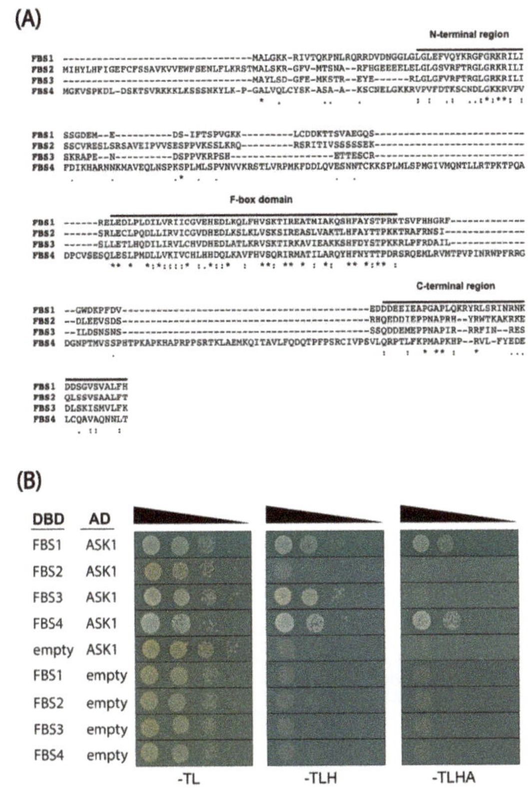

Figure 1. The Arabidopsis F-BOX STRESS INDUCED (FBS) protein family. (**A**) Full-length protein sequence alignment of the four Arabidopsis FBS family members (FBS1–FBS4) created with the T-COFFEE sequence alignment program. Asterisks are fully conserved residues, colons are strongly conserved residue properties, and periods are weakly conserved residue properties. (**B**) FBS family interactions with ASK1 in yeast two-hybrid assays. Diploid yeast strains with indicated test constructs as bait (DBD) and prey (AD) were grown in liquid culture, diluted (OD_{600} = 10^0, 10^{-1}, 10^{-2}, 10^{-3}), and spotted on SD medium minus Trp/Leu (-TL), minus Trp/Leu/His (-TLH), and minus Trp/Leu/His/Ade (-TLHA).

2.2. Identification of a New FBS1 Interactor

In addition to ASK1, the only established FBS1 interacting proteins belong to the 14-3-3 family [20]. However, since the interaction dynamics are not consistent with ubiquitylation of 14-3-3 proteins by SCF^{FBS1} [20], we sought additional FBS1 interactors as candidate targets that could connect FBS proteins to biological processes. Two additional related proteins were identified as partners for FBS1, which we have named FBS INTERACTING PROTEINs (FBIPs). FBIP1 (At3g54190) was identified in the same yeast two-hybrid screen that found 14-3-3 proteins as FBS1 interactors [20]. FBIP1 is also listed as an FBS1 interactor by the SUBA4 database (http://suba.live/, accessed on 16 September 2021) from high-throughput protein-protein interaction (PPI) screening [28,29]. FBIP1 is 467 residues in length and is a member of the transducin/WD40 repeat-like superfamily of proteins. WD40 repeats typically form a β-propeller domain that acts as a scaffold in mediating protein-protein or protein-DNA interactions [30]. Seven putative WD40 repeat-like sequences were predicted in FBIP1 by the WD40-repeat protein Structures Predictor database version 2.0 (WDSPdb 2.0) [31], although these predictions fall into the low confidence category (Figure 2). A second protein highly similar to FBIP1 was identified in the Arabidopsis genome by BLAST search, which we have named FBIP2 (At2g38630). The protein sequence identity and similarity between FBIP1 and FBIP2 are just over 91% and 96%, respectively (Figure 2).

```
FBIP1   -MEGRRITASPRPCSG-RRIVAKKRSRPDGFVNSVKKLQRREISSRKDRA
FBIP2   MMEGRRIIANPRPCSGSRRVIAKKRSRPDGFVNSVKKLQRREISSRMDRA
         ****** *.****** **::************************* ***

FBIP1   FSISTAQERFRNMRLVEQYDTHDPKGHCLVALPFLMKRTKVIEIVAARDI
FBIP2   FSISTAQERFRNMRLVEQYDTHDPKGYCLVSLPNLLKRSKVIEIVAARDI
        **************************:***:** *:**:***********

FBIP1   VFALAHSGVCAAFSRESNKRICFLNVSPDEVIRSLFYNKNNDSLITVSVY
FBIP2   VFALTLSGVCASFSRETNKKVCFLNVSPDEVIRSLFYNKNNDSLITVSVY
        ****: *****:****:**::*****************************

FBIP1   ASDNFSSLKCRSTRIEYILRGQPDAGFALFESESLKWPGFVEFDDVNGKV
FBIP2   ASDNYSSLKCRSTRIEYILRGQADAGFPLFESESLKWPGFVEFDDVNGKV
        ****:*****************.****.**********************

FBIP1   LTYSAQDSVYKVFDLKNYTMLYSISDKNVQEIKISPGIMLLIFKRAASHV
FBIP2   LTYSAQDSVYKVFDLKNYALLYSISDKNVQEIKISPGIMLLIFKRAASHV
        ******************::******************************

FBIP1   PLKILSIEDGTVLKSFNHLLHRNKKVDFIEQFNEKLLVKQENENLQILDV
FBIP2   PLKILSIEDGTLLKSFHHLLHRNKKVDFIEQFNEKLLVKQENENLQILDV
        ***********:****:*********************************

FBIP1   RNAELMEVSRAEFMTPSAFIFLYENQLFLTFRNRNVSVWNFRGELVTSFE
FBIP2   RNAELIEVSRTDFMTPSAFIFLYENQLFLTFRNRNVSVWNFRGELVTSFE
        *****:****::**************************************

FBIP1   DHLLWHPDCNTNNIYITSDQDLIISYCKADTEDQWIEGNAGSINISNILT
FBIP2   DHLLWHPDCNTNNIYITSDQDLIISYCKADTEDQWIEGNAGSINISNILT
        **************************************************

FBIP1   GKCLAKITPSSGPPKDDESSSSNCMGKNSKQRRNAVAEALEDITALFYDE
FBIP2   GKCLAKIKANNGPPKEEDCSSSDL-G-NSSRRRSAVAEALEDITALFYDE
        *******....****:::.***:   * **.:**.***************

FBIP1   ERNEIYTGNRHGLVHVWSN
FBIP2   ERNEIYTGNRHGLLHVWSN
        *************:*****
```

Figure 2. FBS INTERACTING PROTEIN (FBIP) sequence features. Full-length protein sequence alignment of the two Arabidopsis FBIP family members created with the T-COFFEE sequence alignment program. Blue indicates locations of seven WD40-like repeat sequences predicted by the WD40-repeat protein Structure Predictor version 2.0 (WDSPdb 2.0). Asterisks are fully conserved residues, colons are strongly conserved residue properties, and periods are weakly conserved residue properties.

We gained no additional insight on the FBIP function using various bioinformatics resources. Other than putative WD repeat-like sequences, no sequence features were identified using various domain or motif prediction programs. BLAST and PSI-BLAST searches with FBIP1 and FBIP2 sequences failed to identify additional significant hits in Arabidopsis. We did, however, find very highly conserved FBIP protein sequences throughout the plant kingdom, including in bryophytes (the top BLAST hit in *Physcomitrella patens* is about 77% identical and 85% similar to *Arabidopsis* FBIP1). By investigating AtGenExpress ATH1 array datasets [32–34], we found that *FBIP1* is constitutively expressed in most tissues and organs of Arabidopsis, and throughout its life cycle, but we found no conditions where *FBIP1* is more highly expressed compared to the other conditions. *FBIP2* is not represented on the ATH1 array.

2.3. FBS Interactions with FBIPs

We confirmed that the full-length FBS1 and FBIP1 interact with yeast two-hybrid analysis. The interaction between FBS1 and FBIP1 elicited growth in yeast strains on both less stringent (-TLH) and more stringent (-TLHA) nutritional selection, and FBS1 yielded growth with FBIP2 on -TLH (Figure 3A). Family-wide interactions between each FBS protein and the two FBIP proteins were also assessed (Figure S1). Growth was observed for FBS3 and FBIP1, but not with FBS2 or FBS4. No additional interactions were observed with FBIP2. Collectively, the yeast two-hybrid results suggest that FBS1 and FBIP1 might be the primary FBS/FBIP protein interaction pair or possibly bind with the strongest affinity, but that some other family-wide interactions might be possible.

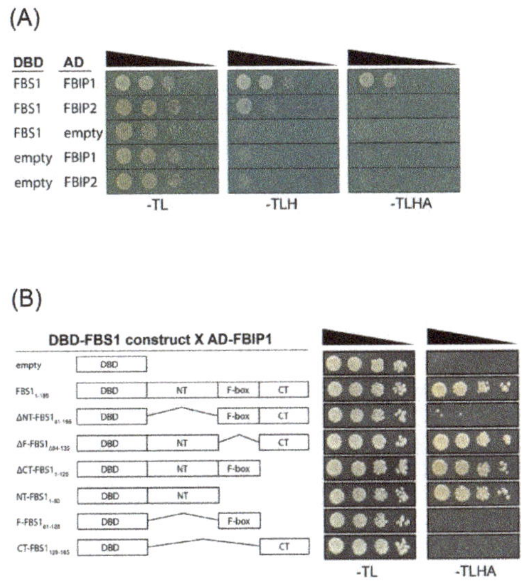

Figure 3. Yeast two-hybrid (Y2H) interactions between FBS1 and FBIP proteins. (**A**) Full-length FBS1 interactions with full-length FBIP1 and FBIP2. Diploid yeast strains with indicated test constructs as bait (DBD) and prey (AD) were grown in liquid culture, diluted (OD_{600} = 10^0, 10^{-1}, 10^{-2}, 10^{-3}), and spotted on SD medium minus Trp/Leu (-TL), minus Trp/Leu/His (-TLH), and minus Trp/Leu/His/Ade (-TLHA). (**B**) Truncated FBS1 bait (DBD) construct interaction with full length FBIP1 prey (AD). Amino acid deletions are indicated on the left.

FBS proteins have two regions of unknown function outside of the F-box domain and, presumably, at least one of these interacts with a target. In order to determine which parts of FBS1 are important for the FBIP1 interaction, we created truncated versions of FBS1 with the N-terminal (NT), F-box or C-terminal (CT) regions removed in different combinations and tested under stringent (-TLHA) selection (Figure 3B). Removing the N-terminal region (ΔNT-$FBS1_{81–185}$) abolished the ability of FBS1 to interact with FBIP1, while removal of the F-box domain (ΔF-$FBS1_{\Delta 84–135}$) or C-terminal region (ΔCT-$FBS1_{1–128}$) did not. The FBS1 N-terminal region (NT-$FBS1_{1–80}$) in combination with the full-length FBIP1 yielded growth on -TLHA, indicating that the FBS1 N-terminal domain alone is sufficient to mediate this interaction.

Near the conserved N-terminal domains of FBS1 and FBS2 we found a LXLXL sequence (Figure 1A), which is the most prominent form of an EAR motif found in many different types of transcriptional regulators [35,36]. The EAR motif mediates the interaction with the WD40 repeat-containing protein TOPLESS (TPL) and TOPLESS RELATED (TPR) co-repressor proteins [37–39]. We considered whether this LXLXL sequence in the N-terminal region of FBS1 might: (1) Function as a canonical EAR motif to interact with TOPLESS, and/or (2) if it could be important for mediating interactions with FBIPs. However, substituting all three leucine residues for alanine in FBS1 did not alter its interaction with FBIP1, and FBS1 did not interact with TPL (both as bait or as prey) in our yeast two-hybrid system.

2.4. FBS Interactions with FBIP Occur in the Nucleus

Next, we used bimolecular fluorescence complementation (BiFC) to test the FBS interaction with FBIP in plants and determine where the interaction occurs in a cell. The FBS and FBIP family proteins were expressed in *Nicotiana benthamiana* leaves as C-terminal fusions to either N-terminal (nYFP) or C-terminal (cYFP) halves of yellow fluorescent protein (YFP). In multiple independent experiments, the YFP fluorescence was observed for pairings between FBS1 and FBIP1 and FBIP2 (Figure 4). This YFP signal co-localized with that of a co-infiltrated H2B-RFP construct, which localizes exclusively in the nucleus [40], and shows that interactions between FBS1 and FBIP proteins also occur in the nucleus. Similar experiments found that FBS2–FBS4 also interact with FBIP1 in the nucleus (Figure S2). We observed interactions for FBS3 and FBS4 with FBIP2 (Figure S3), although we note that these interactions were more variable in the number of YFP positive nuclei across independent replicates, and with consistently fewer interactions for FBS3 and FBIP2. We did not observe any interactions between FBS2 and FBIP2. All FBS and FBIP fusion protein constructs were tested as pairs with empty nYFP or cYFP vectors, and in all pairings we were unable to detect any fluorescent signal similar to the FBS/FBIP test pairs (Figure S4). These findings show that in plants the FBS proteins participate in family-wide interactions in the nucleus.

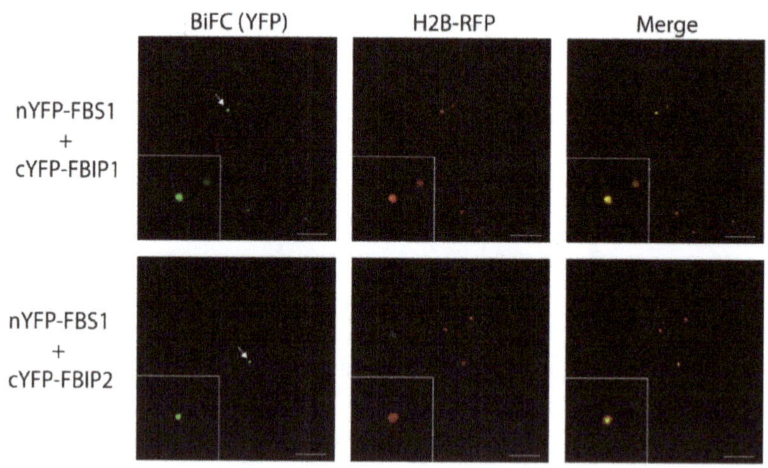

Figure 4. Bimolecular fluorescence complementation (BiFC) interactions between FBS1 and FBIP proteins. Laser-scanning confocal microscopy of *N. benthamiana* epidermal cells expressing N-terminal nYFP- or cYFP-tagged FBS1 and FBIP proteins. FBS1 interactions with FBIP1 (top row) or FBIP2 (bottom row) are visualized on the BiFC yellow channel (YFP, left column). A co-expressed H2B-RFP (as nuclear marker) is visualized on the red channel (RFP, middle column) and YFP/RFP images are overlaid (Merge, right column). Arrow indicates selected nuclei in the expanded inset image. Scale bar = 100 µm.

2.5. FBS1 Destabilizes FBIP1

With the interaction established between multiple FBS and FBIP protein pairs, we next asked if the protein abundance relationship between FBS1 and FBIP1 is consistent with FBIP1 being a ubiquitylation target of SCF^{FBS1}. If a protein is ubiquitylated by a particular SCF complex and subsequently degraded by the 26S proteasome, then increasing the abundance of the F-box component typically increases in vivo targeting and decreases substrate abundance [41]. Therefore, we tested the effects of varying FBS1 protein levels on the FBIP1 abundance in our *N. benthamiana* expression system by co-infiltrating *Agrobacterium* harboring these test constructs in different relative concentrations. Increasing the presence of FBS1 protein resulted in a corresponding decrease in the FBIP1 protein abundance by Western blot analysis (Figure 5). In comparison, when the FBS1 abundance was increased relative to the co-infiltrated 14-3-3λ in an identical setup, we did not observe any decrease in 14-3-3λ abundance as the amount of expressed FBS1 was increased (Figure S5). This finding is congruous with previous observations that FBS1 and 14-3-3 interactions are not consistent with targeting [20]. Therefore, since the abundance of FBIP1 decreases in an FBS1-dependent manner, we conclude that FBIPs are viable candidates for SCF^{FBS1} ubiquitylation targets.

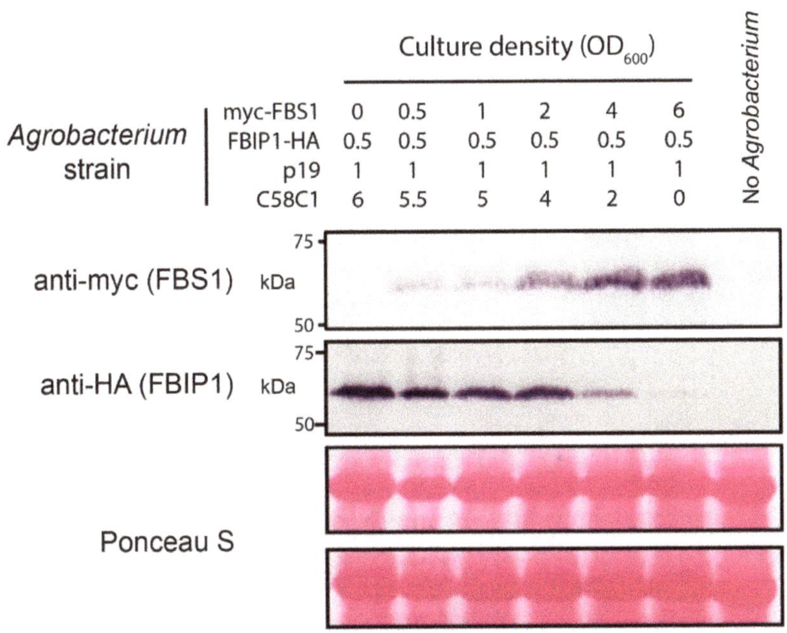

Figure 5. FBS1 influence on FBIP1 protein abundance in plants. *N. benthamiana* leaves were infiltrated with *Agrobacterium* (C58C1) strains to express the tagged proteins. *Agrobacterium* mixes contained varying cell densities of strains harboring expression constructs (myc-FBS1 and/or FBIP1-HA), a suppressor protein (p19) or untransformed cells. Total protein was isolated from leaves 3 days after infiltration, separated by SDS-PAGE, transferred, and probed with antibodies against myc (top row, FBS1) or HA (second row, FBIP1). Bottom two rows show Ponceau S staining of the major subunit of Rubisco from the same two blots as a loading control.

3. Discussion

As substrate adapters for SCF-type E3 ligases, FBX proteins act at the interface between core UPS components and specific cellular outputs, including those that help plant cells mitigate the effects of environmental stress. Previous work with FBS1 strongly alluded to some role in plant stress responses, possibly by regulating the expression of stress genes [19–21], but a more detailed understanding was limited by the unknown identity of ubiquitylation target(s) and by possible redundancy within the *FBS* gene family. Here, we have identified a pair of WD40 repeat-like superfamily proteins, FBIP1 and FBIP2, that both interact with FBS family proteins. These family-wide interactions indicate that functional redundancy within these two families is likely, but at the same time suggest a more robust stress response module. The FBS protein interaction with FBIPs in the nucleus points to a role for these proteins in the regulation of gene expression and/or other chromosomal events. Finally, FBIP proteins are strong candidates for SCF^{FBS} ubiquitylation targets that act in plant stress responses, and they provide new tools with which to investigate unique FBX protein N-terminal events in plants.

The exclusive nuclear localization FBS and FBIP protein interactions under the conditions we tested offer a critical clue as to the molecular functions of both protein families. One hypothesis for the FBIP function stemming from this result is that they regulate gene expression, which is an idea supported by the finding that hundreds of JA/ABA and other stress genes are mis-expressed in the *fbs1-1* background [21]. Some plant nuclear localized WD40 repeat proteins have direct actions in transcription regulation [39,42,43] or chromatin modification [44–46], and in these cases the WD40 repeat proteins are essential components of multi-protein assemblies. For example, TOPLESS (TPL) is a well-studied WD40 repeat-containing co-repressor protein that acts in diverse developmental and environmental-response pathways [39]. TPL interacts with different DNA-bound transcriptional complexes and it recruits chromatin modifying enzymes and/or Mediator to repress gene expression [40,47,48]. TRANSPARENT TESTA GLABRA 1 (TTG1), another WD40 repeat protein, serves as a scaffold and mediates different combinations of bHLH and R2R3-type MYB DNA-binding transcription factors to regulate flavonoid metabolism and various developmental processes [43,49]. The FBIP proteins may function similarly to TPL or TTG1 and act as scaffolds and/or in recruitment roles for complexes that regulate transcription. Knowing additional FBIP interactors, which may include more recognizable proteins with readily inferred functions, will help address this hypothesis.

Future work will also be guided by questions that address interaction dynamics between FBIPs and the N-terminal region of FBS proteins, and the consequences of these associations. There are 13 residue positions in the FBS N-terminal region, ranging from moderately to absolutely conserved, that could be critical for the interaction with FBIPs. Future work will include identification of the exact residue or residues in FBS proteins mediating this interaction. Given numerous FBS connections to stress, but that *FBIP1* appears to be constitutively expressed across different plant organs and environmental conditions, it could be the case that FBIP proteins are components of a stress-response system working at the post-translational level. Next steps include a rigorous assessment of conditions under which SCF^{FBS} complexes form and interact with FBIP proteins in vivo. Furthermore, whether some additional factor (i.e., post-translational modification) stimulates SCF^{FBS} association with FBIP proteins, as in the case of some other SCF targeting events [50], is well worth investigating. The idea that additional in vivo factors or modification mediates the FBS/FBIP interaction is consistent with the finding that we observed more family-wide interactions in plant BiFC experiments compared to yeast two-hybrid. Knowing that SCF complexes in some atypical contexts ubiquitylate targets via the FBX protein N-terminal interactions [8,17], and that FBS1 appears to destabilize FBIP1, a leading hypothesis for future work is that FBIP proteins are bona fide ubiquitylation substrates for SCF^{FBS}. Considering our work here and the general knowledge surrounding SCF action, our current model is that stress stimulates increased SCF^{FBS}-dependent ubiquitylation of FBIP proteins, which are then degraded in response to this environmental trigger, resulting in cellular changes.

The atypical structure of FBS proteins, along with the identification of FBIPs as FBS N-terminal interactors, leads to a few intriguing hypotheses regarding how this SCF complex may impact cellular pathways in plant stress. If FBIP is a bona fide target with a biological function distinct from a more typical C-terminal target, then SCF^{FBS} complexes provide an exciting opportunity to study how plants coordinate more than one cellular pathway related to stress. N- and C-terminal targeting events might be simultaneous under a given condition, and in this situation SCF^{FBS} may integrate a response by ubiquitylating two distinct protein types, each interacting with a different region of the FBS substrate adapter. Alternatively, N- and C-terminal targeting may be asynchronous and condition dependent, in which case SCF^{FBS} may entail a switch that works in or leads to two different cellular states. At this point, however, we cannot completely exclude the possibility that FBIPs are not targets (see above), but instead serve in an alternative capacity that enables or inhibits the FBS action. One idea then is that FBIPs are accessories that help recruit other proteins as ubiquitylation targets. In humans, Cks1 directly associates with the N-terminus of FBX protein Skp2 to direct SCF^{Skp2} interaction with ubiquitylation target p27 in human cell cycle regulation [10,50]. In Arabidopsis, KAI2 and D14 interact with FBX protein MAX2 in SCF^{MAX2} complex to mediate ubiquitylation of SMXL transcription factors [51], though in these cases KAI2 and D14 are also FBX C-terminal interactors. To address these scenarios or others a critical piece of information to learn is the identity of a FBS C-terminal region-interacting protein that we presume to exist. Future work can then investigate higher order SCF^{FBS} complex assembly and action.

The 14-3-3 proteins directly regulate a range of cellular processes in plant cells [52], including core signaling pathways and transcriptional reprogramming events in cold and salt stress responses [22–26]. FBS protein interactions with FBIPs will almost certainly be a vital tool used to fully understand the connections between FBS proteins and the 14-3-3 protein regulatory network. Five of 13 Arabidopsis 14-3-3 proteins interact with FBS1. However, 14-3-3 proteins are unlikely ubiquitylation targets of SCF^{FBS1} and the consequences of these interactions are unknown [20]. One hypothesis regarding this interaction is that 14-3-3 proteins promote dimerization of SCF^{FBS} ligases [20], which in other situations enhances ubiquitylation targeting by SCF complexes [53,54]. As our understanding further develops regarding FBIPs as putative targets, their cellular abundance will be an essential readout in studies that investigate 14-3-3 effects on SCF^{FBS} activity. The 14-3-3 proteins exert regulatory effects through other mechanisms, however, through controlling the subcellular localization of client proteins or by shifting the location themselves [52]. In salt stress, FBS1 interactors 14-3-3λ and 14-3-3κ act at the plasma membrane and release signaling component SOS2 to activate salt stress tolerance [26,55]. Cold temperature triggers the FBS1 interactor 14-3-3λ to translocate from the cytosol into the nucleus where it interacts with and adjusts cold-responsive C-repeat-binding factor (CBF) action [25]. Considering that the FBS1 interaction with FBIPs was exclusively nuclear under the conditions tested here, an investigation of temporal and spatial aspects of 14-3-3/FBS interactions relative to FBS/FBIP interactions in plant cells before and during environmental stress will add more broadly to our understanding of the 14-3-3 stress response network in plant cells.

4. Materials and Methods

Bioinformatics: Gene and protein sequences were obtained from The Arabidopsis Information Resource (http://www.arabidopsis.org, accessed on 16 September 2021). Protein sequences were aligned using T-COFFEE (http://www.ebi.ac.uk/Tools/msa/tcoffee, accessed on 16 September 2021) accessed through the European Bioinformatics Institute (EBI) website (http://www.ebi.ac.uk, accessed on 16 September 2021) [56]. WD40 repeat-like sequences were identified in FBIP1 and FBIP2 using the WD40-repeat protein Structures Predictor database version 2.0 (WDSPdb 2.0; http://www.wdspdb.com/wdsp/, accessed on 16 September 2021) [31]. Basic Local Alignment Search Tool (BLAST) and Position-Specific Iterative (PSI)-BLAST were accessed through the National Center for Biotechnology Information (NCBI) website (http://www.ncbi.nlm.nih.gov, accessed on 16 September 2021) and used to search the RefSeq database. Candidate protein interactors were identified by searching the SUBA4 database (http://suba.live/, accessed on 16 September 2021) [29].

Gateway cloning: Gene-specific primers (Supplementary Table S1) were used with PCR to amplify coding sequences from pooled *Arabidopsis thaliana* (accession Col-0) cDNA. Amplicons were inserted into the pENTR/D-TOPO vector (Thermo Fisher Scientific,

Waltham, MA, USA) according to the manufacturer's protocols. Then, the genes were transferred with the LR Clonase II enzyme mix (Thermo Fisher Scientific, Waltham, MA, USA) into pCL112 or pCL113 [57] destination vectors for BiFC experiments, and into pGBKT7-GW (Addgene plasmid #61703) or pGADT7-GW (Addgene plasmid #61702) destination vectors for yeast two-hybrid experiments. Alternatively (Figure 3B), *FBS1* and *FBIP1* sequences were cloned into pBI770/pBI771 and tested for the interaction, as done previously [20]. Primers used to create *FBS1* truncation constructs are indicated in Supplementary Table S1.

Yeast two-hybrid assays: *Saccharomyces cerevisiae* cells were grown, transformed, mated, and selected by standard yeast protocols. Bait constructs (GAL4 DNA-binding domain, DBD) were transformed into Y2H Gold and prey constructs (GAL4 activation domain, AD) and Y187 strains by the LiAc method (Takara Bio; San Jose, CA, USA). Haploid strains were mated to produce diploid strains to test for the interactions. Diploid strains were grown for 24 h at 30 °C in the liquid synthetic defined (SD) medium minus Trp/Leu (-TL) medium with shaking. Thereafter, cells were washed in sterile water, cell concentrations were adjusted to $OD_{600} = 10^0, 10^{-1}, 10^{-2}, 10^{-3}$, and 10 µL was spotted on SD -TL (control), SD minus Trp/Leu/His (-TLH), and SD minus Trp/Leu/His (-TLHA) selective plates. The plates were incubated for 2 days at 30 °C and then scanned to produce images.

Bimolecular fluorescence complementation (BiFC): Recombinant plasmids were transformed into the *Agrobacterium tumefaciens* strain GV3101 (pMP90) by electroporation and selected under appropriate antibiotics. *A. tumefaciens* seed cultures were grown in LB with the appropriate antibiotic selection for 2 days with shaking at 30 °C. Then, they were used to inoculate 50 mL LB containing the appropriate antibiotics plus 10 µM acetosyringone and grown for an additional 24 h. The cells were pelleted and resuspended in the infiltration medium (10 mM MES, 10 mM $MgCl_2$, 100 µM acetosyringone) and incubated for 5 h with rocking at room temperature. The cells were pelleted a second time, resuspended in the infiltration medium, and the appropriate nYFP/cYFP, H2B-RFP constructs were combined at a final OD_{600} of 1.0 for each test/control construct with suppressor strains (p19, $\gamma\beta$, PtoHA, HcPro) at a final OD_{600} of 0.5. *Nicotiana benthamiana* leaves from 4-week-old plants were infiltrated by a syringe with the *A. tumefaciens* mixes. The underside of whole leaf mounts was visualized using laser-scanning confocal microscopy 3 days after infiltration with a Nikon D-Eclipse C1 Confocal laser scanning microscope (Nikon Instruments) with either: (1) Excitation at 488 nm with an emission band pass filter of 515/30 or (2) excitation at 561 nm with an emission band pass filter of 650 LP.

Co-infiltration: *FBS1*, *FBIP1*, and *14-3-3λ* were cloned into pN-TAPa (9X myc tag), pGWB14 (3X HA tag) or pGWB12 (VSVG tag) vectors [58], respectively, using a Gateway strategy as above. Recombinant plasmids were transformed by electroporation into the *A. tumefaciens* strain C58C1Rif/pGV2260. *A. tumefaciens* was grown to a stationary phase in the LB medium containing the appropriate antibiotics plus 50 µg/mL acetosyringone. Bacteria were pelleted and washed with 10 mM $MgCl_2$, and then resuspended in 10 mM $MgCl_2$ and 150 µg/mL acetosyringone. Cell densities were adjusted to OD_{600} of 0.5. After 3 h of incubation, *A. tumefaciens* strains containing each construct were adjusted to varying concentrations and mixed with the same volume of an *A. tumefaciens* strain containing the viral suppressor p19, treated in the same way, but adjusted to OD_{600} of 1.0. The abaxial side of leaves from 3–4 week-old *N. benthamiana* were infiltrated with this bacterial suspension. After 3 days, the leaf material was collected and immediately frozen in liquid N_2 for protein extraction.

Protein extraction and Western blotting: Approximately 100 µg of frozen tissue was homogenized in 200 µL of 1× Laemmli loading buffer plus 4 M urea, boiled for 5 min, and centrifuged at 10,000× *g* for 5 min. Then, 10 µL of the supernatant were loaded onto 8%, 10% or 15% polyacrylamide gels and subjected to SDS-PAGE using the standard protocols. The separated proteins were blotted onto a Hybond-P+ membrane (Amersham Pharmacia Biotech, Amersham, UK) using the standard protocols, and then the membranes were probed with anti-c-Myc, anti-HA antibody or anti-VSVG antibodies (all from Sigma-

Aldrich, St. Louis, MO, USA). The blots were developed using an alkaline phosphatase kit (BCIP/NBT kit; Invitrogen; Waltham, MA, USA).

AGI numbers: FBS1 (At1g61340), FBS2 (At4g21510), FBS3 (At4g05010), FBS4 (At4g35930), FBIP1 (At3g54190), and FBIP2 (At2g38630).

Supplementary Materials: The following are available online at https://www.mdpi.com/article/10.3390/plants10102228/s1, Figure S1: Yeast two-hybrid FBS1–FBS4 interactions with FBIP1 and FBIP2. Figure S2: Bimolecular fluorescence complementation (BiFC) interactions between FBS1–FBS4 and FBIP1. Figure S3: Bimolecular fluorescence complementation (BiFC) interactions between FBS1–FBS4 and FBIP2. Figure S4: YFP channel positive and negative controls. Figure S5: FBS1 influence on 14-3-3 λ protein abundance in plants. Table S1: Primer sequences used for cloning.

Author Contributions: E.S.-G., E.C.F., E.V.P., L.E.O., A.A.F., A.J.R., M.R.-S., J.M.G. and B.T. designed the experiments. E.S.-G., E.C.F., E.V.P., L.E.O., A.A.F., A.J.R. and B.T. conducted the experiments and analyzed the data. B.T. wrote the manuscript. All authors have read and agreed to the published version of the manuscript.

Funding: This research was funded by grants from the M.J. Murdock Charitable Trust (NS-2016262 and 20141205:MNL:11/20/14) for materials and student summer research stipends, and funds from the University Enrichment Committee (UEC) at the University of Puget Sound for materials and student summer research stipends.

Institutional Review Board Statement: Not applicable.

Informed Consent Statement: Not applicable.

Data Availability Statement: Not applicable.

Acknowledgments: We thank David Somers (The Ohio State University) for the H2B-RFP construct, Frank Harmon (University of California at Berkeley/USDA Plant Gene Expression Center) for the yeast two-hybrid vectors, Faride Unda (University of British Columbia) for the BiFC vectors, and Ruirui Huang and Vivian Irish (Yale University) for the yeast two-hybrid TOPLESS constructs. Moreover, we thank Andreas Madlung (University of Puget Sound) for critical reading of the manuscript and other helpful discussions. Finally, we would like to thank Michal Morrison-Kerr (University of Puget Sound) for her indispensable help in supporting Puget Sound undergraduate research students.

Conflicts of Interest: The authors declare no conflict of interest. The funders had no role in the design of the study; in the collection, analyses, or interpretation of data; in the writing of the manuscript, or in the decision to publish the results.

References

1. Hua, Z.; Vierstra, R.D. The Cullin-RING Ubiquitin-Protein Ligases. *Annu. Rev. Plant Biol.* **2011**, *62*, 299–334. [CrossRef]
2. Gagne, J.M.; Downes, B.P.; Shiu, S.H.; Durski, A.M.; Vierstra, R.D. The F-Box Subunit of the SCF E3 Complex Is Encoded by a Diverse Superfamily of Genes in Arabidopsis. *Proc. Natl. Acad. Sci. USA* **2002**, *99*, 11519–11524. [CrossRef]
3. Sheard, L.B.; Tan, X.; Mao, H.; Withers, J.; Ben-Nissan, G.; Hinds, T.R.; Kobayashi, Y.; Hsu, F.F.; Sharon, M.; Browse, J.; et al. Jasmonate Perception by Inositol-Phosphate-Potentiated COI1-JAZ Co-Receptor. *Nature* **2010**, *468*, 400–405. [CrossRef]
4. Fang, Q.; Zhou, F.; Zhang, Y.; Singh, S.; Huang, C. Degradation of STOP1 Mediated by the F-box Proteins RAH1 and RAE1 Balances Aluminum Resistance and Plant Growth in Arabidopsis Thaliana. *Plant J.* **2021**, *106*, 493–506. [CrossRef] [PubMed]
5. Wang, P.; Nolan, T.M.; Clark, N.M.; Jiang, H.; Montes-Serey, C.; Guo, H.; Bassham, D.C.; Walley, J.W.; Yin, Y. The F-Box E3 Ubiquitin Ligase BAF1 Mediates the Degradation of the Brassinosteroid-Activated Transcription Factor BES1 through Selective Autophagy in Arabidopsis. *Plant Cell* **2021**, koab210. [CrossRef]
6. Jin, J. Systematic Analysis and Nomenclature of Mammalian F-Box Proteins. *Genes Dev.* **2004**, *18*, 2573–2580. [CrossRef] [PubMed]
7. Wang, Z.; Liu, P.; Inuzuka, H.; Wei, W. Roles of F-Box Proteins in Cancer. *Nat. Rev. Cancer* **2014**, *14*, 233–247. [CrossRef]
8. Lee, C.-M.; Feke, A.; Li, M.-W.; Adamchek, C.; Webb, K.; Pruneda-Paz, J.; Bennett, E.J.; Kay, S.A.; Gendron, J.M. Decoys Untangle Complicated Redundancy and Reveal Targets of Circadian Clock F-Box Proteins. *Plant Physiol.* **2018**, *177*, 1170–1186. [CrossRef] [PubMed]
9. Matsumoto, A.; Tateishi, Y.; Onoyama, I.; Okita, Y.; Nakayama, K.; Nakayama, K.I. Fbxw7β Resides in the Endoplasmic Reticulum Membrane and Protects Cells from Oxidative Stress. *Cancer Sci.* **2011**, *102*, 749–755. [CrossRef]
10. Spruck, C.; Strohmaier, H.; Watson, M.; Smith, A.P.L.; Ryan, A.; Krek, W.; Reed, S.I. A CDK-Independent Function of Mammalian Cks1: Targeting of SCFSkp2 to the CDK Inhibitor P27Kip. *Mol. Cell* **2001**, *7*, 12. [CrossRef]

11. Kirk, R.; Laman, H.; Knowles, P.P.; Murray-Rust, J.; Lomonosov, M.; Meziane, E.K.; McDonald, N.Q. Structure of a Conserved Dimerization Domain within the F-Box Protein Fbxo7 and the PI31 Proteasome Inhibitor. *J. Biol. Chem.* **2008**, *283*, 22325–22335. [CrossRef]
12. Nelson, D.E.; Randle, S.J.; Laman, H. Beyond Ubiquitination: The Atypical Functions of Fbxo7 and Other F-Box Proteins. *Open Biol.* **2013**, *3*, 130131. [CrossRef] [PubMed]
13. Kim, W.Y.; Fujiwara, S.; Suh, S.S.; Kim, J.; Kim, Y.; Han, L.; David, K.; Putterill, J.; Nam, H.G.; Somers, D.E. ZEITLUPE Is a Circadian Photoreceptor Stabilized by GIGANTEA in Blue Light. *Nature* **2007**, *449*, 356–360. [CrossRef]
14. Sawa, M.; Nusinow, D.A.; Kay, S.A.; Imaizumi, T. FKF1 and GIGANTEA Complex Formation Is Required for Day-Length Measurement in Arabidopsis. *Science* **2007**, *318*, 261–265. [CrossRef]
15. Yasuhara, M. Identification of ASK and Clock-Associated Proteins as Molecular Partners of LKP2 (LOV Kelch Protein 2) in Arabidopsis. *J. Exp. Bot.* **2004**, *55*, 2015–2027. [CrossRef] [PubMed]
16. Zoltowski, B.D.; Imaizumi, T. Structure and Function of the ZTL/FKF1/LKP2 Group Proteins in Arabidopsis. In *The Enzymes*; Elsevier: Amsterdam, The Netherlands, 2014; Volume 35, pp. 213–239.
17. Más, P.; Kim, W.-Y.; Somers, D.E.; Kay, S.A. Targeted Degradation of TOC1 by ZTL Modulates Circadian Function in Arabidopsis Thaliana. *Nature* **2003**, *426*, 567–570. [CrossRef] [PubMed]
18. Song, Y.H.; Estrada, D.A.; Johnson, R.S.; Kim, S.K.; Lee, S.Y.; MacCoss, M.J.; Imaizumi, T. Distinct Roles of FKF1, GIGANTEA, and ZEITLUPE Proteins in the Regulation of CONSTANS Stability in Arabidopsis Photoperiodic Flowering. *Proc. Natl. Acad. Sci. USA* **2014**, *111*, 17672–17677. [CrossRef] [PubMed]
19. Maldonado-Calderon, M.T.; Sepulveda-Garcia, E.; Rocha-Sosa, M. Characterization of Novel F-Box Proteins in Plants Induced by Biotic and Abiotic Stress. *Plant Sci. Int. J. Exp. Plant Biol.* **2012**, *185*, 208–217. [CrossRef] [PubMed]
20. Sepulveda-Garcia, E.; Rocha-Sosa, M. The Arabidopsis F-Box Protein AtFBS1 Interacts with 14-3-3 Proteins. *Plant Sci. Int. J. Exp. Plant Biol.* **2012**, *195*, 36–47. [CrossRef] [PubMed]
21. Gonzalez, L.E.; Keller, K.; Chan, K.X.; Gessel, M.M.; Thines, B.C. Transcriptome Analysis Uncovers Arabidopsis F-BOX STRESS INDUCED 1 as a Regulator of Jasmonic Acid and Abscisic Acid Stress Gene Expression. *BMC Genom.* **2017**, *18*, 533. [CrossRef]
22. Catala, R.; Lopez-Cobollo, R.; Mar Castellano, M.; Angosto, T.; Alonso, J.M.; Ecker, J.R.; Salinas, J. The Arabidopsis 14-3-3 Protein RARE COLD INDUCIBLE 1A Links Low-Temperature Response and Ethylene Biosynthesis to Regulate Freezing Tolerance and Cold Acclimation. *Plant Cell* **2014**, *26*, 3326–3342. [CrossRef]
23. Van Kleeff, P.J.; Jaspert, N.; Li, K.W.; Rauch, S.; Oecking, C.; de Boer, A.H. Higher Order Arabidopsis 14-3-3 Mutants Show 14-3-3 Involvement in Primary Root Growth Both under Control and Abiotic Stress Conditions. *J. Exp. Bot.* **2014**, *65*, 5877–5888. [CrossRef] [PubMed]
24. Zhou, H.; Lin, H.; Chen, S.; Becker, K.; Yang, Y.; Zhao, J.; Kudla, J.; Schumaker, K.S.; Guo, Y. Inhibition of the Arabidopsis Salt Overly Sensitive Pathway by 14-3-3 Proteins. *Plant Cell* **2014**, *26*, 1166–1182. [CrossRef]
25. Liu, Z.; Jia, Y.; Ding, Y.; Shi, Y.; Li, Z.; Guo, Y.; Gong, Z.; Yang, S. Plasma Membrane CRPK1-Mediated Phosphorylation of 14-3-3 Proteins Induces Their Nuclear Import to Fine-Tune CBF Signaling during Cold Response. *Mol. Cell* **2017**, *66*, 117–128. [CrossRef] [PubMed]
26. Yang, Z.; Wang, C.; Xue, Y.; Liu, X.; Chen, S.; Song, C.; Yang, Y.; Guo, Y. Calcium-Activated 14-3-3 Proteins as a Molecular Switch in Salt Stress Tolerance. *Nat. Commun.* **2019**, *10*, 1199. [CrossRef] [PubMed]
27. Kuroda, H.; Yanagawa, Y.; Takahashi, N.; Horii, Y.; Matsui, M. A Comprehensive Analysis of Interaction and Localization of Arabidopsis SKP1-like (ASK) and F-Box (FBX) Proteins. *PLoS ONE* **2012**, *7*, e50009. [CrossRef] [PubMed]
28. Arabidopsis Interactome Mapping Consortium; Dreze, M.; Carvunis, A.-R.; Charloteaux, B.; Galli, M.; Pevzner, S.J.; Tasan, M.; Ahn, Y.-Y.; Balumuri, P.; Barabasi, A.-L.; et al. Evidence for Network Evolution in an Arabidopsis Interactome Map. *Science* **2011**, *333*, 601–607. [CrossRef]
29. Hooper, C.M.; Castleden, I.R.; Tanz, S.K.; Aryamanesh, N.; Millar, A.H. SUBA4: The Interactive Data Analysis Centre for Arabidopsis Subcellular Protein Locations. *Nucleic Acids Res.* **2017**, *45*, D1064–D1074. [CrossRef]
30. Jain, B.P.; Pandey, S. WD40 Repeat Proteins: Signalling Scaffold with Diverse Functions. *Protein J* **2018**, *37*, 391–406. [CrossRef] [PubMed]
31. Ma, J.; An, K.; Zhou, J.-B.; Wu, N.-S.; Wang, Y.; Ye, Z.-Q.; Wu, Y.-D. WDSPdb: An Updated Resource for WD40 Proteins. *Bioinformatics* **2019**, *35*, 4824–4826. [CrossRef] [PubMed]
32. Schmid, M.; Davison, T.S.; Henz, S.R.; Pape, U.J.; Demar, M.; Vingron, M.; Scholkopf, B.; Weigel, D.; Lohmann, J.U. A Gene Expression Map of Arabidopsis Thaliana Development. *Nat. Genet.* **2005**, *37*, 501–506. [CrossRef]
33. Kilian, J.; Whitehead, D.; Horak, J.; Wanke, D.; Weinl, S.; Batistic, O.; D'Angelo, C.; Bornberg-Bauer, E.; Kudla, J.; Harter, K. The AtGenExpress Global Stress Expression Data Set: Protocols, Evaluation and Model Data Analysis of UV-B Light, Drought and Cold Stress Responses. *Plant J. Cell Mol. Biol.* **2007**, *50*, 347–363. [CrossRef] [PubMed]
34. Goda, H.; Sasaki, E.; Akiyama, K.; Maruyama-Nakashita, A.; Nakabayashi, K.; Li, W.; Ogawa, M.; Yamauchi, Y.; Preston, J.; Aoki, K.; et al. The AtGenExpress Hormone and Chemical Treatment Data Set: Experimental Design, Data Evaluation, Model Data Analysis and Data Access. *Plant J.* **2008**, *55*, 526–542. [CrossRef] [PubMed]
35. Kagale, S.; Rozwadowski, K. EAR Motif-Mediated Transcriptional Repression in Plants: An Underlying Mechanism for Epigenetic Regulation of Gene Expression. *Epigenetics* **2011**, *6*, 141–146. [CrossRef] [PubMed]

36. Shyu, C.; Figueroa, P.; DePew, C.L.; Cooke, T.F.; Sheard, L.B.; Moreno, J.E.; Katsir, L.; Zheng, N.; Browse, J.; Howe, G.A. JAZ8 Lacks a Canonical Degron and Has an EAR Motif That Mediates Transcriptional Repression of Jasmonate Responses in Arabidopsis. *Plant Cell* **2012**, *24*, 536–550. [CrossRef]
37. Long, J.A. TOPLESS Regulates Apical Embryonic Fate in Arabidopsis. *Science* **2006**, *312*, 1520–1523. [CrossRef]
38. Pauwels, L.; Barbero, G.F.; Geerinck, J.; Tilleman, S.; Grunewald, W.; Pérez, A.C.; Chico, J.M.; Bossche, R.V.; Sewell, J.; Gil, E.; et al. NINJA Connects the Co-Repressor TOPLESS to Jasmonate Signalling. *Nature* **2010**, *464*, 788–791. [CrossRef]
39. Causier, B.; Ashworth, M.; Guo, W.; Davies, B. The TOPLESS Interactome: A Framework for Gene Repression in Arabidopsis. *Plant Physiol.* **2012**, *158*, 423–438. [CrossRef]
40. Wang, L.; Kim, J.; Somers, D.E. Transcriptional Corepressor TOPLESS Complexes with Pseudoresponse Regulator Proteins and Histone Deacetylases to Regulate Circadian Transcription. *Proc. Natl. Acad. Sci. USA* **2013**, *110*, 761–766. [CrossRef]
41. Dos Santos Maraschin, F.; Memelink, J.; Offringa, R. Auxin-Induced, SCF TIR1—Mediated Poly-Ubiquitination Marks AUX/IAA Proteins for Degradation. *Plant J.* **2009**, *59*, 100–109. [CrossRef]
42. Ke, J.; Ma, H.; Gu, X.; Thelen, A.; Brunzelle, J.S.; Li, J.; Xu, H.E.; Melcher, K. Structural Basis for Recognition of Diverse Transcriptional Repressors by the TOPLESS Family of Corepressors. *Sci. Adv.* **2015**, *1*, e1500107. [CrossRef]
43. Long, Y.; Schiefelbein, J. Novel TTG1 Mutants Modify Root-Hair Pattern Formation in Arabidopsis. *Front. Plant Sci.* **2020**, *11*, 383. [CrossRef] [PubMed]
44. Li, H.; He, Z.; Lu, G.; Lee, S.C.; Alonso, J.; Ecker, J.R.; Luan, S. A WD40 Domain Cyclophilin Interacts with Histone H3 and Functions in Gene Repression and Organogenesis in Arabidopsis. *Plant Cell* **2007**, *19*, 2403–2416. [CrossRef] [PubMed]
45. Zhu, J.; Jeong, J.C.; Zhu, Y.; Sokolchik, I.; Miyazaki, S.; Zhu, J.-K.; Hasegawa, P.M.; Bohnert, H.J.; Shi, H.; Yun, D.-J.; et al. Involvement of Arabidopsis HOS15 in Histone Deacetylation and Cold Tolerance. *Proc. Natl. Acad. Sci. USA* **2008**, *105*, 4945–4950. [CrossRef] [PubMed]
46. Mehdi, S.; Derkacheva, M.; Ramström, M.; Kralemann, L.; Bergquist, J.; Hennig, L. The WD40 Domain Protein MSI1 Functions in a Histone Deacetylase Complex to Fine-Tune Abscisic Acid Signaling. *Plant Cell* **2016**, *28*, 42–54. [CrossRef] [PubMed]
47. Krogan, N.T.; Hogan, K.; Long, J.A. APETALA2 Negatively Regulates Multiple Floral Organ Identity Genes in Arabidopsis by Recruiting the Co-Repressor TOPLESS and the Histone Deacetylase HDA19. *Development* **2012**, *139*, 4180–4190. [CrossRef]
48. Leydon, A.R.; Wang, W.; Gala, H.P.; Gilmour, S.; Juarez-Solis, S.; Zahler, M.L.; Zemke, J.E.; Zheng, N.; Nemhauser, J.L. Repression by the Arabidopsis TOPLESS Corepressor Requires Association with the Core Mediator Complex. *eLife* **2021**, *10*, e66739. [CrossRef]
49. Lloyd, A.; Brockman, A.; Aguirre, L.; Campbell, A.; Bean, A.; Cantero, A.; Gonzalez, A. Advances in the MYB–BHLH–WD Repeat (MBW) Pigment Regulatory Model: Addition of a WRKY Factor and Co-Option of an Anthocyanin MYB for Betalain Regulation. *Plant Cell Physiol.* **2017**, *58*, 1431–1441. [CrossRef]
50. Skaar, J.R.; Pagan, J.K.; Pagano, M. Mechanisms and Function of Substrate Recruitment by F-Box Proteins. *Nat. Rev. Mol. Cell Biol.* **2013**, *14*, 369–381. [CrossRef]
51. Wang, L.; Xu, Q.; Yu, H.; Ma, H.; Li, X.; Yang, J.; Chu, J.; Xie, Q.; Wang, Y.; Smith, S.M.; et al. Strigolactone and Karrikin Signaling Pathways Elicit Ubiquitination and Proteolysis of SMXL2 to Regulate Hypocotyl Elongation in Arabidopsis Thaliana. *Plant Cell* **2020**, *32*, 2251–2270. [CrossRef]
52. Zhao, X.; Li, F.; Li, K. The 14-3-3 Proteins: Regulators of Plant Metabolism and Stress Responses. *Plant Biol. J.* **2021**, *23*, 531–539. [CrossRef]
53. Welcker, M.; Larimore, E.A.; Swanger, J.; Bengoechea-Alonso, M.T.; Grim, J.E.; Ericsson, J.; Zheng, N.; Clurman, B.E. Fbw7 Dimerization Determines the Specificity and Robustness of Substrate Degradation. *Genes Dev.* **2013**, *27*, 2531–2536. [CrossRef]
54. Barbash, O.; Lee, E.; Diehl, J. Phosphorylation-Dependent Regulation of SCFFbx4 Dimerization and Activity Involves a Novel Component. *Oncogene* **2011**, *30*, 1995–2002. [CrossRef] [PubMed]
55. Tan, T.; Cai, J.; Zhan, E.; Yang, Y.; Zhao, J.; Guo, Y.; Zhou, H. Stability and Localization of 14-3-3 Proteins Are Involved in Salt Tolerance in Arabidopsis. *Plant Mol. Biol.* **2016**, *92*, 391–400. [CrossRef] [PubMed]
56. Madeira, F.; mi Park, Y.; Lee, J.; Buso, N.; Gur, T.; Madhusoodanan, N.; Basutkar, P.; Tivey, A.R.N.; Potter, S.C.; Finn, R.D.; et al. The EMBL-EBI Search and Sequence Analysis Tools APIs in 2019. *Nucleic Acids Res.* **2019**, *47*, W636–W641. [CrossRef] [PubMed]
57. Zhu, D.; Maier, A.; Lee, J.-H.; Laubinger, S.; Saijo, Y.; Wang, H.; Qu, L.-J.; Hoecker, U.; Deng, X.W. Biochemical Characterization of *Arabidopsis* Complexes Containing constitutively photomorphogenic1 and suppressor of phya Proteins in Light Control of Plant Development. *Plant Cell* **2008**, *20*, 2307–2323. [CrossRef]
58. Nakagawa, T.; Kurose, T.; Hino, T.; Tanaka, K.; Kawamukai, M.; Niwa, Y.; Toyooka, K.; Matsuoka, K.; Jinbo, T.; Kimura, T. Development of Series of Gateway Binary Vectors, PGWBs, for Realizing Efficient Construction of Fusion Genes for Plant Transformation. *J. Biosci. Bioeng.* **2007**, *104*, 34–41. [CrossRef] [PubMed]

plants MDPI

Review

Crosstalk between Ca^{2+} and Other Regulators Assists Plants in Responding to Abiotic Stress

Yaoqi Li †, Yinai Liu †, Libo Jin * and Renyi Peng *

Biomedicine Collaborative Innovation Center of Zhejiang Province, Institute of Life Sciences, College of Life and Environmental Science, Wenzhou University, Wenzhou 325035, China; 20461337004@stu.wzu.edu.cn (Y.L.); 21461338012@stu.wzu.edu.cn (Y.L.)

* Correspondence: libo9518@126.com (L.J.); 20170032@wzu.edu.cn (R.P.)

† These authors contributed equally to this work.

Abstract: Plants have evolved many strategies for adaptation to extreme environments. Ca^{2+}, acting as an important secondary messenger in plant cells, is a signaling molecule involved in plants' response and adaptation to external stress. In plant cells, almost all kinds of abiotic stresses are able to raise cytosolic Ca^{2+} levels, and the spatiotemporal distribution of this molecule in distant cells suggests that Ca^{2+} may be a universal signal regulating different kinds of abiotic stress. Ca^{2+} is used to sense and transduce various stress signals through its downstream calcium-binding proteins, thereby inducing a series of biochemical reactions to adapt to or resist various stresses. This review summarizes the roles and molecular mechanisms of cytosolic Ca^{2+} in response to abiotic stresses such as drought, high salinity, ultraviolet light, heavy metals, waterlogging, extreme temperature and wounding. Furthermore, we focused on the crosstalk between Ca^{2+} and other signaling molecules in plants suffering from extreme environmental stress.

Keywords: Ca^{2+}; abiotic stress response; Ca^{2+} sensors; signal transduction; abiotic stress tolerance calcium; heat stress; cold stress

Citation: Li, Y.; Liu, Y.; Jin, L.; Peng, R. Crosstalk between Ca^{2+} and Other Regulators Assists Plants in Responding to Abiotic Stress. *Plants* **2022**, *11*, 1351. https://doi.org/10.3390/plants11101351

Academic Editors: Ewa Muszyńska, Petronia Carillo, Kinga Dziurka and Mateusz Labudda

Received: 25 February 2022
Accepted: 18 May 2022
Published: 19 May 2022

1. Introduction

Calcium ions (Ca^{2+}) are important ions that maintain the normal physiological functions of plant cells and are involved in physiological metabolism in plants [1]. Ca^{2+} also functions as a ubiquitous secondary messenger involved in plant responses to various stresses [2]. Usually, there is a significant increase in the cytosolic Ca^{2+} concentration ($[Ca^{2+}]_{cyt}$) in plant cells that is caused by low temperature [3], salt [4], drought [5] and other abiotic stresses. Ca^{2+} spikes are triggered by Ca^{2+} influx through channels or Ca^{2+} efflux through pumps. This increase is recognized, amplified and transmitted downstream by Ca^{2+}-binding proteins, also known as calmodulin or Ca^{2+} sensors, which regulate plant cell division, cell elongation, stomatal movement, various stress responses and growth and development through a series of conduction cascades [6].

The main function of Ca^{2+} in plant stress resistance is to stabilize plant cell walls and membranes. It can activate or inhibit various ion channels on the membrane to achieve a balance of ion concentrations inside or outside the cell. The activities of specific enzymes are activated or inhibited by Ca^{2+} in cells to regulate biochemical reactions in plants [7,8]. Moreover, the transcriptional expression of multiple anti-stress genes is regulated by changes in calcium signaling to enhance the adaptability of plants exposed to extreme environments [9]. Under abiotic stress conditions, changes in the calcium ion concentration in the plant cytoplasm can be generally recognized as a cellular secondary messenger to distinguish different original signals; it also continues to transmit the signal downstream by interacting with calcium-binding proteins, causing a series of biochemical reactions in the cells to adapt or resist various stresses [10].

Ca^{2+}, considered a secondary messenger for plant signal transduction, transmits extracellular information and regulates many physiological and biochemical responses to primary signals, such as light, hormones, and gravity [11]. Cytosolic Ca^{2+} cannot be maintained at a high level for a long time. If the concentration is too high, the Ca^{2+} will react with phosphoric acid, which is necessary for the metabolism of energy substances, and produce a precipitate that inhibits the normal physiological growth of cells or even causes cell death. In normal plant cells, most Ca^{2+} exists in a bound form, collectively known as the calcium pool, and calcium-storing proteins with high capacity and low affinity for Ca^{2+} can enhance Ca^{2+}-buffering capacity. Due to this low affinity, when Ca^{2+} channels in calcium banks open, Ca^{2+}-binding proteins can be rapidly dissociated from Ca^{2+}, releasing it into the cytoplasm so that Ca^{2+} signals can be accurately and rapidly transmitted [12]. Ca^{2+} enters the cell through the Ca^{2+} channel, which is actually a protein on the plasma membrane that is maintained in the on or off state according to changes in its conformation. This channel rapidly stimulates and induces Ca^{2+} release from the vacuole. There are two major vacuolar uptake mechanisms, including P-type Ca^{2+} pumps and a family of cation/H^+ exchangers, which are responsible for high-affinity Ca^{2+} uptake and low-affinity with high-capacity Ca^{2+} uptake, respectively. Although research on the Ca^{2+} transport pathway mainly focuses on the regulation of $[Ca^{2+}]_{cyt}$ by calmodulin (CaM) on the cell membrane, Ca^{2+} flow through internal membrane systems, such as the endoplasmic reticulum and mitochondrial membrane, is also critical when studying the transport patterns of Ca^{2+} signals [13,14].

Because the distribution and transfer of intracellular Ca^{2+} are the basis for the formation of Ca^{2+} signals, the increase or decrease in intracellular Ca^{2+} concentrations directly affect the generation and termination of Ca^{2+} signals. When there is no external stimulation, cytosolic Ca^{2+} is insufficient to activate CaM, which lacks its own catalytic activity. However, under extreme environmental conditions, $[Ca^{2+}]_{cyt}$ increases rapidly, producing Ca^{2+} signals, and the reaction with CaM transmits the signal downward to allow subsequent physiological and biochemical reactions to occur [15]. Finally, restoration of the normal $[Ca^{2+}]_{cyt}$ levels occurs by reloading calcium stores after completing Ca^{2+} signalling, and through the calcium efflux system, which consists of Ca^{2+}-ATPase pumps and Ca^{2+}/H^+ exchangers, to remove excess Ca^{2+}(Figure 1) [14,16].

Calcium sensors in plants are composed of Ca^{2+}-binding proteins, such as CaMs, calmodulin-like-proteins (CMLs), calcineurin-B-like proteins (CBLs), and Ca^{2+}-dependent protein kinases (CDPKs). CBLs interact with CBL-interacting protein kinases (CIPKs) to form a CBL/CIPK signaling network, which plays a key role in the plant response to abiotic stress. These networks may contain many interactions, with CBLs activating CIPKs and CIPKs phosphorylating CBLs. Phosphorylation is the major mechanism affecting downstream proteins [17].

There are three major elements, influx, efflux and decoding, that affect Ca^{2+}-signal translation. Ca^{2+} influx is mediated by depolarization-activated, hyperpolarization-activated and voltage-independent Ca^{2+}-permeable channels, which are encoded by genes, including *cyclic nucleotide-gated channels* (*CNGCs*), *glutamate receptor-like channels* (*GLRs*), *mechanosensitive channels of small* (*MscS*) and *conductance-like channels* (*MSLs*), *annexins*, *mid1-complementing activity channels* (*MCAs*), *Piezo channels* and *channel 1* (*OSCA1*) [18]. The Ca^{2+}-efflux system, the calcium-dependent protein kinase ZmCDPK7 consisting of autoinhibited Ca^{2+}-ATPases (ACAs), ER-type Ca^{2+}-ATPases (ECAs), and P1-ATPases (HMA1), enables Ca^{2+} efflux to form an informative signature. Specificity in Ca^{2+}-based signaling is achieved via Ca^{2+} signatures with cognate Ca^{2+}-binding proteins. The decoding step is carried out by protein families such as CDPKs, CBL, CIPKs, CaM and CMLs (Figure 2) [19,20].

Plants constantly suffer from various abiotic stresses during their growth and development. Ca^{2+}, acting as a secondary messenger, plays an essential role in the plant response to abiotic stresses; it can not only transmit and recognize various regulatory signals but also participate in gene expression and normal protein functions [21,22]. This review summarizes the biological process of cytosolic Ca^{2+} in response to abiotic stresses, such as drought, high temperature, high salinity, heavy metals, waterlogging, and mechanical damage. Furthermore, we focus on both the crosstalk of cytosolic Ca^{2+} with other signaling molecules and biomacromolecules in plants suffering from extreme environmental stresses.

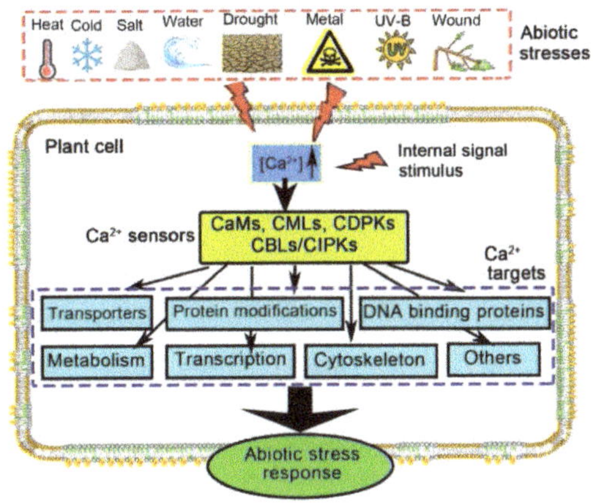

Figure 1. The Ca^{2+} signaling network in plant cells. Abiotic stress, including high-temperature stress (heat), low-temperature stress (cold), salt stress (salt), waterlogging stress (water), drought stress (drought), heavy-metal stress (metal), ultraviolet-B radiation stress (UV-B) and wound stress (wound), gives rise to an increase in [Ca^{2+}], which is subsequently decoded by Ca^{2+} sensors such as Ca^{2+}-dependent protein kinases (CDPKs), calmodulin-like-proteins (CMLs), calmodulins (CaMs), and calcineurin-B like proteins (CBLs) and their interacting protein kinases (CIPKs). These sensors activate various downstream responses that in turn result in an overall response precisely according to the original stimulus.

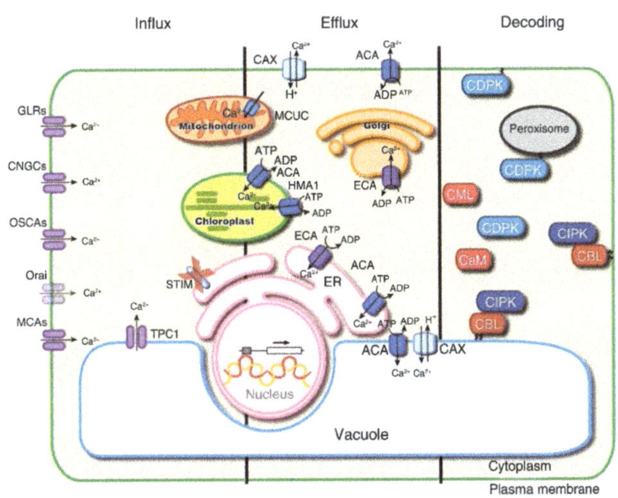

Figure 2. The generation and translation of Ca^{2+} signals in plant cells. Three major processes, including influx, efflux and decoding, can alter the effects of Ca^{2+}-signal translation. GLRs: glutamate receptor-like channels, CNGCs: cyclic nucleotide-gated channels, OSCAs: hyperosmolality-induced Ca^{2+} increase channels, ACAs: Ca^{2+}-ATPases, ECAs: Ca^{2+}-ATPases, HMA1: P1-ATPases, MCUC: mitochondrial calcium uniporter complex, CAX: Ca^{2+} exchangers, CDPKs: calcium-dependent protein kinases, CBL: calcineurin B-like, and CIPKs: protein kinases. Reproduced with permission from [20], copyright 2017 Elsevier.

2. Molecular Mechanisms of Crosstalk between Ca^{2+} and Other Regulators in Response to Abiotic Stresses in Plants

2.1. Drought Stress

Drought is a common adverse factor inhibiting plant growth and development; high levels of drought lead to an increase in the content of reactive oxygen species (ROS) that promote membrane peroxidation and damage membrane structure [23]. Ca^{2+} plays an important regulatory role in the signaling related to the plant drought stress response, reflecting its ability to regulate the activity of some enzymes and improve the ROS-scavenging ability. In addition, damage caused by drought can be reduced with Ca^{2+} channel activation mediating stomatal closure, which reduces transpiration flux to control water loss, thus improving plant water use efficiency [24].

By monitoring the water potential of the root vascular system, plants can transmit stress signals from roots to leaves, regulate stomatal closure and induce the expression of related genes to avoid dehydration [25]. Ca^{2+} efflux was observed in epidermal cells and mesophyll cells of barley roots under drought stress conditions. Extracellular pH affects K^+ absorption, Ca^{2+} outflow and H^+ influx/alkalization in the leaves, which may be a chemical signal in the barley response to drought stress [26]. The application of molybdenum to wheat decreased the transpiration of wheat leaves but increased the Ca^{2+} concentration and other osmotic substances in wheat roots, which increased the osmotic pressure and further enhanced the water absorption capacity of wheat roots [27].

The abscisic acid (ABA)-dependent Ca^{2+} signaling pathway is the main response to drought stress in plants. ABA activates plasma membrane calcium channels in various ways to stimulate the release of Ca^{2+} from intracellular calcium stores, and several secondary messengers, including ROS, nitric oxide (NO), inositol 1,4,5-trisphosphate (IP3) and cyclic ADP-ribose (cADPR), are involved in this process. When water deficit occurs, ABA accumulates in the leaves. On the one hand, it activates phospholipase C and decomposes IP3, which can activate the intracellular calcium pool in guard cells to allow stomatal closure. On the other hand, intracellular Ca^{2+} can also be increased by cADPR, but no receptors for IP3 and cADPR have been identified until now in plants [28,29]. ABA can also rapidly induce an intracellular Ca^{2+} increase through hydrogen peroxide (H_2O_2), leading to plasma membrane hyperpolarization and direct activation of plasma membrane hyperpolarization-activated calcium channels (HACCs) and vacuolar membrane Ca^{2+} channels to achieve stomatal closure regulation [30]. At present, the pathway of NO modulating the crosstalk between ABA and H_2O_2 and activating the calcium signaling pathway has also been further revealed [31]. Wang et al. mentioned that extracellular Ca^{2+} and ABA promote stomatal closure by promoting H_2O_2 to produce calcium signals dependent on NO synthesis [32]. In *Arabidopsis* ABI mutants, H_2O_2 and NO activate calcium signals depending on cyclic guanosine 3′,5′-monophosphate (cGMP), which likely acts upstream of calcium signals. After exogenous calcium treatment, ion channels can be activated by intracellular calcium signaling, and calcium signal production processes mediated by ABA and H_2O_2 may be performed in the following sequence: ABA$\rightarrow H_2O_2\rightarrow$NO$\rightarrow$cGMP$\rightarrow Ca^{2+}$. The upstream calcium-sensing signal is converted to a calcium-receiving signal, and then the downstream calcium signal can produce biological reactions promoting stomatal closure [33].

Ca^{2+} not only acts as a secondary messenger in the rapid response to upstream stimulation, but more importantly, the Ca^{2+} signaling system also contains a large number of different types of calcium signal receptors, such as CDPKs, CaM, CBL, and CIPK, which receive exogenous calcium signals and convert them into endogenous calcium signals. These signals are then phosphorylated and dephosphorylated or eventually interact with other proteins to regulate stomatal movement [34–36]. The interaction between CBL9 and CIPK3 negatively regulates Ca^{2+}-dependent ABA signaling in *Arabidopsis* [37]. It was found that VvK1.1 in grapevine corresponds to the AKT1 channel in *Arabidopsis*, and dominates K^+ uptake in root periphery cells. VvK1.1 and AKT1 have common functions, such as regulation by CIPK23, which occurs independently in grapevine under drought stress; this process is essential for stomatal movement regulated by K^+ flow [38]. During stomatal closure, the relationship between ABA and Ca^{2+} is not a simple upstream and downstream regulatory process. In the early stage of drought stress, Ca^{2+} can rapidly induce ABA biosynthesis and activate Ca^{2+} channels on the plasma membrane by utilizing turgor pressure or pH change to increase the intracellular Ca^{2+} concentration instantly. Then, the expression of related transcription factors and genes, including zeaxanthin epoxidase, nine-cis-epoxy carotenoid dioxygenase, abscisic aldehyde oxidase and molybdenum cofactor sulfurase, is increased by the protein kinase cascade reaction. ABA inhibition of type 2C protein phosphatase leads to phosphorylation and activation of sucrose-nonfermenting-1-related protein kinase 2, which in turn stimulates the expression of ABA-responsive genes, thereby promoting ABA biosynthesis; then, the generated ABA in turn promotes an increase in Ca^{2+} concentration [39]. Another hypothesis is that ABA induces the activation of calcium decoding signal elements, including calcium-permeable ion channels, Ca^{2+}/H^+ antiporters and Ca^{2+}-ATPases, and transduces calcium signals to alter stomatal aperture and transpiration efficiency to regulate water use efficiency in plants. Moreover, calcium channel proteins, such as *Arabidopsis thaliana two-pore channel 1* (*AtTPC1*) and *TaTPC1* from wheat, also regulate stomatal closure [40]. This hypothesis may explain the role of these genes in plant responses to drought and cold stress.

In addition to stomatal closure, plants can increase their water retention capacity by regulating stomatal density and other developmental processes to respond to drought. GT-2like 1 (GTL), a trihelix transcription family member, regulates stomatal motility by regulating the expression of *stomatal density and distribution1* (*SDD1*) genes. When PtaGTL1 identified in *Populus tremula* × *P. alba* was transferred to *Arabidopsis thaliana*, GTL increased *SDD1* gene expression by binding to Ca^{2+}-CaM, thus reducing stomatal density and the transpiration rate and improving water use efficiency under drought stress [41]. Therefore, to adapt to different degrees of water deficit, plants adjust the stomatal number and leaf area through growth and development and balance the relationship between water use efficiency and photosynthesis to achieve the optimal adaptation point, which may be an effective strategy for plants to cope with long-term drought stress [42].

2.2. Salt Stress

Salt stress usually causes ion toxicity, osmotic imbalance and oxidative stress, resulting in limited plant growth and thereby affecting the sustainability of crop yields. In the external environment, hypersaline stress occurs when a high enough salt content significantly changes the water potential, thus affecting the plant [43]. Ca^{2+} also plays a significant regulatory role in plant resistance to salt stress. For example, Ca^{2+} inhibits Na^+ influx by regulating Na^+ entry into the main cell channel nonselective cation channels (NSCCs). Moreover, Ca^{2+} prevents the outflow of K^+ by inhibiting K^+ permeable outwardly rectifying conductance (KORC) channel and initiates the salt overly sensitive (SOS) signal transduction pathway, which regulates the development of plasticity in roots during salt stress adaptation; for example, SOS3 is required for auxin biosynthesis, root polar movement and the formation and maintenance of auxin gradients [44–46].

Usually, Ca^{2+} influx is related to hydroxyl radicals (OH·) and Ca^{2+} influx channels on the plasma membrane in wheat roots. Salt-stress-induced nicotinamide adenine dinucleotide phosphate (NADPH) oxidase on the plasma membrane produces a large number of superoxide anion radicals (O_2^-) extracellularly during electron transfer to O_2, which are then rapidly converted into H_2O_2 and OH·. Notably, both OH· and H_2O_2 can activate the Ca^{2+} channels to induce extracellular Ca^{2+} flow into the cells [19]. Overall, ROS have been identified as key regulators of Ca^{2+} influx.

When suffering from salt stress, roots are the sensory part of plants that initiate the response and adaptive behavior to defend against stress damage as part of first-line defense. The SOS signaling pathway is activated by the increase in Ca^{2+} in the root cytoplasm caused by salt stress, which mediates cell signal transduction by SOS3/SCABp8-SOS2-SOS1 at the cellular level [47]. In this process, SOS3 functions as a Ca^{2+}-binding protein, interacts with SOS2 to form a complex and then activates downstream SOS1 through phosphorylation, thus maintaining K^+ and Na^+ homeostasis inside and outside the cell. Furthermore, SOS3 has also been shown to play a key role in mediating the recombination of Ca^{2+}-dependent actin filaments during salt stress [48].

CDPKs are a large polygenic family whose members contain a serine/threonine protein kinase catalytic domain as an effector region and a calmodulin-like domain for binding to Ca^{2+}. These proteins can directly activate and regulate target proteins when sensing Ca^{2+} signals, thus playing an essential role in a variety of physiological processes in plants. OsCPK12 has been shown to be a crucial factor in salt stress tolerance, acting as a positive regulator of stress tolerance by regulating ABA signaling and reducing ROS accumulation in rice [49]. For example, rice overexpressing OsTPC1 show enhanced tolerance to stress through positive regulation of ABA signaling and salt signaling pathways. Some researchers suggest that ABA receptors may be upstream factors that regulate intracellular Ca^{2+} levels in plants under salt stress conditions [50]. Other experiments have shown that ABA receptors may exist inside cells or outside the plasma membrane. On the surface of the plasma membrane, when ABA acts on its receptor, the activated part interacts with G protein, which binds to the plasma membrane to activate phospholipase C and stimulate the release of Ca^{2+} from the calcium pool [51].

Ca^{2+}-ATPase (PCA1) has been identified as essential for the adjustment of salt tolerance in the moss *Physcomitrella patens*. PCA1 encodes a PIIB-type Ca^{2+}-ATPase, which is a plant-specific Ca^{2+} pump with an N-terminal autoinhibitory calmodulin-binding domain that has been confirmed with in vivo complementation analysis of Ca^{2+} transport-deficient yeast strains. This class of Ca^{2+} pumps may trigger the initiation of stress adaptation mechanisms in Ca^{2+} signaling pathways. In contrast to the transient $[Ca^{2+}]_{cyt}$ increase caused by NaCl in the wild-type, hyperaccumulation of cytosolic Ca^{2+} in PCA1 mutants remained high and did not return to prestimulus $[Ca^{2+}]_{cyt}$ levels. Therefore, Ca^{2+} pumps contribute to the production of stress-induced Ca^{2+} signatures [52]. In addition, based on the isolation of *monocation-induced* $[Ca^{2+}]_i$ *increases 1 Arabidopsis* mutant, which affects Ca^{2+} influx under salt stress, an association between salt sensing and GIPC-gated Ca^{+2+} influx has been inferred. It has been demonstrated that Ca^{2+} channels are gated by GIPCs in plants [53].

2.3. Extreme Temperature Stress

2.3.1. Low-Temperature Stress

A large number of free radicals are produced in plants exposed to low-temperature stress, thereby damaging the membrane system. When plants are subjected to low-temperature stress, Ca^{2+} channels are opened, and intracellular $[Ca^{2+}]_{cyt}$ increases rapidly to induce calcium signaling [54]. Finally, the process is completed after signal transfer from the extramembrane into the membrane. On the one hand, the results of Ca^{2+} treatment of tobacco seedlings subjected to low-temperature stress showed that Ca^{2+} could increase the content of intracellular bound calcium and improve the activities of catalase, superoxide dismutase (SOD), peroxidase (POD) and other antioxidant enzymes, but reduce the content of malondialdehyde [55,56]. Furthermore, the decrease in enzyme activity after Ca^{2+} treatment was lower than that after Ca^{2+}-free treatment, and the membrane permeability of tobacco seedlings also recovered quickly after growth had stopped. Therefore, it is speculated that Ca^{2+} can improve plant cold resistance and maintain the stability of the membrane system [57,58]. Another study demonstrated that Ca^{2+} and CaM could regulate the freezing resistance of citrus protoplasts, while treatment with the exogenous CaM blocker TFP or the Ca^{2+}-chelating agent ethylene glycol diethyl ether diamine tetra-acetic acid (EGTA) could also inhibit the freezing resistance of citrus [59]. CBLs are a special class of Ca^{2+} receptors that specifically interact with CIPK protein kinases to activate downstream target proteins and decode Ca^{2+} signals. The expression of CIPK7 is induced by low temperature, can interact with the CBL1 protein in vitro and may be associated with CBL1 protein in vivo. Compared with wild-type plants, CBL1 mutant plants showed CIPK7 expression is affected by CBL1, suggesting that CIPK7 may bind to the calcium receptor CBL and participate in plants' cold response [60,61].

In contrast to CaM and CBL, which have to couple with Ca^{2+} to change their conformation and be activated, CDPKs, which are constitutively activated and directly phosphorylated, transduce calcium signals by interacting with the site of the calcium receptor or forming a peptide chain [62]. CDPKs are involved in the intermediate process instead of participating in the initial response to low temperature in rice. Moreover, several Ca^{2+}-related genes, such as CDPK13, are regulated by low-temperature stress in plants [63]. In rice, the CDPK13 gene is expressed in leaf sheaths and calli during the initial 2 weeks of growth, and CDPK13 is phosphorylated in response to low temperature and gibberellin (GA) signaling. Simultaneously, low temperature or exogenous GA3 treatment resulted in the elevation of CDPK13 gene expression and protein accumulation. Compared to wild-type and cold-sensitive rice, CDPK13-overexpressing-line rice showed stronger cold tolerance and a higher rate of plant recovery from cold injury, implying that CDPK13 might be a key protein in the rice signaling network responding to low-temperature stress [64,65].

Calcium channels are not only the key to the generation of calcium signals but also the rapid transport pathway and regulatory element for Ca^{2+} across the membrane [66]. At present, *Arabidopsis thaliana* two-pore channel 1 (AtTPC1) is the most studied calcium channel protein. Stomatal closure of *attpc1-2* functional deficient mutants treated with ABA, methyl jasmonate (MeJA) and Ca^{2+} was detected, and the results demonstrated that both ABA and MeJA can induce the accumulation of ROS and NO to cause an increase in $[Ca^{2+}]_{cyt}$ and cytoplasmic alkalization and activate anion channels in both wild-type and mutant plants, thus causing the stomata to be closed. However, compared with that in wild-type *Arabidopsis*, exogenous Ca^{2+} could not induce stomatal closure or activate anion channels on the plasma membrane in *attpc1-2* mutants. Taken together, we can conclude that AtTPC1 protein is involved in both stomatal closure and plasma membrane anion channel activation and is regulated by exogenous calcium signals in guard cells; however, it is not regulated by ABA and MeJA [67]. Stomatal closure is a common adaptive response of plants to low temperature. Stomatal guard cells respond quickly to abiotic stress stimuli, such as low temperature and drought [68].

Studies in eukaryotic cells suggest the overall translation rate can be regulated by an increased AMP/ATP ratio, which leads to activation of 5′-AMP-activated protein kinase and the release of Ca^{2+} from the endoplasmic reticulum, which triggers the phosphorylation of eukaryotic extension factor 2 by its activated specific kinase eukaryotic elongation factor 2 kinase [69].

2.3.2. High-Temperature Stress

High-temperature stress also gives rise to plant cell membrane damage, osmotic regulation imbalance, an accumulation of ROS, an inhibition of photosynthesis, cell aging and death, thus limiting plant distribution, growth and productivity [70]. Exogenous application of Ca^{2+} effectively improves high-temperature stress resistance in laver and tomato [71,72] and alleviates the damage caused by high-temperature stress in ornamental plants such as chrysanthemum [73]. In tomato, spraying calcium chloride on the leaf surface can increase the activities of protective enzymes and soluble protein contents in leaf intima and reduce the malonic acid content, thus enhancing high-temperature-stress adaptability [71]. Further research showed that Ca^{2+} treatment can significantly improve the net photosynthetic rate, transpiration rate and stomatal conductance of tomato leaves suffering from high-temperature stress [74]. On the other hand, significant upregulation of PhCAM1 and PhCAM2 expression is related to the change in $[Ca^{2+}]_{cyt}$ when high-temperature stress occurs, while the expression of PhCAM1 and PhCAM2 is not obviously changed after EGTA is added, implying that the Ca^{2+} signaling system and CAM play a major role in the regulation of resistance to high-temperature stress in *Pyropia haitanensis* [58,75]. Based on the above descriptions, it can be clearly seen that Ca^{2+} can not only stabilize the cell membrane structure but also prevent damage to photosynthetic organs from ROS under high-temperature stress by regulating osmotic balance and the antioxidant system. Additionally, Ca^{2+}, acting as an essential signaling substance, participates in signal transduction when high-temperature stress occurs and enhances high-temperature resistance in plants [76,77].

High-temperature stress also induces heat stress transcription factor (HSP) expression, and many of these factors act as molecular chaperones to prevent protein denaturation and maintain protein homeostasis [78]. Similarly to mammalian heat shock transcription factors (HSFs), plant HSFs are released from the binding and inhibition of HSP70 and HSP90 and combine with misfolded proteins under high-temperature stress. Therefore, HSFs can be used to activate the high-temperature stress response. In contrast, high-temperature stress also activates mitogen-activated protein kinases (MAPKs) and regulates the expression of HSP genes. This may be closely related not only to changes in membrane fluidity but also to calcium signaling induced by high-temperature stress, which is especially required for HSP gene expression and high-temperature stress tolerance acquisition [79,80]. The common features between signals of low- and high-temperature stress are not limited to membrane fluidity changes, calcium signaling and MAPK activation, as they also include ROS, NO, and phospholipid signaling [81,82].

The calcium-dependent protein kinase ZmCDPK7 positively regulates heat stress tolerance in maize. ABA regulates ZmCDPK7 expression by phosphorylation of the respiratory burst oxidase homologue RBOHB in a Ca^{2+}-dependent manner, thus triggering ROS accumulation, which further promotes ZmCDPK7 expression. Moreover, ZmCDPK7 plays a crucial role in maintaining protein quality and reducing heat stress damage by activating the chaperone function of sHSP17.4 through Ca^{2+}-dependent phosphorylation [83].

The Ca^{2+}/calmodulin-dependent phosphatase calcineurin plays a role in morphogenesis and calcium homeostasis during temperature-induced mycelium-to-yeast dimorphism of *Paracoccidioides brasiliensis*. Intracellular Ca^{2+} levels increased immediately after the onset of dimorphism. The extracellular or intracellular chelation of Ca^{2+} inhibits dimorphism, while extracellular Ca^{2+} addition accelerates dimorphism. In addition, the calcineurin inhibitor cyclosporine A disrupts intracellular Ca^{2+} homeostasis and reduces mRNA transcription of the *CCH1* gene in the Ca^{2+} channel of the yeast cell plasma membrane, effectively reducing cell growth or resulting in abnormal growth morphology *P. brasiliensis* [84].

2.4. Heavy-Metal Stress

Increasing the Ca^{2+} content in soil can enhance the heavy-metal tolerance of plants. The accumulation of active Al^{3+} and Mn^{2+}, as well as the lack of nutrients in acidic soil, are important limiting factors for crop growth [85]. Earlier studies showed that Al^{3+} could induce Ca^{2+} loss in plants and inhibit Ca^{2+} absorption and root growth, thereby suppressing plant growth and development. However, salicylic acid (SA) can alleviate Al^{3+}-induced inhibition of soybean root elongation and reduce the Al^{3+} content in plants. The plant response to Al^{3+} stress requires endogenous SA and Ca^{2+} for the transmission and amplification of the Al^{3+} stress signal, which strengthens the subsequent physiological response [86]. In addition, citric acid (CA) secreted from soybean roots can alleviate Al^{3+} toxicity. Both CA secretion and SA content changes are affected by Ca^{2+}, and it has been speculated that SA and Ca^{2+} might be linked to the Al^{3+} tolerance mechanism of soybean. Moreover, both Ca^{2+} and SA can alleviate the physiological reaction of root growth inhibition caused by aluminum, promote the secretion of citric acid, improve the enzyme activities of SOD, POD, ascorbate peroxidase and other antioxidant systems, reduce the accumulation of ROS, and alleviate oxidative stress damage to improve the Al^{3+} tolerance of soybean. Additionally, SA may participate in the Al^{3+} tolerance mechanism by increasing the endogenous Ca^{2+} level [87,88]. Exogenous Ca^{2+} can increase the relative expression levels of PLC and PLD genes, indicating that Ca^{2+} has some effects on the changes in phospholipase in soybean root tip cells, which may be related to changes in microtubule structure [89].

The addition of exogenous calcium can reduce the content of heavy metal ions in plants growing in soils with excessive amounts of heavy metals, such as Cu^{2+}, Cr^{6+} and Pb^{2+}, and improve their ability to resist heavy metal stress [90,91]. According to research findings, when the Cu^{2+} concentration increased, the Ca^{2+} content in plant roots increased, which may be significant for improving plant resistance to Cu^{2+} stress [92]. In addition, Cr^{6+} stress activates plant endogenous hydrogen sulfide (H_2S) synthesis and Ca^{2+} signal transduction. H_2S and Ca^{2+} alone or in combination can significantly reduce the injury caused by Cr^{6+} stress; however, the effect is better when they are used in combination. In contrast, treatment with H_2S synthesis inhibitors or Ca^{2+} chelating agents enhances environment-induced stress. This result suggests the synergistic effects of H_2S and Ca^{2+} in response to Cr^{6+} stress in *Setaria italica* [93].

2.5. Wound Stress

Usually, wounds from mechanical damage caused by harsh weather conditions, such as wind and rain, or by geological disasters, including debris flows and landslides, induce the release of calcium signals to regulate the overall response to stress and further improve the survival ability of plants [94,95]. Wound signaling is required for initiating plant regeneration. Plants promote changes in downstream cell fate due to signal transduction cascades induced by wounds [96]. Wounds also promote changes in cell membrane potential (Vm), fluctuations in Ca^{2+} concentration, ROS bursts, and drastic increases in the concentrations of jasmine, ethylene, SA and other plant hormones [97]. Therefore, Ca^{2+}, as a vital part of wound signaling, may regulate the transcription of downstream genes accompanied by signal transduction and trigger some physiological and biochemical reactions locally or systemically. Studies have shown that the loss of cell membrane integrity at the site of injury may allow cytoplasmic inclusions of damaged cells to enter the intercellular space, thus changing the original ion concentration and composition, which further affects the state of various ion channels on the cell membrane and leads to fluctuations in transmembrane potential and calcium concentration [98,99]. Furthermore, $GdCl_3$, a calcium channel inhibitor, has been shown to inhibit plasma membrane depolarization induced by single-cell injury [100].

Ca^{2+} signals respond to wounds rapidly (often within just 2 s) in plants suffering from mechanical damage and then propagate to specific undamaged distal tissues after 2 min. The ethylene synthesis-related genes ACS2, ACS6, ACS7 and ACS8 were rapidly upregulated within 30 min after leaf injury in *Arabidopsis.* At the same time, wounding rapidly activated the expression of mitogen-activated protein kinase (MPK) along with calcium-dependent protein kinase (CPK) [101]. To some extent, its transmission depends on glutamate receptor-like 3.3/3.6 (glR3.3/3.6) proteins, which are regulated by glutamate concentration. Mutation of both glr3.3 and glr3.6 leads to the long-distance transport of Ca^{2+} being blocked, and the expression of defense genes is subsequently reduced in undamaged areas, while glutamate contents are reduced concurrently [102]. Moreover, Ca^{2+} also functions as an intracellular secondary messenger to regulate the biochemical state of cells near wounds, and Ca^{2+}-dependent MC4 in the cytoplasm has catalytic activity due to the wound-induced $[Ca^{2+}]_{cyt}$ increase. The defense response occurs by catalyzing the elicitor peptide precursors into mature peptides located on the cytoplasmic side of the vacuole membrane; in turn, these peptides are recognized by the cytoplasmic vacuolar membrane-targeted receptor-elicitor peptide receptors [103,104]. Although Ca^{2+} transfer over long distances depends on ROS produced by NADPH oxidase, inhibition of calcium ion signaling can weaken the wound response to jasmonic acid (JA) and ethylene production [97,105]. Therefore, these results indicate that there is a closely linked interaction among various substances related to wounding signals.

Ca^{2+} is directly involved in the generation and propagation of long-distance signals in plants. Under strong local stress, variation potential (VP), a long-distance intercellular electrical signal, is the potential mechanism for coordinating functional responses to different plant cells, which can cause functional changes in unstimulated organs and tissues, namely, systematic responses of plants. Specifically, ligand-dependent or mechanically sensitive Ca^{2+} channels are activated by the propagation of chemical or hydraulic signals or a combination of these potentially distant signals. Subsequent Ca^{2+} influx can trigger VP production, thus inducing H^+-ATPase inactivation and possibly Cl^- channel activation [106].

In long-distance ROS signal transduction, RESPIRATORY BURST HOMOLOG D (RBOHD) is a ferric oxidoreductase that can be activated directly by calcium ions binding to its EF-Hand motif and phosphorylated by various protein kinases, such as CPK5 and CIPK. *Botrytis*-induced kinase 1 (BIK1) is also under Ca^{2+}-dependent regulation by CPK28 and phosphorylates RBOHD. ROS-activated Ca^{2+}-permeable channels on the plasma membrane provide a mechanism for RBOHD to trigger its further activation [107]. The crosstalk between Ca^{2+} and ROS to transmit these signals among cells across long distances, namely, that of RBOH, is activated by Ca^{2+}-dependent protein kinases in the presence of Ca^{2+}. This leads to the accumulation of nonprotoplast ROS, leading to induced Ca^{2+} release from adjacent cells. Then, another Ca^{2+}-dependent protein kinase is activated circularly [99,108]. In this way, signals are transmitted over long distances within plants (Figure 3) [109].

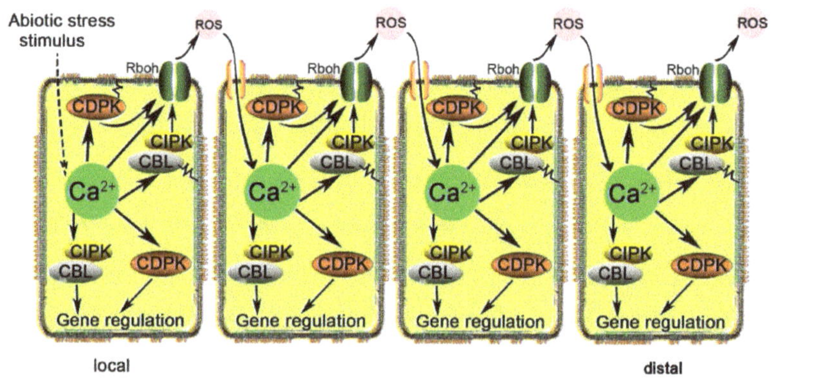

Figure 3. Schematic model of Ca^{2+}- and ROS-mediated cell-to-cell signal propagation over long distances in plants. Stimulating the production of cytosolic Ca^{2+} signals results in the activation of RBOHD by Ca^{2+}-regulated kinases, which produce ROS and then propagate the signal by activating Ca^{2+} channels in neighboring cells.

Generally, Jasmonate-associated VQ domain protein 1 (JAV1) associates with JASMONATE ZIM domain protein 8 (JAZ8) and WRKY51 to form the JAV1-JAZ8-WRKY51 (JJW) complex, which inhibits the expression of jasmonate (JA) synthesis genes. Once the plant sustains an injury, the sudden increase in the concentration of Ca^{2+} causes calmodulin to sense Ca^{2+} and combine with JAV1, thus phosphorylating JAV1, depolymerizing the JJW complex, and releasing the transcriptional inhibition of the JA synthesis gene lipoxygenase 2 (LOX2), which finally results in the accumulation of large quantities of jasmine in response to wound stress [110,111].

2.6. Waterlogging Stress

Plants will be damaged by a lack of sufficient oxygen (O_2) for respiration when they are exposed to waterlogging or submergence stress [112]. Under flooding conditions, when O_2 is lacking, it will likely cause a massive buildup of CO_2 as respiration and metabolism proceed, and when this occurs, intracellular Ca^{2+} in plants is required for the response to waterlogging-induced hypoxia stress in nonphotosynthetic organs [113,114]. Hypoxia promotes a real-time $[Ca^{2+}]_{cyt}$ increase and ROS accumulation, which may be interdependent [115]. For example, ROS in guard cells and root cells can activate Ca^{2+} channels, and Ca^{2+} can also promote ROS accumulation. Furthermore, in mutants with a loss of function of the PM-NAD(P)H oxidase subunits, ROS produced by the defective enzyme can activate Ca^{2+} channels on the cell membrane to achieve Ca^{2+} flow, thus contributing to the promotion of root tip growth. It has been demonstrated that ROS accumulation is coupled with Ca^{2+} dynamics in pollen tubes and root tips; however, relevant and reliable biochemical evidence about whether ROS directly activate NADPH oxidase is necessary [116].

Previous research showed that $[Ca^{2+}]_{cyt}$ acts as a key transducer of hypoxic signals in rice and wheat protoplasm exposed to hypoxia stress [117], and alcohol dehydrogenases (ADH), whose activity is involved in resistance to waterlogging, displayed significant improvement in maize [118]. In corn cells, hypoxic signaling rapidly elevates $[Ca^{2+}]_{cyt}$ by the release of intracellular stores of Ca^{2+}; however, Ca^{2+} is not only involved in hypoxic signal transduction but also affects the activity of related Ca^{2+}-dependent enzymes, such as alcohol dehydrogenase, reflecting tolerance to hypoxia [119]. Studies indicate that Ca^{2+} influx can promote the reduction in H_2S in plants suffering from waterlogging-induced hypoxia stress [120]. H_2S production by CBS is 3.5 times higher in the presence of Ca^{2+}/CaM than in the absence of Ca^{2+}, but it is inhibited by treatment with CaM inhibitors. The application of exogenous Ca^{2+} and its ion carrier A23187 markedly increased H_2S-induced antioxidant activity, while the calcium-chelating agent EGTA, the plasma membrane channel blocker La^{3+}, and calmodulin antagonists attenuated this resistance [58]. During waterlogging, hypoxia stress causes the accumulation of H_2O_2, activates the ROS-induced Ca^{2+} channel and triggers the self-amplifying "ROS-Ca^{2+} hub", which further increases K^+ loss and cell inactivation. The increased content of gamma-aminobutyric acid (GABA) induced by hypoxia is beneficial to the recovery of membrane potential and the maintenance of homeostasis between cytosolic K^+ and Ca^{2+} signaling. In addition, the ROS-Ca^{2+} hub can be better regulated by elevated GABA through transcriptional control of RBOH gene expression, thus preventing the excessive accumulation of H_2O_2 and allowing plants to more easily survive waterlogging [121].

2.7. UV-B Radiation Stress

UV-B radiation stress not only has adverse effects on plant morphology, such as plant dwarfing and leaf thickening, but also harms plant physiological processes, including chloroplast structure damage, photosynthetic rate decreases, and transpiration weakening [122–124]. Studies suggest that there are at least two pathways involved in the cytoplasmic Ca^{2+} response to UV-B radiation stress in plants. On the one hand, enhanced UV-B radiation triggers a significant increase in the free Ca^{2+} concentration in the cytoplasm of wheat mesophyll cells, which may release Ca^{2+} from the intracellular calcium pool or increase intracellular Ca^{2+} influx. UV-B radiation inhibits CaM, leading to it dissociating from the inhibitory region and in turn binding to the active site, which leaves the Ca^{2+} pump in a resting state [125]. On the other hand, UV-B radiation possibly promotes phosphatase dephosphorylation in the inhibitory region and combines with the active site to play an inhibitory role [126]. In addition, the calcium pump is directly activated to change the transport of intracellular Ca^{2+} under UV-B radiation conditions, thereby increasing $[Ca^{2+}]_{cyt}$. Furthermore, a slightly increased $[Ca^{2+}]_{cyt}$ can not only act on the membrane skeleton and significantly reduce the deformability of cells but is also involved in the lipid redistribution of the membrane and the decline in membrane stability [127].

The total phenol content of wheat under UV-B+$CaCl_2$ treatment increased by 10.3% compared with UV-B treatment alone. Most of the genes related to phenolic biosynthesis were upregulated during wheat germination, suggesting that exogenous Ca^{2+} promotes the accumulation of free phenols and bound phenols in germinal wheat exposed to UV-B radiation. In addition, treatment with Ca^{2+} can significantly alleviate membrane lipid peroxidation, activate antioxidant enzymes and regulate plant hormone levels. However, the Ca^{2+} channel blocker $LaCl_3$ significantly reduced TPC and APX activity [128]. These contrasting results suggested that Ca^{2+} was involved in the regulation of phenolic metabolism, antioxidant enzyme activity and endogenous plant hormone levels of germinal wheat in response to UV-B radiation stress [129].

SA is considered to be a synergist of H_2O_2, which may contribute to the generation or maintenance of ROS signaling levels and participate in many signaling responses to abiotic stresses, such as UV-B [130] and heavy metals [86]. Ca^{2+} is essential for H_2O_2- and SA-mediated signal transduction. *Arabidopsis thaliana* BTB and TAZ domain proteins (AtBTs) are Ca^{2+}-dependent CaM-binding proteins. The AtBT family may be a signal transduction center, and the signal transduction chain includes Ca^{2+}, H_2O_2 and SA. These signals may regulate transcription by altering AtBT expression and conformation [131].

3. Calcium Ion Downstream Signaling Response

Under abiotic stress conditions, plants transmit information through a second messenger, allowing cells to transmit external information into the cell interior. The cells then respond by triggering downstream reactions, consisting of transcriptional regulation and protein modification, to influence appropriate adaptive responses [132]. For example, in response to heat stress, altered membrane fluidity is sensed through Ca^{2+} channels and receptor-like kinases. Heat stress transcription factor A1 (HsfA1) transcription factors are the main heat-stress-resistance regulatory factors in plants. When activated by heat, they target downstream transcription factors, microRNAs and *ONSEN* (a copia-like retrotransposon) to induce the expression of heat stress-responsive genes that are critical for ROS clearance, protein homeostasis and heat stress memory [133]. Downstream events of Ca^{2+} signal transduction are mainly mediated by Ca^{2+}-binding proteins. In *Arabidopsis*, membrane hyperpolarization and ROS-activated Ca^{2+}-permeable channels under K^+ deficiency result in an increase in cytoplasmic Ca^{2+}, and Ca^{2+} signals are sensed by specific sensors and transmitted downstream. CBL1/CBL9 recruits the cytoplasmic kinase CIPK23 to the plasma membrane, where CIPK23 activates AKT1-mediated uptake of K^+ through phosphorylation. [134,135].

Calcium regulates the actin cytoskeleton either directly by binding to actin-binding proteins (ABPs) and regulating their activity or indirectly through calcium-stimulated protein kinases, such as CDPKs. The oscillation of the Ca^{2+} concentration gradient in the tip region of the pollen tube affects actin dynamics, and the remodeling of the actin cytoskeleton is associated with pollen tube elongation, showing that the Ca^{2+} concentration gradient may precisely regulate actin dynamics and promote pollen tube growth [136].

4. Conclusions and Perspectives

As one of the most important signaling molecules in cells, the Ca^{2+} signal transduction pathway is widely involved in the regulation of growth and development, abiotic stress response and many other physiological processes. Various studies have confirmed that abiotic stresses such as drought, high salt, ultraviolet light, heavy metal, waterlogging and extreme temperature can lead to a rapid increase in intracellular Ca^{2+} via the regulation of a variety of Ca^{2+} channels and trigger the Ca^{2+} signaling process. Then, the signals are decoded by Ca^{2+} sensors, following a series of physiological reactions through appropriate transduction pathways. Ca^{2+} is involved in crosstalk between other signaling molecules and phytohormone interactions when plants suffer from abiotic stress. In general, calcium, as the central node of the regulatory network, assists other regulators in adapting to adverse abiotic stresses.

Although the many molecular mechanisms behind Ca^{2+} involvement in abiotic stress responses have been elucidated, it remains unclear how plants can accurately distinguish the types and intensities of external stimuli and thus regulate $[Ca^{2+}]_{cyt}$ in a precise and complex way so that they can respond to a series of complex upstream signals accurately and exclusively and ensure signal transduction sensitivity and specificity concurrently. Furthermore, because crosstalk between Ca^{2+} and other signaling molecules is vital for the stress response, the mechanism of stress perception and the system of signal transduction at the biological level should be investigated. Therefore, the next important task for Ca^{2+}

signaling research is to determine which physiological reactions are involved in the various Ca^{2+}-targeted proteins downstream of calcium signaling and which downstream molecules are regulated to affect gene expression. Moreover, with recent advances in techniques and the development of molecular biology, cell biology, genetics and other disciplines, the role of Ca^{2+} signaling will certainly be elucidated more thoroughly.

Author Contributions: R.P. and L.J. contributed to the conception of this review. Y.L. (Yaoqi Li), Y.L. (Yinai Liu) and R.P. (Renyi Peng) designed and produced the figures. R.P. (Renyi Peng) and Y.L. (Yaoqi Li) wrote the manuscript. Y.L. (Yaoqi Li), Y.L. (Yinai Liu), L.J. and R.P. (Renyi Peng) revised the manuscript. All authors have read and agreed to the published version of the manuscript.

Funding: Financial support from the Natural Science Foundation of Zhejiang Province (Grant No. LQ20C020003).

Institutional Review Board Statement: Not applicable.

Informed Consent Statement: Not applicable.

Data Availability Statement: Not applicable.

Conflicts of Interest: The authors declare no conflict of interest.

Abbreviations

$[Ca^{2+}]_{cyt}$	Cytosolic Ca^{2+} concentration
ABA	Abscisic acid
ABPs	Actin-binding proteins
ACAs	Ca^{2+}-ATPases
ACS	1-aminocyclopropane-1-carboxylic acid synthases
ADH	Alcohol dehydrogenases
APX	Ascorbate peroxidase
AtBTs	*Arabidopsis thaliana* BTB and TAZ domain proteins
AtTPC1	*Arabidopsis thaliana* Two pore channel 1
BIK1	*Botrytis*-induced kinase 1
CA	Citric acid
cADPR	Cyclic ADP-ribose
CaMs	Calmodulins
CAX	Ca^{2+} ex-changers
CBLs	Calcineurin-B like proteins
CDPKs	Ca^{2+}-dependent protein kinases
cGMP	Cyclic guanosine 3′,5′-monophosphate
CIPKs	CBL-interacting protein kinases
CMLs	Calmodulin-like-proteins
CNGCs	Cyclic nucleotide-gated channels
CPK	Calcium-dependent protein kinase
ECAs	ER-type Ca^{2+} -ATPases
EGTA	Ethylene glycol diethyl ether diamine tetraacetic acid
ET	Evapotranspiration
GA	Gibberellin
GABA	Gamma-aminobutyric acid
glR3.3/3.6	Glutamate receptor-like 3.3/3.6
GLRs	Glutamate receptor-like channels
GTL	GT-2like 1
H_2O_2	Hydrogen peroxide
H_2S	Hydrogen sulfide
HACCs	Hyperpolarization-activated calcium channels
HMA1	P1-ATPases
HsfA1	Heat stress transcription factor A1
HSFs	Heat shock transcription factors
HSPs	Heat stress transcription factors

IP3	Inositol 1,4,5-trisphosphate
JA	Jasmonic acid
JAV1	Jasmonate-associated VQ domain protein 1
JAZ8	JASMONATE ZIM domain protein 8
JJW	JAV1-JAZ8-WRKY51
KORC	K^+ permeable outwardly rectifying conductance
LOX2	Lipoxygenase 2
MAPKs	Mitogen-activated protein kinases
MCAs	Mid1-complementing activity channels
MCUC	Mitochondrial calcium uniporter complex
MeJA	Methyl jasmonate
MPK	Mitogen-activated protein kinase
MscS	Mechanosensitive channels of small
MSLs	Conductance-like channels
NADPH	Nicotinamide adenine dinucleotide phosphate
NO	Nitric oxide
NSCCs	Nonselective cation channels
O_2^-	Superoxide anion radicals
OH	Hydroxyl radical
OSCAs	Hyperosmolality-induced Ca^{2+} increase channels
PCA1	Ca^{2+}-ATPase
PLC	Phopholipase C
PLD	Phopholipase D
POD	Peroxidase
ROS	Reactive oxygen species
SA	Salicylic acid
SOD	Superoxide dismutase
SOS	Salt overly sensitive
TPC	Total phenolic contents
UV-B	Ultraviolet-B radiation stress
VP	Variation potential

References

1. Poovaiah, B.W.; Reddy, A.S.N.; Leopold, A.C. Calcium messenger system in plants. *CRC Crit. Rev. Plant Sci.* **1987**, *6*, 47–103. [CrossRef] [PubMed]
2. Aldon, D.; Mbengue, M.; Mazars, C.; Galaud, J.-P. Calcium signalling in plant biotic interactions. *Int. J. Mol. Sci.* **2018**, *19*, 665. [CrossRef] [PubMed]
3. Gao, Y.L.; Zhang, G.Z. A calcium sensor calcineurin B-like 9 negatively regulates cold tolerance via calcium signaling in *Arabidopsis thaliana*. *Plant Signal. Behav.* **2019**, *14*, 6. [CrossRef] [PubMed]
4. Zhao, Y.; Pan, Z.; Zhang, Y.; Qu, X.L.; Zhang, Y.G.; Yang, Y.Q.; Jiang, X.N.; Huang, S.J.; Yuan, M.; Schumaker, K.S.; et al. The actin-related protein2/3 complex regulates mitochondrial-associated calcium signaling during salt stress in *Arabidopsis*. *Plant Cell* **2013**, *25*, 4544–4559. [CrossRef]
5. Jing, X.; Cai, C.J.; Fan, S.H.; Wang, L.J.; Zeng, X.L. Spatial and temporal calcium signaling and Its physiological effects in Moso Bamboo under drought stress. *Forests* **2019**, *10*, 224. [CrossRef]
6. Iqbal, Z.; Iqbal, M.S.; Singh, S.P.; Buaboocha, T. Ca^{2+}/Calmodulin complex triggers CAMTA transcriptional machinery under stress in plants: Signaling cascade and molecular regulation. *Front. Plant Sci.* **2020**, *11*, 16. [CrossRef]
7. Liang, C.J.; Zhang, Y.Q.; Ren, X.Q. Calcium regulates antioxidative isozyme activity for enhancing rice adaption to acid rain stress. *Plant Sci.* **2021**, *306*, 10. [CrossRef]
8. Demidchik, V.; Shabala, S.; Isayenkov, S.; Cuin, T.A.; Pottosin, I. Calcium transport across plant membranes: Mechanisms and functions. *New Phytol.* **2018**, *220*, 49–69. [CrossRef]
9. Zeng, H.; Zhao, B.; Wu, H.; Zhu, Y.; Chen, H. Comprehensive in silico characterization and expression profiling of nine gene families associated with calcium transport in soybean. *Agronomy* **2020**, *10*, 1539. [CrossRef]
10. Ranty, B.; Aldon, D.; Cotelle, V.; Galaud, J.P.; Thuleau, P.; Mazars, C. Calcium sensors as key hubs in plant responses to biotic and abiotic stresses. *Front. Plant Sci.* **2016**, *7*, 7. [CrossRef]
11. Michal Johnson, J.; Reichelt, M.; Vadassery, J.; Gershenzon, J.; Oelmüller, R. An *Arabidopsis* mutant impaired in intracellular calcium elevation is sensitive to biotic and abiotic stress. *BMC Plant Biol.* **2014**, *14*, 162. [CrossRef] [PubMed]
12. Matthus, E.; Wilkins, K.A.; Swarbreck, S.M.; Doddrell, N.H.; Doccula, F.G.; Costa, A.; Davies, J.M. Phosphate starvation alters abiotic-stress-induced cytosolic free calcium increases in roots. *Plant Physiol.* **2019**, *179*, 1754–1767. [CrossRef] [PubMed]
13. Barkla, B.J.; Hirschi, K.D.; Pittman, J.K. Exchangers man the pumps. *Plant Signal. Behav.* **2008**, *3*, 354–356. [CrossRef] [PubMed]

14. Bose, J.; Pottosin, I.I.; Shabala, S.S.; Palmgren, M.G.; Shabala, S. Calcium efflux systems in stress signaling and adaptation in plants. *Front. Plant Sci.* **2011**, *2*, 17. [CrossRef]
15. Sun, Q.P.; Guo, Y.; Sun, Y.; Sun, D.Y.; Wang, X.J. Influx of extracellular Ca^{2+} involved in jasmonic-acid-induced elevation of Ca^{2+} (cyt) and JR1 expression in *Arabidopsis thaliana*. *J. Plant Res.* **2006**, *119*, 343–350. [CrossRef]
16. Bouche, N.; Yellin, A.; Snedden, W.A.; Fromm, H. Plant-specific calmodulin-binding proteins. *Annu. Rev. Plant Biol.* **2005**, *56*, 435–466. [CrossRef]
17. Ma, X.; Li, Q.H.; Yu, Y.N.; Qiao, Y.M.; Haq, S.U.; Gong, Z.H. The CBL-CIPK pathway in plant response to stress signals. *Int. J. Mol. Sci.* **2020**, *21*, 5668. [CrossRef]
18. Demidchik, V.; Shabala, S. Mechanisms of cytosolic calcium elevation in plants: The role of ion channels, calcium extrusion systems and NADPH oxidase-mediated 'ROS-Ca^{2+} Hub'. *Funct. Plant Biol.* **2018**, *45*, 9–27. [CrossRef]
19. Yang, Y.L.; Xu, S.J.; An, L.Z.; Chen, N.L. NADPH oxidase-dependent hydrogen peroxide production, induced by salinity stress, may be involved in the regulation of total calcium in roots of wheat. *J. Plant Physiol.* **2007**, *164*, 1429–1435. [CrossRef]
20. Edel, K.H.; Marchadier, E.; Brownlee, C.; Kudla, J.; Hetherington, A.M. The evolution of calcium-based signalling in plants. *Curr. Biol.* **2017**, *27*, R667–R679. [CrossRef]
21. Liu, J.; Niu, Y.; Zhang, J.; Zhou, Y.; Ma, Z.; Huang, X. Ca^{2+} channels and Ca^{2+} signals involved in abiotic stress responses in plant cells: Recent advances. *Plant Cell Tissue Organ Cult.* **2018**, *132*, 413–424. [CrossRef]
22. Wen, F.; Ye, F.; Xiao, Z.L.; Liao, L.; Li, T.J.; Jia, M.L.; Liu, X.S.; Wu, X.Z. Genome-wide survey and expression analysis of calcium-dependent protein kinase (CDPK) in grass Brachypodium distachyon. *BMC Genom.* **2020**, *21*, 53. [CrossRef] [PubMed]
23. Abid, M.; Ali, S.; Qi, L.K.; Zahoor, R.; Tian, Z.W.; Jiang, D.; Snider, J.L.; Dai, T.B. Physiological and biochemical changes during drought and recovery periods at tillering and jointing stages in wheat (*Triticum aestivum* L.). *Sci. Rep.* **2018**, *8*, 15. [CrossRef]
24. de Carvalho, M.H.C. Drought stress and reactive oxygen species production, scavenging and signaling. *Plant Signal. Behav.* **2008**, *3*, 156–165. [CrossRef] [PubMed]
25. Kuromori, T.; Fujita, M.; Takahashi, F.; Yamaguchi-Shinozaki, K.; Shinozaki, K. Inter-tissue and inter-organ signaling in drought stress response and phenotyping of drought tolerance. *Plant J.* **2022**, *109*, 342–358. [CrossRef]
26. Feng, X.; Liu, W.X.; Zeng, F.R.; Chen, Z.H.; Zhang, G.P.; Wu, F.B. K^+ Uptake, H^+-ATPase pumping activity and Ca^{2+} efflux mechanism are involved in drought tolerance of barley. *Environ. Exp. Bot.* **2016**, *129*, 57–66. [CrossRef]
27. Wu, S.W.; Sun, X.C.; Tan, Q.L.; Hu, C.X. Molybdenum improves water uptake via extensive root morphology, aquaporin expressions and increased ionic concentrations in wheat under drought stress. *Environ. Exp. Bot.* **2019**, *157*, 241–249. [CrossRef]
28. Cousson, A. Involvement of phospholipase C-independent calcium-mediated abscisic acid signalling during *Arabidopsis* response to drought. *Biol. Plant.* **2009**, *53*, 53–62. [CrossRef]
29. Jiao, C.F.; Yang, R.Q.; Gu, Z.X. Cyclic ADP-ribose and IP3 mediate abscisic acid-induced isoflavone accumulation in soybean sprouts. *Biochem. Biophys. Res. Commun.* **2016**, *479*, 530–536. [CrossRef]
30. Zou, J.-J.; Li, X.-D.; Ratnasekera, D.; Wang, C.; Liu, W.-X.; Song, L.-F.; Zhang, W.-Z.; Wu, W.-H. *Arabidopsis* CALCIUM-DEPENDENT PROTEIN KINASE8 and CATALASE3 function in abscisic acid-mediated signaling and H_2O_2 homeostasis in stomatal guard cells under drought stress. *Plant Cell* **2015**, *27*, 1445–1460. [CrossRef]
31. Sun, L.R.; Li, Y.P.; Miao, W.W.; Piao, T.T.; Hao, Y.; Hao, F.S. NADK2 positively modulates abscisic acid-induced stomatal closure by affecting accumulation of H_2O_2, Cat$(^{2+})$ and nitric oxide in *Arabidopsis* guard cells. *Plant Sci.* **2017**, *262*, 81–90. [CrossRef] [PubMed]
32. Wang, C.; Deng, Y.; Liu, Z.; Liao, W. Hydrogen sulfide in plants: Crosstalk with other signal molecules in response to abiotic stresses. *Int. J. Mol. Sci.* **2021**, *22*, 12068. [CrossRef] [PubMed]
33. Dubovskaya, L.V.; Bakakina, Y.S.; Kolesneva, E.V.; Sodel, D.L.; McAinsh, M.R.; Hetherington, A.M.; Volotovski, I.D. cGMP-dependent ABA-induced stomatal closure in the ABA-insensitive *Arabidopsis* mutantabi1-1. *New Phytol.* **2011**, *191*, 57–69. [CrossRef] [PubMed]
34. Liu, L.; Xiang, Y.; Yan, J.W.; Di, P.C.; Li, J.; Sun, X.J.; Han, G.Q.; Ni, L.; Jiang, M.Y.; Yuan, J.H.; et al. BRASSINOSTEROID-SIGNALING KINASE 1 phosphorylating CALCIUM/CALMODULIN-DEPENDENT PROTEIN KINASE functions in drought tolerance in maize. *New Phytol.* **2021**, *231*, 695–712. [CrossRef]
35. Huda, K.M.K.; Banu, M.S.A.; Yadav, S.; Sahoo, R.K.; Tuteja, R.; Tuteja, N. Salinity and drought tolerant OsACA6 enhances cold tolerance in transgenic tobacco by interacting with stress-inducible proteins. *Plant Physiol. Biochem.* **2014**, *82*, 229–238. [CrossRef]
36. Mohanta, T.K.; Bashir, T.; Hashem, A.; Abd_Allah, E.F.; Khan, A.L.; Al-Harrasi, A.S. Early events in plant abiotic stress signaling: Interplay between calcium, reactive oxygen species and phytohormones. *J. Plant Growth Regul.* **2018**, *37*, 1033–1049. [CrossRef]
37. Sanyal, S.K.; Kanwar, P.; Yadav, A.K.; Sharma, C.; Kumar, A.; Pandey, G.K. *Arabidopsis* CBL interacting protein kinase 3 interacts with ABR1, an APETALA2 domain transcription factor, to regulate ABA responses. *Plant Sci.* **2017**, *254*, 48–59. [CrossRef]
38. Cuellar, T.; Pascaud, F.O.; Verdeil, J.L.; Torregrosa, L.; Adam-Blondon, A.F.; Thibaud, J.-B.; Sentenac, H.; Gaillard, I. A grapevine shaker inward K^+ channel activated by the calcineurin B-like calcium sensor 1-protein kinase CIPK23 network is expressed in grape berries under drought stress conditions. *Plant J.* **2010**, *61*, 58–69. [CrossRef]
39. Cheng, P.L.; Gao, J.J.; Feng, Y.T.; Zhang, Z.X.; Liu, Y.N.; Fang, W.M.; Chen, S.M.; Chen, F.D.; Jiang, J.F. The chrysanthemum leaf and root transcript profiling in response to salinity stress. *Gene* **2018**, *674*, 161–169. [CrossRef]
40. Song, W.Y.; Zhang, Z.B.; Shao, H.B.; Guo, X.L.; Cao, H.X.; Zhao, H.B.; Fu, Z.Y.; Hu, X.J. Relationship between calcium decoding elements and plant abiotic-stress resistance. *Int. J. Biol. Sci.* **2008**, *4*, 116–125. [CrossRef]

41. Weng, H.; Yoo, C.Y.; Gosney, M.J.; Hasegawa, P.M.; Mickelbart, M.V. Poplar GTL1 is a Ca^{2+}/calmodulin-binding transcription factor that functions in plant water use efficiency and drought tolerance. *PLoS ONE* **2012**, *7*, e32925. [CrossRef] [PubMed]
42. Durand, M.; Cohen, D.; Aubry, N.; Buré, C.; Tomášková, I.; Hummel, I.; Brendel, O.; Le Thiec, D. Element content and expression of genes of interest in guard cells are connected to spatiotemporal variations in stomatal conductance. *Plant Cell Environ.* **2020**, *43*, 87–102. [CrossRef] [PubMed]
43. Marin-Guirao, L.; Sandoval-Gil, J.M.; Garcia-Munoz, R.; Ruiz, J.M. The stenohaline seagrass posidonia oceanica can persist in natural environments under fluctuating hypersaline conditions. *Estuaries Coasts* **2017**, *40*, 1688–1704. [CrossRef]
44. Hryvusevich, P.; Navaselsky, I.; Talkachova, Y.; Straltsova, D.; Keisham, M.; Viatoshkin, A.; Samokhina, V.; Smolich, I.; Sokolik, A.; Huang, X.; et al. Sodium influx and potassium efflux currents in sunflower root cells under high salinity. *Front. Plant Sci.* **2021**, *11*, 10. [CrossRef] [PubMed]
45. Wegner, L.H.; De Boer, A.H. Properties of two outward-rectifying channels in root xylem parenchyma cells suggest a role in K^+ homeostasis and long-distance signaling. *Plant Physiol.* **1997**, *115*, 1707–1719. [CrossRef] [PubMed]
46. Mahajan, S.; Pandey, G.K.; Tuteja, N. Calcium- and salt-stress signaling in plants: Shedding light on SOS pathway. *Arch. Biochem. Biophys.* **2008**, *471*, 146–158. [CrossRef]
47. Ma, L.; Ye, J.; Yang, Y.; Lin, H.; Yue, L.; Luo, J.; Long, Y.; Fu, H.; Liu, X.; Zhang, Y.; et al. The SOS2-SCaBP8 complex generates and fine-tunes an AtANN4-dependent calcium signature under salt stress. *Dev. Cell* **2019**, *48*, 697–709. [CrossRef]
48. Ma, D.-M.; Xu, W.-R.; Li, H.-W.; Jin, F.-X.; Guo, L.-N.; Wang, J.; Dai, H.-J.; Xu, X. Co-expression of the *Arabidopsis* SOS genes enhances salt tolerance in transgenic tall fescue (*Festuca arundinacea Schreb.*). *Protoplasma* **2014**, *251*, 219–231. [CrossRef]
49. Zhang, H.; Zhang, Y.; Deng, C.; Deng, S.; Li, N.; Zhao, C.; Zhao, R.; Liang, S.; Chen, S. The *Arabidopsis* Ca^{2+}-dependent protein kinase CPK12 is involved in plant response to salt stress. *Int. J. Mol. Sci.* **2018**, *19*, 4062. [CrossRef]
50. Kurusu, T.; Sakurai, Y.; Miyao, A.; Hirochika, H.; Kuchitsu, K. Identification of a putative voltage-gated Ca^{2+}-permeable channel (OSTPC1) involved in Ca^{2+} influx and regulation of growth and development in rice. *Plant Cell Physiol.* **2004**, *45*, 693–702. [CrossRef]
51. Cousson, A.; Vavasseur, A. Two potential Ca^{2+}-dependent transduction pathways in stomatal closing in response to abscisic acid. *Plant Physiol. Biochem.* **1998**, *36*, 257–262. [CrossRef]
52. Qudeimat, E.; Frank, W. Ca^{2+} signatures The role of Ca^{2+}-ATPases. *Plant Signal. Behav.* **2009**, *4*, 350–352. [CrossRef] [PubMed]
53. Jiang, Z.H.; Zhou, X.P.; Tao, M.; Yuan, F.; Liu, L.L.; Wu, F.H.; Wu, X.M.; Xiang, Y.; Niu, Y.; Liu, F.; et al. Plant cell-surface GIPC sphingolipids sense salt to trigger Ca^{2+} influx. *Nature* **2019**, *572*, 341–346. [CrossRef] [PubMed]
54. Ma, Y.; Dai, X.Y.; Xu, Y.Y.; Luo, W.; Zheng, X.M.; Zeng, D.L.; Pan, Y.J.; Lin, X.L.; Liu, H.H.; Zhang, D.J.; et al. COLD1 confers chilling tolerance in rice. *Cell* **2015**, *160*, 1209–1221. [CrossRef]
55. Yang, N.; Peng, C.L.; Cheng, D.; Huang, Q.; Xu, G.H.; Gao, F.; Chen, L.B. The over-expression of calmodulin from Antarctic notothenioid fish increases cold tolerance in tobacco. *Gene* **2013**, *521*, 32–37. [CrossRef]
56. Luo, Q.; Wei, Q.; Wang, R.; Zhang, Y.; Zhang, F.; He, Y.; Yang, G.; He, G. Ectopic expression of BdCIPK31 confers enhanced low-temperature tolerance in transgenic tobacco plants. *Acta Biochim. Biophys. Sin.* **2018**, *50*, 199–208. [CrossRef]
57. Wood, N.T.; Allan, A.C.; Haley, A.; Viry-Moussaïd, M.; Trewavas, A.J. The characterization of differential calcium signalling in tobacco guard cells. *Plant J.* **2000**, *24*, 335–344. [CrossRef]
58. Li, Z.G.; Gong, M.; Xie, H.; Yang, L.; Li, J. Hydrogen sulfide donor sodium hydrosulfide-induced heat tolerance in tobacco (*Nicotiana tabacum* L.) suspension cultured cells and involvement of Ca^{2+} and calmodulin. *Plant Sci.* **2012**, *185*, 185–189. [CrossRef]
59. Li, W.; Sun, Z.-H.; Zhang, W.-C.; Ma, X.-T.; Liu, D.-H. Role of Ca^{2+} and calmodulin on freezing tolerance of citrus protoplasts. *Acta Biochim. Biophys. Sin.* **1997**, *23*, 262–266.
60. Cheong, Y.H.; Kim, K.-N.; Pandey, G.K.; Gupta, R.; Grant, J.J.; Luan, S. CBL1, a calcium sensor that differentially regulates salt, drought, and cold responses in *Arabidopsis*. *Plant Cell* **2003**, *15*, 1833–1845. [CrossRef]
61. Cheong, Y.H.; Pandey, G.K.; Grant, J.J.; Batistic, O.; Li, L.; Kim, B.G.; Lee, S.C.; Kudla, J.; Luan, S. Two calcineurin B-like calcium sensors, interacting with protein kinase CIPK23, regulate leaf transpiration and root potassium uptake in *Arabidopsis*. *Plant J.* **2007**, *52*, 223–239. [CrossRef] [PubMed]
62. Zhuang, Q.; Chen, S.; Jua, Z.; Yao, Y. Joint transcriptomic and metabolomic analysis reveals the mechanism of low-temperature tolerance in *Hosta ventricosa*. *PLoS ONE* **2021**, *16*, e0259455. [CrossRef] [PubMed]
63. Wan, B.L.; Lin, Y.J.; Mou, T.M. Expression of rice Ca^{2+}-dependent protein kinases (CDPKs) genes under different environmental stresses. *FEBS Lett.* **2007**, *581*, 1179–1189. [CrossRef] [PubMed]
64. Yang, G.; Shen, S.; Yang, S.; Komatsu, S. OsCDPK13, a calcium-dependent protein kinase gene from rice, is induced in response to cold and gibberellin. *Plant Physiol. Biochem.* **2003**, *41*, 369–374. [CrossRef]
65. Abo-El-Saad, M.; Wu, R. A rice membrane calcium-dependent protein kinase is induced by gibberellin. *Plant Physiol.* **1995**, *108*, 787–793. [CrossRef] [PubMed]
66. Sanders, D.; Brosnan, J.M.; Muir, S.R.; Allen, G.; Crofts, A.; Johannes, E. Ion channels and calcium signalling in plants: Multiple pathways and cross-talk. *Biochem. Soc. Symp.* **1994**, *60*, 183–197.
67. Islam, M.M.; Munemasa, S.; Hossain, M.A.; Nakamura, Y.; Mori, I.C.; Murata, Y. Roles of AtTPC1, vacuolar two pore channel 1, in *Arabidopsis* stomatal closure. *Plant Cell Physiol.* **2010**, *51*, 302–311. [CrossRef]

68. Agurla, S.; Gahir, S.; Munemasa, S.; Murata, Y.; Raghavendra, A.S. Mechanism of stomatal closure in plants exposed to drought and cold stress. In *Survival Strategies in Extreme Cold and Desiccation: Adaptation Mechanisms and Their Applications*; IwayaInoue, M., Sakurai, M., Uemura, M., Eds.; Springer Singapore Pte Ltd.: Singapore, 2018; Volume 1081, pp. 215–232.
69. Knight, J.R.P.; Bastide, A.; Roobol, A.; Roobol, J.; Jackson, T.J.; Utami, W.; Barrett, D.A.; Smales, C.M.; Willis, A.E. Eukaryotic elongation factor 2 kinase regulates the cold stress response by slowing translation elongation. *Biochem. J.* **2015**, *465*, 227–238. [CrossRef]
70. Suzuki, N.; Katano, K. Coordination between ROS regulatory systems and other pathways under heat stress and pathogen attack. *Front. Plant Sci.* **2018**, *9*, 8. [CrossRef]
71. Li, T.-L.; Li, M.; Sun, Z.-P. Regulation effect of calcium and salicylic acid on defense enzyme activities in tomato leaves under sub-high temperature stress. *Yingyong Shengtai Xuebao* **2009**, *20*, 586–590.
72. Hu, Z.J.; Li, J.X.; Ding, S.T.; Cheng, F.; Li, X.; Jiang, Y.P.; Yu, J.Q.; Foyer, C.H.; Shi, K. The protein kinase CPK28 phosphorylates ascorbate peroxidase and enhances thermotolerance in tomato. *Plant Physiol.* **2021**, *186*, 1302–1317. [CrossRef] [PubMed]
73. Xing, X.J.; Ding, Y.R.; Jin, J.Y.; Song, A.P.; Chen, S.M.; Chen, F.D.; Fang, W.M.; Jiang, J.F. Physiological and transcripts analyses reveal the mechanism by which melatonin alleviates heat stress in chrysanthemum seedlings. *Front. Plant Sci.* **2021**, *12*, 17. [CrossRef] [PubMed]
74. Dou, H.O.; Xv, K.P.; Meng, Q.W.; Li, G.; Yang, X.H. Potato plants ectopically expressing *Arabidopsis thaliana* CBF3 exhibit enhanced tolerance to high-temperature stress. *Plant Cell Environ.* **2015**, *38*, 61–72. [CrossRef] [PubMed]
75. Zheng, H.Y.; Wang, W.L.; Xu, K.; Xu, Y.; Ji, D.H.; Chen, C.S.; Xie, C.T. Ca^{2+} influences heat shock signal transduction in *Pyropia haitanensis*. *Aquaculture* **2020**, *516*, 8. [CrossRef]
76. Virdi, A.S.; Singh, S.; Singh, P. Abiotic stress responses in plants: Roles of calmodulin-regulated proteins. *Front. Plant Sci.* **2015**, *6*, 19. [CrossRef]
77. Tiwari, A.; Singh, P.; Khadim, S.R.; Singh, A.K.; Singh, U.; Singh, P.; Asthana, R.K. Role of Ca^{2+} as protectant under heat stress by regulation of photosynthesis and membrane saturation in Anabaena PCC 7120. *Protoplasma* **2019**, *256*, 681–691. [CrossRef]
78. Tripp, J.; Mishra, S.K.; Scharf, K.D. Functional dissection of the cytosolic chaperone network in tomato mesophyll protoplasts. *Plant Cell Environ.* **2009**, *32*, 123–133. [CrossRef]
79. Barbero, F.; Guglielmotto, M.; Islam, M.; Maffei, M.E. Extracellular fragmented self-DNA is involved in plant responses to biotic stress. *Front. Plant Sci.* **2021**, *12*, 17. [CrossRef]
80. Doubnerova, V.; Ryslava, H. *Roles of HSP70 in Plant Abiotic Stress*; CRC Press-Taylor & Francis Group: Boca Raton, FL, USA, 2014; pp. 44–66.
81. Peng, X.; Zhang, X.N.; Li, B.; Zhao, L.Q. Cyclic nucleotide-gated ion channel 6 mediates thermotolerance in *Arabidopsis* seedlings by regulating nitric oxide production via cytosolic calcium ions. *BMC Plant Biol.* **2019**, *19*, 368. [CrossRef] [PubMed]
82. Zhang, Q.T.; Zhang, L.L.; Geng, B.; Feng, J.R.; Zhu, S.H. Interactive effects of abscisic acid and nitric oxide on chilling resistance and active oxygen metabolism in peach fruit during cold storage. *J. Sci. Food Agric.* **2019**, *99*, 3367–3380. [CrossRef]
83. Zhao, Y.L.; Du, H.W.; Wang, Y.K.; Wang, H.L.; Yang, S.Y.; Li, C.H.; Chen, N.; Yang, H.; Zhang, Y.H.; Zhu, Y.L.; et al. The calcium-dependent protein kinase ZmCDPK7 functions in heat-stress tolerance in maize. *J. Integr. Plant Biol.* **2021**, *63*, 510–527. [CrossRef] [PubMed]
84. Campos, C.B.L.; Di Benedette, J.P.T.; Morais, F.V.; Ovalle, R.; Nobrega, M.P. Evidence for the role of calcineurin in morphogenesis and calcium homeostasis during mycelium-to-yeast dimorphism of *Paracoccidioides brasiliensis*. *Eukaryot. Cell* **2008**, *7*, 1856–1864. [CrossRef] [PubMed]
85. Moustaka, J.; Ouzounidou, G.; Baycu, G.; Moustakas, M. Aluminum resistance in wheat involves maintenance of leaf Ca^{2+} and Mg^{2+} content, decreased lipid peroxidation and Al accumulation, and low photosystem II excitation pressure. *Biometals* **2016**, *29*, 611–623. [CrossRef] [PubMed]
86. Liu, Y.; Xu, R.K. The forms and distribution of aluminum adsorbed onto maize and soybean roots. *J. Soils Sediments.* **2015**, *15*, 491–502. [CrossRef]
87. Lan, T.; You, J.F.; Kong, L.N.; Yu, M.; Liu, M.H.; Yang, Z.M. The interaction of salicylic acid and Ca^{2+} alleviates aluminum toxicity in soybean (*Glycine max* L.). *Plant Physiol. Biochem.* **2016**, *98*, 146–154. [CrossRef]
88. Hashimoto, Y.; Smyth, T.J.; Hesterberg, D.; Israel, D.W. Soybean root growth in relation to ionic composition in magnesium-amended acid subsoils: Implications on root citrate ameliorating aluminum constraints. *Soil Sci. Plant Nutr.* **2007**, *53*, 753–763. [CrossRef]
89. Cousson, A. Two calcium mobilizing pathways implicated within abscisic acid-induced stomatal closing in *Arabidopsis thaliana*. *Biol. Plant* **2007**, *51*, 285–291. [CrossRef]
90. Kinraide, T.B. Three mechanisms for the calcium alleviation of mineral toxicities. *Plant Physiol.* **1998**, *118*, 513–520. [CrossRef]
91. Huang, T.L.; Huang, H.J. ROS and CDPK-like kinase-mediated activation of MAP kinase in rice roots exposed to lead. *Chemosphere* **2008**, *71*, 1377–1385. [CrossRef]
92. Maksymiec, W.; Baszynski, T. Are calcium ions and calcium channels involved in the mechanisms of Cu^{2+} toxicity in bean plants? The influence of leaf age. *Photosynthetica* **1999**, *36*, 267–278. [CrossRef]
93. Fang, H.H.; Jing, T.; Liu, Z.Q.; Zhang, L.P.; Jin, Z.P.; Pei, Y.X. Hydrogen sulfide interacts with calcium signaling to enhance the chromium tolerance in *Setaria italica*. *Cell Calcium* **2014**, *56*, 472–481. [CrossRef]

94. Tian, W.; Wang, C.; Gao, Q.F.; Li, L.G.; Luan, S. Calcium spikes, waves and oscillations in plant development and biotic interactions. *Nat. Plants* **2020**, *6*, 750–759. [CrossRef] [PubMed]
95. Pawelek, A.; Duszyn, M.; Swiezawska, B.; Szmidt-Jaworska, A.; Jaworski, K. Transcriptional response of a novel HpCDPK1 kinase gene from *Hippeastrum x hybr.* to wounding and fungal infection. *J. Plant Physiol.* **2017**, *216*, 108–117. [CrossRef] [PubMed]
96. Uemura, T.; Wang, J.Q.; Aratani, Y.; Gilroy, S.; Toyota, M. Wide-field, real-time imaging of local and systemic wound signals in Arabidopsis. *J. Vis. Exp.* **2021**, *172*, e62114. [CrossRef] [PubMed]
97. Mohanta, T.K.; Occhipinti, A.; Zebelo, S.A.; Foti, M.; Fliegmann, J.; Bossi, S.; Maffei, M.E.; Bertea, C.M. Ginkgo biloba responds to herbivory by activating early signaling and direct defenses. *PLoS ONE* **2012**, *7*, e32822. [CrossRef] [PubMed]
98. Costa, A.; Luoni, L.; Marrano, C.A.; Hashimoto, K.; Koster, P.; Giacometti, S.; De Michelis, M.I.; Kudla, J.; Bonza, M.C. Ca^{2+}-dependent phosphoregulation of the plasma membrane Ca^{2+}-ATPase ACA8 modulates stimulus-induced calcium signatures. *J. Exp. Bot.* **2017**, *68*, 3215–3230. [CrossRef]
99. Farmer, E.E.; Gao, Y.Q.; Lenzoni, G.; Wolfender, J.L.; Wu, Q. Wound- and mechanostimulated electrical signals control hormone responses. *New Phytol.* **2020**, *227*, 1037–1050. [CrossRef]
100. Moyen, C.; Hammond-Kosack, K.E.; Jones, J.; Knight, M.R.; Johannes, E. Systemin triggers an increase of cytoplasmic calcium in tomato mesophyll cells: Ca^{2+} mobilization from intra- and extracellular compartments. *Plant Cell Environ.* **1998**, *21*, 1101–1111. [CrossRef]
101. Ludwig, A.A.; Saitoh, H.; Felix, G.; Freymark, G.; Miersch, O.; Wasternack, C.; Boller, T.; Jones, J.D.G.; Romeis, T. Ethylene-mediated cross-talk between calcium-dependent protein kinase and MAPK signaling controls stress responses in plants. *Proc. Natl. Acad. Sci. USA* **2005**, *102*, 10736–10741. [CrossRef]
102. Fichman, Y.; Mittler, R. Integration of electric, calcium, reactive oxygen species and hydraulic signals during rapid systemic signaling in plants. *Plant J.* **2021**, *107*, 7–20. [CrossRef]
103. Hander, T.; Fernandez-Fernandez, A.D.; Kumpf, R.P.; Willems, P.; Schatowitz, H.; Rombaut, D.; Staes, A.; Nolf, J.; Pottie, R.; Yao, P.F.; et al. Damage on plants activates Ca^{2+}-dependent metacaspases for release of immunomodulatory peptides. *Science* **2019**, *363*, eaar7486. [CrossRef] [PubMed]
104. Hou, S.G.; Yin, C.C.; He, P. Cleave and Unleash: Metacaspases Prepare Peps for Work. *Trends Plant Sci.* **2019**, *24*, 787–790. [CrossRef] [PubMed]
105. Matschi, S.; Hake, K.; Herde, M.; Hause, B.; Romeis, T. The calcium-dependent protein kinase CPK28 regulates development by inducing growth phase-specific, spatially restricted alterations in jasmonic acid levels independent of defense responses in *Arabidopsis*. *Plant Cell* **2015**, *27*, 591–606. [CrossRef] [PubMed]
106. Vodeneev, V.; Akinchits, E.; Sukhov, V. Variation potential in higher plants: Mechanisms of generation and propagation. *Plant Signal. Behav.* **2015**, *10*, 7. [CrossRef]
107. Choi, W.G.; Hilleary, R.; Swanson, S.J.; Kim, S.H.; Gilroy, S. Rapid, long-distance electrical and calcium signaling in plants. In *Annual Review of Plant Biology*; Merchant, S.S., Ed.; Annual Reviews: Palo Alto, CA, USA, 2016; Volume 67, pp. 287–307.
108. Takahashi, F.; Mizoguchi, T.; Yoshida, R.; Ichimura, K.; Shinozaki, K. Calmodulin-dependent activation of MAP kinase for ROS homeostasis in *Arabidopsis*. *Mol. Cell* **2011**, *41*, 649–660. [CrossRef]
109. Steinhorst, L.; Kudla, J. Calcium and reactive oxygen species rule the waves of signaling. *Plant Physiol.* **2013**, *163*, 471–485. [CrossRef]
110. Bonaventure, G.; Gfeller, A.; Proebsting, W.M.; Hortensteiner, S.; Chetelat, A.; Martinoia, E.; Farmer, E.E. A gain-of-function allele of TPC1 activates oxylipin biogenesis after leaf wounding in *Arabidopsis*. *Plant J.* **2007**, *49*, 889–898. [CrossRef]
111. Yan, C.; Fan, M.; Yang, M.; Zhao, J.P.; Zhang, W.H.; Su, Y.; Xiao, L.T.; Deng, H.T.; Xie, D.X. Injury activates Ca^{2+}/Calmodulin-dependent phosphorylation of JAV1-JAZ8-WRKY51 complex for jasmonate biosynthesis. *Mol. Cell* **2018**, *70*, 136–149. [CrossRef]
112. Ho, V.T.; Tran, A.N.; Cardarelli, F.; Perata, P.; Pucciariello, C. A calcineurin B-like protein participates in low oxygen signalling in rice. *Funct. Plant Biol.* **2017**, *44*, 917–928. [CrossRef]
113. Ou, L.J.; Liu, Z.B.; Zhang, Y.P.; Zou, X.X. Effects of exogenous Ca^{2+} on photosynthetic characteristics and fruit quality of pepper under waterlogging stress. *Chil. J. Agric. Res.* **2017**, *77*, 126–133. [CrossRef]
114. He, L.Z.; Yu, L.; Li, B.; Du, N.S.; Guo, S.R. The effect of exogenous calcium on cucumber fruit quality, photosynthesis, chlorophyll fluorescence, and fast chlorophyll fluorescence during the fruiting period under hypoxic stress. *BMC Plant Biol.* **2018**, *18*, 180. [CrossRef] [PubMed]
115. Steffens, B.; Kovalev, A.; Gorb, S.N.; Sauter, M. Emerging roots alter epidermal cell fate through mechanical and reactive oxygen species signaling. *Plant Cell* **2012**, *24*, 3296–3306. [CrossRef] [PubMed]
116. Wang, F.F.; Chen, Z.H.; Liu, X.H.; Colmer, T.D.; Shabala, L.; Salih, A.; Zhou, M.X.; Shabala, S. Revealing the roles of GORK channels and NADPH oxidase in acclimation to hypoxia in *Arabidopsis*. *J. Exp. Bot.* **2017**, *68*, 3191–3204. [CrossRef] [PubMed]
117. Yemelyanov, V.V.; Shishova, M.F.; Chirkova, T.V.; Lindberg, S.M. Anoxia-induced elevation of cytosolic Ca^{2+} concentration depends on different Ca^{2+} sources in rice and wheat protoplasts. *Planta* **2011**, *234*, 271–280. [CrossRef] [PubMed]
118. Sedbrook, J.C.; Kronebusch, P.J.; Trewavas, G.G.B.A.J.; Masson, P.H. Transgenic AEQUORIN reveals organ-specific cytosolic Ca^{2+} responses to anoxia in *Arabidopsis thaliana* seedlings. *Plant Physiol.* **1996**, *111*, 243–257. [CrossRef]
119. Peng, R.Y.; Bian, Z.Y.; Zhou, L.N.; Cheng, W.; Hai, N.; Yang, C.Q.; Yang, T.; Wang, X.Y.; Wang, C.Y. Hydrogen sulfide enhances nitric oxide-induced tolerance of hypoxia in maize (*Zea mays* L.). *Plant Cell Rep.* **2016**, *35*, 2325–2340. [CrossRef]

120. Li, Y.Q.; Sun, D.; Xu, K.; Jin, L.B.; Peng, R.Y. Hydrogen sulfide enhances plant tolerance to waterlogging stress. *Plants* **2021**, *10*, 1928. [CrossRef]
121. Wu, Q.; Su, N.N.; Huang, X.; Cui, J.; Shabala, L.; Zhou, M.X.; Yu, M.; Shabala, S. Hypoxia-induced increase in GABA content is essential for restoration of membrane potential and preventing ROS-induced disturbance to ion homeostasis. *Plant Commun.* **2021**, *2*, 12. [CrossRef]
122. Frohnmeyer, H.; Loyall, L.; Blatt, M.R.; Grabov, A. Millisecond UV-B irradiation evokes prolonged elevation of cytosolic-free Ca^{2+} and stimulates gene expression in transgenic parsley cell cultures. *Plant J.* **1999**, *20*, 109–117. [CrossRef]
123. Zhang, X.X.; Tang, X.X.; Wang, M.; Zhang, W.; Zhou, B.; Wang, Y. ROS and calcium signaling mediated pathways involved in stress responses of the marine microalgae *Dunaliella salina* to enhanced UV-B radiation. *J. Photochem. Photobiol. B-Biol.* **2017**, *173*, 360–367. [CrossRef]
124. Yu, G.H.; Li, W.; Yuan, Z.Y.; Cui, H.Y.; Lv, C.G.; Gao, Z.P.; Han, B.; Gong, Y.Z.; Chen, G.X. The effects of enhanced UV-B radiation on photosynthetic and biochemical activities in super-high-yield hybrid rice Liangyoupeijiu at the reproductive stage. *Photosynthetica* **2013**, *51*, 33–44. [CrossRef]
125. Barabas, K.N.; Szegletes, Z.; Pestenacz, A.; Fulop, K.; Erdei, L. Effects of excess UV-B irradiation on the antioxidant defence mechanisms in wheat (*Triticum aestivum* L.) seedlings. *J. Plant Physiol.* **1998**, *153*, 146–153. [CrossRef]
126. Christie, J.M.; Jenkins, G.I. Distinct UV-B and UV-A/blue light signal transduction pathways induce chalcone synthase gene expression in *Arabidopsis* cells. *Plant Cell* **1996**, *8*, 1555–1567. [PubMed]
127. An, L.; Feng, H.; Tang, X.; Wang, X. Changes of microsomal membrane properties in spring wheat leaves (*Triticum aestivum* L.) exposed to enhanced ultraviolet-B radiation. *J. Photochem. Photobiol. B Biol.* **2000**, *57*, 60–65. [CrossRef]
128. Chen, Z.J.; Ma, Y.; Yang, R.Q.; Gu, Z.X.; Wang, P. Effects of exogenous Ca^{2+} on phenolic accumulation and physiological changes in germinated wheat (*Triticum aestivum* L.) under UV-B radiation. *Food Chem.* **2019**, *288*, 368–376. [CrossRef]
129. Gao, L.M.; Wang, X.F.; Shen, Z.H.; Li, Y.F. The application of exogenous gibberellic acid enhances wheat seedlings UV-B tolerance by ameliorating DNA damage and manipulating UV-absorbing compound biosynthesis in wheat seedling leaves. *Pak. J. Bot.* **2018**, *50*, 2167–2172.
130. Wang, J.X.; Ding, H.D.; Zhang, A.; Ma, F.F.; Cao, J.M.; Jiang, M.Y. A novel mitogen-activated protein kinase gene in maize (*Zea mays*), ZmMPK3, is involved in response to diverse environmental cues. *J. Integr. Plant Biol.* **2010**, *52*, 442–452. [CrossRef]
131. Du, L.Q.; Poovaiah, B.W. A novel family of Ca^{2+}/calmodulin-binding proteins involved in transcriptional regulation: Interaction with fsh/Ring3 class transcription activators. *Plant Mol. Biol.* **2004**, *54*, 549–569. [CrossRef]
132. Marcec, M.J.; Gilroy, S.; Poovaiah, B.W.; Tanaka, K. Mutual interplay of Ca^{2+} and ROS signaling in plant immune response. *Plant Sci.* **2019**, *283*, 343–354. [CrossRef]
133. Gong, Z.Z.; Xiong, L.M.; Shi, H.Z.; Yang, S.H.; Herrera-Estrella, L.R.; Xu, G.H.; Chao, D.Y.; Li, J.R.; Wang, P.Y.; Qin, F.; et al. Plant abiotic stress response and nutrient use efficiency. *Sci. China-Life Sci.* **2020**, *63*, 635–674.
134. Wang, Y.; Wu, W.H. Potassium transport and signaling in higher plants. In *Annual Review of Plant Biology*; Merchant, S.S., Ed.; Annual Reviews: Palo Alto, CA, USA, 2013; Volume 64, pp. 451–476.
135. Li, L.G.; Kim, B.G.; Cheong, Y.H.; Pandey, G.K.; Luan, S. A Ca^{2+} signaling pathway regulates a K^+ channel for low-K response in *Arabidopsis*. *Proc. Natl. Acad. Sci. USA* **2006**, *103*, 12625–12630. [PubMed]
136. Qian, D.; Xiang, Y. Actin cytoskeleton as actor in upstream and downstream of calcium signaling in plant cells. *Int. J. Mol. Sci.* **2019**, *20*, 16.

plants

MDPI

Article

A Possible Mode of Action of Methyl Jasmonate to Induce the Secondary Abscission Zone in Stems of *Bryophyllum calycinum*: Relevance to Plant Hormone Dynamics

Michał Dziurka [1,*], Justyna Góraj-Koniarska [2], Agnieszka Marasek-Ciolakowska [2], Urszula Kowalska [2], Marian Saniewski [2], Junichi Ueda [3] and Kensuke Miyamoto [4,*]

1 The Franciszek Górski Institute of Plant Physiology, Polish Academy of Sciences, Niezapominajek 21, 30-239 Krakow, Poland
2 The National Institute of Horticultural Research, Konstytucji 3 Maja 1/3, 96-100 Skierniewice, Poland; justyna.goraj@inhort.pl (J.G.-K.); agnieszkamarasek@wp.pl (A.M.-C.); urszula.kowalska@inhort.pl (U.K.); marian.saniewski@inhort.pl (M.S.)
3 Department of Biological Science, Graduate School of Science, Osaka Prefecture University, 1-1 Gakuen-cho, Naka-ku, Sakai, Osaka 599-8531, Japan; ueda@b.s.osakafu-u.ac.jp
4 Faculty of Liberal Arts and Sciences, Osaka Prefecture University, 1-1 Gakuen-cho, Naka-ku, Sakai, Osaka 599-8531, Japan
* Correspondence: m.dziurka@ifr-pan.krakow.pl (M.D.); miyamoto@las.osakafu-u.ac.jp (K.M.); Tel.: +48-12-425-1833 (M.D.); +81-72-254-9741 (K.M.)

Abstract: Plants can react to environmental stresses through the abscission of infected, damaged, or senescent organs. A possible mode of action of methyl jasmonate (JA-Me) to induce the formation of the secondary abscission zone (SAZ) in the stems of *Bryophyllum calycinum* was investigated concerning plant hormone dynamics. Internode segments were prepared mainly from the second or third internode from the top of plants with active elongation. JA-Me applied to the middle of internode segments induced the SAZ formation above and below the treatment after 5–7 days. At 6 to 7 days after JA-Me treatment, the above and below internode pieces adjacent to the SAZ were excised and subjected to comprehensive analyses of plant hormones. The endogenous levels of auxin-related compounds between both sides adjacent to the SAZ were quite different. No differences were observed in the level of jasmonic acid (JA), but the contents of 12-oxo-phytodienoic acid (OPDA), a precursor of JA, and *N*-jasmonyl-leucine (JA-Leu) substantially decreased on the JA-Me side. Almost no effects of JA-Me on the dynamics of other plant hormones (cytokinins, abscisic acid, and gibberellins) were observed. Similar JA-Me effects on plant hormones and morphology were observed in the last internode of the decapitated growing plants. These suggest that the application of JA-Me induces the SAZ in the internode of *B. calycinum* by affecting endogenous levels of auxin- and jasmonate-related compounds.

Keywords: auxin-related compound; *Bryophyllum calycinum*; indole-3-acetic acid; methyl jasmonate; plant hormone dynamics; secondary abscission

Citation: Dziurka, M.; Góraj-Koniarska, J.; Marasek-Ciolakowska, A.; Kowalska, U.; Saniewski, M.; Ueda, J.; Miyamoto, K. A Possible Mode of Action of Methyl Jasmonate to Induce the Secondary Abscission Zone in Stems of *Bryophyllum calycinum*: Relevance to Plant Hormone Dynamics. *Plants* **2022**, *11*, 360. https://doi.org/10.3390/plants11030360

Academic Editor: Tae-Hwan Kim

Received: 14 December 2021
Accepted: 25 January 2022
Published: 28 January 2022

1. Introduction

Plants encounter plentiful biotic and abiotic stresses, leading to shedding (separation) of no longer needed or damaged organs such as leaves, branches, flowers, and fruits, from the parent plants. This process is known as abscission, and it is strongly associated with plant growth and development [1–9]. In the process of abscission, mechanical weakening of cell walls at the abscission zone is brought about by the degradation of the middle lamella by multiple cell-wall-degrading enzymes such as cellulase, polygalacturonases, pectin methyl esterases, and so forth, resulting in shedding [4,9–15].

The position and the time of the formation of abscission zones are determined genetically in each organ, and abscission zones once formed commonly do not differentiate further. Contrarily, in response to tissue injury or infection, differentiation of abscission zones in abnormal positions on stems, petioles, pedicles, and branches, designated as the secondary abscission zone (SAZ), can occur in vivo [1,16]. The secondary abscission has been observed primarily in various in vitro systems involving pedicels of *Malus sylvestris* [17] and *Pyrus communis* [18], stems of *Impatiens sultani* [10,11,19], *Morus alba* [20], *Citrus sinensis* [21], and *Phaseolus vulgaris* [16], and petiole explants of *P. vulgaris* [22], *Pisum sativum* pedicle, or *Euphorbia pulcherrima* flower [23].

The SAZs are induced by some signals, especially plant hormone cues, between neighboring cells [8,24–26]. According to histological analyses, the formation of the SAZ in the stems of *Bryophyllum calycinum* was characterized by the presence of newly synthesized cell plates resulting from periclinal cell division within one layer of mother cells in stems [27].

Plant hormones are well known to play an essential role in plant growth and development, including the abscission or induction of transdifferentiation in mature cortical cells. Ueda et al. [28,29] reported that jasmonic acid (JA) and methyl jasmonate (JA-Me) as senescence-promoting substances promoted the abscission of bean petiole explants. JA-Me also promotes leaf abscission in intact *Kalanchoe blossfeldiana* [30] and *B. calycinum* plants [31]. Furthermore, Saniewski et al. [31] have reported that JA-Me at a concentration of 0.5% (w/w) applied as a lanolin paste in different stem explants or the debladed petiole induced the formation of the SAZs in *B. calycinum*. These suggest that JA and JA-Me (designated as jasmonates, JAs) have a powerful effect of inducing the SAZ and developing an abscission zone that has already been initiated in plant tissues, resulting in leaf abscission. Ito and Nakano [32] have suggested that a decrease in auxin levels might be considered to provide the first signal for abscission in pedicel abscission in tomatoes. In the stem of *B. calycinum*, indole-3-acetic acid (IAA) applied to a decapitated shoot or internode explants totally prevented the formation of the SAZ in the stems induced by JA-Me [31,33]. However, it should be mentioned that only IAA application substantially induces the formation of the SAZ not only in internode explants, petiole segments, and the petiole after excision of the leaf blade but also in decapitated stems in intact plants of *B. calycinum* [31,33]. It is suggested that in mechanisms of the SAZ formation induced by exogenously applied IAA in the internode of *B. calycinum*, an auxin gradient is vital, and the gradient results from polar IAA transport from the application site [27,31,33]. However, those phenomena induced by JA-Me have not been reported.

As mentioned earlier, plants belonging to the Crassulaceae family show fascinating phenomena, such as leaf abscission and secondary abscission zone formation, easily inducted. This was the reason we chose for experiments the important medicinal plant *Bryophyllum calycinum* (syn. *Kalanchoe pinnata*) [34].

To clarify JA-Me's possible mode of action to induce the formation of the SAZ in terms of its plant hormone dynamics, we focused on differences in plant hormone dynamics between adjacent tissues to the SAZ induced by JA-Me in stem segments, as well as decapitated growing plants of *B. calycinum*. In this paper, comprehensive analyses of plant hormones in JA-Me treated stems, mainly internodes segments, of *B. calycinum* were reported.

2. Results

2.1. The Effect of JA-Me on Induction of the Secondary Abscission in Internode Segments and Decapitated Plants of Bryophyllum calycinum

In *Bryophyllum calycinum*, JA-Me application (0.5%; w/w in lanolin) to the middle of internode segments induced formation of the SAZ, observed at length from 0.5 to 2 cm above and below the JA-Me treatment, 5–7 days after the treatment (Figure 1). JA-Me application induced senescence or loss of chlorophylls in the internode segments in both acropetal and basipetal directions. Treatment with JA-Me (0.5%, w/w in lanolin) at the middle of the last internode in decapitated growing plants also induced the SAZ below the treatment (Supplementary Figure S1).

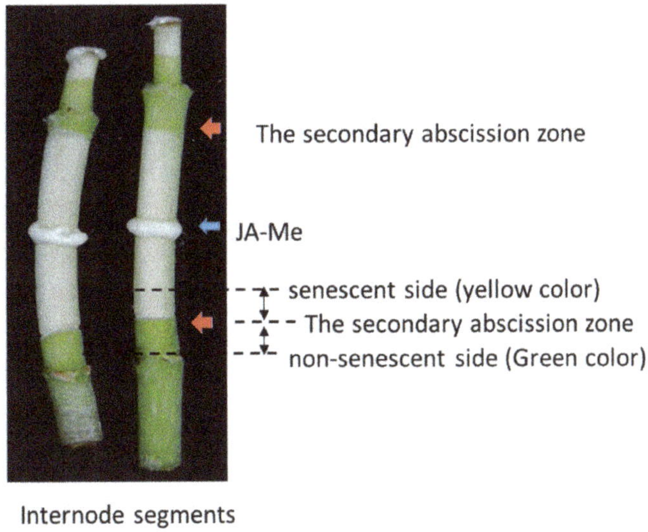

Figure 1. Secondary abscission zone (SAZ) induced by the application of methyl jasmonate (JA-Me) in internode segments of *Bryophyllum calycinum*. The treatment was made in the middle of internode explants. Photograph was taken 8 days after treatment. Red and blue arrows indicate the SAZ and JA-Me treatment place, respectively. Stem pieces (ca. 3–4 mm in length) above and below the SAZ were subjected to comprehensive plant hormone analyses.

2.2. *Changes in the Levels of Endogenous Plant Hormones in Relation to the Formation of the Secondary Abscission Zone Induced by JA-Me*

Comprehensive analyses of the endogenous plant hormones and their related compounds concerning the induction of the SAZ were performed in the internode segments of *B. calycinum*. At the appropriate time or 6 or 7 days after the treatment, small pieces of the internode segments adjacent to the SAZ were harvested for plant hormone analyses, as illustrated in Figure 1. Similar internode pieces above and below the SAZ in decapitated growing plants of *B. calycinum* were also subjected to the plant hormone analyses (Supplementary Figure S1).

2.2.1. Effect of JA-Me on Auxin-Related Compounds

As shown in Figure 2, the following auxins and their related compounds were successfully identified in internode segments of *B. calycinum*: indole-3-acetic acid (IAA), indole-3-acetamide (IAM), indole-3-acetonitrile (IAN), 2-oxindole-3-acetic acid (OxIAA), indole-3-carboxylic acid (ICA), indole-3-acetyl-aspartic acid (IAAsp), indole-3-acetyl-glutamic acid (IAGlu), and indole-3-propionic acid (IPA).

In the internode segment treated with JA-Me, endogenous levels of IAA, IAGlu, IAAsp, OxIAA, and IAM, in the above (senescent side, yellow color) and below pieces (non-senescent side, green color) adjacent to the secondary abscission in internode segments were similar. However, the contents of IAN and ICA were lower in the senescent than in the non-senescent side (Figure 2). These results suggest that the SAZ formation induced by JA-Me is closely related to the modification of IAA biosynthetic pathways via IAM, IAN, and ICA from tryptophan.

It should be mentioned that the endogenous level of IPA is much higher in the senescent than in the non-senescent side, suggesting that IAA metabolism to IPA is possible to be related to the SAZ induced by JA-Me (Figure 2).

Similar results of the effect of JA-Me on the endogenous levels of auxin-related compounds in the internode segments were obtained in the last internode of decapitated growing plants of *B. calycinum* (Supplementary Figure S2). These results suggest that the application of JA-Me substantially affects the IAA metabolism in the internode of *B. calycinum* and then might induce secondary abscission.

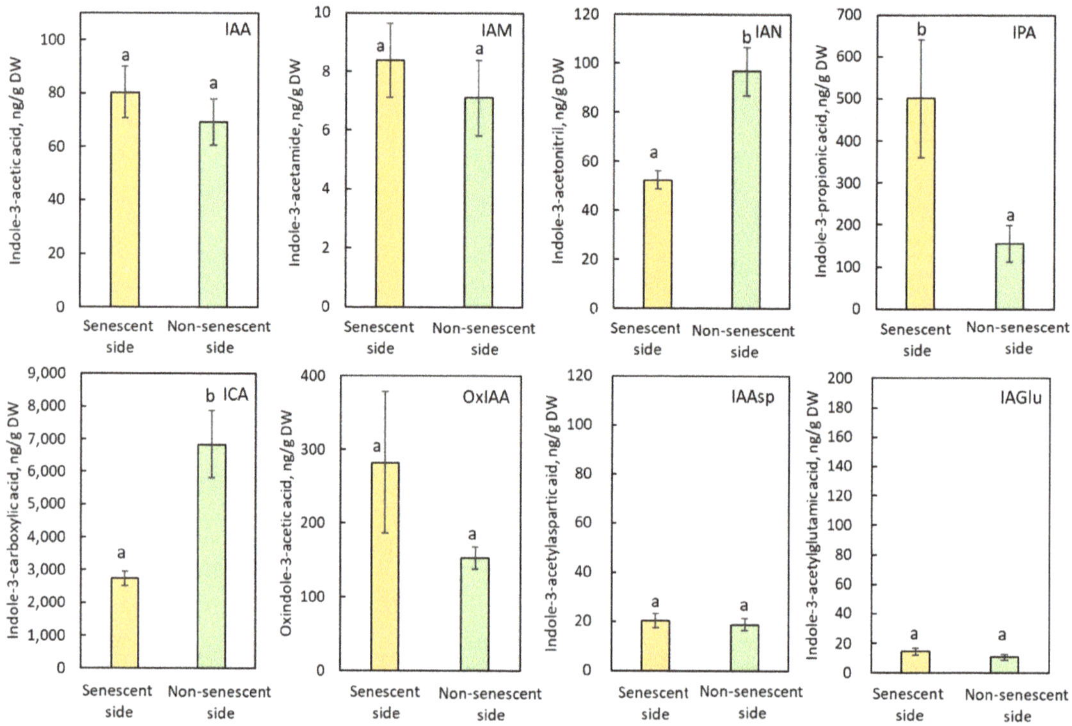

Figure 2. Endogenous levels of auxin-related compounds in the senescent and non-senescent sides of the SAZ induced by JA-Me in the internode explants of *Bryophylum calycinum*. IAA: indole-3-acetic acid; IAM: in-dole-3-acetamide; IAN: indole-3-acetonitrile; IPA: indole-3-propionic acid; ICA: indole-3-carboxylic acid; OxIAA: 2-oxindole-3-acetic acid; IAAsp: indole-3-acetylaspartic acid; IAGlu: indole-3-acetylglutamic acid. Values are the mean with standard error (n = 6). Different letters on the column (a, b) indicated statistically significant at $p < 0.05$ after ANOVA.

2.2.2. Effect of JA-Me on Jasmonate-Relating Compounds, Abscisic Acid, Salicylic Acid and Benzoic Acid

The contents of 12-oxo-phytodienoic acid (OPDA) and *N*-jasmonyl-leucine (JA-Leu) were substantially lower in the stem above the senescent than in the non-senescent side, but the content of jasmonic acid (JA) was similar in the stem pieces below and above the SAZ (senescent and non-senescent) in the internode explants (Figure 3).

An almost similar tendency was observed in the decapitated growing plants of *B. calycinum* (Supplementary Figure S3), suggesting that the application of JA-Me substantially increases endogenous levels of JA.

The endogenous levels of abscisic acid (ABA), salicylic acid (SA), and benzoic acid (BA) occurred in similar amounts in the stem pieces below and above the SAZs induced by JA-Me both in stem explants and in the internode of decapitated growing plants of *B. calycinum* (Figure 4 and Supplementary Figure S4).

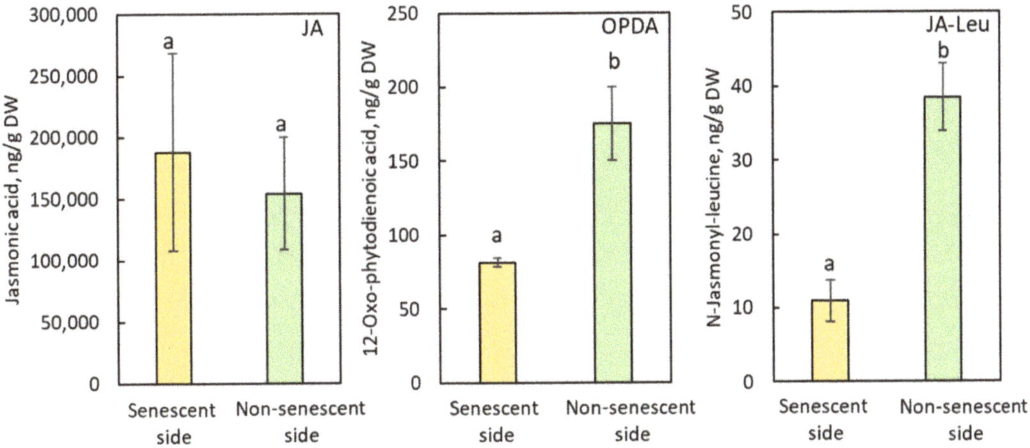

Figure 3. Endogenous levels of jasmonate-related compounds in the senescent and non-senescent sides of the SAZ induced by JA-Me in the internode explants of *Bryophyllum calycinum*. JA: jasmonic acid; OPDA: 12-oxo-phytodienoic acid; JA-Leu: *N*-jasmonyl-leucine. Values are the mean with standard error (n = 6). Different letters on the column (a, b) indicated statistically significant at $p < 0.05$ after ANOVA.

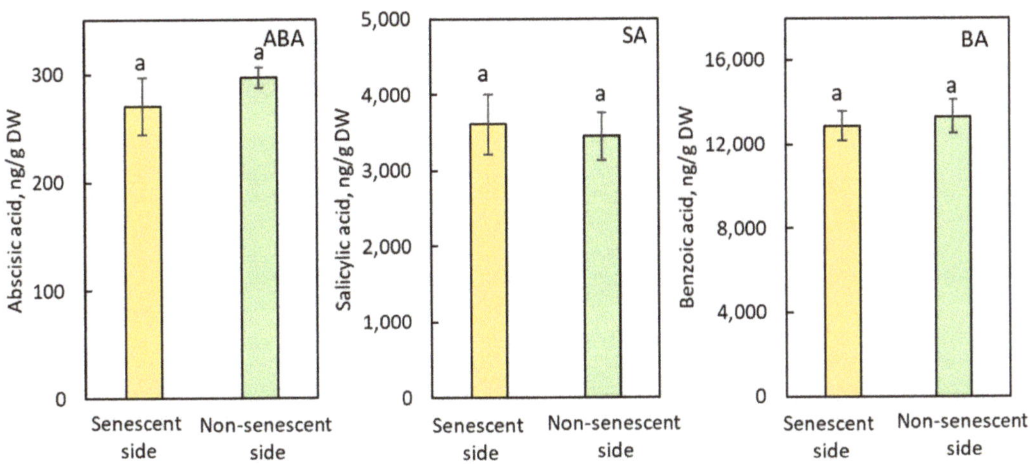

Figure 4. Endogenous levels of abscisic acid (ABA), salicylic acid (SA), and benzoic acid (BA) in the senescent and non-senescent sides of the SAZ induced by JA-Me in the internode explants of *Bryophylum calycinum*. Values are the mean with standard error (n = 6). Different letters on the column (a) indicated statistically significant at $p < 0.05$ after ANOVA.

2.2.3. Effect of JA-Me on Cytokinins

The contents of identified cytokinins such *trans*-zeatin (t-Z), *cis*-zeatin (c-Z), *trans*-zeatin riboside (t-ZR), and *cis*-zeatin riboside (c-ZR) were similar in the senescent and non-senescent sides of SAZ induced by JA-Me both in internode explants and in the internode of decapitated growing plants of *B. calycinum* (Figure 5 and Supplementary Figure S5).

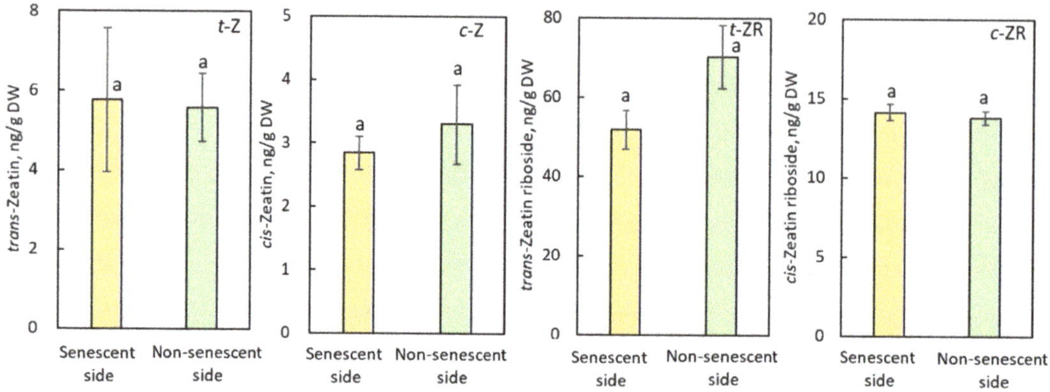

Figure 5. Endogenous levels of cytokinins in the senescent and non-senescent sides of the SAZ induced by JA-Me in the internode explants of *Bryophylum calycinum*. t-Z: *trans*-Zeatin; c-Z: *cis*-Zeatin; t-ZR: *trans*-Zeatin riboside; c-ZR: *cis*-Zeatin riboside. Values are the mean with standard error (n = 6). Different letters on the column (a) indicated statistically significant at $p < 0.05$ after ANOVA.

2.2.4. Effect of JA-Me on Gibberellins

Thirteen gibberellins (GAs), gibberellin A_1 (GA_1), GA_3, GA_4, GA_5, GA_6, GA_7, GA_8, GA_9, GA_{15}, GA_{19}, GA_{20}, GA_{44}, and GA_{53}, were also successfully identified in the internode segments of *B. calycinum*. Similar levels of these GAs were found in both the senescent and non-senescent sides of SAZ induced by JA-Me, except that GA_8 was lower above the SAZ (senescent side; Table 1). A similar tendency was observed in the decapitated growing plants of *B. calycinum* treated with JA-Me (Supplementary Table S1).

Table 1. Endogenous levels of gibberellins in the senescent and non-senescent sides of the SAZ, induced by JA-Me in the internode explants of *Bryophylum calycinum*. Values are the mean with standard error (n = 6). Different letters (a, b) on the column indicated statistically significant at $p < 0.05$ after ANOVA.

	Endogenous Levels (ng/g DW)	
	Senescent Side	**Non-Senescent Side**
Gibberellin A_1	36.26 ± 1.52 a	35.88 ± 1.11 a
Gibberellin A_3	4785.54 ± 382.88 a	5420.84 ± 420.90 a
Gibberellin A_4	54.37 ± 25.21 a	35.49 ± 20.50 a
Gibberellin A_5	61.05 ± 7.16 a	57.94 ± 19.21 a
Gibberellin A_6	547.32 ± 25.36 a	595.66 ± 44.81 a
Gibberellin A_7	54.37 ± 9.60 a	35.49 ± 20.50 a
Gibberellin A_8	23.71 ± 9.47 a	74.10 ± 6.85 b
Gibberellin A_9	65.62 ± 4.20 a	62.66 ± 3.40 a
Gibberellin A_{15}	1.23 ± 0.31 a	2.02 ± 0.42 a
Gibberellin A_{19}	61.30 ± 2.87 a	63.42 ± 4.55 a
Gibberellin A_{20}	83.55 ± 11.62 a	125.63 ± 32.06 a
Gibberellin A_{44}	61.49 ± 2.17 a	59.94 ± 1.97 a
Gibberellin A_{53}	98.90 ± 16.75 a	87.910 ± 12.97 a

3. Discussion

As mentioned in the Introduction (Section 1), many plant species develop the secondary abscission zone that extends between organs and the main body of the plants to shed. Plant hormones may play an essential role in the transdifferentiation in mature cortical cells to induce the SAZ. JA-Me and JA (designated as jasmonates, JAs) show the powerful effect of inducing the SAZ in stems and developing an abscission zone that has already been initiated in plant tissues in *B. calycinum* [31,33]. JAs were applied in lanolin paste, where lanolin alone did not affect morphological changes in the internode segments. This situation was demonstrated in previous works [27,31,35]. In the stem of *B. calycinum*, IAA applied to a decapitated shoot or internode explants prevented the formation of the SAZ induced by JA-Me [31,33]. Contrarily, IAA application has also been demonstrated to substantially induce the formation of the SAZ not only in internode explants, petiole segments, and petiole after excision of the leaf blade in intact plants but also decapitated stems in intact plants of *B. calycinum* [31,33]. A decrease in auxin levels might be considered as providing the first signal for abscission, as suggested in Arabidopsis [36] and tomatoes [32,37]. JAs, together with the disruption of endogenous auxin status by the decapitation or excision, may trigger the formation of the SAZ. The results confirm our previous observations [27,31,33], indicating that JA-Me is translocated in stem explants of *B. calycinum* in both ways, acropetally and basipetally, from the place of treatment. The SAZ development place is considered the final result of the stem's secondary abscission formation and senescence. Thus, it could be asserted that fresh, green tissues of the stem below the SAZ are not affected by JA-Me and can also be treated as a control.

What kinds of hormonal control factors are responsible for the formation of the SAZ induced by JA-Me? The SAZ formation by JA-Me has already been reported to be closely related to auxins [27,31]. Therefore, it is worthwhile to study the dynamics of plant hormones, especially auxins in the senescent and the non-senescent sides of the SAZ induced by JA-Me. Notably, the IAA gradient was not observed in the explants between the induced SAZ on both sides. The same situation occurred in the internode of the decapitated growing plant (Figure 2 and Supplementary Figure S2).

It has been reported that JA-Me is converted into JA and jasmonyl-isoleucine (JA-Ile), activating the jasmonates signaling pathway and emission of volatile organic compounds in *Achyranthes bidentate* [38]. The application of JA-Me resulted in the differences in endogenous levels of auxin-related compounds such IAN, ICA, and IPA in the senescent and non-senescent sides of the SAZ (Figure 2 and Supplementary Figure S2). Endogenous levels of OxIAA, which is one of IAA metabolites, were also different. These results suggest that the SAZ induced by JA-Me is closely related to the disruption of IAA metabolism in the stem adjacent to the SAZ.

IPA and IBA, other auxins that share similar structural scaffolds, are strongly conjugated and hydrolyzed with enzymes with similar or even higher activities than with IAA or IAA conjugates [39]. The occurrence of IBA has been reported in various plants, including *B. calycinum* [27]. In the present study, we report for the first time the occurrence of IPA in *B. calycinum*. The natural occurrence of IPA is scant, and until now, little is known about the physiological activity of IPA compared to IAA [39]. The content of IPA was relatively high in the stem of *B. calycinum*, and evidently, the content of IPA further increased on the stem side of JA-Me treatment, suggesting that IPA is responsible for the SAZ formation in *B. calycinum*.

Jasmonates (JAs) might function as a core signal in the plant hormone signaling network, a signal of JAs interacting with other hormone signaling to regulate plant growth, and abiotic and biotic stress tolerance [40–43]. Evidence for a close functional relationship between JAs signaling and auxin homeostasis has been well documented [44,45]. Du et al. [46] showed that biosynthesis and signaling of JA and IAA are differentially regulated by different abiotic stresses in rice, suggesting that the balance between JA and IAA homeostasis and their signaling are critical for plant development and stress responses. The application of JA-Me substantially induces an increase in the endogenous levels of JA in the stem of explant and internode of the decapitated growing plant of *B. calycinum*, as well as the disruption of auxin metabolism, but negligibly affected dynamics of ABA, cytokinins, and GAs (Figures 3–5, Table 1, Supplementary Figures S3–S5 and Table S1). Thus, cross-talk between JAs and auxin might be essential for the induction of the SAZ formation.

Based on the results of comprehensive analyses of endogenous plant hormones, Marasek-Ciolakowska et al. [27] strongly suggested that GAs and cytokinins did not contribute to the formation of the IAA-induced SAZ in *B. calycinum*. In this experiment, JA-Me also little affected the endogenous level of GAs, ABA, and cytokinins in stems above and below the SAZ (Figures 3 and 4, Table 1, Supplementary Figures S3 and S4, and Table S1). Thus, these plant hormones seem not to contribute to the formation of the JA-Me-induced SAZ in *B. calycinum* as the IAA-induced one [27].

Until now, four tryptophan (Trp)-dependent pathways of IAA biosynthesis, namely the indole-3-acetamide (IAM) pathway, the indole-3-pyruvic acid (IPyr) pathway, the tryptamine pathway, and the indole-3-acetaldoxime (IAOx) pathway, were identified in plants [39,47–50], although biosynthesis pathway(s) of IAA in plants of the Crassulaceae family (succulents) is unknown. The Trp-independent IAA biosynthesis from indole was also documented in some plants [50]. In *Arabidopsis thaliana*, indole-3-carbaldehyde and indole-3-carboxylic acid (ICA) are synthesized from Trp via intermediates such IAOx and IAN, although ICA can also be attributed to the degradation of IAA [51]. Whether ICA can be converted to IPA and vice versa, as indole-3-butyric acid (IBA) and IAA interconversions, has not been shown as yet [39]. ICA has been identified in *Pinus sylvestris* needles, in the leaves of *Ginkgo biloba*, and in the stem of *B. calycinum* [27,31,52].

The occurrence of IAM and IAN in the stem of *B. calycinum* may suggest that biosynthesis of IAA in the plant is going through the IAM and IAOx pathways since IAM and IAN are downstream intermediate metabolites of IAOx [50,53]. The IAOx-dependent IAA biosynthesis pathway was indicated in some plants, but it is not a common pathway [50]. Other pathways of IAA biosynthesis are also possible in *B. calycinum*. Intensive studies on JAs-dependent changes in metabolism or biosynthesis of IAA and the physiological function of ICA, related to the secondary abscission formation, will be needed in the future.

4. Materials and Methods

4.1. Plant Materials and Hormone Treatment

Three- to six-month-old plants of *Bryophyllum calycinum* Salisb. (Crassulaceae), propagated from epiphyllous buds arising in the marginal notches of the leaves, were used in the experiments. Stem segments and decapitated stems of growing *B. calycinum* plants were used in methyl jasmonate (JA-Me) treatment.

Internode segments at the length of ca. 4–5 cm with two nodes (leaves removed) were excised, from mainly the second or third internodes from the top of growing plants with active elongation. The segments were treated with JA-Me at 0.5% (*w*/*w*) in lanolin paste in the middle of the internode and kept vertically in a 50 mL glass chamber with moistened papers at the bottom of these explants under natural light conditions in a greenhouse, as shown in Figure 1. In June, August, and September, experiments were repeated three times with 15 to 20 explants.

A similar experiment with decapitated growing plants was carried out. After decapitation of the apical part of the growing plant shoot, JA-Me (0.5%, *w*/*w* in lanolin) was applied in the middle of the last internode, as shown in Supplementary Figure S1. The experiment was repeated twice from August to October with 20 explants.

4.2. Analyses of Plant Hormones in Relation to the Formation of the Secondary Abscission Zone Induced by Methyl jasmonate

Analyses of plant hormones were performed according to the methods reported previously [27,35,54–56]. At 6 to 7 days after treatment with JA-Me, the below (non-senescent, green) and above (senescent, yellow) parts of ca. 3–4 mm internode pieces adjacent to the SAZ formed by JA-Me application in the stem of *B. calycinum* were excised, respectively. Excised samples were immediately frozen in liquid N_2 and then lyophilized. Lyophilized materials in each piece of internode were combined, and an aliquot of a small amount (ca. 10 mg DW) was used for comprehensive plant hormone analyses. Lyophilized materials with appropriate amounts of a mixture of each stable isotope-labeled plant hormone as an internal standard were extracted with an organic solvent consisting of methanol/water/formic acid = 15: 4: 1 (*v*/*v*/*v*) three times. Respective extracts were combined and then evaporated under N_2. The extract obtained was re-suspended in 3%

methanol in 1 M formic acid and then cleaned up on hybrid SPE cartridges (BondElut Plexa PCX, Agilent, Santa Clara, CA, USA). Qualitative and quantitative analyses of plant hormones and other related compounds were performed on a HPLC-MS/MS system with UHPLC apparatus (Agilent Infinity 1260, Agilent, Waldbronn, Germany) coupled to a triple quadruple mass spectrometer ESI-MS/MS (6410 Triple Quad LC/MS, Agilent, Santa Clara, CA., USA). Plant hormones were separated on an Ascentis Express RP-Amide analytical column (particle size: 2.7 μm; 2.1 mm × 150 mm; Supelco, Bellefonte, PA., USA) at 60 °C, at a linear gradient of water vs. acetonitrile both with 0.01% of formic acid. As internal standards, [$^{15}N_4$] dihydrozeatin, [$^{15}N_4$] kinetin, [$^{2}H_5$] *trans*-zeatin riboside (t-ZR), [$^{2}H_5$] indole-3-acetic acid (IAA), [$^{2}H_4$] indole-3-acetonitrile, [$^{2}H_4$] salicylic acid (SA), [$^{2}H_2$] gibberellin A_1 (GA_1), [$^{2}H_2$] gibberellin A_4 (GA_4), [$^{2}H_2$] gibberellin A_5 (GA_5), [$^{2}H_2$] gibberellin A_6 (GA_6), [$^{2}H_6$] *cis, trans*-abscisic acid (ABA), [$^{2}H_5$] benzoic acid (BA), [$^{2}H_5$] jasmonic acid (JA), and [$^{2}H_5$] dinor-12-oxo-phytodienoic acid (dinor-OPDA) were used. All standards, except for [$^{2}H_5$] JA supplied by CND Isotopes (Quebeck, Canada) and [$^{2}H_5$] dinor OPDA supplied by Cayman Chem. Comp. (Ann Arbor, USA), were from OlChemim (Olomouc, Czech Republic) at the highest available purity. Multiple reaction monitoring (MRM) transitions were used to identify and quantify all compounds of interest. Quantitation was based on calibration curves obtained with each pure standard compound taking account of the recovery rates of an internal standard used. Further technical details are given by the references cited above.

4.3. Statistical Analysis

The analysis of variance (ANOVA) was conducted using STATISTICA software (StatSoft, Kraków, Poland). To compare the means, Duncan's multiple range test was used. Values of $p < 0.05$ were considered to be statistically significant. Values are expressed as the mean with standard error. Different letters in the columns in the figures and tables indicate statistical differences.

5. Conclusions

A comprehensive study of the dynamics of plant hormones in the stem pieces above and below the SAZ induced by the application of JA-Me in *B. calycinum* revealed that the application of JA-Me substantially affected auxin metabolism and the endogenous status of JAs. However, it negligibly affected the endogenous IAA levels. These suggest that the mode of JA-Me action to induce the SAZ is different from that of IAA, whereas IAA also induces the SAZ. JA-Me functions as a trigger modifying metabolism of IAA and JAs to induce the formation of the SAZ in the stem of *B. calycinum*.

Supplementary Materials: The following are available online at https://www.mdpi.com/article/10.3390/plants11030360/s1, Figure S1: The secondary abscission zone induced by the application of methyl jasmonate (JA-Me) in the last internode of decapitated growing plants of *Bryophyllum calycinum*, Figure S2: Endogenous levels of auxin-related compounds in the stem pieces above and below the secondary abscission zone induced by JA-Me in the last internode of decapitated growing plants of *Bryophyllum calycinum*, Figure S3: Endogenous levels of jasmonate-related compounds in the stem pieces above and below the secondary abscission zone induced by JA-Me in the last internode of decapitated growing plants of *Bryophyllum calycinum*. Figure S4: Endogenous levels of abscisic acid, salicylic acid, and benzoic acid in the stem pieces above and below the secondary abscission zone induced by JA-Me in the last internode of decapitated growing plants of *Bryophyllum calycinum*. Figure S5: Endogenous levels of cytokinins in the stem pieces above and below the secondary abscission zone induced by JA-Me in the last internode of decapitated growing plants of *Bryophyllum calycinum*. Table S1: Endogenous levels of gibberellins in the stem pieces above and below the secondary abscission zone induced by JA-Me in the last internode of decapitated growing plants of *Bryophyllum calycinum*.

Author Contributions: Conceptualization, A.M.-C., K.M. and M.S.; Methodology, M.D., A.M.-C. and M.S.; Software, J.G.-K.; Investigation, M.D., A.M.-C. and U.K.; Writing—Original Draft Preparation, A.M.-C., J.G.-K., K.M. and M.S.; Writing—Review and Editing, M.D., A.M.-C., K.M., J.U. and M.S.; Visualization, A.M.-C., K.M. and M.S.; Supervision, M.S.; Funding Acquisition A.M.-C. All authors have read and agreed to the published version of the manuscript.

Funding: This work was partly supported by the Polish Ministry of Science and Higher Education through statutory funds of the Research Institute of Horticulture, Skierniewice, Poland (Grant ZBS/7/2021).

Data Availability Statement: The data sets generated for this study are available in this article and Supplementary Material.

Conflicts of Interest: The authors declare no conflict of interest.

References

1. Addicott, F.T. *Abscission*; University of California Press: Berkeley, CA, USA, 1982.
2. Osborne, D.J. Morphogenetic signals and markers in vitro and in vivo. In *Morphogenesis in Plants Molecular Approaches*; Roubelakis-Angelakis, K.A., Van Thanh, K., Eds.; Springer Science & Business Media: New York, NY, USA, 1993; Volume 253, pp. 1–17.
3. Van Doorn, W.G.; Stead, A.D. Abscission of flowers and floral parts. *J. Exp. Bot.* **1997**, *48*, 821–837. [CrossRef]
4. Roberts, J.A.; Whitelaw, C.A.; Gonzalez-Carranza, Z.H.; McManus, M.T. Cell Separation Processes in Plants—Models, Mechanisms and Manipulation. *Ann. Bot.* **2000**, *86*, 223–235. [CrossRef]
5. Taylor, J.E.; Whitelaw, C.A. Signals in abscission. *New Phytol.* **2001**, *151*, 323–340. [CrossRef]
6. Tucker, M.L.; Kim, J. Abscission research: What we know and what we still need to study. *Stewart Postharvest Rev.* **2015**, *11*, 7. [CrossRef]
7. Patharkar, O.R.; Walker, J.C. Advances in abscission signaling. *J. Exp. Bot.* **2018**, *69*, 733–740. [CrossRef] [PubMed]
8. Meir, S.; Sundaresan, S.; Riov, J.; Agarwal, I.; Philosoph-Hadas, S. Role of auxin depletion in abscission control. *Stewart Postharvest Rev.* **2015**, *11*, 15. [CrossRef]
9. Lee, Y. More than cell wall hydrolysis: Orchestration of cellular dynamics for organ separation. *Curr. Opin. Plant Biol.* **2019**, *51*, 37–43. [CrossRef]
10. Warren Wilson, J.; Warren Wilson, P.M.; Walker, E.S. Abscission sites in nodal explants of *Impatiens sultani*. *Ann. Bot.* **1987**, *60*, 693–704. [CrossRef]
11. Warren Wilson, J.; Walker, E.S.; Warren Wilson, P.M. The role of basipetal auxin transport in the positional control of abscission sites induced in *Impatiens sultani* stem explants. *Ann. Bot.* **1988**, *62*, 487–495. [CrossRef]
12. Warren Wilson, J.; Palni, L.M.S.; Warren Wilson, P.M. Auxin concentrations in nodes and internodes of *Impatiens sultani*. *Ann. Bot.* **1999**, *83*, 285–292. [CrossRef]
13. Bleecker, A.B.; Patterson, S.E. Last exit: Senescence, abscission, and meristem arrest in *Arabidopsis*. *Plant Cell* **1997**, *9*, 1169. [CrossRef] [PubMed]
14. Roberts, J.A.; Elliott, K.A.; Gonzalez-Carranza, Z.H. Abscission, dehiscence, and other cell separation processes. *Annu. Rev. Plant Biol.* **2002**, *53*, 131–158. [CrossRef] [PubMed]
15. Patterson, S.E. Cutting Loose. Abscission and Dehiscence in *Arabidopsis*. *Plant Physiol.* **2001**, *126*, 494–500. [CrossRef] [PubMed]
16. Webster, B.D.; Leopold, A.C. Stem Abscission in *Phaseolus vulgaris* Explants. *Bot. Gaz.* **2015**, *133*, 292–298. [CrossRef]
17. Pierik, R.L.M. Induction of secondary abscission in apple pedicels in vitro. *Physiol. Plant.* **1977**, *39*, 271–274. [CrossRef]
18. Pierik, R.L.M. Hormonal regulation of secondary abscission in pear pedicels in vitro. *Physiol. Plant.* **1980**, *48*, 5–8. [CrossRef]
19. Warren Wilson, P.M.; Warren Wilson, J.; Addicott, F.T.; Mckenzie, R.H. Induced abscission sites in internodal explants of *Impatiens sultani*: A new system for studying positional control: With an appendix: A mathematical model for abscission sites. *Ann. Bot.* **1986**, *57*, 511–530. [CrossRef]
20. Suzuki, T. Shoot-tip abscission and adventitious abscission of internodes in mulberry (*Morus alba*). *Physiol. Plant.* **1991**, *82*, 483–489. [CrossRef]
21. Plummer, J.A.; Vine, J.H.; Mullins, M.G. Regulation of stem abscission and callus growth in shoot explants of sweet orange [*Citrus sinensis* (L.) Osbeck]. *Ann. Bot.* **1991**, *67*, 17–22. [CrossRef]
22. McManus, M.T.; Thompson, D.S.; Merriman, C.; Lyne, L.; Osborne, D.J. Transdifferentiation of mature cortical cells to functional abscission cells in bean. *Plant Physiol.* **1998**, *116*, 891–899. [CrossRef]
23. Hvoslef-Eide, A.K.; Munster, C.M.; Mathiesen, C.A.; Ayeh, K.O.; Melby, T.I.; Rasolomanana, P.; Lee, Y. Primary and secondary abscission in *Pisum sativum* and *Euphorbia pulcherrima*-how do they compare and how do they differ? *Front. Plant Sci.* **2016**, *6*, 1204. [CrossRef] [PubMed]
24. Pang, Y.; Zhang, J.; Cao, J.; Yin, S.Y.; He, X.Q.; Cui, K.M. Phloem transdifferentiation from immature xylem cells during bark regeneration after girdling in *Eucommia ulmoides* Oliv. *J. Exp. Bot.* **2008**, *59*, 1341–1351. [CrossRef] [PubMed]

25. Yamaguchi, M.; Goué, N.; Igarashi, H.; Ohtani, M.; Nakano, Y.; Mortimer, J.C.; Nishikubo, N.; Kubo, M.; Katayama, Y.; Kakegawa, K.; et al. VASCULAR-RELATED NAC-DOMAIN6 and VASCULAR-RELATED NAC-DOMAIN7 effectively induce transdifferentiation into xylem vessel elements under control of an induction system. *Plant Physiol.* **2010**, *153*, 906–914. [CrossRef]
26. Reusche, M.; Thole, K.; Janz, D.; Truskina, J.; Rindfleisch, S.; Drübert, C.; Polle, A.; Lipka, V.; Teichmanna, T. Verticillium infection triggers VASCULAR-RELATED NAC DOMAIN7-dependent de novo xylem formation and enhances drought tolerance in *Arabidopsis*. *Plant Cell* **2012**, *24*, 3823–3837. [CrossRef] [PubMed]
27. Marasek-Ciolakowska, A.; Saniewski, M.; Dziurka, M.; Kowalska, U.; Góraj-Koniarska, J.; Ueda, J.; Miyamoto, K. Formation of the secondary abscission zone induced by the interaction of methyl jasmonate and auxin in *Bryophyllum calycinum*: Relevance to auxin status and histology. *Int. J. Mol. Sci.* **2020**, *21*, 2784. [CrossRef]
28. Ueda, J.; Miyamoto, K.; Aoki, M.; Momotani, Y.; Kato, J.; Kamisaka, S. The mode of actions of jasmonic acid and its methyl ester on the growth and the abscission. In Proceedings of the 14th International Conference on Plant Growth Substances, Amsterdam, The Netherlands, 21–25 July 1991; p. 80.
29. Ueda, J.; Miyamoto, K.; Hashimoto, M. Jasmonates promote abscission in bean petiole explants: Its relationship to the metabolism of cell wall polysaccharides and cellulase activity. *J. Plant Growth Regul.* **1996**, *15*, 189–195. [CrossRef]
30. Saniewski, M.; Wegrzynowicz-Lesiak, E. Methyl jasmonate-induced leaf abscision in *Kalanchoe blosfediana*. *Acta Hortic.* **1995**, *394*, 315–324. [CrossRef]
31. Saniewski, M.; Ueda, J.; Miyamoto, K. Methyl jasmonate induces the formation of secondary abscission zone in stem of *Bryophyllum calycinum* Salisb. *Acta Physiol. Plant.* **2000**, *22*, 17–23. [CrossRef]
32. Ito, Y.; Nakano, T. Development and regulation of pedicel abscission in tomato. *Front. Plant Sci.* **2015**, *6*, 442. [CrossRef]
33. Saniewski, M.; Góraj-Koniarska, J.; Gabryszewska, E.; Miyamoto, K.; Ueda, J. Auxin effectively induces the formation of the secondary abscission zone in *Bryophyllum calycinum* Salisb. (Crassulaceae). *Acta Agrobot.* **2016**, *69*, 3. [CrossRef]
34. Monroy, L.A.V.; Cauich, J.R.C.; Ortega, A.M.M.; Campos, M.R.S. Medicinal plants as potential functional foods or resources for obtaining anticancer activity metabolites. In *Oncological Functional Nutrition*; Campos, M., Ortega, A., Eds.; Academic Press: London, UK, 2021; pp. 161–194. [CrossRef]
35. Marasek-Ciolakowska, A.; Dziurka, M.; Kowalska, U.; Góraj-koniarska, J.; Saniewski, M.; Ueda, J.; Miyamoto, K. Mode of action of 1-naphthylphthalamic acid in conspicuous local stem swelling of succulent plant, *Bryophyllum calycinum*: Relevance to the Aspects of Its Histological Observation and Comprehensive Analyses of Plant Hormones. *Int. J. Mol. Sci.* **2021**, *22*, 3118. [CrossRef] [PubMed]
36. Basu, M.M.; González-Carranza, Z.H.; Azam-Ali, S.; Tang, S.; Shahid, A.A.; Roberts, J.A. The manipulation of auxin in the abscission zone cells of *Arabidopsis* flowers reveals that indoleacetic acid signaling is a prerequisite for organ shedding. *Plant Physiol.* **2013**, *162*, 96–106. [CrossRef] [PubMed]
37. Meir, S.; Philosoph-Hadas, S.; Sundaresan, S.; Selvaraj, K.S.V.; Burd, S.; Ophir, R.; Kochanek, B.; Reid, M.S.; Jiang, C.Z.; Lers, A. Microarray analysis of the abscission-related transcriptome in the tomato flower abscission zone in response to auxin depletion. *Plant Physiol.* **2010**, *154*, 1929–1956. [CrossRef] [PubMed]
38. Tamogami, S.; Noge, K.; Abe, M.; Agrawal, G.K.; Rakwal, R. Methyl jasmonate is transported to distal leaves via vascular process metabolizing itself into JA-Ile and triggering VOCs emission as defensive metabolites. *Plant Signal. Behav.* **2012**, *7*, 1378. [CrossRef] [PubMed]
39. Ludwig-Müller, J. Synthesis and hydrolysis of auxins and their conjugates with different side-chain lengths: Are all products active auxins? *Period. Biol.* **2020**, *121*, 81–96. [CrossRef]
40. Yang, J.; Duan, G.; Li, C.; Liu, L.; Han, G.; Zhang, Y.; Wang, C. The crosstalks between jasmonic acid and other plant hormone signaling highlight the involvement of jasmonic acid as a core component in plant response to biotic and abiotic stresses. *Front. Plant Sci.* **2019**, *10*, 1349. [CrossRef] [PubMed]
41. Liu, H.; Timko, M.P. Jasmonic Acid signaling and molecular crosstalk with other phytohormones. *Int. J. Mol. Sci.* **2021**, *22*, 2914. [CrossRef]
42. Huang, H.; Liu, B.; Liu, L.; Song, S. Jasmonate action in plant growth and development. *J. Exp. Bot.* **2017**, *68*, 1349–1359. [CrossRef]
43. Yu, X.; Zhang, W.; Zhang, Y.; Zhang, X.; Lang, D.; Zhang, X. The roles of methyl jasmonate to stress in plants. *Funct. Plant Biol.* **2019**, *46*, 197–212. [CrossRef]
44. Hentrich, M.; Böttcher, C.; Düchting, P.; Cheng, Y.; Zhao, Y.; Berkowitz, O.; Masle, J.; Medina, J.; Pollmann, S. The jasmonic acid signaling pathway is linked to auxin homeostasis through the modulation of YUCCA8 and YUCCA9 gene expression. *Plant J.* **2013**, *74*, 626–637. [CrossRef]
45. Pérez, A.C.; Goossens, A. Jasmonate signalling: A copycat of auxin signalling? *Plant. Cell Environ.* **2013**, *36*, 2071–2084. [CrossRef] [PubMed]
46. Du, H.; Liu, H.; Xiong, L. Endogenous auxin and jasmonic acid levels are differentially modulated by abiotic stresses in rice. *Front. Plant Sci.* **2013**, *4*, 397. [CrossRef] [PubMed]
47. Woodward, A.W.; Bartel, B. Auxin: Regulation, action, and interaction. *Ann. Bot.* **2005**, *95*, 707–735. [CrossRef] [PubMed]
48. Chandler, J.W. Auxin as compère in plant hormone crosstalk. *Planta* **2009**, *231*, 1–12. [CrossRef]
49. Ludwig-Müller, J. Auxin conjugates: Their role for plant development and in the evolution of land plants. *J. Exp. Bot.* **2011**, *62*, 1757–1773. [CrossRef]

50. Di, D.W.; Zhang, C.; Luo, P.; An, C.W.; Guo, G.Q. The biosynthesis of auxin: How many paths truly lead to IAA? *Plant Growth Regul.* **2016**, *78*, 275–285. [CrossRef]
51. Böttcher, C.; Chapman, A.; Fellermeier, F.; Choudhary, M.; Scheel, D.; Glawischnig, E. Thebiosynthetic pathway of indole-3-carbaldehyde and indole-3-carboxylic acid derivatives in *Arabidopsis*. *Plant Physiol.* **2014**, *165*, 841–853. [CrossRef]
52. Sandberg, G.; Jensen, E.; Crozier, A. Analysis of 3-indole carboxylic acid in Pinus sylvestris needles. *Phytochemistry* **1984**, *23*, 99–102. [CrossRef]
53. Sugawara, S.; Hishiyama, S.; Jikumaru, Y.; Hanada, A.; Nishimura, T.; Koshiba, T.; Zhao, Y.; Kamiya, Y.; Kasahara, H. Biochemical analyses of indole-3-acetaldoxime-dependent auxin biosynthesis in Arabidopsis. *Proc. Natl. Acad. Sci. USA* **2009**, *106*, 5430–5435. [CrossRef]
54. Dziurka, M.; Janeczko, A.; Juhász, C.; Gullner, G.; Oklestková, J.; Novák, O.; Saja, D.; Skoczowski, A.; Tóbiás, I.; Barna, B. Local and systemic hormonal responses in pepper leaves during compatible and incompatible pepper-tobamovirus interactions. *Plant Physiol. Biochem.* **2016**, *109*, 355–364. [CrossRef]
55. Płażek, A.; Dubert, F.; Kopeć, P.; Dziurka, M.; Kalandyk, A.; Pastuszak, J.; Wolko, B. Seed hydropriming and smoke water significantly improve low-temperature germination of *Lupinus angustifolius* L. *Int. J. Mol. Sci.* **2018**, *19*, 992. [CrossRef] [PubMed]
56. Dziurka, K.; Dziurka, M.; Muszyńska, E.; Czyczyło-Mysza, I.; Warchoł, M.; Juzoń, K.; Laskoś, K.; Skrzypek, E. Anatomical and hormonal factors determining the development of haploid and zygotic embryos of oat (*Avena sativa* L.). *Sci. Rep.* **2022**, *12*, 548. [CrossRef] [PubMed]

plants

MDPI

Article

Cold Acclimation in *Brachypodium* Is Accompanied by Changes in Above-Ground Bacterial and Fungal Communities

Collin L. Juurakko [1,*], George C. diCenzo [1] and Virginia K. Walker [1,2]

[1] Department of Biology, Queen's University, Kingston, ON K7L 3N6, Canada; george.dicenzo@queensu.ca (G.C.d.); walkervk@queensu.ca (V.K.W.)

[2] Department of Biomedical and Molecular Sciences, School of Environmental Studies, Queen's University, Kingston, ON K7L 3N6, Canada

* Correspondence: 11cj10@queensu.ca; Tel.: +1-613-533-6000 (ext. 77360)

Abstract: Shifts in microbiota undoubtedly support host plants faced with abiotic stress, including low temperatures. Cold-resistant perennials prepare for freeze stress during a period of cold acclimation that can be mimicked by transfer from growing conditions to a reduced photoperiod and a temperature of 4 °C for 2–6 days. After cold acclimation, the model cereal, *Brachypodium distachyon*, was characterized using metagenomics supplemented with amplicon sequencing (16S ribosomal RNA gene fragments and an internal transcribed spacer region). The bacterial and fungal rhizosphere remained largely unchanged from that of non-acclimated plants. However, leaf samples representing bacterial and fungal communities of the endo- and phyllospheres significantly changed. For example, a plant-beneficial bacterium, *Streptomyces* sp. M2, increased more than 200-fold in relative abundance in cold-acclimated leaves, and this increase correlated with a striking decrease in the abundance of *Pseudomonas syringae* (from 8% to zero). This change is of consequence to the host, since *P. syringae* is a ubiquitous ice-nucleating phytopathogen responsible for devastating frost events in crops. We posit that a responsive above-ground bacterial and fungal community interacts with *Brachypodium*'s low temperature and anti-pathogen signalling networks to help ensure survival in subsequent freeze events, underscoring the importance of inter-kingdom partnerships in the response to cold stress.

Keywords: *Brachypodium distachyon*; cold acclimation; microbiome; amplicon and shotgun sequencing; metagenomics; *Pseudomonas*; *Streptomyces*

Citation: Juurakko, C.L.; diCenzo, G.C.; Walker, V.K. Cold Acclimation in *Brachypodium* Is Accompanied by Changes in Above-Ground Bacterial and Fungal Communities. *Plants* **2021**, *10*, 2824. https://doi.org/10.3390/plants10122824

Academic Editor: Ewa Muszyńska

Received: 30 November 2021
Accepted: 16 December 2021
Published: 20 December 2021

1. Introduction

As sessile organisms, plants are at the mercy of an array of abiotic stresses, and, as winter approaches in mid- to high-latitudes and altitudes, one such stress is low temperature. Plants employ various strategies that allow them to recognise and cope with the cold [1]. As autumn progresses, perennials undergo a period of cold acclimation, which in a few days of low temperature exposure allows them to physiologically prepare for freezing conditions. Such preparations include changed levels of hundreds of proteins, the accumulation of fatty acids, lipid remodelling for plasma membrane protection, increased production of cryoprotective metabolites, such as soluble sugars and amino acids, as well as chaperones and reactive oxygen scavengers [2]. This acclimation process also appears to coincide with changes in host-associated microbial communities. Such a turnover in microbiota could assist plants in preparing for sub-zero temperature conditions and their vulnerability to psychrophilic pathogens. Indeed, winter seasonality in the plant microbiome has been previously reported [3–5]. Although the impact of cold acclimation on the microbiomes of perennial grass has not been hitherto explored, the identification of their bacterial and fungal communities offers the promise of understanding how the battle against coming winter conditions can be won by partnerships.

The perennial grass and model cereal, *Brachypodium distachyon* (hereinafter, *Brachypodium*), is capable of cold acclimation, reaching peak freezing tolerance after two days at 4 °C, and is associated with changes in the abundance of multiple plasma membrane proteins at 2–6 days [6]. In turn, these proteins are involved in complex crosstalk networks that prime the *Brachypodium* defensive response to a variety of abiotic and pathogenic stresses. Studies of cold acclimation have, for the most part, ignored the host-associated microbiota [1,7,8]. Nevertheless, the plant microbiome is emerging as an important factor in stress responses, including symbiont-mediated tolerance [9,10].

The general beneficial effects of microbes on plant fitness under a variety of stressful conditions have recently come to be known as the "Defence Biome" [5,10–17]. Symbiont-mediated fitness benefits may be a collective result of microbial exudates and function, for example, by facilitating early stress sensing and more efficient nutrient uptake and transfer, as well as by the induction of plant stress genes [9,10]. Specifically, symbiont-mediated cold tolerance has been directly demonstrated with some plant species and plant growth promoting bacteria (PGPBs) [9]. For example, *Burkholderia phytofirmans*-inoculated grape vines expressed cold stress-responsive genes earlier than non-inoculated vines [18] and *Streptomyces neyagawaensis* J6-inoculated turfgrass showed enhanced cold tolerance over non-inoculated plants [19]. Microbes thus have a demonstrated role in plant protection. They excrete a variety of products to benefit host plants, including anti-pathogenic microbial compounds and osmolytes, including proline and trehalose, as well as scavengers of reactive oxygen species, such as superoxide dismutase, catalase, and peroxidases [9,10,20]. Taken together, plant-associated microbial communities undoubtedly help plants survive cold stress.

The identification of host-associated microbiota that enhance freezing tolerance may lead the way to the development of synthetic cocktails of species that could eventually be used to inoculate crops or seeds to enhance cold tolerance [21]. Here, shotgun sequencing and metagenomic analysis of the phyllosphere/endosphere and rhizosphere in cold-acclimated *Brachypodium* is an important first step towards this goal. Our experimental inoculation of a commercial growing mix with old pasture soil allowed for the exposure and subsequent identification of bacterial and fungal taxa that thrived after transfer of the growing plants to low temperatures and thus are prospective native partners in the cold acclimation process. In addition, we contribute to the general appreciation of the robustness of the plant abiotic stress response, which employs communities of diverse organisms for survival.

2. Materials and Methods

2.1. Soil Inoculation and Preparation

Commercial potting soil (Sun Gro Horticulture, Agawam, MA, USA) was autoclaved twice and sealed in a double layer of plastic autoclave bags before being inoculated with bulk field soil (5% *w*/*v*). Bulk field soil was sampled using a sterilized trowel from the active layer (3–7 cm depth) in autumn (29 October 2020) after 96 h of day and night temperatures of ~5 °C and ~0 °C, respectively. The sampled fallow field had been left unfertilized and unplowed for 26 years and without domestic grazing animals for 15 years (Figure S1). It was characterized by grasses, including orchard grass, brome, and timothy (*Dactylis*, *Bromus*, and *Phleum* species, respectively) on clay soils and was located north of Sydenham, Ontario, Canada (44°24′26″ N, 76°36′1″ W). Soils were thoroughly mixed for 15 min using a cement mixer that had been rinsed with 70% ethanol, with the inoculated soil then stored in a lidded container that had also been rinsed with 70% ethanol. The inoculated soil mixture was kept at room temperature until use.

2.2. Plant Material and Growth Conditions

Surface-sterilized *Brachypodium* seeds of an inbred line (ecotype: *Bd*21) (RIKEN, Wakō, Japan) were sown in the inoculated potting soil and grown in a temperature-controlled chamber (Conviron GEN2000, Queen's University Phytotron, Kingston, ON, Canada) on a 20 h light (~100 µmol m^{-2} s^{-1}; 22 °C) and 4 h dark (22 °C) light cycle. *Brachypodium* that had been grown under standard conditions for three weeks (Figure S2) were then cold acclimated by transferring the plants to a low temperature chamber (Coldmatic Refrigeration, Etobicoke, ON, Canada) (4 °C, 12 h light as indicated above; 12 h dark) for 6 days [6]. Plants maintained at standard conditions until time of use were considered the non-acclimated controls.

2.3. Microbiome Extraction and Preparation

Microbiome extractions were performed under sterile conditions. Above-ground extractions were from tissue excised from the tips of primary leaves. Phyllosphere microbes are found on the leaf surface and endosphere microbiota include communities that enter the plant through the leaves, as well as those that circulate within the xylem. Rather than separate these, we reasoned that both phyllosphere and endosphere communities would be driven by the changing environmental conditions, in addition to plant interactions. Accordingly, these leaf microbiota were extracted together using a DNeasy Plant Pro Kits (Qiagen, Hilden, Germany), following the manufacturer's recommended directions, using 10 mg of leaf tissue per plant (10 plants per replicate for a total of 100 mg of tissue) and three replicates.

Extractions of the below-ground, tightly bound root soil of the rhizosphere (Figure S3) were performed as previously described [22] using a DNeasy PowerSoil Pro Kit (Qiagen, Hilden, Germany), following the manufacturer's recommendations. Adhering root soil (25 mg per plant) was released from the roots following careful removal of the plants from the pots and gentle shaking. Extra care was taken to remove any root tissue, or non-soil material from samples, such as wood or perlite. Three replicates were performed, each using 10 individual plants. DNA purity and concentration was quantified using a Synergy H1 microplate reader with a Take3 Micro-Volume Plate (both BioTek Instruments Inc., Winooski, VT, USA).

2.4. Shotgun Metagenomics Library Preparation and Sequencing

Libraries were prepared using an Illumina DNA Prep (M) Tagmentation library preparation kit (Illumina Inc., San Diego, CA, USA), following the manufacturer's user guide. Initial DNA concentration was evaluated using the Qubit dsDNA HS Assay Kit (Life Technologies, Carlsbad, CA, USA). Eukaryotic DNA was depleted in leaf tissue samples using an NEBNext Microbiome DNA Enrichment Kit (New England Biolabs, Ipswich, MA, USA), following the manufacturer's user guide to decrease the probability of recovery of host genomic, chloroplast, and mitochondrial DNA sequences [23]. DNA (500 ng) was used for depletion of the eukaryotic DNA, as recommended by Molecular Research LP (MR DNA; Shallowater, TX, USA). The enriched microbial DNA was quantified using the Qubit dsDNA HS Assay Kit (Life Technologies, Carlsbad, CA, USA) (Table S1). Subsequently, 50 ng of DNA was used to prepare the libraries. The samples underwent simultaneous fragmentation and addition of adapter sequences, which were utilized during a limited-cycle polymerase chain reaction in which unique indices were added to the sample. Following library preparation, library concentration and mean library size were determined using the Qubit dsDNA HS Assay Kit (Life Technologies, Carlsbad, CA, USA) and the Agilent 2100 Bioanalyzer (Agilent Technologies, Santa Clara, CA, USA), respectively. Libraries were pooled in equimolar ratios (0.6 nM), and sequencing was performed on a NovaSeq 6000 platform (Illumina Inc., San Diego, CA, USA) to a depth of 10 million 2 $\times$ 150 bp reads.

2.5. Preprocessing and Quality Control

Analysis of sequencing data was performed following the Sunbeam pipeline (v2.1.0) [24] with 26 available cores (15.425 Gb of memory each) on Ubuntu (v18.04.05). Raw fastq files of paired-end reads were quality controlled to remove adapter sequences using Cutadapt (v3.4.0) [25] and Trimmomatic (v0.3.9) [26], following which read quality was assessed using FastQC (v0.11.9) [27]. Low-complexity sequences were masked using Komplexity (v0.3.6) [24] and contaminating plant host reads were removed by Sunbeam following mapping of reads to the *Brachypodium* genome (RefSeq assembly accession GCF_000005505.3) using BWA (v0.7.17) [28]. Following initial host read decontamination, individual reads were interrogated using the National Center for Biotechnology Information (NCBI) BLAST (*blastn;* available at https://blast.ncbi.nlm.nih.gov/Blast.cgi; accessed on 18 August 2021), revealing numerous hits to mitochondrial genomic sequences. Subsequently, several mitochondrial genomic sequences (detailed below) were subsequently downloaded and added to the host genome path for removal of contaminating mitochondrial sequences. This process was repeated until a subset of individual reads did not return any mitochondrial genomes with high coverage.

Most mitochondrial genomes used to filter contaminating sequences were retrieved from NCBI from the following species with GenBank IDs: *Saccharum officinarum* cv. Khon Kaen 3 (NC_031164.1), *Sorghum bicolor* (NC_008360.1), *Triticum aestivum* cv. Chinese Yumai (NC_036024.1), *Oryza sativa* (NC_011033.1), *Zea mays* (NC_007982.1), *Lolium perenne* (JX999996.1), *Oryza coarctata* (MG429050.1), *Sporobolus alterniflorus* (MT471321.1), *Aegilops speltoides* (AP013107.1), *Stipa capillata* (MZ161090.1, MZ161091.1, MZ161093.1, MZ161092.1), *Bambusa oldhamii* (EU365401.1), and a *Brachypodium* sequence (AC276583.1), suggesting a partial *Brachypodium* mitochondrial draft genome. In addition, the *Hordeum vulgare* mitochondria genome sequence was downloaded from Ensembl Plants (ID: IBSC_v2, chromosome Mt). Pre-processing and quality control data is summarized in Table S2.

2.6. Taxonomic Classification

Taxonomic assignment was performed on the quality-controlled and host-decontaminated reads using a Kraken2 (v2.1.2) [29] database containing RefSeq libraries [30] of archaea (628 sequences), bacteria (58,811 sequences), fungi (1579 sequences), and protozoa (11,151 sequences) for a total of 72,217 sequences and ~110 billion bp (as of 24 June 2021). A Bayesian re-estimation of abundance with the Kraken (Bracken) (v2.6) [31] database was subsequently built with the Kraken2 database using the default 35 k-mer length and 150 bp read lengths. Kraken2 was run as an integrated module of Sunbeam using the development branch. Bracken was run on the Kraken2 output files, and the Bracken outputs were combined using the combine_bracken_outputs.py function for downstream analysis. Barplots were produced using the thresholds indicated in the legends to group together low abundant taxa for visual presentation. For diversity analysis, the kraken-biom tool (v1.0.1) (https://github.com/smdabdoub/kraken-biom; accessed on 27 September 2021) was used to convert Bracken outputs at the species level into .biom files for use with the Phyloseq (v1.36.0) [32] and Vegan (v2.5.7) [33] R packages.

2.7. Core and Functional Microbiome

To further characterize the microbiomes, PAST (Paleontological Statistics, v4.08, available at https://www.nhm.uio.no/english/research/infrastructure/past/; accessed on 15 November 2021) [34] was used for similarity percentage (SIMPER) analyses using the Bray–Curtis similarity matrix to compare leaf and rhizosphere-associated microbiota and to facilitate the identification of a core microbiome [35–37]. Core microbiomes were calculated based on species and ASVs present in 100% of the tissue-specific samples with >5% relative abundance.

Paired-end quality-controlled and decontaminated reads outputted by Sunbeam were concatenated using the command "cat sample_R1.fq sample_R2.fq > merged_sample.fq" and inputted into HUMAnN (v3.0.0) [38] running MetaPhlan (v3.0) [38], Bowtie2 (v2.4.4) [39], DIAMOND (v2.0.11) [40], and SAMtools (v1.13) [41,42]. Sequences were processed using the default UniRef90 database and the following parameters for MetaPhlAn: –stat_q 0, –bt2_ps very-sensitive-local; the following parameters for HUMAaN 3: –nucleotide-subject-coverage-threshold 5.0, –translated-subject-coverage-threshold 5.0; and the following parameters for and Bowtie 2: -D 20 -R 3 -N 1 -L 20 -i S,1,0.50 –local.

Gene families were regrouped and renamed to the uniref90_Pfam database using the humann_regroup_table and humann_rename_table commands. Special features, including ungrouped genes and unintegrated pathways, were retained by skipping normalization in favour of downstream normalization using MaAsLin2 (v1.6.0) [43]. The final renamed gene family and unnormalized pathway abundance tables were joined using the humann_join_table command and split into the stratified and unstratified tables using the humann_split_table command, the latter of which was used for differential abundance testing. Standard HUMAnN3 MetaCyc assigned metabolic pathways were used for analysis and were assigned classes based on the respective associated MetaCyc pathway superclasses. All scripts can be found in Supplementary File S1.

2.8. Amplicon Sequencing

Aliquots of the DNA extractions used for shotgun sequencing were sent to MR DNA for amplification and barcoded amplicon sequencing of the 16S rRNA V4 region using primers 515F (5′-GTGYCAGCMGCCGCGGTAA-3′) [44] and 806R (5′-GGACTACNVGGG TWTCTAAT-3′) [45], and of the ITS region using primers ITS1F (5′-CTTGGTCATTTAGAGG AAGTAA-3′) and ITS2R (5′-GCTGCGTTCTTCATCGATGC-3′) [46]. Peptide nucleic acid clamps pP01 (5′-GGCTCAACCCTGGACAG-3′), as previously described [47], were used to reduce amplification of *Brachypodium*-contaminating sequences during the amplification of the 16S rRNA V4 regions. Blank kit controls for both Plant Pro and PowerSoil Pro kits were performed in triplicate and subjected to the same amplification and sequencing as the corresponding samples. Sequencing was performed on a MiSeq platform (Illumina Inc., San Diego, CA, USA) for ITS and NovaSeq 6000 platform (Illumina Inc., San Diego, CA, USA) for 16S.

2.9. Amplicon Sequence Processing

Sequences were processed using QIIME2 (v2021.4) [48]. Raw .fastq files were demultiplexed and non-biological sequences were removed, including primers, adapters, spacers, and linkers, using FASTqProcessor (v20.11.19). Sequences were trimmed and denoised to remove any chimeras and singletons using DADA2 (v1.18) [49] before being grouped into amplicon single variants (ASVs). ASVs were used for taxonomic classification with SILVA (v138) for 16S rRNA sequences and UNITE (v8) for ITS sequences [50–55]. In the leaf samples, any taxa classified as eukaryota, chloroplast, mitochondria, archaea, or unclassified were filtered out of the 16S rRNA feature tables. Shannon's diversity index was used as a measure for alpha diversity and Bray–Curtis dissimilarity distance was used as a measure for community dissimilarity. Principal coordinate analysis (PCoA) was performed using Bray–Curtis dissimilarity matrices and plots made in R using ggplot2. Differential abundance between cold-acclimated and non-acclimated samples and between blank kit controls and samples was also assessed at the genus taxonomic levels using ANCOM-BC in R (v1.2.2) [56]. All commands and codes used can be found in Supplementary File S1.

2.10. Statistical Analysis

All statistical analyses were performed in RStudio (v1.3.1073) running R (v4.1.1) and all scripts used are available in Supplementary File S1. All plots, when necessary, were cleaned up using Inkscape (v0.92.2). Alpha and beta diversity analysis was performed using the Vegan and Phyloseq packages and PCoA plots were performed using ggplot2 (v3.3.5). To find differentially abundant taxa between the two temperature conditions, ANCOM-BC was run on Bracken outputs with default parameters for shotgun data and feature tables for amplicon data. Output coefficients representing the natural log fold-change model were converted to log2 fold changes. ANCOM-BC outputs were parsed to remove any low abundant taxa from differential abundance results.

3. Results

3.1. Pre-Processing, Shotgun Sequencing, and Kit Controls

Initial DNA samples representing the cold-acclimated (CA) leaf and rhizosphere were sent for shotgun sequencing without eukaryotic depletion, revealing high host contamination in the leaves (not shown). Subsequent replicate samples undergoing eukaryotic depletion proved successful as the classification of processed reads showed a full order of magnitude better recovery of microbial sequences. DNA and library concentrations and average size, quality control, host read decontamination, and Kraken2 classification results are summarized in the Supplementary Materials (Figure S4, Tables S1 and S2). Although shotgun DNA library construction was attempted on the blank kit controls, a lack of sufficient DNA resulted in no results for this sequencing method. However, the same control samples were subject to amplicon marker gene sequencing. Following QIIME2 processing, it was determined through diversity analysis and PCoA using Bray–Curtis dissimilarities that the microbial compositions associated with the kits were significantly different than the *Brachypodium* leaf ($p < 0.001$ 16S, $p < 0.05$ ITS, pairwise PERMANOVA) and rhizosphere microbiomes ($p < 0.001$ 16S, $p < 0.05$ ITS, pairwise PERMANOVA) (Figure S5).

3.2. Compatible Results with Shotgun and Amplicon Sequencing

The correlation between taxa identified in both the shotgun data and the amplicon data was assessed at the genus level in order to compare the two methods. In the CA rhizosphere, the genera identified by shotgun metagenomic and 16S rRNA amplicon sequences, as well as shotgun metagenomics and ITS amplicon sequencing, were well correlated ($R^2 = 0.93$ and $R^2 = 0.88$, respectively) (Figure S6). The non-acclimated (NA) rhizosphere shotgun and 16S rRNA, and the shotgun and ITS amplicon results ($R^2 = 0.91$ and $R^2 = 0.45$, respectively) also correlated, but less well. It is notable that for the leaf microbiome, bacterial taxa in the CA shotgun and 16S rRNA samples, as well as for the NA leaf samples, showed mixed correlations ($R^2 = 0.31$ and $R^2 = 0.75$, respectively). Insufficient fungal reads in the leaves following Bracken re-estimation resulted in no correlation between the shotgun and ITS reads in the leaves.

3.3. Cold Acclimation and the Rhizosphere Microbiome

In total, 4646 microbial species were identified in the rhizosphere shotgun data with $45 \pm 3\%$ of reads remaining unclassified. The majority of identified reads, $99.70 \pm 0.06\%$, represented bacterial microbes with $0.15 \pm 0.03\%$ and $0.13 \pm 0.02\%$ representing fungi and archaea, respectively. Alpha diversity, assessed using Shannon's diversity index, across all rhizosphere samples was 4.98 ± 0.21 and was not significantly different between conditions with 5.07 ± 0.29 in the CA and 4.91 ± 0.94 in the NA samples. The rhizosphere was dominated by *Streptomyces* sp. M2, a PGPB, accounting for approximately one-third of the taxa in all samples. Rounding out the top abundant species across the rhizosphere samples were taxa present at 1–10% abundance, which included *Actinocatenispora sera, Actinocatenispora thailandica, Rhodanobacter denitrificans*, and *Rhodanobacter* sp. *FDA-ARGOS* 1247 (Figure 1A; Table S3). Nearly half of all species in the rhizosphere shotgun data were below a cut-off value (0.2%) for low relative abundance leaving a balance of 53% and 56% of species found in NA and CA samples, respectively.

The amplicon analysis identified 651 distinct ASVs at the genus level. Alpha diversity appeared similar in the NA and CA samples (6.79 ± 0.25 and 6.40 ± 0.16, respectively) and differences were not significant. Both conditions were dominated by the genera *Streptomyces, Actinocatenispora*, and *Rhodanobacter* (Figure 1B; Table S3). After CA, low abundant taxa (<1% relative abundance) remained equal at 29%. Again, a similar number of ASVs were considered at low abundance under NA and CA conditions (20% and 15%, respectively). ITS analysis showed 25 distinct ASVs at the genus level (Figure 1C). *Ascomycota* and *Apiotrichum* each represented a third of the ASVs in the rhizosphere irrespective of conditions (Figure 1C; Table S3). Alpha diversity was significantly different ($p < 0.05$, two-tailed t-test) at 3.43 ± 0.06 in the CA and 3.05 ± 0.17 in the NA.

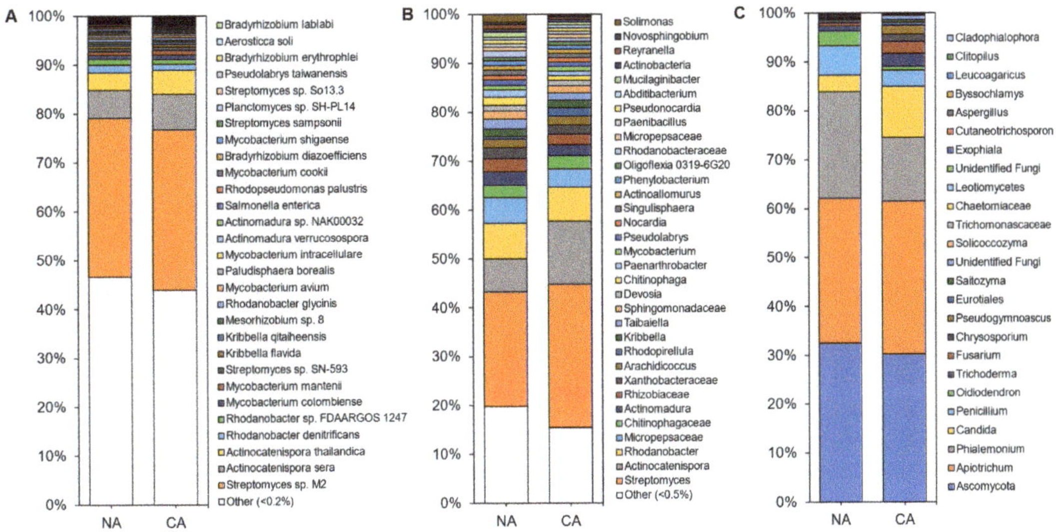

Figure 1. Average relative abundance of the taxonomies of the non-acclimated and cold-acclimated *Brachypodium distachyon* rhizosphere microbiomes: (**A**) species identified from shotgun sequencing and metagenomics classified using a custom Kraken2 database, (**B**) distinct amplicon sequence variants assigned down to the genus or lowest possible level by QIIME2 using the SILVA database for 16S rRNA sequences amplified using the V4 region of prokaryotes, and (**C**) distinct amplicon sequence variants assigned down to the genus or lowest possible level by QIIME2 using the UNITE database for ITS regions of eukaryotes.

Although there were few changes in the rhizosphere community following 6 days at 4 °C, differential abundance testing using ANCOM with bias control and parsed for taxa above the assigned low relative abundance thresholds (Figure 1) identified two modestly differentially abundant species (out of 143; 1.4%) in the shotgun data. *Kribbella qitaiheensis* (log2 fold change: 0.37) and *Kribbella flavida* (log2 fold change: 0.38) increased in relative abundance after CA (Figure 2A). In addition, the relative abundance of three fungal genera (out of 25; 12%) changed following CA, including a decrease in *Penicillium* (log2 fold change: −1.8) and *Phialemonium* (log2 fold change: −1.7) and a more substantial relative increase in *Pseudogymnoascus* (log2 fold change: 8.43) (Figure 2B).

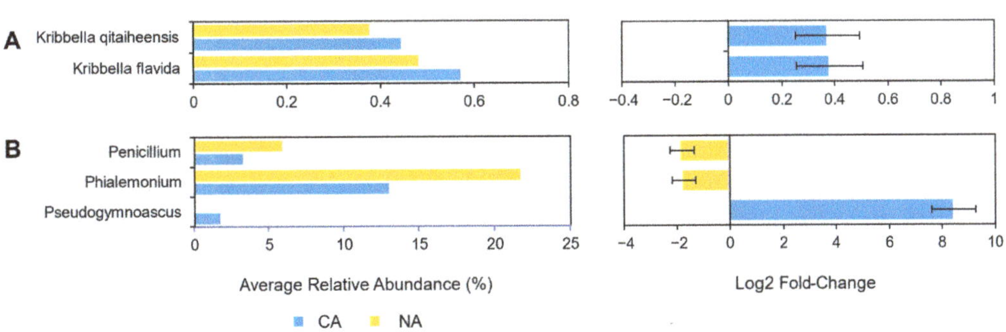

Figure 2. Differentially abundant taxa between the non-acclimated and cold-acclimated *Brachypodium distachyon* rhizosphere microbiomes as determined by ANCOM-BC and showing their average relative abundance in both conditions and log2 fold changes with error bars representing standard error: (**A**) species identified by Kraken2 from shotgun sequencing data that are differentially abundant and above an average relative abundance threshold of 0.2%, and (**B**) ITS amplicon sequence variants that are differentially abundant. Only statistically significant changes are shown, as determined by ANCOM-BC.

Although shifts in the rhizosphere community appeared modest, the Bray–Curtis dissimilarity analysis showed that the shotgun rhizosphere communities were significantly different under the two temperature regimes ($p < 0.01$, pairwise PERMANOVA) (Figure 3A). In contrast, there were no differences in Bray–Curtis dissimilarity for the amplicon analysis, either for 16S (Figure 3B) or ITS data (Figure 3C). Taking all the results together, it appears that overall, the CA regime resulted in only a very minor shift in the rhizosphere microbial community. We speculate that a longer period of low temperature with concomitant changes in root exudates would be required for a more dramatic change in the root-associated microbiota.

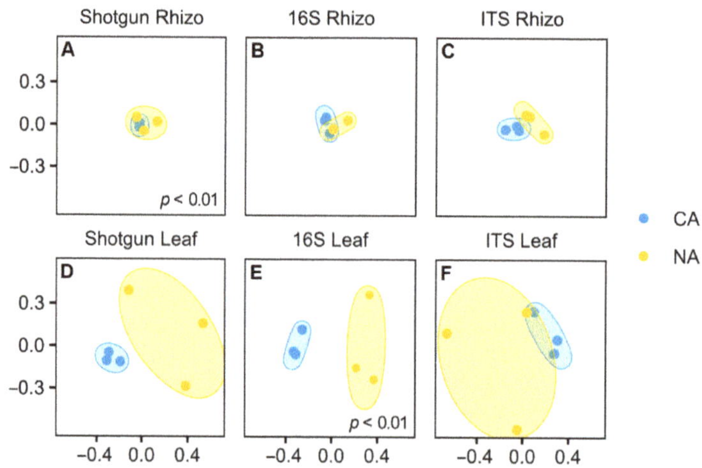

Figure 3. Principal coordinate analysis comparing non-acclimated and cold-acclimated conditions in each sample type for each sequencing method, for the following samples: (**A**) shotgun sequencing in the rhizosphere, (**B**) 16S rRNA sequencing of the V4 region in the rhizosphere, (**C**) ITS sequencing of the rhizosphere samples, (**D**) shotgun sequencing of the leaf samples, (**E**) 16S rRNA sequencing of the V4 region in the leaf samples, and (**F**) ITS sequencing of the leaf samples. Pairwise PERMANOVAs were conducted between conditions with significance as noted.

3.4. Cold Acclimation and the Leaf Microbiome

Although shotgun sequencing of the leaf, representing the endosphere and phyllosphere microbiomes, identified 143 microbial species with the most abundant taxa shown (Figure 4A; Table S4), an average of 92 ± 4% of the reads remained unclassified, with a portion of these likely attributable to as yet unsequenced host mitochondrial sequences (Figure S4C). Bacteria accounted for ~100% of the microbiota except in a couple of samples from which a few fungal sequences were recovered. Overall, alpha diversity was significantly lower ($p < 5 \times 10^{-6}$, two-tailed *t*-test) in leaf samples (3.18 ± 0.36) compared to rhizosphere samples (4.99 ± 0.21).

Leaf alpha diversity did not significantly change after CA treatment (mean Shannon indices at 3.30 ± 0.29 in NA samples and 3.06 ± 0.47 in CA samples). However, the taxa profile changed with the cyanobacteria *Microcystis aeruginosa*, decreasing from ~27% to ~13% relative abundance after CA. *Streptomyces* sp. M2 showed the opposite profile, increasing from ~4% to ~15% average relative abundance after transfer to 4 °C. NA leaves were dominated by the plant pathogens *Pseudomonas syringae* and '*Candidatus* Liberibacter africanus', as well as the plant beneficial *Rhodococcus qingshengii*, whose levels substantially decreased in the CA conditions. Lower abundant reads (<1%) made up about a quarter of the taxa, similar to the CA samples.

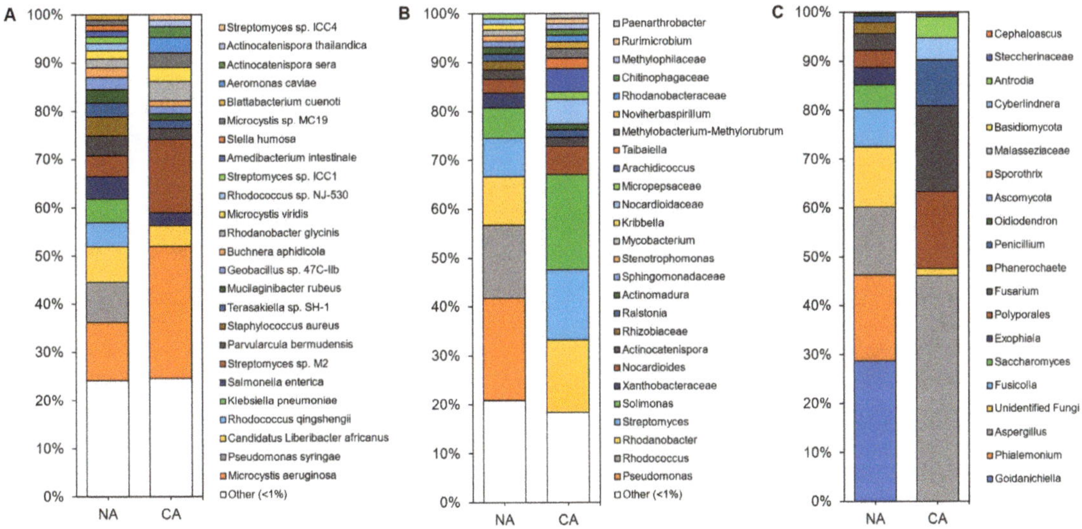

Figure 4. Average relative abundance of the taxonomies of the non-acclimated and cold-acclimated *Brachypodium distachyon* leaf microbiomes representing the endosphere and phyllosphere: (**A**) species identified from shotgun sequencing and metagenomics classified using a custom Kraken2 database, (**B**) distinct amplicon sequence variants assigned down to the genus or lowest possible level by QIIME2 using the SILVA database for 16S rRNA sequences, amplified using the V4 region of prokaryotes, and (**C**) distinct amplicon sequence variants assigned down to the genus or lowest possible level by QIIME2 using the UNITE database for ITS regions of eukaryotes.

Amplicon sequencing of the 16S rRNA from the leaves identified 188 distinct ASVs at the genus level (with the most abundant shown in Figure 4B and Table S4). Again, alpha diversity was not significantly different between conditions (5.04 ± 0.25 and 4.60 ± 0.70 in the CA and NA samples, respectively). Taxa present under both conditions included the genera *Solimonas*, *Rhodanobacter*, and *Streptomyces*. *Pseudomonas* and *Rhodococcus* were abundant (21% and 15% average relative abundance, respectively) in NA conditions, but decreased in relative abundance after transfer of the plants to 4 °C with log2 fold changes of −4.18 and −5.41, respectively. The cereal growth-promoting genus *Nocardioides* and an unidentified genus from the same family, *Nocardioidaceae*, both increased in abundance to represent 11% of the taxa in CA plants. ASVs at low relative abundance (<1%) made up a similar 18% and 21% of CA and NA 16S samples, respectively. ITS analysis resulted in 20 distinct ASVs at the genus level (Figure 4C).

After shotgun sequence analysis, 3.5% (5/143) of the taxa were identified as differentially abundant between the NA and CA conditions (Figure 5A). After transfer to 4 °C, reads attributed to *P. syringae* (log2 fold change: −8.68) and *R. qingshengii* (log2 fold change: −8.33) decreased so that there was a change in the estimated average relative abundance of *P. syringae* and *R. qingshengii* from 8.2% and 5.0% to 0%, respectively. At the same time there was a corresponding increase in the relative abundance of *Streptomyces* sp. M2 (log2 fold change: 2.81), *A. sera* (log2 fold change: 3.20), and *A. thailandica* (log2 fold change: 3.87). In 16S CA samples, nine other taxa increased, including the genus *Solimonas*, which increased in relative abundance but was below the low abundance threshold. In total, 5.9% (11/188) of the identified sequences above the threshold were found to be differentially abundant. For the ITS analysis, the genus *Phialemonium* represented 5% (1/20) of the ASVs and decreased in relative abundance (log2 fold change: −10.6) (Figure 5C).

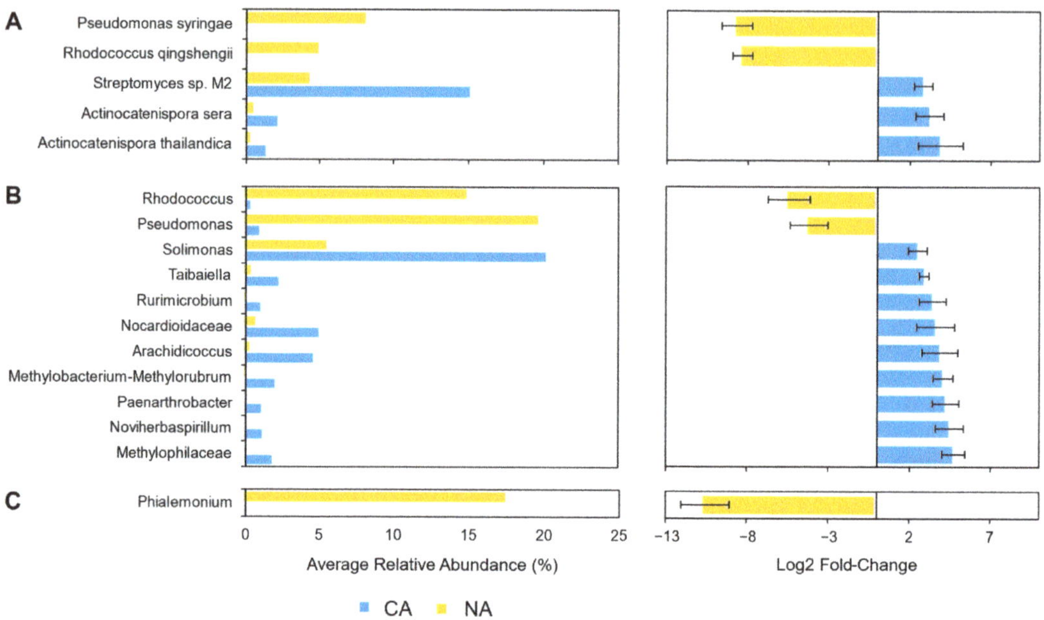

Figure 5. Differentially abundant taxa between the non-acclimated and cold-acclimated *Brachypodium distachyon* leaf microbiomes representing the endosphere and phyllosphere as determined by ANCOM-BC and showing their average relative abundance in both conditions and log2 fold changes with error bars representing standard error: (**A**) species identified by Kraken2 from shotgun sequencing data that are differentially abundant and above an average relative abundance threshold of 1%, (**B**) distinct 16S rRNA amplicon sequence variants assigned by QIIME2 and the SILVA database to the genus level that are differentially abundant, and (**C**) distinct ITS amplicon sequence variants assigned by QIIME2 and the UNITE database to the genus level that are differentially abundant. Only statistically significant changes are shown as determined by ANCOM-BC.

Despite the apparent community differences, Bray–Curtis dissimilarity analysis suggested that the microbial communities identified with the shotgun sequencing approach were not significantly different, undoubtedly due to the low number of sequences (Figure 3D), similar to the leaf ITS communities. Supporting that conclusion, 16S rRNA communities were shown to be significantly different between conditions ($p < 0.01$, pairwise PERMANOVA) with the analysis supported by high ASV numbers (Figure 3E).

3.5. Dissimilarity Comparisons and Core Microbiome

The root and leaf-associated microbiomes were further independently characterized with SIMPER to identify taxa that contributed the most dissimilarity between NA and CA regimes (Table 1). For microbiota isolated from the rhizosphere, the taxa contributing to the top ~25% of dissimilarity were *Streptomyces* sp. M2, *A. sera*, and *A. thailandica* for the shotgun data, the genera *Actinocatenispora* and *Streptomyces* for the 16S data, and the genera *Phialemonium* and *Apiotrichum* for the ITS data. For leaf samples, taxa contributing to the top ~25% dissimilarity were *M. aeruginosa* and *Streptomyces* sp. M2 for the shotgun data, the genera *Pseudomonas* and *Rhodococcus* for the 16S data, and the genera *Aspergillus* and *Goidanichiella* for the ITS data.

Highly conserved taxa that are present in most samples, typically ~70%, can be considered part of the "core" microbiome that orchestrates the interactions between the host and the microbiota [57]. As described in the methods, we employed strict criteria that the taxa must appear in all of the samples for each condition (Table 2). In the rhizosphere, the core microbiota identified in the shotgun analysis included *Streptomyces* sp. M2 and

Actinocatenispora sera. Core taxa in the leaves included *Streptomyces* sp. M2 and '*Candidatus* Liberibacter africanus', both of which persisted across the two different conditions and all samples. The larger number of taxa associated with the rhizosphere ASVs were consistent with the microbes identified by shotgun analysis and indicated bacterial (*Streptomyces*, *Actinocatenispora*, and *Rhodanobacter*) as well as fungal taxa (*Ascomycota*, *Apiotrichum*, *Phialemonium*, and *Candida*) as contributors to the core microbiome. Leaf ASVs revealed that bacteria (*Streptomyces*, *Rhodanobacter*, and *Solimonas*), as well as a single unidentified fungal sequence, comprised the core.

Table 1. Similarity of percentage (SIMPER) analysis of microbiota contributing to the top ~25% of dissimilarity (Bray–Curtis) between non-acclimated (NA) and cold-acclimated (CA) samples (showing average relative abundance in %) in both leaf tissue and rhizosphere performed in PAST (v4.08).

Taxa	NA (%)	CA (%)	Average Dissimilarity	Contribution (%)	Cumulative (%)
Shotgun Rhizo (Overall Average Dissimilarity 7.1%)					
Streptomyces sp. M2	32.5	32.8	0.9	12.8	12.8
Actinocatenispora sera	5.7	7.3	0.8	11.2	24.0
Actinocatenispora thailandica	3.6	4.9	0.6	9.1	33.1
16S Rhizo (Overall Average Dissimilarity 10.7%)					
Actinocatenispora	8.1	11.4	1. 7	15.7	15.7
Streptomyces	24.2	26.6	1.4	13.4	29.2
ITS Rhizo (Overall Average Dissimilarity 19.0%)					
Phialemonium	21.3	13.4	4.0	21.1	21.1
Apiotrichum	29.9	30.8	4.0	21.1	42.2
Shotgun Leaf (Overall Average Dissimilarity 52. 6%)					
Microcystis aeruginosa	12.1	27.4	9.6	18.2	18.2
Streptomyces sp. M2	4.4	15.1	5.4	10.2	28.4
16S Leaf (Overall Average Dissimilarity 60.9%)					
Pseudomonas	19.7	1.0	9.4	15.4	15.4
Rhodococcus	15.0	0.4	7.3	12.0	27.4
ITS Leaf (Overall Average Dissimilarity 80.3%)					
Aspergillus	15.9	47.5	16.6	20.7	20.7
Goidanichiella	25.1	0.0	12.6	15.7	36.4

Table 2. Core microbiota taxa (species or distinct ASVs as indicated) present in 100% of samples for each sequencing and analysis method of shotgun, 16S rRNA, and ITS sequencing methodologies with an average relative abundance >5%.

Phyla	Class	Order	Family	Genus	Species
Core rhizosphere species (shotgun)					
Actinobacteria	Actinomycetia	*Streptomycetales*	*Streptomycetaceae*	*Streptomyces*	*Streptomyces* sp. M2
Actinobacteria	Actinomycetia	*Micromonosporales*	*Micromonosporaceae*	*Actinocatenispora*	*Actinocatenispora sera*
Core rhizosphere genera (16S)					
Actinobacteria	Actinomycetia	*Streptomycetales*	*Streptomycetaceae*	*Streptomyces*	
Actinobacteria	Actinomycetia	*Micromonosporales*	*Micromonosporaceae*	*Actinocatenispora*	
Proteobacteria	Gammaproteobacteria	*Xanthomonadales*	*Rhodanobacteraceae*	*Rhodanobacter*	
Core rhizosphere genera (ITS)					
Ascomycota					
Basidiomycota	Tremellomycetes	*Trichosporonales*	*Trichosporonaceae*	*Apiotrichum*	
Ascomycota	Sordariomycetes	*Sordariales*	*Cephalothecaceae*	*Phialemonium*	

Table 2. *Cont.*

Phyla	Class	Order	Family	Genus	Species
Ascomycota	Saccharomycetes	*Saccharomycetales*	*Saccharomycetaceae*	*Candida*	
		Core leaf species (shotgun)			
Actinobacteria	Actinomycetia	*Streptomycetales*	*Streptomycetaceae*	*Streptomyces*	*Streptomyces* sp. M2
Proteobacteria	Alphaproteobacteria	*Hyphomicrobiales*	*Rhizobiaceae*	*Liberibacter*	* '*Candidatus* L. a.'
		Core leaf genera (16S)			
Actinobacteria	Actinomycetia	*Streptomycetales*	*Streptomycetaceae*	*Streptomyces*	
Proteobacteria	Gammaproteobacteria	*Xanthomonadales*	*Rhodanobacteraceae*	*Rhodanobacter*	
Proteobacteria	Gammaproteobacteria	*Salinisphaerales*	*Solimonadaceae*	*Solimonas*	
		Core leaf genera (ITS)			
Unidentified Fungi					

* '*Candidatus* Liberibacter africanus'.

4. Discussion

The plant-microbiome partnership is responsive to stress, with the details of the signalling between the kingdoms of Eubacteria, Fungi, and Planta only beginning to be investigated [9,10,58,59]. Sub-zero temperatures are a particular challenge, resulting in cellular dehydration, membrane rupture, and increased vulnerability to psychrophilic pathogens and death, but some perennials respond to earlier non-freezing temperatures, and/or shortened day lengths to initiate a signalling response. This CA stress triggers changes in plant metabolism, resulting in cold-hardening and survival during subsequent freeze events and is accompanied by significant changes in the leaf microbiome community profile, but with less substantial community shifts in the rhizosphere (Figures 1 and 3).

4.1. Little Change in Rhizosphere Communities after Cold Acclimation

The different sequencing methodologies employed, either amplicon or shotgun analyses, generally yielded compatible results. As indicated, there were few changes in the rhizosphere community after the shift to low temperatures, as shown by the overlapping PCoA groupings with rare exceptions, and for the most part these did not make up a large proportion of the taxa. The rhizosphere communities from both NA and CA plants contained taxa previously reported in bound soils associated with *Brachypodium* and similar to those found in wheat [22]. Some species of the order *Burkholderiales* have been isolated from ryegrass rhizospheres and are associated with nutrient acquisition such that there is interest in their potential as beneficial probiotics for crop enhancement [60]. Ascomycota is dominant in grassland soils, which can be low in organic matter and nutrients, playing key roles in cyanobacteria-dominated soils as well as having important roles in cycling carbon and nitrogen in addition to nutrient transport [61]. The fact that these taxa are shared in wheat and *Brachypodium* underscores the co-evolution of the plant–host relationship, since microbiota in the dicot, *Arabidopsis*, is distinct [22]. As noted, neither the *Brachypodium* bacterial nor fungal communities changed significantly after the plants were moved to 4 °C, suggesting that there was insufficient time for the soil to reach that temperature. Indeed, investigations of cold-responsive rhizosphere microbiota in maize used 5 weeks exposure to "chilling" conditions compared to our 6-day treatment [17]. In addition, it is notable that the myriad of CA-dictated changes made in the above-ground portion of *Brachypodium* are not apparently signalled to the rhizosphere during the treatment regimen.

4.2. Shifts in Leaf Communities Accompany Cold Acclimation

Compared to the rhizosphere, which is relatively protected from rapid abiotic and biotic stresses, leaves are exposed to daily temperature fluctuations, visible and ultraviolet light, herbivore and mechanical damage, and arguably more pathogens. Within two days of the shift to CA conditions, the *Brachypodium* leaf membrane is protected from freeze-induced electrolyte leakage, contains elevated levels of soluble sugars, and shows changes in the abundance profiles of hundreds of proteins [6]. The leaf community response was also rapid, as revealed by numerous abundance changes in the bacterial and fungal microbiota, as well as in the proportion of individual core taxa, as supported by the distinct groupings shown in PCoAs (Figures 3 and 5; Table 2). Similarly, cold-associated shifts occurred in leaves from European grasslands over winter while the rhizosphere was relatively unchanged [4]. As in the rhizosphere data, results from the two sequencing methods were generally consistent. However, a notable exception was for sequences corresponding to the toxic cyanobacteria *Microcystis aeruginosa*, which were abundant in NA and increased after CA, but only when using the shotgun methodology. It is possible that these sequences were misclassified as chloroplast DNA and were mistakenly filtered from the amplicon data. We speculate that the increase in relative abundance of cyanobacteria after CA is likely due to the reduction in evaporation on the leaf surfaces at low temperatures, consistent with their preference for aquatic habitats, and their known colonization of the phyllosphere [62].

For other taxa, there was clear evidence of a change in relative abundance after CA that was generally consistent irrespective of the sequencing methodology. This included three prominent *Actinobacteria* species that increased in relative read numbers, including the grassland-associated *Actinocatenispora thailandica* and *Actinocatenispora sera*, as well as the mycelium-producing *Streptomyces* sp. M2, a known PGPB [63]. Although present in the rhizosphere samples under both conditions, *Streptomyces* sp. M2 increased 216-fold in relative abundance following CA in leaves. Presumably, it promotes plant growth with its extensive repertoire of antibiotics, plant growth hormones, siderophores, and insecticides [63–65]. Strikingly, this *Streptomyces* strain can inhibit the plant pathogen *P. syringae*, perhaps due to siderophores that chelate iron required by *Pseudomonas* [63]. Such inhibition could explain the disappearance of *P. syringae* after CA treatment, representing a log2 fold change of −8.7.

Other bacteria also showed inverse abundance profiles depending upon the condition, as described in the Results section. Fungal ascomycete taxa similarly exchanged their relative abundance, with a decrease in the genus *Goidanichiella* and an increase in the genus *Aspergillus* detected after CA. These changes may be related to the temperature regime since *Goidanichiella* was reported to dominate summer-collected wheat leaves whereas cold-tolerant *Aspergillus* are of interest as growth promoters likely due to their ability to solubilize phosphates [66,67].

4.3. Leaf Cold Acclimation Associated with Low Temperature and Pathogen Responses

After transfer to 4 °C, the leaf microbiome was impacted by the temperature shift and also showed changes in the relative abundance of potential pathogens. These observations reflect the results of network analysis of hundreds of plasma membrane proteome changes after CA that showed crosstalk between pathways for low temperature stress and disease and defence [6]. *Brachypodium* responds to CA by diverting resources away from growth and to the stress response. It appears then that the host–microbiome works together in a joint effort to prepare for the worsening conditions associated with winter.

One of the most obvious examples of the connection between low temperature and disease is found in the ice nucleation-active plant pathogen *P. syringae*, which can facilitate the formation of ice at temperatures just below 0 °C, presumably to lyse plant cells and thus access nutrients [68]. In NA leaves, *P. syringae* was a large contributor to the bacterial taxa (8% of the shotgun reads). However, as the temperature drops, such a large proportion of *P. syringae* in the leaf microbiota would surely present a grave risk to the host plant. Remarkably, after CA there was no evidence of this bacteria. This disappearance is undoubtedly fostered by *Brachypodium*'s defence pathways that lead to the production of multiple proteins, including antifreeze proteins, that target the ice nucleator, but we propose that the microbiome also supports this protective strategy.

Coincident with the collapse of the *P. syringae* population, there was a 216-fold increase in the relative abundance of *Streptomyces* sp. M2 (0.1% to 15.1%). It is important to note that this increase after CA cannot be explained by sensitivity to the NA growth conditions since it is routinely cultured at 30 °C [69]. Thus, the change in its abundance is independent of the temperature shift and may be fostered by *Brachypodium*. As mentioned, this PGPB secretes antibiotics and siderophores and is known to inhibit *P. syringae* [63]. *Rhodococcus* also decreased 40-fold in relative abundance, but to date there is no information on its interaction with *Streptomyces* or other plant beneficials. Nevertheless, as well as directly targeting *P. syringae*, it is likely that *Streptomyces* alerts plant defences against other phytopathogens since the inoculation of *Streptomyces* spp. induces the expression of defense-related genes—at least, so it was found to do in a pea crop [70]. This ability could also explain why *Streptomyces* spp. are not limited to inhibition of bacterial species but also inhibit fungal phytopathogens in planta [71,72].

Therefore, in addition to combating the cold-associated pathogen *P. syringae*, *Streptomyces* sp. M2 likely contributes to the overall cold tolerance of *Brachypodium* and thus would be central to the cold-acclimated microbiome. *Streptomyces* spp. have a variety of adaptations for cold resistance, including the production of cold shock proteins and small solutes for cryoprotection [73–75]. These products may assist host survival, since a strain of *Streptomyces* was shown to alleviate the effects of cold stress in turfgrass [19] and drought stress in maize [76]. In addition, BioCyc genome-wide predictions indicate that *Streptomyces* sp. M2 produces key oxidative stress enzymes that can be secreted in *Streptomyces* spp. [77–79]. In addition, *Streptomyces* sp. M2 synthesizes cryoprotective soluble sugars that coincidentally increase rapidly in CA *Brachypodium* [6,80]. The synthesis of the osmoprotectant proline may also benefit host plants, as inoculation of sugarcane with *Streptomyces* increased proline content and drought tolerance [81]. *Streptomyces* spp. are also reported to increase drought tolerance in maize and aid in the accumulation of soluble sugars [76].

Another bacterial taxon, the genus *Solimonas*, increased 3.3-fold after CA, and although these species have a wide temperature optimum, they are characterized by polar lipids and fatty acids, which are known to contribute to cold tolerance [82]. In parallel findings, *Brachypodium* shows changes in metabolic pathways leading to restructuring of the plasma membrane after CA, a common vulnerability for both microbes and their hosts [6,83,84]. Already mentioned was the cold tolerance of the plant-beneficial fungus *Aspergillus*. More insight could be revealed by an investigation of the functional microbiomes of CA *Brachypodium*. However, due to low reads and sequencing depths, our results can only be considered preliminary (see Supplementary File S2 and Figures S7–S9). Nevertheless, in parallel with the CA *Brachypodium* plasma membrane proteome [6], microbial proteins involved in pathways that intersect with low temperature tolerance, such as the synthesis of soluble cryoprotectants, oxidative stress, and pathogen resistance, were detected in the microbiome in response to cold stress. Again, this underscored the inter-dependent and symbiotic character of the CA response.

4.4. Prospects and Conclusions

Taken together, both the changes in microbial community profiles following CA and the functional role of these plant beneficials suggest that commercial growers could see some benefit from the inoculation of mixed community strains, including *Streptomyces* sp. M2, for protection against *P. syringae* and other phytopathogens, while at the same time benefiting from other plant growth-promoting characteristics as well as enhancing cold resilience. With the presentation of this first CA *Brachypodium* microbiome, it is hoped that the insights gained will inspire treatment options to enhance cold tolerance and other intersecting stresses tailored toward specific agriculturally important grain crops [1,9,85,86].

This special issue of *Plants* asks, "What makes the life of stressed plants a little easier?" The answer for *Brachypodium* undergoing acclimation to low temperature in preparation for the coming winter is very clear. It is the strong partnership with a shifting above-ground bacterial and fungal community that works in concert with plant networks that intersect cold-, drought-, and antipathogen-signalling pathways to ensure that within only a few days host plants survive freeze events. Not only does it make the life of plants a "little easier", we also argue that it may very well be essential for survival. Therefore, we propose that the battle against winter condition stresses is won by important inter-kingdom partnerships.

Supplementary Materials: The following are available online at https://www.mdpi.com/article/10.3390/plants10122824/s1. Supplementary File S1: Contains all scripts and commands used. Supplementary File S2: Contains all Supplemental Tables, Supplemental Figures, and Supplemental Text describing functional classification of shotgun data. Figure S1: Bulk soil collection from a farm field. Figure S2: Representative three-week-old *Brachypodium distachyon*. Figure S3: Image showing an example of the tightly bound root soil still attached to the plant. Figure S4: Read statistics of the shotgun sequencing processing for averages of the cold-acclimated and non-acclimated leaf and rhizosphere samples. Figure S5: Principal coordinate analysis plots comparing the taxonomic communities from amplicon sequencing blank kit controls. Figure S6: Correlation plots comparing shotgun sequencing to amplicon sequencing results under both non-acclimated and cold-acclimated conditions in the leaf and rhizosphere samples. Figure S7: Heatmaps showing the average relative abundance of the Pfam domains. Figure S8: Heatmap showing the top 50 most abundant MetaCyc pathways in rhizosphere samples. Figure S9: Heatmap showing the top 50 most abundant MetaCyc pathways in leaf samples. Table S1: DNA, final library concentration, and average library size. Table S2: Summary of quality control and preprocessing of metagenomic reads from shotgun sequencing. Table S3: Summary of the top ten average relative abundant taxa for rhizosphere samples showing average relative abundance for each non-acclimated and cold-acclimated conditions of shotgun, 16S, and ITS sequencing. Table S4: Summary of the top ten average relative abundant taxa for leaf samples showing average relative abundance for each non-acclimated and cold-acclimated conditions of shotgun, 16S, and ITS sequencing.

Author Contributions: C.L.J. conducted all experiments, analyzed all data, and produced all figures. C.L.J. wrote the initial draft of the manuscript and all authors contributed to manuscript revision. G.C.d. and V.K.W. supervised the work. All authors have read and agreed to the published version of the manuscript.

Funding: This research was funded by a Natural Sciences and Engineering Research Council of Canada Discovery grant to V.K.W and funding from Queen's University to G.C.d.

Institutional Review Board Statement: This study did not involve humans or animals. *Brachypodium distachyon* plants were used in this study. Seeds of an inbred line, ecotype Bd21, were kindly provided by RIKEN, Wakō, Japan.

Data Availability Statement: All raw sequences were deposited in the National Centre for Biotechnology Information (NCBI) Sequence Read Archive (SRA) under BioProject ID: PRJNA782211, available at https://www.ncbi.nlm.nih.gov/bioproject/782211 (accessed on 21 November 2021).

Acknowledgments: We acknowledge Kristy Moniz for her technical support early in the project and MiGS and MR DNA for their preparation of libraries and sequencing, as well as Scot Dowd of MR DNA for his suggestions and advice.

Conflicts of Interest: The authors declare no conflict of interest.

References

1. Juurakko, C.L.; di Cenzo, G.C.; Walker, V.K. Cold acclimation and prospects for cold-resilient crops. *Plant Stress* **2021**, *2*, 100028. [CrossRef]
2. Suzuki, N.; Mittler, R. Reactive oxygen species and temperature stresses: A delicate balance between signaling and destruction. *Physiol. Plant.* **2006**, *126*, 45–51. [CrossRef]
3. Grady, K.L.; Sorensen, J.W.; Stopnisek, N.; Guittar, J.; Shade, A. Assembly and seasonality of core phyllosphere microbiota on perennial biofuel crops. *Nat. Commun.* **2019**, *10*, 4135. [CrossRef]
4. Bei, Q.; Moser, G.; Müller, C.; Liesack, W. Seasonality affects function and complexity but not diversity of the rhizosphere microbiome in European temperate grassland. *Sci. Total Environ.* **2021**, *784*, 147036. [CrossRef]
5. Chialva, M.; De Rose, S.; Novero, M.; Lanfranco, L.; Bonfante, P. Plant genotype and seasonality drive fine changes in olive root microbiota. *Curr. Plant Biol.* **2021**, *28*, 100219. [CrossRef]

6. Juurakko, C.L.; Bredow, M.; Nakayama, T.; Imai, H.; Kawamura, Y.; di Cenzo, G.C.; Walker, V.K. The *Brachypodium distachyon* cold-acclimated plasma membrane proteome is primed for stress resistance. *G3* **2021**, *11*, jkab198. [CrossRef]
7. Thomashow, M.F. Plant cold acclimation: Freezing tolerance genes and regulatory mechanisms. *Annu. Rev. Plant Biol.* **1999**, *50*, 571–599. [CrossRef]
8. Guo, X.; Liu, D.; Chong, K. Cold signaling in plants: Insights into mechanisms and regulation. *J. Integr. Plant Biol.* **2018**, *60*, 745–756. [CrossRef] [PubMed]
9. Acuña-Rodríguez, I.S.; Newsham, K.K.; Gundel, P.E.; Torres-Díaz, C.; Molina-Montenegro, M.A. Functional roles of microbial symbionts in plant cold tolerance. *Ecol. Lett.* **2020**, *23*, 1034–1048. [CrossRef] [PubMed]
10. Liu, H.; Brettell, L.E.; Qiu, Z.; Singh, B.K. Microbiome-mediated stress resistance in plants. *Trends Plant Sci.* **2020**, *25*, 733–743. [CrossRef]
11. Saikkonen, K.; Faeth, S.H.; Helander, M.; Sullivan, T.J. Fungal endophytes: A continuum of interactions with host plants. *Annu. Rev. Ecol. Syst.* **1998**, *29*, 319–343. [CrossRef]
12. Porras-Alfaro, A.; Bayman, P. Hidden fungi, emergent properties: Endophytes and microbiomes. *Annu. Rev. Phytopathol.* **2011**, *49*, 291–315. [CrossRef]
13. Rho, H.; Kim, S.H. Endophyte effects on photosynthesis and water use of plant hosts: A meta-analysis. In *Functional Importance of the Plant Microbiome*; Springer: Cham, Switzerland, 2017; pp. 43–69.
14. Yadav, S.K. Cold stress tolerance mechanisms in plants. A review. *Agron. Sustain. Dev.* **2010**, *30*, 515–527. [CrossRef]
15. Hardoim, P.R.; Van Overbeek, L.S.; Berg, G.; Pirttilä, A.M.; Compant, S.; Campisano, A.; Sessitsch, A. The hidden world within plants: Ecological and evolutionary considerations for defining functioning of microbial endophytes. *Microbiol. Mol. Biol. Rev.* **2015**, *79*, 293–320. [CrossRef]
16. Rho, H.; Hsieh, M.; Kandel, S.L.; Cantillo, J.; Doty, S.L.; Kim, S.H. Do endophytes promote growth of host plants under stress? A meta-analysis on plant stress mitigation by endophytes. *Microb. Ecol.* **2018**, *75*, 407–418. [CrossRef]
17. Beirinckx, S.; Viaene, T.; Haegeman, A.; Debode, J.; Amery, F.; Vandenabeele, S.; Goormachtig, S. Tapping into the maize root microbiome to identify bacteria that promote growth under chilling conditions. *Microbiome* **2020**, *8*, 54. [CrossRef]
18. Theocharis, A.; Bordiec, S.; Fernandez, O.; Paquis, S.; Dhondt-Cordelier, S.; Baillieul, F.; Barka, E.A. *Burkholderia phytofirmans* PsJN primes *Vitis vinifera* L. and confers a better tolerance to low nonfreezing temperatures. *Mol. Plant Microbe Interact.* **2012**, *25*, 241–249. [CrossRef]
19. Jeon, C.W.; Kim, D.R.; Bae, E.J.; Kwak, Y.S. Changes in bacterial community structure and enriched functional bacteria associated with turfgrass monoculture. *Front. Bioeng. Biotechnol.* **2021**, *8*, 1495. [CrossRef]
20. Liu, X.M.; Xu, Q.L.; Li, Q.Q.; Zhang, H.; Xiao, J.X. Physiological responses of the two blueberry cultivars to inoculation with an arbuscular mycorrhizal fungus under low-temperature stress. *J. Plant Nutr.* **2017**, *40*, 2562–2570. [CrossRef]
21. Rocha, I.; Ma, Y.; Souza-Alonso, P.; Vosátka, M.; Freitas, H.; Oliveira, R.S. Seed coating: A tool for delivering beneficial microbes to agricultural crops. *Front. Plant Sci.* **2019**, *10*, 1357. [CrossRef]
22. Kawasaki, A.; Donn, S.; Ryan, P.R.; Mathesius, U.; Devilla, R.; Jones, A.; Watt, M. Microbiome and exudates of the root and rhizosphere of *Brachypodium distachyon*, a model for wheat. *PLoS ONE* **2016**, *11*, e0164533. [CrossRef]
23. Feehery, G.R.; Yigit, E.; Oyola, S.O.; Langhorst, B.W.; Schmidt, V.T.; Stewart, F.J.; Pradhan, S. A method for selectively enriching microbial DNA from contaminating vertebrate host DNA. *PLoS ONE* **2013**, *8*, e76096.
24. Clarke, E.L.; Taylor, L.J.; Zhao, C.; Connell, A.; Lee, J.J.; Fett, B.; Bittinger, K. Sunbeam: An extensible pipeline for analyzing metagenomic sequencing experiments. *Microbiome* **2019**, *7*, 46. [CrossRef] [PubMed]
25. Martin, M. Cutadapt removes adapter sequences from high-throughput sequencing reads. *EMBnet. J.* **2011**, *17*, 10–12. [CrossRef]
26. Bolger, A.M.; Lohse, M.; Usadel, B. Trimmomatic: A flexible trimmer for Illumina sequence data. *Bioinformatics* **2014**, *30*, 2114–2120. [CrossRef]
27. Andrews, S. FastQC: A Quality Control Tool for High Throughput Sequence Data. 2010. Available online: http://www.bioinformatics.babraham.ac.uk/projects/fastqc (accessed on 20 August 2021).
28. Li, H.; Durbin, R. Fast and accurate short read alignment with Burrows–Wheeler transform. *Bioinformatics* **2009**, *25*, 1754–1760. [CrossRef]
29. Wood, D.E.; Lu, J.; Langmead, B. Improved metagenomic analysis with Kraken 2. *Genome Biol.* **2019**, *20*, 257. [CrossRef]
30. O'Leary, N.A.; Wright, M.W.; Brister, J.R.; Ciufo, S.; Haddad, D.; McVeigh, R.; Pruitt, K.D. Reference sequence (RefSeq) database at NCBI: Current status, taxonomic expansion, and functional annotation. *Nucleic Acids Res.* **2016**, *44*, D733–D745. [CrossRef] [PubMed]
31. Lu, J.; Breitwieser, F.P.; Thielen, P.; Salzberg, S.L. Bracken: Estimating species abundance in metagenomics data. *PeerJ Comput. Sci.* **2017**, *3*, e104. [CrossRef]
32. McMurdie, P.J.; Holmes, S. Phyloseq: An R package for reproducible interactive analysis and graphics of microbiome census data. *PLoS ONE* **2013**, *8*, e61217.
33. Oksanen, J.; Blanchet, F.G.; Friendly, M.; Kindt, R.; Legendre, P.; McGlinn, D. *Vegan: Community Ecology Package. R Package Version 2.5-7*. 2020. Available online: https://rdrr.io/cran/vegan/ (accessed on 15 December 2021).
34. Hammer, Ø.; Harper, D.A.; Ryan, P.D. PAST: Paleontological statistics software package for education and data analysis. *Palaeontol. Electron.* **2001**, *4*, 9.
35. Clarke, K.R. Non-parametric multivariate analyses of changes in community structure. *Aust. J. Ecol.* **1993**, *18*, 117–143. [CrossRef]

36. Lazcano, C.; Boyd, E.; Holmes, G.; Hewavitharana, S.; Pasulka, A.; Ivors, K. The rhizosphere microbiome plays a role in the resistance to soil-borne pathogens and nutrient uptake of strawberry cultivars under field conditions. *Sci. Rep.* **2021**, *11*, 3188. [CrossRef] [PubMed]
37. Marsh, R.; Gavillet, H.; Hanson, L.H.; Ng, C.; Mitchell-Whyte, M.; Major, G.; van der Gast, C. Intestinal function and transit associate with gut microbiota dysbiosis in cystic fibrosis. *MedRxiv* **2021**. Available online: https://www.medrxiv.org/content/10.1101/2021.08.24.21262265v2 (accessed on 20 October 2021). [CrossRef]
38. Beghini, F.; McIver, L.J.; Blanco-Míguez, A.; Dubois, L.; Asnicar, F.; Maharjan, S.; Segata, N. Integrating taxonomic, functional, and strain-level profiling of diverse microbial communities with bioBakery 3. *Elife* **2021**, *10*, e65088. [CrossRef] [PubMed]
39. Langmead, B.; Salzberg, S.L. Fast gapped-read alignment with Bowtie 2. *Nat. Methods* **2012**, *9*, 357–359. [CrossRef]
40. Buchfink, B.; Xie, C.; Huson, D.H. Fast and sensitive protein alignment using DIAMOND. *Nat. Methods* **2015**, *12*, 59–60. [CrossRef] [PubMed]
41. Li, H.; Handsaker, B.; Wysoker, A.; Fennell, T.; Ruan, J.; Homer, N.; Durbin, R. The sequence alignment/map format and SAMtools. *Bioinformatics* **2009**, *25*, 2078–2079. [CrossRef]
42. Danecek, P.; Bonfield, J.K.; Liddle, J.; Marshall, J.; Ohan, V.; Pollard, M.O.; Li, H. Twelve years of SAMtools and BCFtools. *Gigascience* **2021**, *10*, giab008. [CrossRef]
43. Mallick, H.; Tickle, T.L.; McIver, L.J.; Rahnavard, G.; Nguyen, L.H.; Weingart, G.; Subramanian, A. Multivariable Association in Population-Scale Meta'omic Surveys. Submission. 2020. Available online: https://huttenhower.sph.harvard.edu/maaslin2/ (accessed on 6 September 2021).
44. Parada, A.E.; Needham, D.M.; Fuhrman, J.A. Every base matters: Assessing small subunit rRNA primers for marine microbiomes with mock communities, time series and global field samples. *Environ. Microbiol.* **2016**, *18*, 1403–1414. [CrossRef]
45. Apprill, A.; McNally, S.; Parsons, R.; Weber, L. Minor revision to V4 region SSU rRNA 806R gene primer greatly increases detection of SAR11 bacterioplankton. *Aquat. Microb. Ecol.* **2015**, *75*, 129–137. [CrossRef]
46. Gardes, M.; Bruns, T.D. ITS primers with enhanced specificity for basidiomycetes-application to the identification of mycorrhizae and rusts. *Mol. Ecol.* **1993**, *2*, 113–118. [CrossRef] [PubMed]
47. Tannenbaum, I.; Kaur, J.; Mann, R.; Sawbridge, T.; Rodoni, B.; Spangenberg, G. Profiling the *Lolium perenne* microbiome: From seed to seed. *Phytobiomes J.* **2020**, *4*, 281–289. [CrossRef]
48. Bolyen, E.; Rideout, J.R.; Dillon, M.R.; Bokulich, N.A.; Abnet, C.C.; Al-Ghalith, G.A.; Caporaso, J.G. Reproducible, interactive, scalable and extensible microbiome data science using QIIME 2. *Nat. Biotechnol.* **2019**, *37*, 852–857. [CrossRef]
49. Callahan, B.J.; McMurdie, P.J.; Rosen, M.J.; Han, A.W.; Johnson, A.J.A.; Holmes, S.P. DADA2: High-resolution sample inference from Illumina amplicon data. *Nat. Methods* **2016**, *13*, 581–583. [CrossRef]
50. Quast, C.; Pruesse, E.; Yilmaz, P.; Gerken, J.; Schweer, T.; Yarza, P.; Glöckner, F.O. The SILVA ribosomal RNA gene database project: Improved data processing and web-based tools. *Nucleic Acids Res.* **2012**, *41*, D590–D596. [CrossRef]
51. Yilmaz, P.; Parfrey, L.W.; Yarza, P.; Gerken, J.; Pruesse, E.; Quast, C.; Glöckner, F.O. The SILVA and "all-species living tree project (LTP)" taxonomic frameworks. *Nucleic Acids Res.* **2014**, *42*, D643–D648. [CrossRef]
52. Glöckner, F.O.; Yilmaz, P.; Quast, C.; Gerken, J.; Beccati, A.; Ciuprina, A.; Ludwig, W. 25 years of serving the community with ribosomal RNA gene reference databases and tools. *J. Biotechnol.* **2017**, *261*, 169–176. [CrossRef]
53. Nilsson, R.H.; Larsson, K.H.; Taylor, A.F.S.; Bengtsson-Palme, J.; Jeppesen, T.S.; Schigel, D.; Abarenkov, K. The UNITE database for molecular identification of fungi: Handling dark taxa and parallel taxonomic classifications. *Nucleic Acids Res.* **2019**, *47*, D259–D264. [CrossRef] [PubMed]
54. Abarenkov, K.; Zirk, A.; Piirmann, T.; Pöhönen, R.; Ivanov, F.; Nilsson, R.H.; Kõljalg, U. UNITE General FASTA Release for Fungi. Version 04.02. 2020, UNITE Community. Available online: https://search.datacite.org/works/10.15156/bio/786385 (accessed on 15 December 2021).
55. Kõljalg, U.; Nilsson, H.R.; Schigel, D.; Tedersoo, L.; Larsson, K.H.; May, T.W.; Abarenkov, K. The taxon hypothesis paradigm—On the unambiguous detection and communication of taxa. *Microorganisms* **2020**, *8*, 1910. [CrossRef]
56. Lin, H.; Peddada, S.D. Analysis of compositions of microbiomes with bias correction. *Nat. Commun.* **2020**, *11*, 3514. [CrossRef] [PubMed]
57. Kumar, M.; Brader, G.; Sessitsch, A.; Mäki, A.; van Elsas, J.D.; Nissinen, R. Plants assemble species specific bacterial communities from common core taxa in three arcto-alpine climate zones. *Front. Microbiol.* **2017**, *8*, 12. [CrossRef] [PubMed]
58. Santos-Medellín, C.; Liechty, Z.; Edwards, J.; Nguyen, B.; Huang, B.; Weimer, B.C.; Sundaresan, V. Prolonged drought imparts lasting compositional changes to the rice root microbiome. *Nat. Plants* **2021**, *7*, 1065–1077. [CrossRef]
59. Song, Y.; Haney, C.H. Drought dampens microbiome development. *Nat. Plants* **2021**, *7*, 994–995. [CrossRef] [PubMed]
60. Castanheira, N.; Dourado, A.C.; Kruz, S.; Alves, P.I.L.; Delgado-Rodríguez, A.I.; Pais, I.; Fareleira, P. Plant growth-promoting *Burkholderia* species isolated from annual ryegrass in Portuguese soils. *J. Appl. Microbiol.* **2016**, *120*, 724–739. [CrossRef]
61. Challacombe, J.F.; Hesse, C.N.; Bramer, L.M.; McCue, L.A.; Lipton, M.; Purvine, S.; Kuske, C.R. Genomes and secretomes of *Ascomycota* fungi reveal diverse functions in plant biomass decomposition and pathogenesis. *BMC Genom.* **2019**, *20*, 976. [CrossRef] [PubMed]
62. Lee, S.; Kim, J.; Lee, J. Colonization of toxic cyanobacteria on the surface and inside of leafy green: A hidden source of cyanotoxin production and exposure. *Food Microbiol.* **2021**, *94*, 103655. [CrossRef]

63. Worsley, S.F.; Newitt, J.; Rassbach, J.; Batey, S.F.; Holmes, N.A.; Murrell, J.C.; Hutchings, M.I. *Streptomyces* endophytes promote host health and enhance growth across plant species. *Appl. Environ. Microbiol.* **2020**, *86*, e01053-20. [CrossRef]
64. Hutchings, M.I.; Truman, A.W.; Wilkinson, B. Antibiotics: Past, present and future. *Curr. Opin. Microbiol.* **2019**, *51*, 72–80. [CrossRef]
65. Gupta, S.; Pandey, S. ACC deaminase producing bacteria with multifarious plant growth promoting traits alleviates salinity stress in French bean (*Phaseolus vulgaris*) plants. *Front. Microbiol.* **2019**, *10*, 1506. [CrossRef]
66. Rinu, K.; Malviya, M.K.; Sati, P.; Tiwari, S.C.; Pandey, A. Response of cold-tolerant *Aspergillus* spp. to solubilization of Fe and Al phosphate in presence of different nutritional sources. *Int. Sch. Res. Not.* **2013**, *2013*, 598541. [CrossRef]
67. Granzow, S.; Kaiser, K.; Wemheuer, B.; Pfeiffer, B.; Daniel, R.; Vidal, S.; Wemheuer, F. The effects of cropping regimes on fungal and bacterial communities of wheat and faba bean in a greenhouse pot experiment differ between plant species and compartment. *Front. Microbiol.* **2017**, *8*, 902. [CrossRef]
68. Pearce, R.S. Plant freezing and damage. *Ann. Bot.* **2001**, *87*, 417–424. [CrossRef]
69. Daffonchio, D.; Zanardini, E.; Vatta, P.; Sorlini, C. Cometabolic degradation of thiocarbamate herbicides by Streptomyces sp. strain M2 and effects on the cell metabolism. *Ann. Microbiol. Enzimol.* **1999**, *49*, 13–22.
70. Singh, S.P.; Gaur, R. Endophytic *Streptomyces* spp. underscore induction of defense regulatory genes and confers resistance against *Sclerotium rolfsii* in chickpea. *Biol. Control.* **2017**, *104*, 44–56. [CrossRef]
71. Taechowisan, T.; Peberdy, J.F.; Lumyong, S. Isolation of endophytic actinomycetes from selected plants and their antifungal activity. *World J. Microbiol. Biotechnol.* **2003**, *19*, 381–385. [CrossRef]
72. Human, Z.R.; Moon, K.; Bae, M.; De Beer, Z.W.; Cha, S.; Wingfield, M.J.; Venter, S.N. Antifungal *Streptomyces* spp. associated with the infructescences of *Protea* spp. in South Africa. *Front. Microbiol.* **2016**, *7*, 1657. [CrossRef]
73. Beales, N. Adaptation of microorganisms to cold temperatures, weak acid preservatives, low pH, and osmotic stress: A review. *Compr. Rev. Food Sci. Food Saf.* **2004**, *3*, 1–20. [CrossRef]
74. Kim, M.J.; Lee, Y.K.; Lee, H.K.; Im, H. Characterization of cold-shock protein A of Antarctic *Streptomyces* sp. AA8321. *Protein J.* **2007**, *26*, 51–59. [CrossRef]
75. Manullang, W.; Chuang, H.W. *Streptomyces* sp. mitigates abiotic stress response and promotes plant growth. *J. Plant Prot. Res.* **2020**, 263–274.
76. Warrad, M.; Hassan, Y.M.; Mohamed, M.S.; Hagagy, N.; Al-Maghrabi, O.A.; Selim, S.; Abd-Elgawad, H. A bioactive fraction from *Streptomyces* sp. enhances maize tolerance against drought stress. *J. Microbiol. Biotechnol.* **2020**, *30*, 1156–1168. [CrossRef]
77. Zou, P.; Schrempf, H. The heme-independent manganese-peroxidase activity depends on the presence of the C-terminal domain within the *Streptomyces reticuli* catalase-peroxidase CpeB. *Eur. J. Biochem.* **2000**, *267*, 2840–2849. [CrossRef] [PubMed]
78. Folcher, M.; Gaillard, H.; Nguyen, L.T.; Nguyen, K.T.; Lacroix, P.; Bamas-Jacques, N.; Thompson, C.J. Pleiotropic functions of a *Streptomyces pristinaespiralis* autoregulator receptor in development, antibiotic biosynthesis, and expression of a superoxide dismutase. *J. Biol. Chem.* **2001**, *276*, 44297–44306. [CrossRef]
79. Akanuma, G.; Hara, H.; Ohnishi, Y.; Horinouchi, S. Dynamic changes in the extracellular proteome caused by absence of a pleiotropic regulator AdpA in *Streptomyces griseus*. *Mol. Microbiol.* **2009**, *73*, 898–912. [CrossRef]
80. Killham, K.; Firestone, M.K. Proline transport increases growth efficiency in salt-stressed *Streptomyces griseus*. *Appl. Environ. Microbiol.* **1984**, *48*, 239–241. [CrossRef]
81. Wang, Z.; Solanki, M.K.; Yu, Z.X.; Yang, L.T.; An, Q.L.; Dong, D.F.; Li, Y.R. Draft genome analysis offers insights into the mechanism by which *Streptomyces chartreusis* WZS021 increases drought tolerance in sugarcane. *Front. Microbiol.* **2019**, *9*, 3262. [CrossRef] [PubMed]
82. Zhou, Y.; Lai, R.; Li, W.J. The Family Solimonadaceae. In *The Prokaryotes*; Rosenberg, E., DeLong, E.F., Lory, S., Stackebrandt, E., Thompson, F., Eds.; Springer: Berlin/Heidelberg, Germany, 2014.
83. Panoff, J.M.; Thammavongs, B.; Guéguen, M.; Boutibonnes, P. Cold stress responses in mesophilic bacteria. *Cryobiology* **1998**, *36*, 75–83. [CrossRef]
84. Stokes, J.M.; French, S.; Ovchinnikova, O.G.; Bouwman, C.; Whitfield, C.; Brown, E.D. Cold stress makes *Escherichia coli* susceptible to glycopeptide antibiotics by altering outer membrane integrity. *Cell Chem. Biol.* **2016**, *23*, 267–277. [CrossRef]
85. Tiryaki, D.; Aydın, İ.; Atıcı, Ö. Psychrotolerant bacteria isolated from the leaf apoplast of cold-adapted wild plants improve the cold resistance of bean (*Phaseolus vulgaris* L.) under low temperature. *Cryobiology* **2019**, *86*, 111–119. [CrossRef] [PubMed]
86. Howard, M.M.; Muñoz, C.A.; Kao-Kniffin, J.; Kessler, A. Soil microbiomes from fallow fields have species-specific effects on crop growth and pest resistance. *Front. Plant Sci.* **2020**, *11*, 1171. [CrossRef]

plants

MDPI

Article

Bonactin and Feigrisolide C Inhibit *Magnaporthe oryzae Triticum* Fungus and Control Wheat Blast Disease

S. M. Fajle Rabby [1,†], Moutoshi Chakraborty [1,†], Dipali Rani Gupta [1], Mahfuzur Rahman [2], Sanjoy Kumar Paul [1], Nur Uddin Mahmud [1], Abdullah Al Mahbub Rahat [1], Ljupcho Jankuloski [3] and Tofazzal Islam [1,*]

1 Institute of Biotechnology & Genetic Engineering (IBGE), Bangabandhu Sheikh Mujibur Rahman Agricultural University, Gazipur 1706, Bangladesh

2 Extension Service, Davis College of Agriculture, West Virginia University, Morgantown, WV 26506, USA

3 Plant Breeding and Genetics Section, Joint FAO/IAEA Centre, International Atomic Energy Agency, 1400 Vienna, Austria

* Correspondence: tofazzalislam@bsmrau.edu.bd

† These authors contributed equally to this work.

Citation: Rabby, S.M.F.; Chakraborty, M.; Gupta, D.R.; Rahman, M.; Paul, S.K.; Mahmud, N.U.; Rahat, A.A.M.; Jankuloski, L.; Islam, T. Bonactin and Feigrisolide C Inhibit *Magnaporthe oryzae Triticum* Fungus and Control Wheat Blast Disease. *Plants* **2022**, *11*, 2108. https://doi.org/10.3390/plants11162108

Academic Editor: Tika Adhikari

Received: 17 May 2022
Accepted: 18 July 2022
Published: 12 August 2022

Abstract: Wheat blast caused by the *Magnaporthe oryzae Triticum* (MoT) pathotype is one of the most damaging fungal diseases of wheat. During the screening of novel bioactive secondary metabolites, we observed two marine secondary metabolites, bonactin and feigrisolide C, extracted from the marine bacteria *Streptomyces* spp. (Act 8970 and ACT 7619), remarkably inhibited the hyphal growth of an MoT isolate BTJP 4 (5) in vitro. In a further study, we found that bonactin and feigrisolide C reduced the mycelial growth of this highly pathogenic isolate in a dose-dependent manner. Bonactin inhibited the mycelial development of BTJP 4 (5) more effectively than feigrisolide C, with minimal concentrations for inhibition being 0.005 and 0.025 μg/disk, respectively. In a potato dextrose agar (PDA) medium, these marine natural products greatly reduced conidia production in the mycelia. Further bioassays demonstrated that these secondary metabolites could inhibit the MoT conidia germination, triggered lysis, or conidia germinated with abnormally long branched germ tubes that formed atypical appressoria (low melanization) of BTJP 4 (5). Application of these natural products in a field experiment significantly protected wheat from blast disease and increased grain yield compared to the untreated control. As far as we are aware, this is the first report of bonactin and feigrisolide C that inhibited mycelial development, conidia production, conidial germination, and morphological modifications in the germinated conidia of an MoT isolate and suppressed wheat blast disease in vivo. To recommend these compounds as lead compounds or biopesticides for managing wheat blast, more research is needed with additional MoT isolates to identify their exact mode of action and efficacy of disease control in diverse field conditions.

Keywords: antifungal secondary metabolites; biocontrol; abnormal germ tube suppression of appressoria; *Streptomyces* sp.

1. Introduction

Wheat is an essential staple dietary source for approximately 2.5 billion individuals in 89 different nations in the world. In low- and middle-income nations, it outperforms maize or rice as a source of protein. Wheat ranks second only to rice in the context of calorie supply. It is a primary food source in North Africa and West and Central Asia, accounting for up to half of the calories consumed (https://wheat.org/; accessed on 16 May 2022). Nonetheless, wheat is prone to various fungal diseases; the most notorious one is a wheat blast, caused by the pathogenic filamentous fungus *Magnaporthe oryzae Triticum* (MoT) pathotype. In 1985, the first case of the wheat blast was recorded in Brazil [1,2]. In 2016, Bangladesh experienced an alarming epidemic of the wheat blast that was the first incidence of the disease in Asia [3]. That epidemic destroyed 15,000 hectares of wheat fields with a yield loss of up to 100% [4]. Wheat blast is causing concerns among seed scientists as it has the potential to spread to important wheat-growing areas in South Asian and African countries [5]. Plant pathologists have warned that the disease might spread to India, Pakistan, and China, which are the second-, seventh-, and first-highest wheat producers in the world, respectively [4,6–8].

The MoT is a filamentous haploid ascomycete fungus. Its infection cycle has previously been described [9,10]. Briefly, MoT's three-celled hyaline airborne conidium lands on a wheat leaf and attaches to it using adhesive. It then begins to grow, developing into a slender germ tube with an appressorium at the tip. A tiny penetrating peg develops at the base of the appressorium, compressing the cuticle and allowing entry into the wheat epidermis. Wheat plasma membranes are penetrated by bulky, virulent mycelium, which then enters epidermal cells to complete tissue invasion [10–12]. It affects the aerial parts of the wheat plant, specifically the leaves, stems, nodes, and kernels encompassing all growth phases [7,13,14]. MoT usually affects spikes and bleaches the infected spikes, which results in malformed grains or producing no grain at all [4,15]. Wheat heads with severe infection may die, resulting in a considerable decrease in productivity. The early bleaching of spikelets above the infection point and the whole panicle is the most common symptom [4,7,16]. Contaminated seeds or grains as well as airborne conidia spread this disease, and the pathogen may persist in infected crop residues and seeds [17].

There is an ongoing demand for new plant chemotherapeutic agents that are unique from frequently used fungicides in their underlying mechanisms for advanced plant disease management. Another important reason for these needs is the occurrence of fungicide-resistant pathogens, which results from the requirement of using many synthetic fungicides at high rates, with adverse environmental repercussions [18,19]. Several microorganisms have been authorized as biocontrol agents in many countries including the EU to date due to their relatively low toxic residues, environment-friendly properties, and low manufacturing cost [20]. However, scientific research suggests that these benefits are not always achieved as biological pesticides are mostly living organisms, and their performance varies owing to the influence of numerous biotic (nutritional requirements, host species, and pathogenic microbes) and abiotic (moisture, temperature, relative humidity) factors, which limit their fitness under field conditions [21,22]. In addition, some biological control microbes, such as *Bacillus cereus*, are known to cause human diseases, precluding their release in the environment. In this regard, microbial metabolites can be another suitable alternative to live microbes or synthetic fungicides that are also capable of controlling plant diseases with low detrimental effects on human health and the environment [23]. The versatility of biological activity and chemical structure of microbial metabolites as a pesticide is worth considering due to the potential benefits [24]. The second aspect of microbial metabolites as agricultural fungicides is the requirement of a relatively short period for biodegradation. According to Tanaka and Omura [25], they often decay within a month or even a few days, leaving low residue that should be less harmful to the environment. Metabolites derived from diverse microorganisms have been utilized extensively to address commercially important diseases of several plants [26].

Secondary metabolites extracted from *Streptomyces* species have shown a broad range of biological functions by blocking particular enzymes or proteins in signaling cascades [27–30]. Wheat blast management research of our working group took a comprehensive strategy including biologicals and biorational approaches. During the screening of new bioactive natural products against MoT, we discovered that a few metabolites of *Streptomyces* spp. inhibited the growth of MoT mycelia [31]. Two natural secondary metabolites, bonactin and feigrisolide C, extracted from marine *Streptomyces* spp., Act 8970 and ACT 7619, respectively, exhibited substantial growth inhibitory effects against a MoT isolate among many different compounds tested. The first acyclic ester of nonactic acid is bonactin, whereas feigrisolide C is a non-symmetric lactone associated with the nactic acid group [32,33]. Bonactin has shown antimicrobial properties against both bacteria and fungi. Many different microbes including *Bacillus megaterium, Klebsiella pneumoniea, Escherichia coli, Micrococcus luteus,*

Staphylococcus aureus, *Saccharomyces cerevisiae*, and *Alicagenes faecalis* are sensitive to bonactin [33]. In antiviral, antibacterial, and enzyme inhibition tests, feigrisolides were found to have varying degrees of effectiveness. Synthesis of feigrisolide C has been achieved [34]. Nonactic acid esters are in general environmentally benign since soil microbes convert them to H_2O and CO_2 [35]. Inhibitory effects of bonactin and feigrisolide C on zoosporogenesis and motility of phytopathogenic Peronosporomycete zoospores have been reported [29]. A few studies have documented the toxicity level of these compounds to date. Bonactin is reported as non-carcinogenic and non-toxic to aquatic model organisms. It has been reported as a suitable natural compound for schizophrenia disorder, suggesting little or no toxicity to humans [36]. However, further research is needed to ascertain their safety for humans and the environment before using them as a potential lead component for the synthesis of agricultural fungicides for controlling wheat blast. There is currently no information available about the use of nonactic acid esters' antimicrobial activities to control wheat blast disease. To our best knowledge, this is the first report of marine natural antibiotics bonactin and feigrisolide C from *Streptomyces* spp. inhibiting a destructive wheat blast causing a MoT isolate and suppressing the disease in field conditions. The major targets of the current study were to: (i) assess the inhibitory effects of bonactin and feigrisolide C on the mycelia growth of BTJP 4 (5); (ii) evaluate the influences of these marine natural products on conidia production, germination, and the developmental transitions of conidia of BTJP 4 (5); (iii) assess the effect of these compounds on the suppression of wheat blast disease development caused by BTJP 4 (5) on leaves and spikes; and (iv) compare the disease inhibition efficiencies of these natural compounds with a commercialized fungicide Nativo®75WG.

2. Results

2.1. Mycelial Growth Inhibition and Morphological Alteration of Hyphae

Both bonactin and feigrisolide C considerably inhibited MoT mycelium development in the PDA medium in a dose-dependent manner (Figure 1). Bonactin inhibited mycelium development of the MoT isolate BTJP 4 (5) more efficiently than feigrisolide C. When bonactin and feigrisolide C were applied separately at 2 μg/disk, mycelial growth inhibition was 70.8 ± 0.8% and 68.1 ± 1.0%, respectively (Figure 2). Both bonactin and feigrisolide C demonstrated slightly lower inhibitory capacity than Nativo® WG 75 (82.7 ± 0.6% at 2 μg/disk). The inhibitory effects of these natural compounds enhanced as concentrations were raised from 0.005 to 2 μg/disk, reaching up to 71% for bonactin (Figure 2). Bonactin had more inhibitory efficacy than feigrisolide C but was slightly less effective than Nativo®WG 75 against the BTJP 4 (5) isolate. Both substances were ineffective against MoT at quantities lower than 0.005 μg.

Bonactin extensively impeded BTJP 4 (5) hyphal growth at 2 μg/disk (70.8 ± 0.8%), 1.5 μg/disk (65.4 ± 1.0%), and 1 μg/disk (58.6 ± 1.3%), showing that inhibition and accelerated concentrations had a positive correlation. At 2, 1.5, and 1 μg/disk, feigrisolide C inhibited 68.1 ± 1.0%, 63.4 ± 1.3%, and 54.2 ± 1.0% hyphal growth of BTJP 4 (5). Bonactin and feigrisolide C had minimum suppressive concentrations of 0.005 and 0.025 μg/disk, respectively, and these compounds suppressed mycelial growth by 9.81 ± 1.3% and 13.3 ± 0.8% at 0.005 and 0.025 μg/disk, respectively. However, the minimal inhibitory concentration of Nativo® WG 75 was 0.05 μg/disk, but at higher doses starting at 0.1 μg/disk it outperformed the suppression percentage of the two other test compounds at equal concentrations. It is worth noting that at less than 0.1 μg/disk concentration, bonactin and feigrisolide C inhibited mycelial development more efficiently than Nativo® WG 75, and bonactin inhibited BTJP 4 (5) at a 10-fold lower dose.

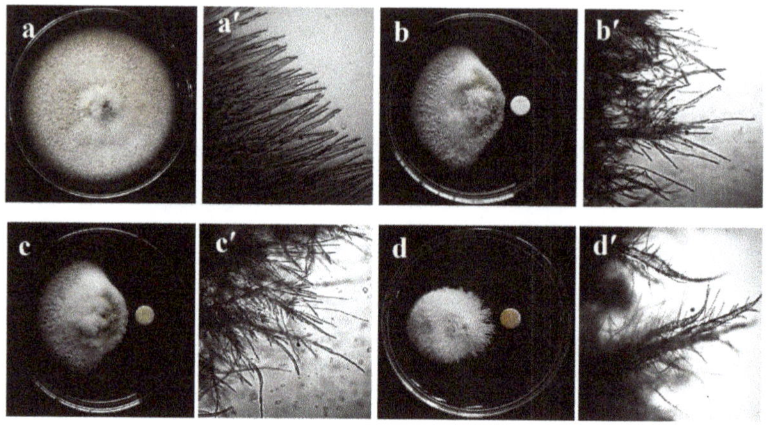

Figure 1. Mycelial growth suppression and morphological changes of hyphae of a wheat blast fungus, *Magnaporthe oryzae Triticum* (MoT) isolate BTJP 4 (5) approaching the paper disks containing two marine natural products, bonatin and feigrisolide C, and Nativo® WG75 (20 μg/disk), a commercial fungicide known to growers as local standard in Bangladesh. Normal mycelial growth (**a**) of BTJP 4 (5) on PDA plate (10 days) and microscopic view of the growing typical tubular hyphal tips (**a′**) in the untreated control. Mycelial growth inhibition (**b**) and abnormal hyphal tips (**b′**) closer to the paper disk containing bonactin. Inhibited mycelia (**c**) and curly and irregular growth of hyphal tips (**c′**) by feigrisolide C. Mycelial growth inhibition (**d**) and severely damaged hyphal tips (**d′**) by the Nativo® WG75. Bar = 50 μm. The micrographs shown in panels A and B were captured with a digital camera (CAMEDIA C-3040 zoom; Olympus Optical Co. Ltd., Tokyo, Japan), and those in panels C and D were taken from a light microscope (IX70-S1F2; Olympus) by using the same digital camera connected to it.

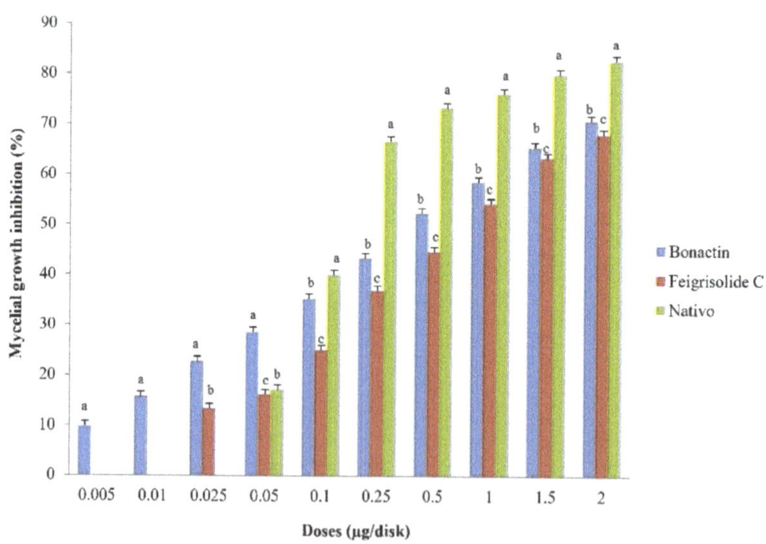

Figure 2. Suppression effects of bonactin, feigrisolide C, and Nativo® WG75 on mycelial growth of *Magnaporthe oryzae Triticum* (MoT) isolate BTJP 4 (5) in PDA media. The data represents the mean ± standard errors of three replications for each rate of the test compound based on the Tukey HSD (honest significance difference) test at a 5% level.

Microscopic examinations of untreated BTJP 4 (5) revealed polar, cylindrical growth with smooth, hyaline, branching, plump, septate, and unbroken hyphae (Figure 1a,a′). Hyphae treated with bonactin and feigrisolide C grew irregularly and exhibited a higher frequency of branching per unit of the hyphal length. Cell walls of the hyphae were not smooth but had ridges that gave them a crinkled look as well as causing irregular cell swelling (Figure 1b,b′,c,c′). Similar effects of the fungicide Nativo®WG75 on hyphal growth were observed. Mycelia closer to the filter disk of Nativo®WG75 showed a comparable modification of MoT hyphae (Figure 1d,d′). However, compared to Nativo®WG75, the two natural products generated slightly different morphological aberrations in MoT, suggesting a possibly different mode of action.

2.2. Conidiogenesis Inhibition

Bonactin, feigrisolide C, and Nativo®WG75 considerably decreased the conidia production of BTJP 4 (5) at concentrations of 1, 5, and 10 μg/mL, respectively, and suppression increased with increasing concentrations from 1 to 5 to 10 g/mL (Figure 3). Almost no or only a few conidia were produced at 10 μg/mL in media amended separately with all three compounds. Microscopic examination also revealed broken hyphal tips and complete suppression of conidiophore formation in fungal colonies in Petri plates that were treated with these three compounds at 10 μg/mL.

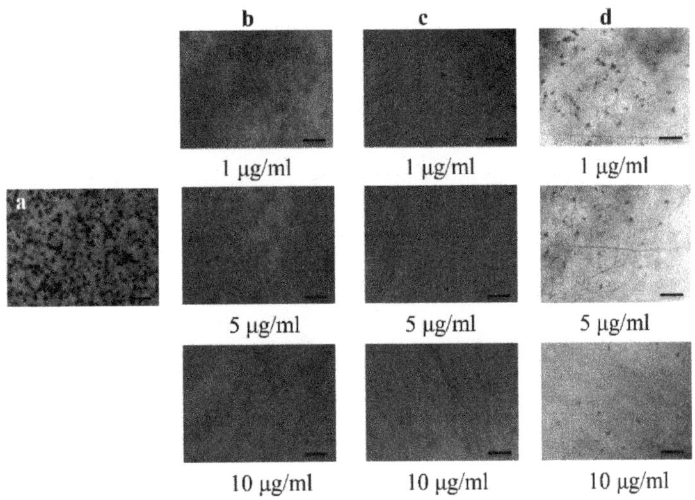

Figure 3. Effects of bonactin, feigrisolide C, and Nativo® WG75 on suppression of conidiogenesis of *M. oryzae Triticum* isolate BTJP 4 (5) in the 96-multiwell plates at 1 μg/mL, 5 μg/mL, and 10 μg/mL. Image (**a**) control. Images in panels (**b–d**) are bonactin, feigrisolide C, and Nativo® WG75, respectively. Bar = 50 μm.

2.3. Inhibition of Conidia Germination and Morphological Aberrations in Germinated Conidia

To determine the MoT isolate BTJP 4(5)'s conidial germination inhibition capacity of test products, bonactin, feigrisolide C, and Nativo® WG75 were added to the multi-well plates at concentrations of 0.5 μg/mL. The rate of conidial germination was recorded after 6, 12, and 24 h of incubation at 25 °C (Table 1). Bonactin and Nativo® WG75 treatments dramatically inhibited conidia germination compared to the control, while no conidia germinated in feigrisolide C-treated plates after 6 h of incubation. All (100%) conidia germinated in the water while it was 49.7 ± 0.6% in Nativo® WG75-treated plates. The BTJP 4 (5)′ conidia germination rates with bonactin and feigrisolide C were 79.1 ± 0.6% and 0 ± 0%, respectively, at 0.5 μg/mL.

Table 1. In vitro effects of bonactin and feigrisolide C on conidia germination and the developmental transitions of *M. oryzae Triticum* (MoT) isolate BTJP 4 (5) at 0.5 μg/mL.

Compound	Time (h)	Effects of Secondary Metabolites on the Developmental Alterations of Conidia of a MoT Isolate	
		Germinated Conidia (% ± SE [a])	Major Morphological Changes Occurred in the Treated Conidia
Water	0	0 ± 0^{e}	No germination
	6	100 ± 0^{a}	Normal germ tube and development of normal appressoria
	12	100 ± 0^{a}	Hyphal growth was observed
	24	100 ± 0^{a}	Huge hyphal growth occurred
Bonactin	0	0 ± 0^{e}	Zero germination
	6	79.1 ± 0.6^{b}	Germinated conidia had short germ tube
	12	79.1 ± 0.6^{b}	12.7 ± 0.4% Normal germ tube and 66.5 ± 0.5% of germ tube formed unusually elongated branches
	24	69.6 ± 0.5^{b}	9.5 ± 0.2% Normal appressoria and 60.1 ± 0.3% abnormal appressoria (low melanization) but no hyphal growth
Feigrisolide C	0	0 ± 0^{e}	No germination
	6	7.4 ± 0.5^{d}	7.4 ± 0.5% conidia lysed; No germination took place
	12	0 ± 0^{d}	No germination took place
	24	0 ± 0^{c}	No germination took place
Nativo® WG75	0	0 ± 0^{e}	Zero germination
	6	49.7 ± 0.6^{c}	Germinated, but germ tube was very short
	12	49.7 ± 0.6^{c}	Normal germ tube formed
	24	0 ± 0^{c}	Zero appressoria formed; zero hyphal growth

[a] Data are mean value ± SE of three replications in each natural compound. Means within a column followed by a different letter(s) are significantly different according to Tukey's HSD (honest significance difference) post-hoc ($p \leq 0.05$).

In the dark at 25 °C, 100% of conidia germination occurred in water during all incubation periods (6 h, 12 h, and 24 h), with normal germ tube and mycelial growth (Table 1, Figure 4a). At 0.5 μg/mL, both bonactin (panel b) and feigrisolide C reduced on the germination of the conidia and the post-germination developmental processes, resulting in abnormal transitions from one stage to another. During 6 h of incubation in the presence of bonactin, the conidia germination rate was 79.1 ± 0.6%, which had short germ tubes. After 12 h, 12.7 ± 0.4% of normal germ tubes were observed, whereas 66.5 ± 0.5% had abnormally long branched germ tubes. After 24 h, there were 9.5 ± 0.2% normal appressoria and 60.1 ± 0.3% atypical appressoria (low melanization), without any hyphal development (Table 1, Figure 4b).

In the case of feigrisolide C, 7.4 ± 0.5% of the conidia lysed after 6 h, and no germination occurred between 6 h and 24 h (Table 1, Figure 5c). In the presence of Nativo® WG75, 49.7 ± 0.6% of conidia germinated with normal germ tubes after 6 and 12 h, but no appressorial development took place. Nativo® WG75 also inhibited sporulation similar to feigrisolide C to prevent further mycelial growth after 24 h (Table 1, Figure 5A). It is worth mentioning that these compounds resulted in excessively long branching in germ tubes and conidia lysis, whereas Nativo® WG75 had no such effect.

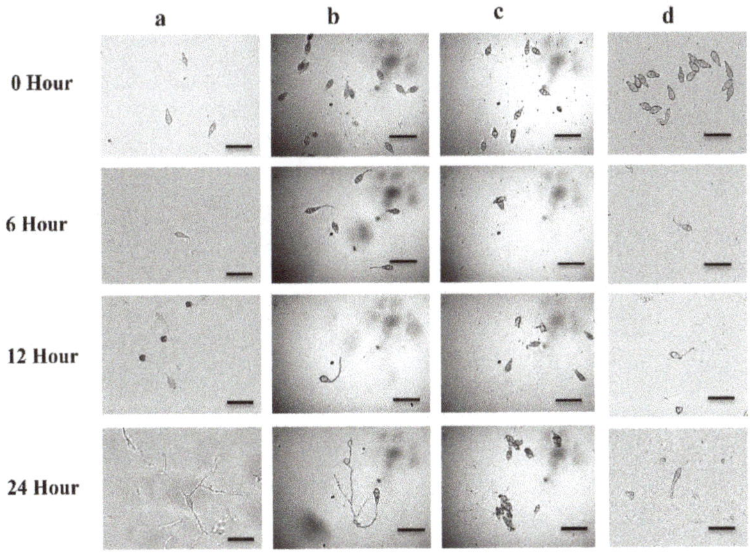

Figure 4. Micrographs showing the changes in germination and developmental transitions of MoT conidia with time-course in the untreated control (panel (**a**)) and the presence of bonactin (panel (**b**)), feigrisolide C (panel (**c**)), and a commercial fungicide Nativo® WG75 (panel (**d**)) at 0.5 µg/mL. Bar = 10 µm.

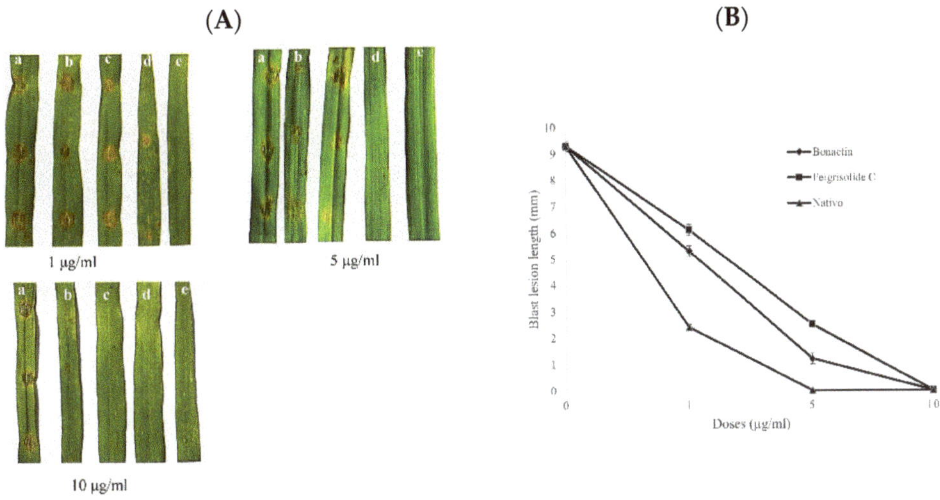

Figure 5. (**A**). Representative images showing wheat blast disease (symptoms) suppression by varying doses (1–10 µg/mL) of bonactin, feigrisolide C, and Nativo® WG75. The compounds were liquefied in 1% DMSO and applied on the detached leaves of wheat (cv. BARI Gom 26) 24 h before artificial point inoculation with 20 µL/point of suspension of conidia containing 1×10^5 conidia/mL. (a) Control, 1% DMSO, (b) bonatin, (c) feigrisolide C, (d) Nativo® WG75, and (e) uninoculated and untreated leaf. (**B**) Average lengths of blast lesions on detached wheat leaves pretreated with bonatin, feigrisolide C, and Nativo®WG75 compared to 1% DMSO treatment control. The data are the means ± standard errors of at least five replications for each dosage of the compounds at $p \leq 0.05$. Vertical bars represent ± standard error.

2.4. Wheat Blast Progression on Excised Wheat Leaves

The two metabolites applied at 1, 5, and 10 µg/mL considerably decreased the wheat blast disease symptoms in excised leaves of wheat infected with BTJP 4 (5). The lesion lengths in the leaves pretreated with bonactin were 5.3 ± 0.2 mm at 1 µg/mL and 1.2 ± 0.2 mm at 5 µg/mL, respectively (Figure 5A,B). The blast lesion lengths with feigrisolide C were 6.1 ± 0.2 mm and 2.5 ± 0.1 mm at 1 µg/mL and 5 µg/mL, respectively (Figure 5A,B). Leaves of wheat treated with bonactin, feigrisolide C, and Nativo®WG75 at 10 µg/mL did not show any blast symptoms (Figure 5A,B). Normal blast lesions were visible on the water-treated leaves with average lengths of 9.3 ± 0.2 mm (Figure 5A,B). In comparison to both compounds, the fungicide effectively reduced lesion progression at 1 and 5 µg/mL.

2.5. Wheat Blast Disease Suppression in the Field at the Heading Stage

To determine the efficacy of these compounds in suppressing blast disease in artificially infected wheat spikes by BTJP 4 (5), a field experiment was conducted by using a commercial fungicide Nativo®75WG at 50 µg/mL as a local standard. In the field, bonactin and feigrisolide C considerably reduced wheat blast disease incidences (41% and 51.3%, respectively) (Figure 6c,d, Table 2), compared to 87.3% disease incidence in the untreated control (Figure 6b, Table 2).

Figure 6. Inhibition of wheat blast disease with bonactin, feigrisolide C at 5 µg/mL, and Nativo® 75WG at 50 µg/mL; (**a**) Uninoculated, untreated spike, (**b**) BTJP 4 (5) inoculation + water control + (**c**) bonactin + BTJP 4 (5) inoculation, (**d**) feigrisolide C + BTJP 4 (5) inoculation, (**e**) Nativo®75WG + BTJP 4 (5) inoculation.

Table 2. Effect of bonactin and feigrisolide C on wheat (variety-BARI Gom-26) yield and yield components in field conditions following the artificial inoculation with BTJP 4 (5).

Treatment	Yield/1 m^2 Plot (gm) *	1000-Grain Weight (gm) *	Blast Incidence (%) *	Blast Severity (%) *
Healthy control	133.07 ± 2.33a	46.63 ± 1.57a	0.00 ± 0.00e	0.00 ± 0.00d
Untreated control	64.60 ± 1.71c	31.77 ± 1.29c	87.33 ± 3.18a	82.67 ± 3.53a
Bonactin	112.97 ± 2.26b	40.09 ± 1.72ab	41.00 ± 1.15c	32.33 ± 2.40b
Feigrisolide C	106.40 ± 2.58b	38.78 ± 3.16b	51.33 ± 3.53b	38.67 ± 1.20b
Nativo® 75WG	126.10 ± 2.70a	43.28 ± 2.52ab	24.00 ± 4.04d	14.33 ± 2.33c

* Yield data are the mean ± SE collected from five replications of each treatment of the test compounds. Data followed by the same letter in a column are not significantly different according to Tukey HSD (honest significance difference) post-hoc statistic at the 5% level.

Furthermore, 32.3 ± 2.40% and 38.6 ± 1.20% blast severities were recorded in wheat plants pretreated with these compounds in comparison to 82.6% in the untreated control. Bonactin (112.9 ± 2.26 gm), feigrisolide C (106.4 ± 2.58 gm), and Nativo® 75WG (126.1 ± 2.70 gm) had significantly increased grain yields compared to the untreated control (64.6 ± 1.71 gm). Grain yields in the Nativo® 75WG were statistically similar to the healthy control (133.1 ± 2.33 gm). Nevertheless, both bioactive natural compounds' treatments had statistically lower but similar grain yields compared to the Nativo® 75WG fungicide and healthy control (Table 2).

Thousand-grain weights for Nativo® 75WG, feigrisolide C, bonactin, and the negative control were 43.2 ± 2.52, 38.7 ± 3.16, 40.1 ± 1.72, and 46.6 ± 1.57 gm, respectively. Grain yields in treated plots were considerably greater than the yield of the untreated control plot (31.7 ± 1.29 gm) (Table 2).

3. Discussion

In this study, we demonstrated for the first time that two nonactic acid esters extracted from marine *Streptomyces* spp. and named bonactin and feigrisolide C inhibited the growth and development of a destructive wheat blast pathogen *M. oryzae Triticum* (MoT) isolate BTJP 4 (5). Additionally, we discovered that these natural compounds were comparable to the commercial fungicide Nativo® WG75 in their efficacy in successfully reducing wheat blast disease in wheat leaves and spikes that had been artificially inoculated by BTJP 4 (5). These treatments also resulted in a modest increase in grain yield although the highest yield was obtained from fungicide treatment followed by two test compounds. Formation of conidia asexually in hyphal conidiophore and germination of conidia are critical for plant infection by the blast fungus [37–40]. Suppression of hyphal growth, conidia formation, and germination of many fungi, such as rice and wheat blast fungi, by various natural products, have been reported [29,31,33,41–45]. The nonactic acid esters are precursors of macrotetrolide antibiotics which have a broad spectrum of antimicrobial, anticancer, acaricidal, insecticidal, immunosuppressive, antiprotozoan (coccidiostatic), and antiparasitic properties [35,46–48]. In the current study, we did not focus on unraveling the underlying molecular mechanism associated with in vitro growth inhibition of wheat blast causing fungal pathogen and suppression of the disease in vivo. However, from a similar study, Islam et al. [29] found that the hydrolysis of mitochondrial ATP via increased ATPase function was likely associated with the mode of action of antimicrobial activities of macrotetrolides against phytopathogenic Peronosporomycete zoospores. Despite having outstanding biological properties, macrotetrolides have received extremely less attention in plant protection studies. To the best of our knowledge, it is the first report of two natural bioactive nonactic acid esters and precursors of macrotetrolides (bonactin and feigrisolide C) originated from marine *Streptomyces* spp. suppressing the highly aggressive wheat blast pathogen MoT isolate BTJP 4 (5) in vitro and in vivo. Additional study is needed to test the efficacy of these compounds against other strains of MoT as well as whether their antiblast activities are linked with the induction of increased mitochondrial ATPase activity in the asexual spores and hyphae of MoT.

One of the key discoveries of this study is that at almost equal concentrations of Nativo® WG75, both bonactin and feigrisolide C dramatically reduced hyphal growth, conidia production, and germination, and also caused morphological changes in germinated conidia. Our findings indicate that these natural substances inhibited conidial germination and mycelium growth, which consequently suppressed wheat blast disease in vivo.

The swelling phenomenon by these compounds on BTJP 4 (5) hyphae is another remarkable observation from our study (Figure 1b'–d'). We utilized doses ranging from 0.005 to 2 μg/disk in our experiment. Swelling increased with increased concentrations, showing a positive correlation of swelling with concentrations. Tensin [49], fengycin [50], gageopeptides, gageotetrin [44], and oligomycins [31] have all been reported to induce developmental aberrations in the tubular growth of the fungal hyphae. Developmental transitions in *Aphanomyces cochlioides* hyphae, such as increased swelling and excessive branching, have been observed in response to xanthobaccin A from *Lysobacter* sp. SB-K88 or m *Pseudomonas fluorescence* phloroglucinols [51–54]. According to Schumacher et al. [33], bonactin from *Streptomyces* sp. greatly suppressed the hyphal development of *Saccharomyces cerevisiae*, but no data on the mycelial growth inhibitory activity of feigrisolide C has been documented to date. So far, this is known to be the first report of some nonactic acid and nonactic acid ester exhibiting swollen-like abnormal hyphae against a destructive wheat pathogen.

Conidiogenesis is the process of producing conidia, which are fungal spores that are grown asexually on the conidiophore [39]. The majority of fungal plant pathogens attack plants by these asexual spores. Inhibiting or preventing conidiogenesis and conidia germination can reduce the likelihood of host infection by fungal pathogens [55,56]. Future plant protection strategies should explore and rely on similar natural compounds that interfere with these processes. Therefore, another noteworthy finding from this study was that these compounds greatly decreased conidiogenesis (Figure 3), and conidial germination, and also triggered morphological alterations of BTJP 4 (5)'s conidia (Table 1, Figure 4).

Lysis of conidia and uneven branching of germ tube tips as well as unusually long hypha-like germ tubes were among the other distinct and interrelated phenomena found in this work (Figure 4B,C). Dame and co-workers [57] discovered a similar occurrence when they found that oligomycins derived from a marine *Streptomyces* sp. triggered lysis of phytopathogenic *Plasmopara viticola* zoospores that causes grapevine downy mildew disease. Homma and colleagues reported that lecithin induced abnormal branching in germ tube tips of rice blast fungus, and prevented the development of appressoria [58]. Similarly, *A. cochlioides*' cystospores germinated with hyperbranched germ tubes by the effects of diacetylphloroglucinol (DAPG) [54]. Bonactin caused atypical appressoria (low melanization), which restricted MoT fungal infection since appressorium melanization is essential for *M. oryzae* pathogenicity [11]. This compound may affect the gene expression related to the synthesis of melanin. This is also the first study to show that two esters impeded conidiogenesis, germination, and the development of appressoria of BTJP 4 (5) conidia. Future research should concentrate on the mechanisms by which these compounds suppress conidia formation, germination, and appressorium formation of MoT, as well as the impact of these natural bioactive compounds on the expression of genes associated with conidia germination and appressorium formation of BTJP 4 (5) or similar MoT isolates.

Nonactic acid esters are relatively safe for the environment since soil microorganisms can quickly convert them to H_2O and CO_2 [59]. Plant growth stimulation and specific insecticidal actions of nonactin antibiotic precursors have been documented [35,60]. Bonactin was reported to have antibacterial action and also antifungal action [33]. In a lab investigation, we noticed that nonactin had remarkable antifungal properties against MoT both in vivo and in vitro (our unpublished data). According to Schumacher et al. [33], antimicrobial activity can be achieved without the requirement for a macrotetrolide ring structure, such as the non-asymmetric lactone feigrisolide C, which has antibacterial and antiviral properties [61]. Islam and his colleagues [29] discovered that bonactin and feigrisolide C with other known macroletrolides suppress zoosporogenesis, hamper motility, as well as trigger lysis of *Plasmopara viticola* zoospores. The findings of the current work do not elucidate the detailed mechanism of action, but they do suggest that stimulation of ATPase activities in mitochondria or/and imbalance/translocation of cell cations could inhibit hyphal development and impede conidia germination. Identifying the role of ATPase in inhibiting hyphal growth, conidiogenesis, conidial germination, and appressoria formation may aid in our understanding of the biology and pathogenesis of filamentous plant pathogens. This naturally occurring ATPase inducer may thus be a promising pioneer ingredient for developing novel, efficient agrochemicals to fight this aggressive fungal pathogen.

In this study, wheat leaves pretreated with the test compounds showed shorter lesions than untreated checks (Figure 5). The majority of those lesions were small, and appeared as brown patches with spots of a pinhead size (scale 1) to roundish and fairly expanded grey dots that ranged in size from 1–2 mm in diameter (scale 3). The untreated control leaves had typical blast lesions covering 26–50% of the leaf surface (scale 7), according to the 9-scale blast disease assessment system developed by the IRRI SES (standard evaluation system) [55,56]. However, in the Nativo® WG75 treatment, no visible blast lesions were present. When disease control studies were conducted at the wheat heading stage, similar results were obtained. In artificially infected wheat spikes, blast disease progression was dramatically inhibited by bonactin and feigrisolide C (Figure 6). A popular systemic fungicide, Nativo® WG75, was used in this study as a local standard and positive control. In terms of suppression efficacy of the MoT fungus, the two marine natural compounds evaluated in the current study were comparable to that of the commercial fungicide. The active components of Nativo® WG75 are tebuconazole and trifloxystrobin. Belonging to the systemic triazole fungicide group, tebuconazole's mode of action is known as demethylase inhibitor (DMI). The development of the fungus is slowed down and can eventually be killed as DMI fungicides interfere with the production of sterol in fungal cell walls [62]. Trifloxystrobin, a fungicide in the strobilurin group, suppresses the spore germination of phytopathogenic fungi by disrupting energy production through blocking mitochondrial electron transport [62]. The modes of action of bonactin and feigrisolide C are possibly distinct from Nativo®WG75, despite the observation of a similar disease inhibition response. More research is needed to determine the fundamental mechanism through which these compounds suppress wheat blast. Before acknowledging these compounds as prospective fungicides for wheat blast, a large-scale field evaluation of their efficiency in preventing wheat blast infection is necessary. Recently, it has been found that secondary metabolites from both marine and terrestrial species can biologically suppress the wheat blast disease [55,56].

In today's agriculture, the development of fungicide resistance across pathogenic microorganisms is a major concern. Due to the inappropriate use of fungicides with a single-site active mode of action such as triazole and strobilurin (QoI), some resistant MoT mutant species have been found widely distributed [19]. Investigators are actively searching for new, effective antifungal chemicals possessing alternative modes of action to protect wheat plants against this lethal pathogenic fungus due to the risk of resistance development in conventional fungicides. The marine natural products, bonactin and feigrisolide C, exhibited almost equivalent bioactivity to the commercialized fungicide Nativo®WG75. The effectiveness of these compounds as inhibitors of the MoT isolate BTJP 4 (5) has suggested using them as candidates for agrochemical with a novel mode of action towards this wheat pathogenic fungus provided they are equally effective on other MoT strains under various agro-ecological regions. Further studies are needed with structurally diverse nactic acids, their esters, and macrotetrolides for understanding the structure-activity relationship of bonactin and feigrisolide C. However, very few reports have been published yet regarding their impacts on humans and the environment, and more research is also required to assess their toxicity level before using them to produce fungicides.

4. Materials and Methods

4.1. Fungal Isolate, the Revival of a Synthetic Medium, and Host Plant Materials

In 2016, during the first wheat blast epidemic in Jhenaidah, Bangladesh, we collected many MoT strains including BTJP 4 (5) from wheat cv. Prodip (BARI Gom-24) that showed blast infection on spikelets These isolates were preserved at 4 °C on dried filter paper for later use. We revived five isolates on a potato dextrose agar (PDA) medium and tested them in the lab for their normal colony characters and aggressiveness to select a representative one (BTJP 4) for this work (4). We also tested isolates collected from the field-infected wheat from 2016 to 2022 and found that they were equally sensitive to the commercial fungicide, Nativo® WG 75 (Figure 7). It appeared the clonal population introduced in Bangladesh from South America had not been mutated [37]. Therefore, we chose BTJP 4(5) for the whole study. On a potato dextrose agar (PDA) medium, the selected isolate was grown for seven days at 25 °C. Ten-days-old PDA-grown fungi fungal colonies were washed in an aseptic environment in a laminar flow hood with 500 mL of deionized water to remove aerial mycelia, and then kept at ambient temperature (25–30 °C) for 2–3 days to induce abundant conidia production [4,40,55,56]. The conidia were scraped out from each plate using a glass slide after adding 15 mL of water to each plate. Two-layer cheesecloth was used to filter out the hyphal mass, and the dilution was conducted to achieve 1×10^5 conidia/mL. Conidial germination was examined under a compound microscope by counting the number. Seedlings of blast disease-susceptible wheat variety Prodip (BARI Gom-24) at the five-leaf stage were used for the bioassay on leaves [31,55,56]. For assessing the wheat blast disease suppression efficacy of bonactin and feigrisolide C, these compounds were sprayed on field-grown wheat spikes at the flowering stage one day before inoculation of the plants with MoT conidia. The detailed method of artificial inoculation of wheat plants by MoT conidia was described recently by Paul et al. [40].

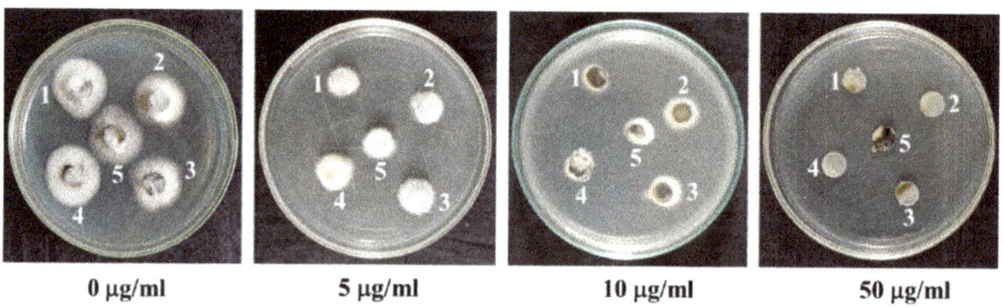

Figure 7. Sensitivity of different strains (**1–5**) of wheat blast fungus *Mahnaporthe oryzae Triticum* obtained from the field-infected spikes to various doses of a commercial fungicide Nativo. **1**, BTKP 22(3) collected in 2022; **2**, BTJP 194-2 collected in 2019; **3**, BTJP 1910-3 collected in 2019; **4**, BTJP 2 g collected in 2017; and **5**, BTJP 4 (5) collected in 2016. The PDA plates were cultured at 25 °C for 3 days after inoculation of the plates by various wheat blast strains.

4.2. Chemicals

Bonactin and feigrisolide C (Figure 8) were derived from the marine bacteria *Streptomyces* spp. Act 8970 and ACT 7619. Dr. Hartmut Laatsch, a Professor of Georg-August-Universitaet Goettingen in Germany, generously provided these pure chemicals as gifts [29]. The fungicide Nativo® WG 75 (50:50 mixtures of trifloxystrobin and tebuconazole) was purchased in Dhaka, Bangladesh from Bayer Crop Science Ltd. Stock solutions of test compounds were prepared using small amounts of DMSO (dimethyl sulfoxide), and then the solutions were diluted with water. The final mixture included a maximum of 1% (v/v) DMSO, which had no impact on the development or sporulation of BTJP 4 (5) mycelium [55,56].

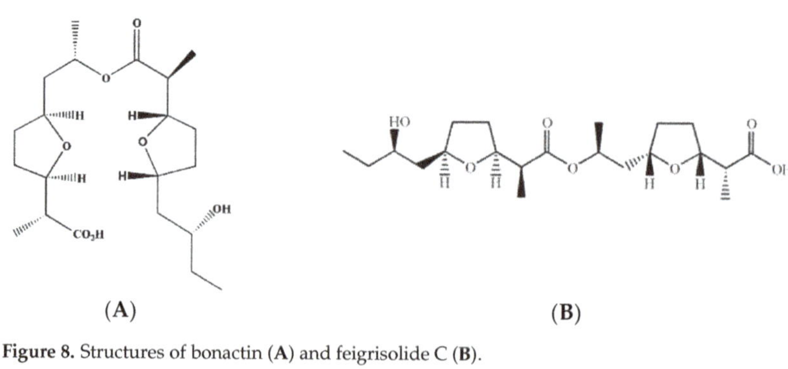

Figure 8. Structures of bonactin (**A**) and feigrisolide C (**B**).

4.3. Suppression of Mycelial Growth and Hyphal Morphological Alteration

Using a modified disk diffusion technique as reported by Chakraborty et al. [31], the mycelial growth inhibition of MoT isolate BTJP 4(5) was determined by the application of bonactin, feigrisolide C, and the commercial fungicide Nativo®WG75 on filter paper disks. To prepare a range of concentrations from 0.005 to 2 μg/disk, the required amounts of natural compounds and the fungicide Nativo® WG75 were dissolved in ethyl acetate and water. Nine-millimeter-diameter filter paper disks (Sigma-Aldrich Co., St. Louis, MO, USA) were used to absorb the test compound solutions. In 9 cm-diameter Petri dishes with 10 mL of PDA, the treated disks were placed 2 cm apart from one side. The filter paper disks containing the test chemicals were placed on the opposite side of the actively growing 5 mm-diameter, 7-days-old mycelial plugs of BTJP 4(5). Petri dishes with fungal hyphal plugs against filter paper disks with Nativo®WG75 were used as a control. As a negative control, filter paper disks were coated with ethyl acetate and then allowed to evaporate at ambient room temperature. A fungal hyphal development reduction was observed after 10 days of culture. The Petri plates used as untreated controls were incubated at 25 °C until the fungus had colonized and covered the whole surface of the agar. The test was conducted five times with five replications for each concentration. Using a ruler and two perpendicular lines drawn on the lower side of each plate, the radial growth of the fungal culture was measured in centimeters. Measurements were also recorded for the inhibition zone and associated fungal colony diameter influenced by the test compounds and the fungicide. Inhibition percentage radial growth (RGIP) [55,56] was calculated as:

$$\text{RGIP \%} = \frac{\text{Control plate radial growth} - \text{Treated plate radial growth}}{\text{Control plate radial growth}} \times 100$$

Results including radial growth suppression from the disk diffusion test were captured using a digital camera of CAMEDIA C-3040 zoom. At 40× and 100× magnification, an Olympus IX70-S1F2 microscope was used to study the mycelial morphology at the sharp end of the cultures approaching the control and treated disks. The mycelial growth including aberration was photographed using the same digital camera attached to the microscope.

4.4. Suppression of Conidiogenesis

The stock solutions of each compound were prepared in 10 μL of DMSO and then diluted with distillate water to obtain concentrations of 1, 5, and 10 μg/ml. The final mixtures had a maximum of 1% (v/v) of DMSO, which had no impact on BTJP 4 (5) sporulation or hyphal development. A 5 mL solution of Nativo®WG75 was prepared to achieve each 1, 5, and 10 μg/mL concentrations by dissolving the required amount of formulation in distilled water that was used as a positive control. A conidiogenesis inhibition test of a MoT isolate was established in our lab and used for this work [31,39,56]. Briefly, to deplete nutrients and promote conidiogenesis, the mycelium of a 10-day-old BTJP 4 (5) Petri plate culture was rinsed [4,39]. After being treated, ten mm BTJP 4 (5) hyphal agar blocks were treated with 50 μL of each test compound and Nativo®WG75 at the aforementioned doses and then placed on Nunc multi-well plates. The mycelial agar block of MoT with 1% DMSO in the same amount of sterile water was used as a negative control. Treated BTJP 4 (5) mycelial plugs were incubated at 28 °C and >90% RH under alternating light and dark cycles for 14 and 10 h, respectively. After 24 h, conidiogenesis was observed under a 40× Zeiss Primo Star microscope for analysis, and pictures were taken with a Zeiss Axiocam ERc 5s. With five replications for each treatment, the test was repeated five times.

4.5. Suppression of Conidial Germination and Morphological Changes in Germinated Conidia

Each natural compound was first liquefied in 10 μL of DMSO before being diluted with distilled water to a concentration of 0.1 μg/mL. As a positive control, a 0.1 μg/mL solution of Nativo®WG75 was prepared in distilled water. We used the methodology developed previously by us for MoT isolate conidial germination investigations [31,55,56]. Briefly, a

100 μL solution containing 1×10^5 conidia/mL of BTJP 4 (5) was directly mixed with a 100 μL solution containing 0.1 μg/mL of product to obtain a 200 μL final solution in the well of a 96-multiwell plate containing test compounds comprising 0.5 μg/mL. Immediately after blending with a glass rod, the suspension was incubated for 6, 12, and 24 h at 25 °C in a Ziploc plastic bag with layers of moist paper towel. Sterile water that contained 1% DMSO was employed as a control. A total of 100 conidia from each of the five replications were examined with a Zeiss Primo Star microscope at a 100× magnification. The photographs were acquired with a Zeiss Axiocam ERc 5s, and the percentage of conidia germination, and the morphological alterations of spore germ tubes and appressoria, were determined. The experiment was repeated five times, with at least five replications for each treatment. The conidia germination percentage was calculated as: $CG\% = (C - T)/C \times 100$; where %CG = conidia germination, C = average conidia germination percentage in control, and T = average conidia germination percentage in treated samples.

4.6. *Wheat Blast Progression on Detached Wheat Leaves*

Bonactin and feigrisolide C stock solutions were made using a small quantity of DMSO. The final DMSO content never exceeded 1% when the natural substances were dissolved in sterile distilled water to obtain concentrations of 1, 5, and 10 μg/mL. Nativo®WG75 was prepared in concentrations of 1, 5, and 10 μg/mL as well. As a negative control, sterilized water that contained 1% DMSO was utilized. This experiment was carried out according to the procedures outlined by Chakraborty et al. [31,55,56]. The first step was to separate wheat leaves from seedlings at the five-leaf stage and place them on plates covered with wet paper towels. Each leaf was treated with three 20 μL drops of the appropriately prepared test compound at the aforementioned concentrations, and the leaves were left to dry for 15 min. Following that, inoculation was conducted on each spot with 1 μL conidial solution containing 1×10^5 BTJP 4 (5) conidia/mL, and the plates were incubated at 28 °C in the darkness for the first 30 h, then under constant lighting for the following two days. The experiment was repeated five times with five different samples each time. For each treatment and compound concentration, the diameter of blast lesions induced by MoT was measured on three leaves per experiment.

4.7. *Determination of Wheat Blast Control Efficacy of Bonactin and Feigrisolide C under Field Conditions*

4.7.1. Soil Preparation and Seed Sowing

The experiment was carried out in the research field of the Bangabandhu Sheikh Mujibur Rahman Agricultural University (BSMRAU) in Gazipur, Bangladesh. The trial site was situated 8.4 m above sea level at a latitude of 24.09° north, and a longitude of 90.26° east. Weeds and stubbles were pulled out of the soil after it had been gently plowed. During soil preparation, adequate quantities of well-decomposed cow dung were applied. Gypsum, muriate of potash, triple super phosphate, and urea were applied as chemical fertilizers at a rate of 11-50-28-70 kg/ha. [63]. 3 to 4 days before seed sowing, the final soil preparation included the application of additional fertilizers along with two-thirds of the urea as a baseline dose. 20 days after the first irrigation, the final one-third of the urea was applied. In the first week of December, BARI Gom-26 wheat seeds were sown. Before sowing, the seeds were treated with Vitavex 200 (3 g/kg seed). There were three replications per treatment and the size of the experimental plot was 1 m^2. All of the plots were properly labeled. The required irrigation work was performed, along with additional cross-cultural tasks. The experiment was conducted using a randomized complete block design (RCBD).

4.7.2. Infection Assay in the Wheat Reproductive Phase

The test compounds were applied at a concentration of 5 μg/mL in each plot and allowed to dry overnight, whereas sterile water containing 1% DMSO served as a negative control. BTJP 4 (5) spore suspension was sprayed to wheat fields immediately after flowering. The positive control was the fungicide Nativo® 75WG, whereas the negative control was deionized distillate water. To establish a humid atmosphere suitable for spore germination, polyethylene sheets were placed over plots before inoculation.

4.7.3. Data Collection and Analysis for Disease Severity

During the reproductive phase, data were recorded on the total number of tillers, productive tillers, infected tillers per hill, the full length of the spikes, diseased area of the spikes, seeds per spike, 1000-grain weight, and grain production per hill. During the vegetative phase, data were collected on the total number of seedlings, the number of infected seedlings per pot, the overall length of the leaves, and the infected area of the leaves. The disease intensity (DI) was calculated using the formula:

$$DI = \frac{\text{Total infected plants}}{\text{Total plants observed}} \times 100$$

A 5-point scale was used to assess the severity of blast disease, with % infection accounting for the length of the spike that was infected by blast. The scales were 0 for the absence of lesions, 1 for infection rates between 1% and 25%, 2 for infection rates between 26% and 50%, 3 for infection rates between 51% and 75%, and 4 for infection rates between 76% and 100% on the length of damaged leaves. Blast severity was measured by the following formula:

$$DS = \frac{n \times v}{N \times V} \times 100\%$$

where DS = disease severity
n = number of blast-infected leaves
v = value score for blast severity
N = number of observed leaves
V = value of highest score.

4.8. Statistical Analysis, Experimental Design, and Replications

The efficacy of the pure compounds was examined in the laboratory and the field, respectively, using completely randomized design (CRD) and randomized complete block design (RCBD). All statistical analyses were performed using Microsoft Office Excel 2015 and IBM SPSS Statistics 25. Tukey's HSD (honest significance difference) test was used to compare the treatment means. The tables and figures utilized the mean value ± standard error and there were five replications per treatment.

5. Conclusions

In this study, we demonstrated for the first time that marine natural products, bonactin and feigrisolide C, from *Streptomyces* species, suppressed the mycelial growth and asexual development of an isolate of MoT fungus and inhibited the progression of wheat blast disease caused by that isolate in vivo. Large-scale in vitro and field testing of these compounds with multiple isolates is necessary to determine whether they are potential candidates or lead compounds for developing an effective fungicide against wheat blast disease. More investigation is also needed to determine their level of toxicity towards humans and the environment, as well as their specific method of action and the structure–activity association between these bioactive natural compounds and the wheat blast fungus *M. oryzae Triticum*.

Author Contributions: Conceptualization and methodology, T.I.; methodology, S.M.F.R., A.A.M.R. and M.C.; formal analysis, M.C., S.M.F.R., S.K.P., A.A.M.R. and N.U.M.; software, M.C., S.M.F.R., S.K.P. and N.U.M.; validation, T.I.; writing—original draft preparation, M.C., S.M.F.R. and S.K.P.; writing—review and editing, T.I., D.R.G., M.R. and L.J.; supervision, T.I. and D.R.G.; project administration, T.I.; funding acquisition, T.I., M.R. and L.J. All authors have read and agreed to the published version of the manuscript.

Funding: The Krishi Gobeshona Foundation (KGF) of Bangladesh provided funding for this study to Tofazzal Islam of the Institute of Biotechnology and Genetic Engineering (IBGE) at BSMRAU, Bangladesh, under project Nos. KGF TF50-C/17 and TF 92-FNS/21.

Institutional Review Board Statement: This research content included no human participants and/or animals.

Informed Consent Statement: This manuscript has not been published or presented elsewhere in part or entirety and is not under consideration by another journal. We have read and understood your journal policies and we believe that neither the manuscript nor the study violates any of these. All the authors have been personally and actively involved in substantive work leading to the manuscript and will hold themselves jointly and individually responsible for its content. All co-authors agreed to this submission.

Data Availability Statement: The manuscript includes all the data.

Conflicts of Interest: The authors disclose that they have no competing interests.

References

1. Igarashi, S.; Utiamada, C.M.; Igarashi, L.C.; Kazuma, A.H.; Lopes, R.S. Pyricularia emtrigo. 1. Ocorrência de *Pyricularia* sp. no estado do Paraná. *Phytopathol. Bras.* **1986**, *11*, 351–352.
2. Kohli, M.M.; Mehta, Y.R.; Guzman, E.; Viedma, L.; Cubilla, L.E. *Pyricularia* blast-a threat to wheat cultivation. *Czech. J. Genet. Plant Breed.* **2011**, *47*, 130–134. [CrossRef]
3. Callaway, E. Devastating wheat fungus appears in Asia for first time. *Nature* **2016**, *532*, 421–422. [CrossRef] [PubMed]
4. Islam, M.T.; Croll, D.; Gladieux, P.; Soanes, D.M.; Persoons, A.; Bhattacharjee, P.; Hossain, M.; Gupta, D.R.; Rahman, M.; Mahboob, M.G.; et al. Emergence of wheat blast in Bangladesh was caused by a South American lineage of *Magnaporthe oryzae*. *BMC Biol.* **2016**, *14*, 84. [CrossRef] [PubMed]
5. CIMMYT. *Wheat Blast Disease: A Deadly and Baffling Fungal Foe. International Maize and Wheat Improvement Center*; CIMMYT: Texcoco, Mexico, 2016.
6. Mundi. Agricultural Production, Supply, and Distribution: Wheat Production by Country in 1000 MT. 2016. Available online: https://www.indexmundi.com/agriculture/?country=bd&commodity=wheat&graph=production (accessed on 16 May 2022).
7. Islam, M.T.; Kim, K.H.; Choi, J. Wheat blast in Bangladesh: The current situation and future impacts. *Plant Pathol. J.* **2019**, *35*, 1–10. [CrossRef] [PubMed]
8. Kamoun, S.; Talbot, N.J.; Islam, M.T. Plant health emergencies demand open science: Tackling a cereal killer on the run. *PLoS Biol.* **2019**, *17*, e3000302. [CrossRef]
9. Chakraborty, M.; Mahmud, N.U.; Ullah, C.; Rahman, M.; Islam, T. Biological and biorational management of blast diseases in cereals caused by *Magnaporthe oryzae*. *Crit. Rev. Biotechnol.* **2021**, *41*, 994–1022. [CrossRef]
10. Eseola, A.B.; Ryder, L.S.; Osés-Ruiz, M.; Findlay, K.; Yan, X.; Cruz-Mireles, N.; Molinari, C.; Garduño-Rosales, M.; Talbot, N.J. Investigating the cell and developmental biology of plant infection by the rice blast fungus *Magnaporthe oryzae*. *Fungal. Gen. Biol.* **2021**, *18*, 103562. [CrossRef]
11. Wilson, R.A.; Talbot, N.J. Under pressure: Investigating the biology of plant infection by *Magnaporthe oryzae*. *Nat. Rev. Microbiol.* **2009**, *7*, 185–195. [CrossRef]
12. Tufan, H.A.; McGrann, G.R.D.; Magusin, A.; Morel, J.B.; Miche, L.; Boyd, L.A. Wheat blast: Histopathology and transcriptome reprogramming in response to adapted and nonadapted *Magnaporthe* isolates. *New Phytol.* **2009**, *184*, 473–484. [CrossRef]
13. Inoue, K.; Suzuki, T.; Ikeda, K.; Jiang, S.; Hosogi, N.; Hyon, G.S.; Hida, S.; Yamada, T.; Park, P. Extracellular matrix of *Magnaporthe oryzae* may have a role in host adhesion during fungal penetration and is digested by matrix metalloproteinases. *J. Gen. Plant Pathol.* **2007**, *73*, 388–398. [CrossRef]
14. Islam, M.T.; Gupta, D.R.; Hossain, A.; Roy, K.K.; He, X.; Kabir, M.R.; Singh, P.K.; Khan, M.; Rahman, A.; Rahman, M.; et al. Wheat blast: A new threat to food security. *Phytopathol. Res.* **2020**, *2*, 28. [CrossRef]
15. Surovy, M.Z.; Mahmud, N.U.; Bhattacharjee, P.; Hossain, M.; Mehebub, M.; Rahman, M.; Majumdar, B.C.; Gupta, D.R.; Islam, T. Modulation of nutritional and biochemical properties of wheat grains infected by blast fungus *Magnaporthe oryzae Triticum* pathotype. *Front. Microbiol.* **2020**, *11*, 1174. [CrossRef]
16. Igarashi, S. Update on wheat blast (*Pyricularia oryzae*) in Brazil. In Proceedings of the International Conference—Wheat for the Nontraditional Warm Areas, Foz do Iguaçu, Brazil, 29 July–3 August 1990; CIMMYT: Mexico, Mexico, 1990; pp. 480–483.
17. Urashima, A.S.; Hashimoto, Y.; Le Don, D.; Kusaba, M.; Tosa, Y.; Nakayashiki, H.; Mayama, S. Molecular analysis of the wheat blast population in Brazil with a homolog of retrotransposon MGR583. *Jpn. J. Phytopathol.* **1999**, *65*, 429–436. [CrossRef]

18. Knight, S.C.; Anthony, V.M.; Brady, A.M.; Greenland, A.J.; Heaney, S.P.; Murray, D.C.; Powell, K.A.; Schulz, M.A.; Sinks, C.A.; Worthington, P.A.; et al. Rationale and perspectives on the development of fungicides. *Annu. Rev. Phytopathol.* **1997**, *35*, 349–372. [CrossRef]
19. Dorigan, A.F.; Carvalho, G.D.; Poloni, N.M.; Negrisoli, M.M.; Maciel, J.L.N.; Ceresini, P.C. Resistance to triazole fungicides in *Pyricularia* species associated with invasive plants from wheat fields in Brazil. *Acta Sci. Agron.* **2019**, *41*, 39332. [CrossRef]
20. Cook, R.J.; Baker, K.F. *The Nature and Practice of Biological Control of Plant Pathogens*; American Phytopathological Society: St. Paul, MN, USA, 1983; p. 1.
21. Lugtenberg, B.; Leveau, J. Biocontrol of plant pathogens: Principles, promises, and pitfalls. In *The Rhizosphere: Biochemistry and Organic Substances at the Soil–Plant Interface*, 2nd ed.; Pinton, R., Varanini, Z., Nannipieri, P., Eds.; CRC Press: Boca Raton, FL, USA, 2007; pp. 267–296.
22. Sundin, G.W.; Werner, N.A.; Yoder, K.S.; Aldwinckle, H.S. Field evaluation of biological control of fire blight in the eastern United States. *Plant Dis.* **2009**, *93*, 386–394. [CrossRef]
23. Porter, N. Physicochemical and biophysical panel symposium biologically active secondary metabolites. *Pestic. Sci.* **1985**, *16*, 422–427.
24. Vining, L.C. Function of secondary metabolites. *Annu. Rev. Microbiol.* **1990**, *44*, 395–427. [CrossRef]
25. Tanaka, Y.T.; Omura, S. Agroactive compounds of microbial origin. *Annu. Rev. Microbiol.* **1993**, *47*, 57–87. [CrossRef]
26. Verma, C.; Jandaik, S.; Gupta, B.K.; Kashyap, N.; Suryaprakash, V.S.; Kashyap, S.; Kerketta, A. Microbial metabolites in plant disease management: Review on biological approach. *Int. J. Chem. Stud.* **2020**, *8*, 2570–2581. [CrossRef]
27. Õmura, S.; Ikeda, H.; Ishikawa, J.; Hanamoto, A.; Takahashi, C.; Shinose, M.; Takahashi, Y.; Horikawa, H.; Nakazawa, H.; Osonoe, T.; et al. Genome sequence of an industrial microorganism *Streptomyces avermitilis*: Deducing the ability of producing secondary metabolites. *Proc. Natl. Acad. Sci. USA* **2001**, *98*, 12215–12220. [CrossRef] [PubMed]
28. Islam, M.T.; von Tiedemann, A.; Laatsch, H. Protein kinase C is likely to be involved in zoosporogenesis and maintenance of flagellar motility in the Peronosporomycete zoospores. *Mol. Plant-Microbe Interact.* **2011**, *24*, 938–947. [CrossRef] [PubMed]
29. Islam, M.T.; Laatsch, H.; von Tiedemann, A. Inhibitory effects of macrotetrolides from *Streptomyces* spp. on zoosporogenesis and motility of Peronosporomycete zoospores are likely linked with enhanced ATPase activity in mitochondria. *Front. Microbiol.* **2016**, *7*, 1824. [CrossRef]
30. Islam, M.T.; von Tiedemann, A. 2,4-Diacetylphloroglucinol suppresses zoosporogenesis and impairs motility of Peronosporomycete zoospores. *World J. Microb. Biot.* **2011**, *27*, 2071–2079. [CrossRef]
31. Chakraborty, M.; Mahmud, N.; Muzahid, A.N.M.; Rabby, S.M.F.; Islam, T. Oligomycins inhibit *Magnaporthe oryzae Triticum* and suppress wheat blast disease. *PLoS ONE* **2020**, *15*, e0233665. [CrossRef]
32. Tang, Y.-Q.; Sattler, I.; Thiericke, R.; Grabley, S.; Feng, X.-Z. Feigrisolides A, B, C and D, new lactones with antibacterial activities from *Streptomyces griseus*. *J. Antibiot.* **2000**, *53*, 934–943. [CrossRef]
33. Schumacher, R.W.; Talmage, S.C.; Miller, S.A.; Sarris, K.E.; Davidson, B.S.; Goldberg, A. Isolation and structure determination of an antimicrobial ester from a marine sediment-derived bacterium. *J. Nat. Prod.* **2003**, *66*, 1291–1293. [CrossRef]
34. Kim, W.H.; Jung, J.H.; Sung, L.T.; Lim, S.M.; Lee, E. Synthesis of the Proposed Structure of Feigrisolide C. *Org. Lett.* **2005**, *7*, 1085–1087. [CrossRef]
35. Zizka, Z. Biological effects of macrotetrolide antibiotics and nonactic acids. *Folia Microbiol.* **1998**, *43*, 7–14. [CrossRef]
36. Thiyagarajamoorthy, D.K.; Arulanandam, C.D.; Dahms, H.U.; Murugaiah, S.G.; Krishnan, M.; Rathinam, A.J. Marine bacterial compounds evaluated by in silico studies as antipsychotic drugs against schizophrenia. *Mar. Biotechnol.* **2018**, *20*, 639–653. [CrossRef] [PubMed]
37. Latorre, S.M.; Were, V.M.; Foster, A.J.; Langner, T.; Malmgren, A.; Harant, A.; Asuke, S.; Reyes-Avila, S.; Gupta, D.R.; Jensen, C.; et al. A pandemic clonal lineage of the wheat blast fungus. *bioRxiv* **2022**. bioRxiv:2022.06.06.494979.
38. Urashima, A.S.; Igarashi, S.; Kato, H. Host range, mating type, and fertility of *Pyricularia grisea* from wheat in Brazil. *Plant Dis.* **1993**, *77*, 1211–1216. [CrossRef]
39. Gupta, D.R.; Surovy, M.Z.; Mahmud, N.U.; Chakraborty, M.; Paul, S.K.; Hossain, M.; Bhattacharjee, P.; Mehebub, M.; Rani, K.; Yeasmin, R.; et al. Suitable methods for isolation, culture, storage, and identification of wheat blast fungus *Magnaporthe oryzae Triticum* pathotype. *Phytopathol. Res.* **2020**, *2*, 30. [CrossRef]
40. Paul, S.K.; Mahmud, N.U.; Gupta, D.R.; Rani, K.; Kang, H.; Wang, G.L.; Jankuloski, L.; Islam, T. *Oryzae* pathotype of *Magnaporthe oryzae* can cause typical blast disease symptoms on both leaves and spikes of wheat under a growth room condition. *Phytopathol. Res.* **2022**, *4*, 9. [CrossRef]
41. Sobolevskaya, M.P.; Fotso, S.; Havash, U.; Denisenko, V.A.; Helmke, E.; Prokofeva, N.G.; Kuznetsova, T.A.; Laatsch, H.; Elyakov, G.B. Metabolites of the sea isolate of bacteria *Streptomyces* sp. 6167. *Chem. Nat. Comp.* **2004**, *40*, 282–285. [CrossRef]
42. Prokofeva, N.G.; Kalinovskaya, N.I.; Lukyanov, P.A.; Kuznetsova, T.A. Membranotropic effects of cyclic lipopeptides produced by a marine isolate of the bacteria *Bacillus pumilus*. *Rus. J. Mar. Biol.* **1996**, *22*, 167–170.
43. He, Y.; Zhu, M.; Huang, J.; Hsiang, T.; Zheng, L. Biocontrol potential of a *Bacillus subtilis* strain BJ-1 against the rice blast fungus *Magnaporthe oryzae*. *Can. J. Plant Pathol.* **2019**, *41*, 47–59. [CrossRef]
44. Chakraborty, M.; Mahmud, N.U.; Gupta, D.R.; Tareq, F.S.; Shin, H.J.; Islam, T. Inhibitory effects of linear lipopeptides from a marine *Bacillus subtilis* on the wheat blast fungus *Magnaporthe oryzae Triticum*. *Front. Microbiol.* **2020**, *11*, 665. [CrossRef]

45. Riungu, G.M.; Muthorni, J.W.; Narla, R.D.; Wagacha, J.M.; Gathumbi, J.K. Management of *Fusarium* head blight of wheat and deoxynivalenol accumulation using antagonistic microorganisms. *Plant Pathol. J.* **2008**, *7*, 13–19. [CrossRef]
46. Meyers, E.; Pansy, F.E.; Perlman, D.; Smith, D.A.; Weisenborn, F.L. The in vitro activity of nonactin and its homologs: Monactin, dinactin, and trinactin. *J. Antibiot.* **1965**, *18*, 128–129.
47. Borrel, M.N.; Pereira, E.; Fiallo, M.; Garnier-Suillerot, A. Mobile ionophores are a novel class of P-glycoprotein inhibitors. The effects of ionophores on 49-O-tetrahydropyranyl- adriamycin incorporation in K562 drug-resistant cells. *Eur. J. Biochem.* **1994**, *223*, 125–133. [CrossRef] [PubMed]
48. Kusche, B.R.; Smith, A.E.; McGuirl, M.A.; Priestley, N.D. Alternating pattern of stereochemistry in the nonactin macrocycle is required for antibacterial activity and efficient ion binding. *J. Am. Chem. Soc.* **2009**, *131*, 17155–17165. [CrossRef] [PubMed]
49. Nielsen, T.H.; Thrane, C.; Christophersen, C.; Anthoni, U.; Sørensen, J. Structure, production characteristics and fungal antagonism of tensin–A new antifungal cyclic lipopeptide from *Pseudomonas fluorescens* strain 96.578. *J. Appl. Microbiol.* **2000**, *89*, 992–1001. [CrossRef] [PubMed]
50. Zhang, L.; Sun, C. Fengycins, cyclic lipopeptides from marine *Bacillus subtilis* strains, kill the plant-pathogenic fungus *Magnaporthe grisea* by inducing reactive oxygen species production and chromatin condensation. *Appl. Environ. Microbiol.* **2018**, *84*, e00445-18. [CrossRef]
51. Islam, M.T.; Hashidoko, Y.; Deora, A.; Ito, T.; Tahara, S. Suppression of damping-off disease in host plants by the rhizoplane bacterium *Lysobacter* sp. strain SB-K88 is linked to plant colonization and antibiosis against soilborne Peronosporomycetes. *Appl. Environ. Microb.* **2005**, *71*, 3786–3796. [CrossRef]
52. Islam, M.T. Disruption of ultrastructure and cytoskeletal network is involved with biocontrol of damping-off pathogen *Aphanomyces cochlioides* by *Lysobacter* sp. strain SB-K88. *Biol. Control* **2008**, *46*, 312–321. [CrossRef]
53. Islam, M.T. Mode of antagonism of a biocontrol bacterium *Lysobacter* sp. SB-K88 toward a damping-off pathogen *Aphanomyces cochlioides*. *World J. Microb. Biot.* **2010**, *26*, 629–637. [CrossRef]
54. Islam, M.T.; Fukushi, Y. Growth inhibition and excessive branching in *Aphanomyces cochlioides* induced by 2,4-diacetylphloroglucinol is linked to disruption of filamentous actin cytoskeleton in the hyphae. *World J. Microb. Biot.* **2010**, *26*, 1163–1170. [CrossRef]
55. Paul, S.K.; Chakraborty, M.; Rahman, M.; Gupta, D.R.; Mahmud, N.U.; Rahat, A.A.M.; Sarker, A.; Hannan, M.A.; Rahman, M.M.; Akanda, A.M.; et al. Marine natural product antimycin A suppresses wheat blast disease caused by *Magnaporthe oryzae Triticum*. *J. Fungi* **2022**, *8*, 618. [CrossRef]
56. Chakraborty, M.; Rabby, S.M.F.; Gupta, D.R.; Rahman, M.; Paul, S.K.; Mahmud, N.U.; Rahat, A.A.M.; Jankuloski, L.; Islam, T. Natural protein kinase inhibitors, staurosporine, and chelerythrine suppress wheat blast disease caused by *Magnaporthe oryzae Triticum*. *Microorganisms* **2022**, *10*, 1186. [CrossRef] [PubMed]
57. Dame, Z.T.; Islam, M.T.; Helmke, E.; von Tiedemann, A.; Laatsch, H. Oligomycins and pamamycin homologs impair motility and induce lysis of zoospores of the grapevine downy mildew pathogen, *Plasmopara viticola*. *FEMS Microbiol. Lett.* **2016**, *363*, fnw167. [CrossRef] [PubMed]
58. Homma, Y.; Takahashi, H.; Arimoto, Y. Studies on the Mode of Action of Soybean Lecithin Part 3. Effects on the Infection Process of Rice Blast Fungus, Pyricularia oryzae. *Jpn. J. Phytopathol.* **1992**, *58*, 514–521. [CrossRef]
59. Sasaki, H.; Suzuki, K.; Ichikawa, T.; Sawada, M.; Iwane, Y.; Ando, K. Microbial degradation of a macrotetrolide miticide in soil. *Appl. Environ. Microbiol.* **1980**, *40*, 264–268. [CrossRef]
60. Jizba, J.; Přikrylová, V.; Ujhelyiová, L.; Varkonda, Š. Insecticidal properties of nonactic acid homononactic acid, the precursors of macrotetrolide antibiotics. *Folia Microbiol.* **1992**, *37*, 299–303. [CrossRef]
61. Kilbourn, B.T.; Dunitz, J.D.; Pioda, A.R.; Simon, W. Structure of the K^+ complex with nonactin, a macrotetrolide antibiotic possessing highly specific K^+ transport properties. *J. Mol. Biol.* **1967**, *30*, 559. [CrossRef]
62. Sauter, H.; Steglich, W.; Anke, T. Strobilurins: Evolution of a new class of active substances. *Angew. Chem. Int. Ed. Engl.* **1999**, *38*, 1328–1349. [CrossRef]
63. *Fertilizer Recommendation Guide (FRG)*; Bangladesh Agricultural Research Council (BARC): Farmgate, Dhaka, 2012; pp. 1–265.

plants

MDPI

Review

An Insight into the Abiotic Stress Responses of Cultivated Beets (*Beta vulgaris* L.)

Seher Yolcu [1,*,†], Hemasundar Alavilli [2,*,†], Pushpalatha Ganesh [3], Muhammad Asif [1], Manu Kumar [4] and Kihwan Song [2,*]

1 Faculty of Engineering and Natural Sciences, Sabanci University, Istanbul 34956, Turkey; muhammad.asif@sabanciuniv.edu
2 Department of Bioresources Engineering, Sejong University, Seoul 05006, Korea
3 Department of Plant Biotechnology, M. S. Swaminathan School of Agriculture, Centurion University of Technology and Management, Odisha 761211, India; pushpabhagyalakshmi@gmail.com
4 Department of Life Science, College of Life Science and Biotechnology, Dongguk University, Seoul 10326, Korea; manukumar007@gmail.com
* Correspondence: seher808@gmail.com (S.Y.); alavilli.sundar@gmail.com (H.A.); khsong@sejong.ac.kr (K.S.)
† These authors contributed equally to this work.

Abstract: Cultivated beets (sugar beets, fodder beets, leaf beets, and garden beets) belonging to the species *Beta vulgaris* L. are important sources for many products such as sugar, bioethanol, animal feed, human nutrition, pulp residue, pectin extract, and molasses. *Beta maritima* L. (sea beet or wild beet) is a halophytic wild ancestor of all cultivated beets. With a requirement of less water and having shorter growth period than sugarcane, cultivated beets are preferentially spreading from temperate regions to subtropical countries. The beet cultivars display tolerance to several abiotic stresses such as salt, drought, cold, heat, and heavy metals. However, many environmental factors adversely influence growth, yield, and quality of beets. Hence, selection of stress-tolerant beet varieties and knowledge on the response mechanisms of beet cultivars to different abiotic stress factors are most required. The present review discusses morpho-physiological, biochemical, and molecular responses of cultivated beets (*B. vulgaris* L.) to different abiotic stresses including alkaline, cold, heat, heavy metals, and UV radiation. Additionally, we describe the beet genes reported for their involvement in response to these stress conditions.

Keywords: beet cultivation; abiotic stress; alkaline; cold; heat; heavy metals; stress tolerance; ultraviolet radiation

Citation: Yolcu, S.; Alavilli, H.; Ganesh, P.; Asif, M.; Kumar, M.; Song, K. An Insight into the Abiotic Stress Responses of Cultivated Beets (*Beta vulgaris* L.). *Plants* **2022**, *11*, 12. https://doi.org/10.3390/plants11010012

Academic Editors: Ewa Muszyńska, Kinga Dziurka and Mateusz Labudda

Received: 6 November 2021
Accepted: 14 December 2021
Published: 21 December 2021

1. Introduction

Economically important cultivated beets such as fodder beets, sugar beets, garden beets (e.g., red beet), and leaf beets (e.g., Swiss chard) belong to the sub-species *Beta vulgaris* L. ssp. *vulgaris* [1,2]. All beets originate from a halophytic plant, *Beta vulgaris* L. ssp. *maritima* (sea beet or wild beet), also known as *Beta maritima* L. [3]. Among them, leaf beets and garden beets are used as vegetables [2,4], fodder beets as animal feed [1,2], and sugar beets serve as the source of sucrose, bioethanol, biodegradable polymers, and biofertilizers [5–8]. In addition to these advantages, beets such as Swiss chard and red beet are a rich source of pigments, termed betalains [9–12]. Cultivation of beets is widely distributed throughout Turkey and Mediterranean and European countries [13]. Fodder beet plants, which grow at a temperature between 8 °C and 25 °C [1], are cultivated in coastal areas of many countries [14] as well as continental habitats [15]. Wild beet (*Beta maritima* L.) is especially distributed along the coasts of Mediterranean Sea and the European North Atlantic Ocean [3], and it shows significantly higher salt tolerance during germination and seedling stages when compared to other beet varieties [15–19]. Although previous reports have shown genetic diversity in beet species, due to insufficient genetic variation in cultivated beets [15,20,21], the use of wild beet can provide a remarkable source of genetic variability for crop improvement under stressful conditions [20].

Crop plants are subjected to various abiotic stresses, resulting in loss of yield or decreased productivity. Plants have different adaptive and protective strategies at morphological, physiological and molecular levels to cope with environmental stress conditions [21]. Although stress conditions negatively affect beet growth, yield, and quality, the beet cultivars are able to tolerate abiotic stress conditions such as salinity, drought, cold, heat, and heavy metals [18,22–28]. Sugar beets exhibit tolerance to cadmium (Cd) and are capable of accumulating heavy metals such as Cd and nickel (Ni) [27]. The improvement of beet varieties with better heat tolerance is also an important task due to climate change and global warming [29]. Therefore, we need breeding techniques and agronomic practices for better tolerance to biotic and abiotic stresses in beets [30]. Thus, cultivated beets and their wild ancestor are important genetic sources for crop breeding programs and studying abiotic stress tolerance [15,31]. In the present review, we summarize the morpho-physiological, biochemical, and molecular alterations in cultivated beets (*B. vulgaris* L.) under alkaline, cold, heat, heavy metal, and UV stresses.

2. Responses of Cultivated Beets (*B. vulgaris* L.) to Different Abiotic Stresses Including Alkaline, Temperature, Heavy Metal, and UV

Although several studies report different responses of beet cultivars to environmental stresses, research articles and reviews mostly focus on salt and drought response mechanisms in beets [22,23,32–34]. However, a comprehensive review describing the responses of cultivated beets to several abiotic stress factors including cold, heat, alkaline, heavy metal, and UV is lacking. Therefore, this review focuses on the responses of cultivated beets (*B. vulgaris* L.) to alkaline, cold, heat, heavy metal, and UV stresses at morpho-physiological, biochemical, and molecular levels. In Table 1, we demonstrate the list of beet genes known for their involvement in response to alkaline, cold, and heavy metal stress.

Table 1. Beet genes known to be involved in response to alkaline, cold, and heavy metal stresses.

Type of Abiotic Stress	Gene Name	References
Alkaline stress	WRKY transcription factor family (*WRKY10* and *16*)	[35]
Alkaline stress	Metal Tolerance Protein 11 (*MTP11*)	[36]
Alkaline stress	Ethylene-insensitive protein 2 (*EIN2*)	[36]
Alkaline stress	Polyphenol Oxidase (*PPO*)	[36]
Cold stress	Integral membrane protein (*IMP*)	[37]
Cold stress	A novel ER-located aquaporin gene (*COLD1*)	[38]
Cold stress	Raffinose synthase 1 and 2 (*RS1* and *RS2*)	[39]
Freezing	Galactinol synthase 2 and 3 (*GOLS2* and *GOLS3*) Raffinose synthase 2 and 5 (*RS2* and *RS5*)	[40]
Heavy metal	Metal tolerance protein (*BmMTP10* and *BmMTP11*)	[41]
Heavy metal	Toxic nickel concentration (*NIC3*, *NIC6* and *NIC8*)	[42]
Heavy metal	Natural resistance-associated macrophage protein 3 (*NRAMP3*)	[43]

2.1. Alkaline Stress

Alkaline stress (high pH) is one of the abiotic constraints of plants, which co-exists with salt stress and elicits severe detrimental damages to global agricultural production [44]. Over 954 million hectares of land on the globe is affected by salinity [45]. Salt stress results from a neutral salt such as NaCl. Although alkaline salt stress is a type of salt stress, it is caused by alkaline salts such as $NaHCO_3$ and Na_2CO_3, which is shortly called alkaline stress and causes more damage than neutral salt [46,47]. Numerous research groups across the globe have been perusing tolerance mechanisms to understand the salt stress responses in various crops and model land plants [48,49]. However, the studies focused on high salinity together with alkaline stress are minuscule [44,46]. Apparently, when the plants simultaneously encounter high salinity and high pH, their cumulative damage is more severe than their single occurrence [44]. Several previous reports determined that sugar beet can sustain moderate exposure to saline and alkaline conditions [35,37,48]. However, only a few reports investigated the responses of beets under alkaline stress conditions [47,50]. Hence, to alleviate the alkaline stress-induced damages in commercially important crops such as beets, we need to build a comprehensive knowledge repository that helps devise better strategies for generating stress-tolerant cultivars to attain sustainable agriculture [35,51,52]. Furthermore, developing high salinity-resistant cultivars will efficiently and rationally utilize salinity-affected areas in cultivated lands [45].

Although alkaline stress and salt stress share many common features, such as osmotic stress and ion toxicity, the alkaline condition has unique differences to consider as a different stress form [51]. The alkaline stress includes three principle factors that negatively impact plant growth and development: high soil pH, Na^+ toxicity, and water deficiency [51]. For example, it has been shown that alkaline stress-induced Na^+ toxicity and oxidative stress decreased photosynthesis and growth in tomato plants. Moreover, alkaline stress led to higher Na^+/K^+ ratio and lower K^+ content in tomato seedlings [51], and the expression of genes encoding Na^+ transporters such as *SlNHX1*, *SlNHX2*, *SlSOS1*, *SlHKT1,1*, and *SlHKT1,2* were found to increase in tomato roots exposed to $NaHCO_3$ [52,53]. However, we still do not know how sugar beet plants maintain Na^+-K^+ homeostasis under alkaline stress conditions and whether Na^+ transporters contribute to the alkaline stress response in beets. High alkaline pH causes the occurrence of oxidative stress through reactive oxygen species (ROS) and the production of malondialdehyde (MDA), which damage the membrane integrity and intracellular components in plants [47]. To decrease the ROS-induced oxidative stress, plants use several enzymatic and non-enzymatic antioxidants [54]. Enzymatic antioxidants including superoxide dismutase (SOD), catalase (CAT), peroxidase (POX) and ascorbate peroxidase (APX) are involved in scavenging of superoxide radicals and hydrogen peroxide (H_2O_2) [54–56]. Under salt stress, cultivated beets and wild beet show higher antioxidant enzyme activities [57–59]. Similarly, Zou et al. [30] reported that the alkaline stress-tolerant beet cultivar KWS0143 displayed higher antioxidant enzyme activities such as CAT and APX than the sensitive cultivar Beta464 under the same growth conditions [30]. This implies that the tolerant plants are bestowed with durable antioxidant defense equipped with APX, CAT and SOD enzymes to circumvent the cellular damages under salt-alkaline stress [30]. Hence, we need to identify genetic resources with a strong innate antioxidant defense system to fortify beet cultivars with alkaline stress tolerance. In addition to oxidative stress, soils with high pH perturb the macro and micronutrient balance in the soil, which drives the plant to a physiological depression [50]. Previously, Oster et al. [60] classified the alkaline stress into three categories based on the alkaline salt percentage in soil. According to this classification, the alkalinity is considered as mild (3% salt content and pH 7.1–8.5), moderate (3–6% salt and the pH is 8.5–9.5), and severe (>3–6% salt and the pH over 9.5) [60]. In contrast to the detrimental effects of alkaline stress, mild alkaline stress can help the plants to grow bigger and healthier [50,61]. Likewise, in a recent report, Geng et al. [50] examined the differential proteomic responses of sugar beet seedlings by treating them with pH 5, pH 7.5, and pH 9.5 (acidic, neutral, and alkaline) conditions. In the study, they found that the acidic pH caused more growth retardation and enzymatic aberrations than that of neutral and alkaline pH conditions [50]. In contrast to other reports, the alkaline conditions (pH 9.5) significantly improved plant height, fresh weight, total leaf and root area, net photosynthetic rate, stomatal conductance, intercellular CO_2 concentration, and chlorophyll contents compared to neutral and acidic soils [50]. Moreover, a few more reports found that mild alkaline stress caused better growth, leaf chlorophyll contents, photosynthetic index, and antioxidant activities in sugar beet seedlings [30,61]. Geng et al. [61] found that neutral salt ($NaCl:Na_2SO_4$, 1:1, Na^+ 100 mM) remarkably decreased growth and photosynthesis when compared with mild neutral salt ($NaCl:Na_2SO_4$, 1:1, Na^+ 25 mM) and alkaline conditions (Na_2CO_3, Na^+ 25 mM) in sugar beet plants. In contrast, plants displayed a significant increase in total biomass, leaf area, and photosynthesis under mild neutral salt and alkaline conditions [61]. Interestingly, sugar beet plant growth was not impacted by high alkaline salt (Na_2CO_3, Na^+ 100 mM) as compared to control [61]. We speculate that by virtue of being tolerant to mild saline–alkaline stress, the sugar beet cultivars might display better growth, and we need further experimental evidence to learn the growth patterns of different beet cultivars altered under mild alkalinity. Nevertheless, the growth retardation of plants is found to be proportionately elevating along with the increase in alkaline stress severity [62]. Additionally, alkaline stress responses in plants are usually governed by a multigenic effect, but not by a single gene expression, which implies the intricate stress signaling mechanism [36,63,64].

Numerous reports suggest that under alkaline stress, several physiological parameters, including stomatal conductance (Gs), transpiration rate (Tr), relative water content (RWC), water use efficiency (WUE), accumulation of photosynthetic pigments, and the net photosynthetic rate (Pn), were dropped [47,62]. Specifically, the photosystem-II (PSII) quantum efficiency (Fv/Fm) ratios are negatively affected by alkaline stress, which reduce the electron transport rate [65]. Furthermore, high alkaline conditions dampen the leaf area (LA) and chlorophyll contents (Chl a and b), specifically Chl b, which lowers the photosynthetic rate and WUE [66]. All these physiological parameters will eventually curtail the seedling growth and seedling emergence under alkaline stress [30,62]. In another study, Liu et al. [66] assessed the physiological responses of white Swiss chard under saline and alkaline conditions. Their study identified that although Swiss chard retains higher RWC under alkaline stress, the seedlings suffered from alkaline stress in terms of plant growth. The growth retardation was likely caused by high pH, $CO_3^{\ 2-}$, and HCO_3^- toxicity [66]. Additionally, the physiological indicators such as chlorophyll contents, WUE, and the ionic balance were also perturbed in Swiss chard under 50–100 mM alkaline stress [66]. While comparing the glycine betaine (GB) and proline levels, they found that the GB levels in sugar beet were lower in 50 mM alkaline stress than that of 50 mM salt stress, whereas they did not find any significant alterations in proline levels [66]. This bolsters the notion that the GB plays a more critical role in mediating the alkaline stress tolerance than proline for Swiss chard [66]. It is a well-known fact that compatible solutes including GB, proline, and soluble sugars are remarkably increased under salt stress conditions to maintain photosynthesis and stomatal conductance in beets [67–69].

In addition to physiological and biochemical responses of beets under alkaline stress, only few genes have been reported to be involved in alkaline stress response in beets. For example, Wu et al. [35] identified 58 putative *WRKY* genes in the sugar beet genome, and among them, nine genes were found to be responsive to the alkaline stress stimulus (~15 mM to 100 mM $NaCHO_3$) in both root and shoot tissues [35]. In the study, they found augmented expression of the *BvWRKY10* gene in shoots and *BvWRKY16* expression in root tissues under alkaline stress [35]. The differential expression of *BvWRKY* genes in different tissues implies their functional roles in mediating the alkaline stress responses in different tissues and needs further experimental attention. The WRKY family of transcription factors is plant-specific and plays many critical roles in diverse aspects of plant physiological processes, including abiotic stress responses [70]. Through a transcriptomic approach, some of the differentially expressed genes (DEGs) were shown in alkaline stress-treated beets. Recently, Zou et al. [36] identified differential expression of 1270 genes in alkaline stress-tolerant cultivar KWS0143 in response to alkaline stress. They irrigated the plants with 75 mM alkaline solution ($Na_2CO_3:NaHCO_3$, 1:2, pH 9.67) and harvested the leaf tissues three (short-term) and seven days (long-term) after the treatments [36]. Compared to the control groups, the short-term and long-term treatments induced the expression of *'Ethylene-insensitive protein 2'* (*LOC104884677*) and *'Metal tolerance protein 11'* (*LOC104886952*) genes, respectively [36]. The results suggest that some of these DEGs would be useful for developing alkaline-tolerant beet cultivars. In another report, Zou et al. [47] assessed the roles of long non-coding RNAs (lncRNAs) in sugar beets under different alkaline stress conditions as previously described in Zou et al. [36] by high-throughput RNA sequencing [47]. In this study, they identified 93 differentially expressed alkaline stress-responsive lncRNAs. Furthermore, additional functional attribution of candidate target genes revealed their association with diverse biological processes, including kinase activity, ribosomal and ribonucleoprotein constituents, and protein metabolic activity, and denotes the association of specific target genes with lncRNAs [47]. In addition, Zou et al. [71] treated the sugar beet seedlings with an alkaline solution and performed small RNA sequencing [71]. They found 53 novel microRNAs (miRNAs) responsive to long-term and short-term alkaline stresses [71]. Similarly, the gene ontology (GO) analysis uncovered enrichment of miRNAs related to the "redox process" and they reported the involvement of *'polyphenol oxidase'* (LOC04900758) gene as the target of alkali-responsive miRNAs. In addition to this, the other 29 miRNAs responsive to long-term alkaline stress can be useful as potential targets to fortify crops with alkaline stress resistance. In Table 2, we summarize the alkaline stress responses in sugar beet varieties.

Table 2. Alkaline stress responses in cultivated beets.

Beet Variety	Stress Treatments	Experimental Results	Reference
B. vulgaris, KWS0143	$NaHCO_3$:Na_2CO_3 (0.5%, 0.7%, 0.9%)	High activity levels of antioxidant enzymes, such as CAT and APX	[30]
B. vulgaris, H004	pH 5, pH 7.5, and pH 9.5	Acidic pH resulted in more growth retardation, photosynthesis, and enzymatic aberrations than neutral and alkaline pH	[50]
B. vulgaris, KWS0143	75 mM alkaline solution ($NaHCO_3$:Na_2CO_3, 2:1, pH 9.67)	Significant inhibition of plant growth	[47]
		A decrease in stomatal conductance (Gs), transpiration rate (Tr), and net photosynthetic rate (Pn)	
		Identification of 93 differentially expressed alkaline stress-responsive lncRNAs	
B. vulgaris, H004	Neutral salt (NaCl:Na_2SO_4, 1:1) and alkaline salt (Na_2CO_3)	Mild neutral salt and alkaline conditions led to a significant increase in total biomass, leaf area, and photosynthesis	[61]
B. vulgaris, KWS0143 and Beta464	0, 25, 50, 75 and 100 mM of mixed (Na_2CO_3:$NaHCO_3$, 1:2) alkaline conditions	The levels of photosynthetic pigments were remarkably diminished by high alkaline stress (75 and 100 mM)	[62]
		Sugar beet displayed resistance to alkaline stress through osmotic adjustment and antioxidant enzymes under mild alkaline stress	
B. vulgaris L. var. *cicla*	50 and 100 mM alkaline salt ($NaHCO_3$ and Na_2CO_3, 9:1)	Growth retardation due to high pH, $CO_3{}^{2-}$, and $HCO_3{}^-$ toxicity	[66]
		Lower GB levels under 50 mM alkaline stress than 50 mM salt stress, whereas no significant alterations in proline levels	
B. vulgaris, Gantang7	0, 15, 25, 50 and 100 mM $NaHCO_3$	Among 58 putative *WRKY* genes, 9 genes were found to be responsive to alkaline stress (~15 mM–100 mM $NaCHO_3$) in both root and shoot	[35]
		Enhanced expression of *BvWRKY10* gene in shoots and *BvWRKY16* expression in roots under alkaline conditions	
B. vulgaris, KWS0143	75 mM alkaline solution (Na_2CO_3:$NaHCO_3$, 1:2, pH 9.67)	Differential expression of 1270 genes in alkaline stress-tolerant cultivar KWS0143 under alkaline stress	[36]
B. vulgaris, KWS0143	75 mM alkaline solution (Na_2CO_3:$NaHCO_3$, 1:2, pH 9.67) for short-term (3 d), and long-term (7 d)	53 novel miRNAs responsive to long-term and short-term alkaline stress	[71]

2.2. Cold and Heat Stresses

Because plants are sessile organisms, the ambient temperature has a profound impetus on their entire life cycle, reflecting on their spatial distribution and seasonal behaviors [72]. Their surrounding temperatures also influence the plant growth rate and development, and each plant system has its own set of minimum, optimum, and maximum range of temperatures for survival [73]. Crop production varies depending on the severity of temperatures [74]. Furthermore, plants differentially respond to cold or heat stress according to their developmental stage. Hence, to circumvent the yield damages associated with capricious climates, we need to accumulate the morpho-physiological responses for individual crop varieties. Furthermore, more studies should be performed in order to characterize stress-responsive genes and determine the molecular mechanisms under low and high-temperature stresses in beets, as we have limited knowledge on beet responses to temperature changes.

2.2.1. Cold Stress

Low temperature is one of the most important constraints, impeding plant growth, distribution, biological activity, production, and, ultimately, economic yield [75]. The sensitivity and responses of sugar beet to cold temperatures depend on its developmental stage. Cold is known to drive several developmental events in sugar beet in early and later stages, such as germination, growth, bolting, and accumulation of molassigenic products in the roots [76]. In sugar beets, exposure to cold temperatures at the early seedling stages causes severe root growth retardation and reduced sugar yield [75,77]. Although cold temperatures (i.e.,−2 °C) result in loss of cotyledon viability, the seedlings at 3–4 leaf stage can withstand freezing temperatures up to −10 °C [78,79]. Furthermore, sugar beet roots and shoots show differential responses to cold stress. For instance, in three sugar beet genotypes (GT1, GT2, and GT3), cold temperatures impacted taproot growth more than the shoot growth [80]. It has been reported that there are variations in cold stress tolerance and sensitivity among *Beta* germplasms [81]. Hence, to generate cold-tolerant varieties in commercially essential crops such as beets, knowledge pertaining to their responses to cold conditions is the most important prerequisite [82]. In some geographical sections, sugar beet seeds are sown in early autumn to expose them to shallow winter temperatures (below 0 °C). This practice helps protect the sugar beets from pathogen *Cercospora* attacks and drought stress [75]. Such an early seed sowing in fall, also known as "autumn sowing", was reported to produce sugar beets with better field emergence than the spring-sown beets [76]. Nevertheless, prolonged exposure of sugar beets at the young seedling stage to extreme cold temperatures seriously limits the yield [75]. Cold-treated sugar beet plants displayed a decrease in photosynthetic efficiency, quantum yield of PSII, leaf CO_2 concentration, CO_2 assimilation rate, and leaf transpiration rate [40,80]. Moreover, compatible solutes such as glucose, fructose, and raffinose in leaves were increased by 0 °C and 4 °C cold treatments [40,80], but decreased in taproots in response to freezing temperature [40]. Consistently, under freezing conditions, the sucrose content decreased in roots, followed by leakage of the root sap due to cell alteration in membrane permeability and infection with microbes. Water infiltration due to rapid freezing/thawing can also lead to softening of the root tissue and gradual rotting [83]. Rodrigues et al. [80] reported an interesting finding for the first time. Vernalization (long-term cold treatment at 4–15 °C) leads to a reversal of phloem translocation from taproots (sink tissue) to shoots (source tissue). Redirection of sugar flux is required for induction of flowering in sugar beet. This process might be the reason for the sugar beet sensitivity to freezing temperatures [80]. In a very recent work, three sugar beet genotypes (GT1, GT2, and GT3) were evaluated for freezing tolerance. Freezing temperatures caused the production of ROS, raffinose accumulation, and transcription of genes involved in raffinose metabolism in leaves and taproots [40]. These results suggest that raffinose metabolism has a protective role against freezing injury in sugar beet. Moreover, ROS-scavenging enzymes including SOD and CAT significantly enhanced in response to 4 °C [40]. Consistently, the maximum expression levels of genes encoding antioxidant enzymes such as CAT, APX, ascorbate reductase, and glutathione peroxidase (GPX) were seen at 4 °C, but the expression was reduced at 0 °C. The findings indicate the temperature-dependent ROS production in sugar beet plants.

To date, very few sugar beet genes that function in cold stress response have been functionally characterized under cold stress conditions. In some reports, the transcript levels of genes involved in photosynthesis and compatible solute biosynthesis were investigated in cold-treated beets. For example, Rodrigues et al. [80] reported a sharp increase in the expression of photosynthesis-related genes encoding rubisco activase, rubisco small subunit, a chlorophyll a/b binding protein, and plastocyanin under cold stress. Kito et al. [39] isolated and characterized two sugar beet genes, *B. vulgaris RS1* and *RS2* (*BvRS1* and *BvRS2*), encoding raffinose synthase, which is involved in raffinose biosynthesis. The transcript levels of *BvRS1* and *BvRS2* genes were induced by cold stress in sugar beet leaves and roots [39]. Similarly, in a very recent study, the transcript abundances of galactinol synthase encoding genes, *GOLS2* and *GOLS3*, and two *RS* genes, *BvRS2* and *BvRS5*, were increased by freezing temperature [40]. Surprisingly, the expression of *BvRS5* gene and raffinose amounts remarkably induced in the taproots of freezing-tolerant beet cultivars, GT2 and GT3, but not in the sensitive one, GT1. As compared to other beet genotypes, the GT2 showed the maximum expression levels of *GOLS* and *RS* genes and raffinose levels in taproots, indicating the highest freezing tolerance in GT2 [40]. These findings suggest that the survival of taproot tissue under cold stress might depend on the accumulation of raffinose. As compatible solutes and antioxidants, raffinose family oligosaccharides have important roles in plant response to abiotic stress and stabilizing membranes and proteins [84,85]. In addition to genes involved in raffinose metabolism, the *B. vulgaris Integral Membrane Protein (BvIMP)* gene is the closest homolog of *A. thaliana early response to dehydration-like 6 (AtERDL6)*, which was previously reported for its cold stress-responsive function [86]. Cold stress may lead to elevations in the transcription of *BvIMP* gene and vacuolar sugar trafficking in sugar beet leaves, which is critical for cold stress response and seed germination [37]. Ectopic overexpression of *BvIMP* in *Arabidopsis* resulted in altered glucose concentration during cold conditions, lower accumulation of monosaccharides, and cold-sensitive phenotype compared to the wild-type [37]. In a recent study, Porcel et al. [38] uncovered and isolated a novel endoplasmic reticulum-located aquaporin gene, *B. vulgaris COLD1 (BvCOLD1)*, which is specific to the *Chenopodiaceae* subfamily. The *BvCOLD1* gene is ubiquitously expressed in all tissues of sugar beet [38]; however, its expression was not changed by cold stress [38,75]. In contrast to the wild-type plants, overexpression of *BvCOLD1* restored the membrane fluidity in transgenic *Arabidopsis* lines under cold temperatures and rendered tolerance to cold stress, suggesting that it could be a useful gene for developing biotechnological strategies in order to generate cold-tolerant beet cultivars [38].

2.2.2. Heat Stress

Elevated temperatures and water deficit conditions tend to elicit similar impacts on plant water content where the evaporation exceeds the water intake, eventually leading to the plant wilting [87]. Across the globe, we face rapid climate changes and adverse weather problems; hence, developing heat-tolerant crops is the need of the hour. High temperatures impede many vital developmental events such as seed germination and impact seed vigor and viability and seedling emergence, and eventually challenge their survival [88,89]. Critical physiological processes, including photosynthesis and PSII activity, were also affected due to electron transport chain block under heat stress [90,91]. Of late, sugar beet cultivation is also expanding to the tropical and sub-tropical areas, and more people pay attention to cultivation of the sugar beets in summer [29,92]. Ironically, there are few studies aimed to select the heat-tolerant sugar beet cultivars. For the identification of the heat-tolerant beet genotypes, currently, there are no universally approved criteria. Different research groups used different parameters to evaluate the heat stress tolerance in different beet cultivars. For instance, Malmir et al. [92] considered the seed vigor index and root length as evaluation parameters of heat stress tolerance in the early growth stage [92]. To investigate the effects of heat on early growth in sugar beet, they compared 31 sugar beet genotypes under heat stress conditions. Among all the variants tested, the tolerant genotype displayed relatively higher germination, seed vigor, plumule length, and seedling length compared to other genotypes, suggesting that the tolerant one is a prospective cultivar to expand the sugar beet cultivation to tropical areas [92]. Under high temperatures, the leaf temperature, which is associated with vapor pressure deficit (VPD) and stomatal conductance, is known to be enhanced [93]. Moreover, another recent study showed the stress tolerance index (STI) and average root and recoverable sugar yields as selection parameters to identify heat-tolerant lines among 18 sugar beet breeding lines [29]. Among them, six lines were found to have the highest yield, and two lines can sustain under heat stress [29]. In a previous work, two fodder beet cultivars (Ecdogelb and Ecdorot) were used to reveal the impacts of different light intensities and temperatures on fodder beet physiology [94]. High temperature affected root weight ratio (RWR), dry leaf weight (DLW), dry root weight (DRW), total dry weight (TDW), specific leaf area (SLA), net assimilation rate (NAR), and relative growth rate (RGR) in both cultivars at low light intensity [94]. For example, under high light intensity and temperature (20 °C), the cultivar Ecdorot exhibited enhancements of leaf weight ratio (LWR). The highest RGR, RWR, and DLW levels were recorded in response to high temperature and low light intensity in both cultivars. High temperatures result in increments of the growth in root crops, but adversely impact the final biomass [95]. When the temperature was increased from 14 °C to 19.6 °C, an increase in the SLA was also observed [94]. Leaf area, which is used as a selection parameter of drought-tolerant beet cultivars, determines the plant growth rate during initial phase of development [96] and is associated with root and sugar yield [32]. Thus, we assume that the leaf area could be an important parameter to enhance sucrose yield of beets under high temperature conditions.

Unfortunately, so far, no beet genes have been functionally characterized under high temperature conditions. Moreover, the knowledge on beet physiological and biochemical responses is very limited. Hence, comprehensive studies should be performed in different beet cultivars under heat conditions to gain a better understanding of heat tolerance mechanisms in beets at different developmental stages. In Table 3, we summarize the low and high temperature stress responses in cultivated beets.

2.3. Heavy Metal Stress

Generally, heavy metals are a group of metals and metalloids with atomic density more than 5 g cm^{-3}, or five times or more, greater than water [97], including lead (Pb), cadmium (Cd), nickel (Ni), cobalt (Co), iron (Fe), zinc (Zn), chromium (Cr), arsenic (As), silver (Ag), and the platinum group elements. Mining and smelting operations and agriculture have caused heavy metal contamination of soils with Cd, copper (Cu), and Zn in many areas of the world [98]. Moreover, due to vigorous mining and industrial activities, the metal pollution in soils is becoming prevalent day by day and posing a severe threat to ecological balance [99,100]. For example, in 2002, 22,000 t of Cd, 93,900 t of Cu, 783,000 t of Pb, and 1,350,000 t of Zn were released into the environment on the global scale [101,102]. The buildup of heavy metals in arable lands results in contamination of soils, making them unsuitable for cultivation of plants, including beets. Therefore, the need for collecting scientific information regarding effects of various heavy metals on plants, response mechanisms of plants to heavy metal stress, and agronomic management of this stress can not be overemphasized.

Table 3. Cold and heat stress responses in cultivated beets.

Beet Variety	Stress Treatments	Experimental Results	Reference
B. vulgaris, Merak, and Antic cultivars	Cold stress (0 °C, 5 °C and 10 °C)	Some parameters, such as proline content, F_v/F_m ratio, and root dry matter, were higher in cold-tolerant varieties than sensitive ones	[77]
		Genetic diversity in cold tolerance of sugar beet cultivars was observed at seedling stage	
B. vulgaris, Bianca	Cold stress (−2 °C)	Prolonged exposure of sugar beets at the young seedling stage to the cold stress seriously limits the yield	[75]
		After short-term cold stress, transcription factors and genes involved in metabolic pathways were expressed in sugar beet leaves and roots	
B. vulgaris	Cold stress (−2 °C and −10 °C)	Sugar beet plantlets at the cotyledon stage completely died at −2 °C; however, at the 3–4 leaf stages, the plants can survive up to −10 °C	[78,79]
B. vulgaris	Cold stress (−5 °C)	Freezing injury results in an increase in tonoplast permeability for sucrose	[83]
		Under freezing conditions, the sucrose content decreased in roots, followed by leakage of the root sap due to cell alteration in membrane permeability and infection with microbes	
B. vulgaris, NK-210 mm-0	Cold stress (4 °C)	The transcript levels of two sugar beet genes, *B. vulgaris RS1* and *RS2* (*BvRS1* and *BvRS2*), encoding raffinose synthase, were induced by cold stress in sugar beet leaves and roots	[39]
B. vulgaris genotypes; GT1, GT2, and GT3	Cold stress (12 °C, 4 °C, and 0 °C)	Raffinose accumulation and transcription of genes involved in raffinose metabolism in leaves and taproots have been observed under low temperature	[40]
B. vulgaris, belladonna	Cold stress (4 °C)	Ectopic overexpression of *BvIMP* in *Arabidopsis* led to altered glucose concentration under cold conditions, lower accumulation of monosaccharides	[37]
B. vulgaris	Cold stress (10 °C)	Overexpression of *BvCOLD1* restored the membrane fluidity in transgenic *Arabidopsis* lines under cold stress and rendered tolerance to cold	[38]
B. vulgaris var. *altissima* Döll	Heat stress (20 °C and 30 °C)	Among 31 sugar beet genotypes, the tolerant genotype exhibited higher germination, seed vigor, plumule length, and seedling length under heat stress	[92]
B. vulgaris, USKPS25 and USC944-6-68 breeding lines	High temperature conditions in the field experiments	The stress tolerance index (STI) showed positive correlation with average root and sugar yields, which were used as selection parameters to identify heat-tolerant lines	[29]
B. vulgaris var. crassa Mansf. Fodder beet cv. Ecdogelb and Ecdorot	Heat and cold stress (18.28, 19.58, 18.26, 17.61, and 14.1 °C)	Two fodder beet cultivars showed the highest levels of RGR, RWR, and DLW under high temperature and low light intensity	[94]

Exposure of plants to toxic levels of heavy metals causes various metabolic and physiological alterations depending on the metal of concern, level of stress, plant species, cultivar, and other biotic and abiotic factors [103–105]. Most of the mineral ions such as Zn, Ni, manganese (Mn), etc., are required for all metabolic activities in plants at miniscule amounts. However, if the metal ion presence exceeds the threshold, they tend to exert detrimental effects on plant metabolism, resulting in leaf chlorosis, necrosis, turgor loss, a decrease in the rate of seed germination, and a crippled photosynthetic apparatus, which could cause plant death [106–108]. Among the heavy metal ions, Cd, Zn, and Cu are reported as the most toxic metals, with serious health hazards to humans when they infiltrate the food chain [109]. Like other plants, heavy metals adversely affect the sugar beet as they proscribe various metabolic activities [27,110,111]. For example, heavy metals such as Pb damage the vacuolar membrane in red beet taproots [112]. Lead is one of the most toxic metals for plant cells, and it negatively affects plant growth, photosynthesis, respiration, and membrane transport [113]. Cd treatment in *B. vulgaris* caused growth retardation, leaf chlorosis, and increased root/whole plant ratio [114] with decreased root-tip respiration and photosynthesis [110,114]. As compared to control plants, Cd-treated plants exhibited lower shoot dry weights, photosynthetic pigments, and reduction in water content of shoots and fine roots, dramatically [114]. Direct application of Cd on isolated leaves, protoplasts, and chloroplasts inhibited CO_2 fixation without affecting the PSI or PSII and dark respiration rate, whereas indirect Cd application through the culture medium decreased the maximal quantum yield of CO_2 assimilation [110]. Papazoglou and Fernando [27] tested the growth and heavy metal tolerance of sugar beet plants in Cd- and Ni-contaminated soil [27]. They found that the highest Ni concentration (20 g) was lethal to the plants, and an interesting fact they found was that the single application of Ni caused higher toxic effects than the combination of Ni and Cd [27]. Nevertheless, the combination of Cd (5 g) and Ni (10 g) treatment resulted in a drastic reduction in fresh and dry biomass of aerial parts and beets, and a decrease in plant height [27]. Very recently, Haque et al. [43] found that toxic levels of Cd cause growth retardation of sugar beet plants because of low iron levels resulting in photosynthetic inefficiency, and cellular oxidative stress [43]. Cd-treated plants displayed sensitivity to oxidative stress, leading to an increase in levels of O_2^- and H_2O_2 in roots and shoots. In addition, Haque et al. [43] examined the antioxidant defense system in sugar beet under heavy metal stress and found that Cd stress caused an enhancement of CAT enzyme activity in the shoots, whereas the activities of other antioxidant enzymes such as SOD, APX, and GR did not change in neither roots nor shoots. Furthermore, the results from a previous study indicated reduced uptake of N, P, Mg, K, Mn, Cu, and Zn upon Cd toxicity [114]. Similar to Cd stress, Zn toxicity decreased macronutrient concentrations (N, K, and Mg), whereas it enhanced the P level in shoots as well as roots [115]. In sugar beets, Cu and Zn treatments also significantly reduced plant growth, shoot and root lengths, and dry weight [116]. At high Cu concentrations, the shoots showed turgor loss, but lower Cu concentration did not affect plant growth [116]. Sagardoy et al. [115] reported that the toxic level of Zn reduced water content, leaf numbers, and root/shoot ratio, along with wrinkled and chlorotic leaves in sugar beet [115]. Root proteome analysis of sugar beet showed slight changes in metabolism under low and mild Zn levels, but higher levels of Zn led to cell death and cessation of metabolism through decreasing aerobic respiration and damaging defense systems required for oxidative stress response. Thus, the results showed that toxic Zn levels caused damages to the oxidative stress defense mechanisms due to Zn competition with divalent cations such as Fe, which might strengthen the symptoms of Zn toxicity in plants [117]. In summary, the results denote that the degree of toxicity of heavy metals on plant metabolism depends on plant species, the duration of stress, and type and concentration of heavy metals they were exposed to [111].

Several studies highlighted foliar uptake of heavy metals and their effects on the membrane permeability through the cuticle and percentage of open stomata in sugar beet [118,119]. A previous study demonstrated that sugar beet seedlings grown in nutrient solution containing high concentrations of $CdCl_2$ showed an increased leaf transpiration rate and a decreased stomatal aperture area. Thus, higher Cd concentrations affected the permeability of the leaf cuticle [119]. Apart from seedlings, Cd stress was also shown to negatively influence sugar beet taproot growth. For instance, long-term Cd exposure caused decreased sucrose uptake and diminished dry weight in taproots, but the direct addition of Cd^{2+} to the medium enhanced the sucrose uptake at the tonoplast [120]. Increased accumulation of Cd lowered the contents of glucose, fructose, and sucrose in both shoots and roots of sugar beet [121] and inhibited the activity of plasma membrane H^+-ATPase (PM H^+-ATPase) [122]. Additionally, in several studies, changes in the activity of enzymes related to metal homeostasis and nitrate metabolism were investigated in heavy metal-treated sugar beets. For instance, the activity of ferric chelate reductase (FCR) involved in iron homeostasis was decreased under short-term exposure of Pb and Cd, but prolonged exposure increased the FCR activity in sugar beet roots [123]. Recently, Haque et al. [43] reported that the reduction in FCR activity and expression of *iron-regulated transporter 1 (BvIRT1)* gene suggested a negative impact of Cd in Fe acquisition. In another study, the Pb-treated sugar beet plants exhibited altered Cu deficiency levels and increased FCR activities [114]. When sugar beet plants were exposed to the highest concentrations of heavy metals (Ni and Cd), the nitrate content and nitrate reductase (NR) activity dramatically dropped in the leaves [111].

To cope with heavy metal stress, plants have developed certain strategies involving two type of mechanisms, i.e., avoidance and tolerance [124]. The avoidance mechanisms emphasize on limiting the uptake of heavy metals (e.g., Cd) into the plant, whereas tolerance refers to storing (e.g., in vacuoles) and accumulation of heavy metals by binding it to peptides, amino acids, and proteins [125,126]. To limit uptake of heavy metals and detoxify them, plants have developed certain mechanisms, including the development of morphological structures such as thick cuticle and cell walls, mycorrhizal symbiosis, and biologically active tissues such as trichomes [127–129]. Sugar beet, like canola, is a non-mycorrhizal plant species, and therefore has a limited ability to phytostabilize heavy metals and has been suggested as a source of phytoremediation of heavy metals [130,131] despite the negative effects of heavy metals on beet growth, physiology, and metabolism. For instance, among different crop plants tested, red beets have the capacity of removing Cd from soils [130]. It has been reported that sugar beet plants have the ability to accumulate Ni, Pb, and Cd [27,132]. Papazoglou and Fernando [27] suggested that sugar beet could be a suitable crop for phytoextraction of Cd as it can accumulate Cd and produce biomass. Similarly, Yadav et al. [132] compared several crops for their capacity to accumulate heavy metals and found that sugar beets accumulated the highest amount of Cd and Pb among the studied crops. These findings clearly suggest that sugar beet could be an efficient source for phytoremediation of heavy metal-contaminated soils. Since heavy metals such as Cd and Pb have serious effects on human and animal health, sugar beets grown on heavy metal-contaminated soils must not be used for food and feed purpose, but only for industrial purposes such as bioethanol production. Due to the hazardous nature of heavy metals, heavy metal-contaminated areas are of limited use, and removal strategies of excessive heavy metals from soils are required [133]. Phytoremediation is a promising approach to dampen the toxic effects of heavy metal pollution by utilizing the artificial hyperaccumulators. Transgenic plants, which can take up the persistent heavy metals, serve as artificial hyperaccumulators. For instance, Liu et al. [131] found an important role of glutathione (synthesized by γ-glutamylcysteine synthetase-glutathione synthetase) in cellular tolerance of heavy metal stress. Overexpression of *γ-glutamylcysteine synthetase-glutathione synthetase (StGCS-GS)* gene from *Streptococcus thermophilus* in sugar beet plants showed the explicit role of *StGCS-GS* in enhancing Cd, Zn, and Cu tolerance and accumulation of these metals in shoots of transgenic sugar beets [131]. Transgenic lines also displayed resistance to different heavy metal combinations, i.e., 50 µM Cd-Zn, Cd-Cu, Zn-Cu, and Cd-Zn-Cu, and had higher levels of glutathione (GSH) and phytochelatin (PC) compared to the WT [131]. Moreover, a study by Dronnet et al. [134] concluded that the sugar beet pulp is economical and highly selective in binding of divalent metal cations such as Cd^{2+}, Cu^{2+}, Ni^{2+}, Pb^{2+} and Zn^{2+}; thus, it could be useful as a substrate to entrap heavy metals in aqueous solution. Surprisingly, it was reported that the intake of juice extracted from red beet roots protects the chickens from Cd-induced oxidative stress with enhanced immune power [135]. However, it is unfortunate that the response mechanisms of cultivated beets and wild beet to heavy metal stress is yet to be investigated in detail. Further comprehensive studies are necessary to examine the influences of heavy metal contamination on different beet cultivars, and yield and quality of bioethanol [27]. In addition, only few genes have been reported for their involvement in heavy metal response in beets. For instance, two MTP genes, *BmMTP10* and *BmMTP11* encoding metal-tolerant proteins from wild beet (*B. maritima*), were found to render tolerance to high concentrations of Mn^{2+} when expressed in yeast cells. Transcript level of *BmMTP10* gene was augmented by the presence of excessive Mn^{2+}, but *BmMTP11* transcription was not altered, suggesting that BmMTP10 and BmMTP11 proteins have non-redundant functions in Mn detoxification [41]. Thus, the study demonstrated that the BmMTP10 protein, which is localized to the Golgi apparatus, is specific to Mn^{2+} transport and decreased Mn^{2+} levels in yeast cells [42]. Ni detoxification was regulated by a couple of genes in *B. maritima* named as toxic nickel concentration (NIC), i.e., *NIC3*, *NIC6*, and *NIC8* [42]. It was speculated that all three genes are involved in tolerance to Ni toxicity. Yeast cells expressing a cDNA clone (NIC6) from *B. maritima* showed substantially high tolerance to Ni but not to the other heavy metals such as Co, Cd, and Zn [42]. Even though the excess Ni accumulation is toxic to plants, *B. maritima* plants overcome the Ni-induced toxicity by internal sequestration, but not by effluxing Ni [42]. In a very recent study, under Cd stress, sugar beet roots displayed higher levels of putative *inactive Cd/Zn-transporting ATPase (BvHMA3)* and *natural resistance-associated macrophage protein 3* (*BvNRAMP3*) gene expression, suggesting that these genes might participate in Cd uptake [43]. Interestingly, in response to Cd application, no significant changes have been observed in the expression of *phytochelatin 3 (BvPC3)* gene encoding PCs [43], which are involved in the detoxification of Cd [136].

Further studies on sugar beet are needed to investigate the physiological, cellular, and molecular alterations induced by heavy metals to help plant biologists develop breeding strategies to improve sugar beet cultivars with efficient phytoremediation ability and ability to grow in heavy metal stress-affected fields [43]. In Table 4, we summarize the heavy metal stress responses in beets.

Table 4. Heavy metal stress responses in beets.

Beet Variety	Stress Treatments	Experimental Results	Reference
B. vulgaris, red beet	0.1–100 μM trimethyllead chloride (Met_3PbCl)	Lead (Pb) damage the vacuolar membrane in red beet taproots	[112]
B. vulgaris, Monohill	10 μM and 50 μM Cd-EDTA or $CdCl_2$	As compared to control plants, Cd-treated plants showed lower shoot dry weights, photosynthetic pigments, and reduction in water content of shoots and fine roots	[114]
		The reduction in uptake of N, P, Mg, K, Mn, Cu, and Zn due to Cd stress	
B. vulgaris, Monohill	Direct Cd application (1, 5, 20, 50, 2000 μM $CdCl_2$) Indirect Cd application (5, 10, 20 μM $CdCl_2$)	Direct application of Cd on isolated leaves, protoplasts and chloroplasts inhibited CO_2 fixation, whereas indirect Cd application through the culture medium decreased the maximal quantum yield of CO_2 assimilation	[110]
B. vulgaris	0; 0.5; 5; 10 g Cd 0 1; 10; 20 g Ni Cd + Ni (0 + 0, 0.25 + 0.5, 2.5 + 5, 5 + 10)	The highest Ni concentration (20 g) is lethal to the plants	[27]
		The single application of Ni causes higher toxic effects than the combination of Ni and Cd	
B. vulgaris	10 μM $CdSO_4$	Cd stress causes growth retardation in sugar beets because of low iron levels resulting in photosynthetic inefficiency, and oxidative damage	[43]
		Sugar beet roots displayed higher levels of *BvHMA3* and *BvNRAMP3* gene expression, whereas the reduction in ferric chelate reductase (FCR) activity and expression of *iron-regulated transporter 1 (BvIRT1)* gene was observed	
B. vulgaris, Orbis	50, 100, and 300 μM $ZnSO_4$	Zn toxicity decreased macronutrient concentrations (N, K, and Mg), whereas it enhanced the P level in shoots as well as roots	[115]
		The toxic level of Zn reduced water content, leaf numbers, and root/shoot ratio along with wrinkled and chlorotic leaves	

Table 4. *Cont.*

Beet Variety	Stress Treatments	Experimental Results	Reference
B. vulgaris, Orbis	50, 100, and 300 μM $ZnSO_4$	High levels of Zn led to cell death and cessation of metabolism through decreasing aerobic respiration and damaging defense systems required for oxidative stress response	[117]
B. vulgaris, Qaweterna	0.1, 1, 10, 100 μM $CuSO_4$, or $ZnSO_4$	Cu and Zn treatments significantly reduced plant growth, shoot and root lengths, and dry weight	[116]
		At high Cu concentrations, the shoots showed turgor loss, but lower Cu concentration did not affect plant growth	
B. vulgaris, Monohill	0 to 10 μM $CdCl_2$	Sugar beet seedlings grown in nutrient solution containing high concentrations of $CdCl_2$ showed an increased leaf transpiration rate and a decreased stomatal aperture area. Thus, higher Cd concentrations affected the permeability of the leaf cuticle.	[119]
B. vulgaris, Monohill	0, 1, 5 or 20 μM Cd^{2+}	Long-term Cd exposure caused decreased sucrose uptake and diminished dry weight in taproots, but direct addition of Cd^{2+} to the medium enhanced the sucrose uptake at the tonoplast	[120]
B. vulgaris, Monohill	0, 5 or 50 μM Cd^{2+}	Increased accumulation of Cd lowered the contents of glucose, fructose, and sucrose in both shoots and roots	[121]
B. vulgaris, Monohill	Short-term application: 10 and 50 μM $CdCl_2$/Cd-EDTA, or 1 and 2 mM Pb-EDTA for 30 min and 1 h Long-term application: 10 and 50 μM $CdCl_2$, /Cd-EDTA, or $PbCl_2$, and 10; 50; 500; 1000 and 2000 μM Pb-EDTA for 7–10 days	The activity of FCR involved in iron homeostasis was decreased under short-term exposure of Pb and Cd, but a prolonged exposure increased the FCR activity in roots	[123]
B. vulgaris, hybrid NS Hy-11	10^{-4}, 10^{-2}, 1 mM $NiSO_4$, or $CdCl_2$	When sugar beet was exposed to the highest concentrations of heavy metals (Ni and Cd), the nitrate content and nitrate reductase (NR) activity dramatically dropped in the leaves	[111]
B. vulgaris, US-8916	0, 50, 100, 200 μM $CdCl_2$, $ZnCl_2$, or $CuCl_2$	Overexpression of *StGCS-GS* from *S. thermophilus* in sugar beets showed the explicit role of this gene in enhancing Cd, Zn, and Cu tolerance and accumulation of these metals in transgenic sugar beets	[131]
B. maritima	75 μM $NiCl_2$	Yeast cells expressing a cDNA clone (NIC6) from *B. maritima* showed high tolerance to Ni	[42]
		B. maritima plants overcome the Ni-induced toxicity by internal sequestration, but not by effluxing Ni	
B. maritima, TR 51196	8 mM Mn^{2+} for yeast cells 2 mM Mn^{2+} for gene expression analyses	Two *MTP* genes, *B. maritima MTP10* and *MTP11* encoding metal-tolerant proteins, were found to render tolerance to high concentrations of Mn^{2+} in yeast cells	[41]
		Transcript level of *BmMTP10* gene was augmented by the excessive Mn^{2+}, but *BmMTP11* transcription was not altered	

2.4. Ultraviolet (UV) Stress

Ultraviolet radiation (UV) causes various changes in metabolic activities of plants, imposing malfunctions and retarded overall growth. The key processes in plants affected by UV radiation include photosynthesis, biomass, respiration, transpiration, etc. UV-B (280–320 nm) radiation becomes a serious threat to the organisms because of the reduction in stratospheric ozone [137]. The stress triggers changes at molecular level by protein degradation, altering the double helical structure of DNA and antioxidant contents, etc. However, under UV stress conditions, plants adopt defensive tolerant mechanisms [138,139].

We have very limited information about the physiological and biochemical responses of beets to UV stress. Moreover, there are no reports on the molecular mechanisms and genes involved in UV stress response in beets. A report by Panagopoulos et al. [140] demonstrated that the leaves of sugar beets curled inwards and positioned towards light source with 68% growth reduction over control (ROC) under yellow light, whereas the plants were dead under a combination of yellow light and UV-B after three weeks [140]. They found that some parameters such as leaf area, fresh and dry weights, and total chlorophyll levels in sugar beet were decreased under UV radiation [140]. On the other hand, carotenoid concentrations showed different patterns upon imposition of UV radiation. For example, yellow light and a combination of white light + UV-B resulted in higher carotenoid contents, suggesting the protective role of these pigments against photo-oxidation [140]. The study also showed an increase in leaf peroxidase activity under the combination of white light and UV-B [140]. The increased peroxidase activity and ultraweak luminescence upon UV-B exposure and ascorbic acid incubated leaves represents a strong correlation in *Hibiscus* leaves [141] and sugar beet [140]. In a recent study, the most widely cultivated Iranian sugar beet variety, BR1, was used to analyze biochemical and physiological responses against different doses (3.042, 6.084, and 9.126 $kJm^{-2}d^{-1}$) of UV-B radiation [142]. The UV-B-treated sugar beet plants showed a drastic growth retardation with reduction in fresh weight, dry weight, and height. Moreover, total chlorophyll and carotenoid contents and photochemical efficiency of PSII were reduced in UV-treated plants. Interestingly, no significant raise in the proline levels was noticed. Betalain levels increased by 8%, 28%, and 34% with increased UV-B radiation of 3.042, 6.084, and 9.126 $kJm^{-2}d^{-1}$, respectively, indicating that these water-soluble pigments possess tolerant metabolic function in sugar beet varieties against UV-B radiation. Hence, it is likely that the BR1 variety is a suitable plant material for areas with UV-B irradiation [142].

Levall and Bornman [143] showed the establishment of a reproducible regeneration technique in sugar beet, wherein production of somaclonal variations was observed and UV-B-tolerant plants were selected. After additional UV-B treatment, unselected somaclones displayed significantly higher UV damage and lower carotenoid levels than the selected plants [143]. The UV irradiation exposure in in vitro conditions exhibited more tolerant callus parts than the protoplasts, paving the way for the selection of UV-tolerant sugar beet somaclones [143]. In another study, Levall and Bornman [137] showed differences between *Cercospora*-sensitive and -tolerant sugar beet plants upon the combined biotic (*Cercospora* fungal infection) and abiotic (UV radiation) stresses. The line tolerant to fungal infection was shown to be tolerant to UV-B alone and combined UV-B and biotic stresses; however, the photosynthetic yield significantly reduced in the sensitive line [137]. A report by Bornman et al. [144] showed that the UV-B radiation was not capable of penetrating organelles such as chloroplasts, resulting in intact thylakoids [144]. On the other hand, the ultrastructural image of sugar beet leaves showed prominent damages due to UV-B radiation (290–320 nm), whereas UV-C (254 nm)-treated sugar beet plants showed fewer structural changes, leading to a higher quantity of starch in chloroplasts, grana stacks fused to each other, and decreased damage to the leaf surface [145].

The results described above suggest that beet plants are adversely affected by UV stress conditions at the morpho-physiological level. However, molecular mechanisms and UV stress-responsive genes in beets are still elusive. Further studies are needed to better understand the UV stress response mechanisms at the morpho-physiological, biochemical, and molecular levels in different beet cultivars. In Table 5, we summarize the responses of cultivated beets to UV radiation.

Table 5. UV stress responses in cultivated beets.

Beet Variety	Stress Treatments	Experimental Results	Reference
B. vulgaris, inbred genotype no. 22	Yellow light (350–450 nm) Yellow light + UV-B (350–450 nm + 280–320 nm)	The leaves curled inwards and positioned towards light source with a 68% growth reduction over control under yellow light, whereas the plants were dead under the combination of yellow light and UV-B	[140]
		Yellow light and a combination of white light and UV-B led to higher carotenoid levels	
B. vulgaris, BR1	3.042, 6.084 and 9.126 $kJm^{-2}d^{-1}$ of UV-B	The UV-B-treated sugar beets showed a drastic growth retardation with reduction in fresh weight, dry weight, and height	[142]
		Total chlorophyll and carotenoid contents and photochemical efficiency of PSII were reduced, but the betalain levels were increased under UV-B	
B. vulgaris, inbred lines S (CCA 242) and T (GGO 480)	13 $kJ\ m^{-2}\ d^{-1}$ of UV-B *Cercospora beticola*	The sugar beet line tolerant to *Cercospora* fungal infection was shown to be tolerant to UV-B alone and combined UV-B and biotic stresses, but the photosynthetic yield significantly reduced in sensitive line	[137]
B. vulgaris, Primahill, derivative 9164	UV-B (290–320 nm) UV-C (254 nm)	The ultrastructural image of sugar beet leaves showed prominent damages due to UV-B (290–320 nm), whereas UV-C (254 nm)-treated plants showed fewer structural changes, leading to a higher quantity of starch in chloroplasts, grana stacks fused to each other, and decreased damage to the leaf surface	[144,145]

3. Concluding Remarks

As an economically important crop plant, cultivated beets have multifarious industrial applications ranging from food and nutrition to sugar and bioethanol production. Despite beet tolerance to different abiotic stresses [16,24], the cultivation of beets is often challenged by various adverse environmental factors [34]. These climatic abnormalities are anticipated to be more aggravated due to human industrial activities as well as global warming effects. Hence, to meet the global food security demands, developing stress-resilient plant genotypes is one of the most important topics for crop production in stress-affected fields. However, selection of the suitable beet genotypes tolerant to environmental conditions is an arduous task for plant breeders [29] as there is no clear and comprehensive understanding about the stress signaling pathways and tolerance mechanisms in different climatic regions. Even though our understanding of the heavy metal accumulation ability of beets is limited, sugar beet plants have been suggested as a candidate for phytoremediation [28,126,135]. Sugar beets grown in contaminated soils pose a serious threat to human and animal health. Therefore, use of sugar beets grown for phytoremediation must be limited to industrial purposes, such as bioethanol production. Furthermore, we have limited experimental data showing the molecular mechanisms underlying the stress response of *B. vulgaris* genotypes under extreme temperatures (cold and heat), UV radiation, high pH, and heavy metals. Although the beet cultivars show some degree of stress resistance, persistent exposure to these abiotic constraints takes a toll of their development and growth potential. On the other hand, the wild beet (*B. maritima*) displays better stress tolerance compared to the modern beet cultivars as it is rich in allelic diversity [18,34]. Most likely, the modern cultivars lost some of their stress tolerance traits during progressive domestication. While utilizing the genetic variability in wild beet and stress-tolerant beets, we can ameliorate the allelic diversity, which further eases the improvement of tolerant varieties.

Since several beet cultivars were introduced and acclimated to tropical and sub-tropical climates, it would be thus essential to establish the pan-genomic studies of beet cultivars to uncover the precise genetic modifications responsible for the ecological adaptations. Establishing the phenotypic and genotypic diversity of various beet cultivars grown in different climatic zones by utilizing the modern bioinformatic advents can enable us to generate stress-resistant crops. Consequently, further investigations are necessary to design breeding strategies under abiotic stress, and compare stress response mechanisms and signaling pathways between cultivated beets and wild beet. In Figure 1, we summarize morpho-physiological, biochemical, and molecular changes in beets under different abiotic stresses including alkaline, cold, heat, heavy metals, and UV radiation.

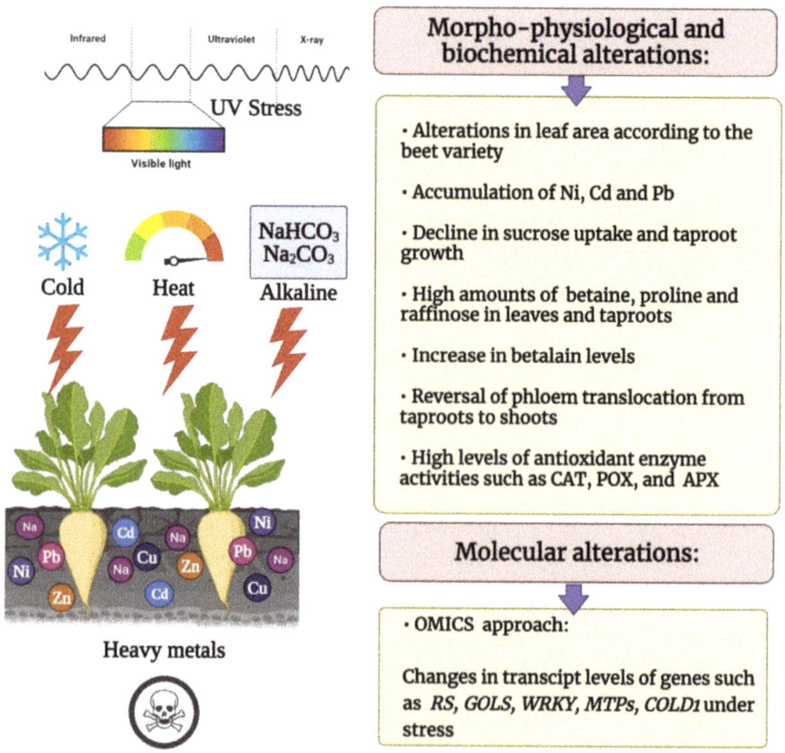

Figure 1. Schematic representation of morpho-physiological, biochemical, and molecular alterations in beets under alkaline, cold, heat, heavy metal, and UV conditions. This figure was created via BioRender.com (accessed on 12 December 2021).

Author Contributions: Conceptualization, S.Y.; writing—original draft preparation, S.Y., H.A., P.G. and M.A.; writing—review and editing, S.Y. and M.K.; preparation of Figure 1 and Tables, S.Y.; funding assistance, K.S. All authors have read and agreed to the published version of the manuscript.

Funding: This research did not receive any specific grant from funding agencies in the public, commercial, or not-for-profit sectors.

Institutional Review Board Statement: Not applicable.

Informed Consent Statement: Not applicable.

Conflicts of Interest: The authors declare no competing interests.

References

1. Henry, K. Fodder Beet. In *Root and Tuber Crops. Handbook of Plant Breeding*; Bradshaw, J.E., Ed.; Springer: New York, NY, USA, 2010; Volume 7, pp. 221–243.
2. Lange, W.; Brandenburg, W.A.; De Bock, T.S.M. Taxonomy and cultonomy of beet (*Beta vulgaris* L.). *Bot. J. Linn. Soc.* **1999**, *130*, 81–96. [CrossRef]
3. Biancardi, E.; Panella, L.; Lewellen, R. *Beta maritima: The Origin of Beets*, 1st ed.; Springer: New York, NY, USA, 2012.
4. Lee, J.H.; Son, C.W.; Kim, M.Y.; Kim, M.H.; Kim, H.R.; Kwak, E.S.; Kim, S.; Kim, M.R. Red beet (*Beta vulgaris* L.) leaf supplementation improves antioxidant status in C57BL/6J mice fed high fat high cholesterol diet. *Nutr. Res. Pract.* **2009**, *3*, 114–121. [CrossRef]
5. Zhang, Y.; Nan, J.; Yu, B. OMICS Technologies and Applications in Sugar Beet. *Front. Plant Sci.* **2016**, *7*, 900. [CrossRef] [PubMed]
6. Hussein, H.-A.A.; Mekki, B.B.; El-Sadek, M.E.A.; El Lateef, E.E. Effect of L-Ornithine application on improving drought tolerance in sugar beet plants. *Heliyon* **2019**, *5*, e02631. [CrossRef] [PubMed]
7. Magaña, C.; Núñez-Sánchez, N.; Fernández-Cabanás, V.M.; García, P.; Serrano, A.; Pérez-Marín, D.; Peman, J.M.; Alcalde, E. Direct prediction of bioethanol yield in sugar beet pulp using near infrared spectroscopy. *Bioresour. Technol.* **2011**, *102*, 9542–9549. [CrossRef] [PubMed]
8. Ahmed, S. Improving biogas production by sugar beet silage co-fermentation: An approach for on-demand biogas energy. *Environ. Sci.* **2018**. [CrossRef]
9. Kugler, F.; Stintzing, F.C.; Carle, R. Identification of Betalains from Petioles of Differently Colored Swiss Chard (*Beta vulgaris* L. ssp. cicla [L.] Alef. Cv. Bright Lights) by High-Performance Liquid Chromatography−Electrospray Ionization Mass Spectrometry. *J. Agric. Food Chem.* **2004**, *52*, 2975–2981. [CrossRef] [PubMed]
10. Cai, Y.; Sun, M.; Corke, H. Antioxidant activity of betalains from plants of the amaranthaceae. *J. Agric. Food Chem.* **2003**, *51*, 2288–2294. [CrossRef] [PubMed]
11. Lee, E.J.; An, D.; Nguyen, C.T.T.; Patil, B.S.; Kim, J.; Yoo, K.S. Betalain and Betaine Composition of Greenhouse- or Field-Produced Beetroot (*Beta vulgaris* L.) and Inhibition of HepG2 Cell Proliferation. *J. Agric. Food Chem.* **2014**, *62*, 1324–1331. [CrossRef]
12. Mzoughi, Z.; Chahdoura, H.; Chakroun, Y.; Cámara, M.; Fernández-Ruiz, V.; Morales, P.; Mosbah, H.; Flamini, G.; Snoussi, M.; Majdoub, H. Wild edible Swiss chard leaves (*Beta vulgaris* L. var. cicla): Nutritional, phytochemical composition and biological activities. *Food Res. Int.* **2019**, *119*, 612–621. [CrossRef] [PubMed]
13. Kumar, S.; Brooks, M.S.-L. Use of Red Beet (*Beta vulgaris* L.) for Antimicrobial Applications—A Critical Review. *Food Bioprocess Technol.* **2018**, *11*, 17–42. [CrossRef]
14. Chakwizira, E.; Meenken, E.D.; Maley, S.; George, M.; Hubber, R.; Morton, J.; Stafford, A. Effects of potassium, sodium and chloride fertiliser rates on fodder beet yield and quality in Canterbury. *Proc. N. Z. Grassl. Assoc.* **2013**, *75*, 261–270. [CrossRef]
15. Ribeiro, I.C.; Pinheiro, C.; Ribeiro, C.M.; Veloso, M.M.; Simoes-Costa, M.C.; Evaristo, I.; Paulo, O.S.; Ricardo, C.P. Genetic Diversity and Physiological Performance of Portuguese Wild Beet (*Beta vulgaris* spp. maritima) from Three Contrasting Habitats. *Front. Plant Sci.* **2016**, *7*, 1293. [CrossRef] [PubMed]
16. Pinheiro, C.; Ribeiro, I.C.; Reisinger, V.; Planchon, S.; Veloso, M.M.; Renaut, J.; Eichacker, L.; Ricardo, C.P. Salinity effect on germination, seedling growth and cotyledon membrane complexes of a Portuguese salt marsh wild beet ecotype. *Theor. Exp. Plant Physiol.* **2018**, *30*, 113–127. [CrossRef]
17. Pakniyat, H.; Armion, M. Sodium and proline accumulation as osmoregulators in tolerance of sugar beet genotypes to salinity. *Pak. J. Biol. Sci.* **2007**, *22*, 4081–4086. [CrossRef]
18. Skorupa, M.; Golebiewski, M.; Kurnik, K.; Niedojadlo, J.; Kesy, J.; Klamkowski, K.; Wojcik, K.; Treder, W.; Tretyn, A.; Tyburski, J. Salt stress vs. salt shock—The case of sugar beet and its halophytic ancestor. *BMC Plant Biol.* **2019**, *19*, 57. [CrossRef] [PubMed]
19. Mostafavi, K. Effect of Salt Stress on Germination and Early Seedling Growth Stage of Sugar Beet Cultivars. *Am.-Eurasian J. Sustain. Agric.* **2012**, *6*, 120–125.
20. Van Geyt, J.P.C.; Lange, W.; Oleo, M.; De Bock, T.S.M. Natural variation within the genus Beta and its possible use for breeding sugar beet: A review. *Euphytica* **1990**, *49*, 57–76. [CrossRef]
21. Choudhary, A.K.; Sultana, R.; Vales, M.I.; Saxena, K.B.; Kumar, R.R.; Ratnakumar, P. Integrated physiological and molecular approaches to improvement of abiotic stress tolerance in two pulse crops of the semi-arid tropics. *Crop J.* **2018**, *6*, 99–114. [CrossRef]
22. Rozema, J.; Cornelisse, D.; Zhang, Y.; Li, H.; Bruning, B.; Katschnig, D.; Broekman, R.; Ji, B.; van Bodegom, P. Comparing salt tolerance of beet cultivars and their halophytic ancestor: Consequences of domestication and breeding programmes. *AoB Plants* **2015**, *7*, plu083. [CrossRef] [PubMed]
23. Niazi, B.H.; Rozema, J.; Amin, R.; Salim, M.; Rashid, A. Physiological Characteristics of Fodderbeet Grown on Saline Sodic Soils of Pakistan. *Pak. J. Biol. Sci.* **1999**, *2*, 595–598. [CrossRef]
24. Wisniewska, A.; Andryka-Dudek, P.; Czerwinski, M.; Choluj, D. Fodder beet is a reservoir of drought tolerance alleles for sugar beet breeding. *Plant Physiol. Biochem.* **2019**, *145*, 120–131. [CrossRef] [PubMed]
25. Stagnari, F.; Galieni, A.; Speca, S.; Pisante, M. Water stress effects on growth, yield and quality traits of red beet. *Sci. Hortic.* **2014**, *165*, 13–22. [CrossRef]
26. Subbarao, G.V.; Wheeler, R.M.; Levine, L.H.; Stutte, G.W. Glycine betaine accumulation, ionic and water relations of red-beet at contrasting levels of sodium supply. *J. Plant Physiol.* **2001**, *158*, 767–776. [CrossRef] [PubMed]

27. Papazoglou, E.G.; Fernando, A.L. Preliminary studies on the growth, tolerance and phytoremediation ability of sugarbeet (*Beta vulgaris* L.) grown on heavy metal contaminated soil. *Ind. Crop. Prod.* **2017**, *107*, 463–471. [CrossRef]
28. Vastarelli, P.; Moschella, A.; Pacifico, D.; Mandolino, G. Water Stress in *Beta vulgaris*: Osmotic Adjustment Response and Gene Expression Analysis in ssp. vulgaris and maritima. *Am. J. Plant Sci.* **2013**, *4*, 11–16. [CrossRef]
29. Abou-Elwafa, S.F.; Amin, A.E.A.; Eujayl, I. Genetic diversity of sugar beet under heat stress and deficit irrigation. *Agron. J.* **2020**, *112*, 3579–3590. [CrossRef]
30. Zou, C.; Sang, L.; Gai, Z.; Wang, Y.; Li, C. Morphological and Physiological Responses of Sugar Beet to Alkaline Stress. *Sugar Tech* **2018**, *20*, 202–211. [CrossRef]
31. Monteiro, F.; Romeiras, M.M.; Batista, D.; Duarte, M.C. Biodiversity assessment of sugar beet species and its wild relatives: Linking ecological data with new genetic approaches. *Am. J. Plant Sci.* **2013**, *4*, 21–34. [CrossRef]
32. Chołuj, D.; Wisniewska, A.; Szafranski, K.M.; Cebula, J.; Gozdowski, D.; Podlaski, S. Assessment of the physiological responses to drought in different sugar beet genotypes in connection with their genetic distance. *J. Plant Physiol.* **2014**, *171*, 1221–1230. [CrossRef] [PubMed]
33. Shaw, B.; Thomas, T.H.; Cooke, D.T. Responses of sugar beet (*Beta vulgaris* L.) to drought and nutrient deficiency stress. *Plant Growth Regul.* **2002**, *37*, 77–83. [CrossRef]
34. Yolcu, S.; Alavilli, H.; Ganesh, P.; Panigrahy, M.; Song, K. Salt and Drought Stress Responses in Cultivated Beets (*Beta vulgaris* L.) and Wild Beet (*Beta maritima* L.). *Plants* **2021**, *10*. [CrossRef] [PubMed]
35. Wu, G.Q.; Li, Z.Q.; Cao, H.; Wang, J.L. Genome-wide identification and expression analysis of the WRKY genes in sugar beet (*Beta vulgaris* L.) under alkaline stress. *PeerJ* **2019**, *7*, e7817. [CrossRef]
36. Zou, C.; Liu, D.; Wu, P.; Wang, Y.; Gai, Z.; Liu, L.; Yang, F.; Li, C.; Guo, G. Transcriptome analysis of sugar beet (*Beta vulgaris* L.) in response to alkaline stress. *Plant Mol. Biol.* **2020**, *102*, 645–657. [CrossRef]
37. Klemens, P.A.W.; Patzke, K.; Trentmann, O.; Poschet, G.; Büttner, M.; Schulz, A.; Marten, I.; Hedrich, R.; Neuhaus, H.E. Overexpression of a proton-coupled vacuolar glucose exporter impairs freezing tolerance and seed germination. *New Phytol.* **2014**, *202*, 188–197. [CrossRef]
38. Porcel, R.; Bustamante, A.; Ros, R.; Serrano, R.; Mulet Salort, J.M. BvCOLD1: A novel aquaporin from sugar beet (*Beta vulgaris* L.) involved in boron homeostasis and abiotic stress. *Plant Cell Environ.* **2018**, *41*, 2844–2857. [CrossRef] [PubMed]
39. Kito, K.; Yamane, K.; Yamamori, T.; Matsuhira, H.; Tanaka, Y.; Takabe, T. Isolation, functional characterization and stress responses of raffinose synthase genes in sugar beet. *J. Plant Biochem. Biotechnol.* **2018**, *27*, 36–45. [CrossRef]
40. Keller, I.; Müdsam, C.; Martins Rodrigues, C.; Kischka, D.; Zierer, W.; Sonnewald, U.; Harms, K.; Czarnecki, O.; Fiedler-Wiechers, K.; Koch, W.; et al. Cold-triggered induction of ROS- and raffinose metabolism in freezing-sensitive taproot tissue of sugar beet. *Front. Plant Sci.* **2021**, *12*, 715767. [CrossRef]
41. Erbasol, I.; Bozdag, G.O.; Koc, A.; Pedas, P.; Karakaya, H.C. Characterization of two genes encoding metal tolerance proteins from *Beta vulgaris* subspecies maritima that confers manganese tolerance in yeast. *Biometals* **2013**, *26*, 795–804. [CrossRef] [PubMed]
42. Bozdag, G.O.; Kaya, A.; Koc, A.; Noll, G.A.; Prüfer, D.; Karakaya, H.C. Characterization of a cDNA from *Beta maritima* that confers nickel tolerance in yeast. *Gene* **2014**, *538*, 251–257. [CrossRef] [PubMed]
43. Haque, A.M.; Tasnim, J.; El-Shehawi, A.M.; Rahman, M.A.; Parvez, M.S.; Ahmed, M.B.; Kabir, A.H. The Cd-induced morphological and photosynthetic disruption is related to the reduced Fe status and increased oxidative injuries in sugar beet. *Plant Physiol. Biochem.* **2021**, *166*, 448–458. [CrossRef]
44. Fang, S.; Hou, X.; Liang, X. Response Mechanisms of Plants Under Saline-Alkali Stress. *Front. Plant Sci.* **2021**, *12*. [CrossRef]
45. Liu, L.; Wang, Y.; Gai, Z.; Liu, D.; Wu, P.; Wang, B.; Zou, C.; Li, C.; Yang, F. Responses of Soil Microorganisms and Enzymatic Activities to Alkaline Stress in Sugar Beet Rhizosphere. *Pol. J. Environ. Stud.* **2020**, *29*, 739–748. [CrossRef]
46. Yu, S.; Yu, L.; Hou, Y.; Zhang, Y.; Guo, W.; Xue, Y. Contrasting Effects of NaCl and NaHCO3 Stresses on Seed Germination, Seedling Growth, Photosynthesis, and Osmoregulators of the Common Bean (*Phaseolus vulgaris* L.). *Agronomy* **2019**, *9*, 409. [CrossRef]
47. Zou, C.; Wang, Y.; Wang, B.; Liu, D.; Liu, L.; Gai, Z.; Li, C. Long non-coding RNAs in the alkaline stress response in sugar beet (*Beta vulgaris* L.). *BMC Plant Biol.* **2020**, *20*, 227. [CrossRef] [PubMed]
48. Alavilli, H.; Awasthi, J.P.; Rout, G.R.; Sahoo, L.; Lee, B.-H.; Panda, S.K. Overexpression of a Barley Aquaporin Gene, HvPIP2;5 Confers Salt and Osmotic Stress Tolerance in Yeast and Plants. *Front. Plant Sci.* **2016**, *7*, 1566. [CrossRef] [PubMed]
49. Alavilli, H.; Lee, H.; Park, M.; Lee, B.-H. Antarctic Moss Multiprotein Bridging Factor 1c Overexpression in Arabidopsis Resulted in Enhanced Tolerance to Salt Stress. *Front. Plant Sci.* **2017**, *8*, 1206. [CrossRef] [PubMed]
50. Geng, G.; Wang, G.; Stevanato, P.; Lv, C.; Wang, Q.; Yu, L.; Wang, Y. Physiological and Proteomic Analysis of Different Molecular Mechanisms of Sugar Beet Response to Acidic and Alkaline pH Environment. *Front. Plant Sci.* **2021**, *12*, 682799. [CrossRef]
51. Gong, B.; Li, X.; Bloszies, S.; Wen, D.; Sun, S.; Wei, M.; Li, Y.; Yang, F.; Shi, Q.; Wang, X. Sodic alkaline stress mitigation by interaction of nitric oxide and polyamines involves antioxidants and physiological strategies in Solanum lycopersicum. *Free Radic Biol. Med.* **2014**, *71*, 36–48. [CrossRef] [PubMed]
52. Olias, R.; Eljakaoui, Z.; Li, J.; Morales, P.A.D.; Marín-Manzano, M.C.; Pardo, J.M.; Belver, A. The plasma membrane Na^+/H^+ antiporter SOS1 is essential for salt tolerance in tomato and affects the partitioning of Na+ between plant organs. *Plant Cell Environ.* **2009**, *32*, 904–916. [CrossRef]
53. Blumwald, E.; Aharon, G.S.; Apse, M.P. Sodium transport in plant cells. *Biochim. Biophys. Acta* **2000**, *1465*, 140–151. [CrossRef]

54. Gill, S.S.; Tuteja, N. Reactive oxygen species and antioxidant machinery in abiotic stress tolerance in crop plants. *Plant Physiol. Biochem.* **2010**, *48*, 909–930. [CrossRef] [PubMed]
55. Asada, K.; Takahashi, M. Production and scavenging of active oxygen in chloroplasts. In *Photoinhibition*; Kyle, D.J., Osmond, C.B., Arntzen, C.J., Eds.; Elsevier: Amsterdam, The Netherlands, 1987; pp. 227–287.
56. Mittler, R. Oxidative stress, antioxidants and stress tolerance. *Trends Plant Sci.* **2002**, *7*, 405–410. [CrossRef]
57. Wang, Y.; Stevanato, P.; Yu, L.; Zhao, H.; Sun, X.; Sun, F.; Li, J.; Geng, G. The physiological and metabolic changes in sugar beet seedlings under different levels of salt stress. *J. Plant Res.* **2017**, *130*, 1079–1093. [CrossRef] [PubMed]
58. Bor, M.; Özdemir, F.; Türkan, I. The effect of salt stress on lipid peroxidation and antioxidants in leaves of sugar beet *Beta vulgaris* L. and wild beet *Beta maritima* L. *Plant Sci.* **2003**, *164*, 77–84. [CrossRef]
59. Li, J.; Cui, J.; Dai, C.; Liu, T.; Cheng, D.; Luo, C. Whole-Transcriptome RNA Sequencing Reveals the Global Molecular Responses and CeRNA Regulatory Network of mRNAs, lncRNAs, miRNAs and circRNAs in Response to Salt Stress in Sugar Beet (*Beta vulgaris*). *Int. J. Mol. Sci.* **2021**, *22*, 289. [CrossRef] [PubMed]
60. Oster, J.D.; Shainberg, I.; Abrol, I.P. Reclamation of Salt-Affected Soils. In *Agricultural Drainage*; American Society of Agronomy, Inc. Crop Science Society of America, Inc. Soil Science Society of America, Inc.: Madison, WI, USA, 1999; Volume 38, pp. 659–691.
61. Geng, G.; Li, R.; Stevanato, P.; Lv, C.; Lu, Z.; Yu, L.; Wang, Y. Physiological and Transcriptome Analysis of Sugar Beet Reveals Different Mechanisms of Response to Neutral Salt and Alkaline Salt Stresses. *Front. Plant Sci.* **2020**, *11*, 571864. [CrossRef]
62. Zou, C.L.; Wang, Y.B.; Liu, L.; Liu, D.; Wu, P.R.; Yang, F.F.; Wang, B.; Tong, T.; Liu, X.M.; Li, C.F. Photosynthetic capacity, osmotic adjustment and antioxidant system in sugar beet (*Beta vulgaris* L.) in response to alkaline stress. *Photosynthetica* **2019**, *57*, 350–360. [CrossRef]
63. Gong, B.; Wen, D.; VandenLangenberg, K.; Wei, M.; Yang, F.; Shi, Q.; Wang, X. Comparative effects of NaCl and $NaHCO_3$ stress on photosynthetic parameters, nutrient metabolism, and the antioxidant system in tomato leaves. *Sci. Hortic.* **2013**, *157*, 1–12. [CrossRef]
64. Xu, D.; Tuyen, D.D. Genetic studies on saline and sodic tolerances in soybean. *Breed Sci* **2012**, *61*, 559–565. [CrossRef] [PubMed]
65. Wu, Z.H.; Yang, C.W.; Yang, M.Y. Photosynthesis, photosystem II efficiency, amino acid metabolism and ion distribution in rice (*Oryza sativa* L.) in response to alkaline stress. *Photosynthetica* **2014**, *52*, 157–160. [CrossRef]
66. Liu, L.; Ueda, A.; Saneoka, H. Physiological responses of white Swiss chard (*Beta vulgaris* L. subsp. cicla) to saline and alkaline stresses. *Aust. J. Crop Sci.* **2013**, *7*, 1046–1052.
67. Ghoulam, C.; Foursy, A.; Fares, K. Effects of salt stress on growth, inorganic ions and proline accumulation in relation to osmotic adjustment in five sugar beet cultivars. *Environ. Exp. Bot.* **2002**, *47*, 39–50. [CrossRef]
68. Taghizadegan, M.; Toorchi, M.; Moghadam Vahed, M.; Khayamim, S. Evaluation of sugar beet breeding populations based morpho-physiological characters under salinity stress. *Pak. J. Bot.* **2019**, *51*. [CrossRef]
69. Liu, L.; Wang, B.; Liu, D.; Zou, C.; Wu, P.; Wang, Z.; Wang, Y.; Li, C. Transcriptomic and metabolomic analyses reveal mechanisms of adaptation to salinity in which carbon and nitrogen metabolism is altered in sugar beet roots. *BMC Plant Biol.* **2020**, *20*, 138. [CrossRef] [PubMed]
70. Li, W.; Pang, S.; Lu, Z.; Jin, B. Function and Mechanism of WRKY Transcription Factors in Abiotic Stress Responses of Plants. *Plants* **2020**, *9*, 1515. [CrossRef]
71. Zou, C.; Wang, Y.; Wang, B.; Liu, D.; Liu, L.; Li, C.; Chen, F. Small RNA Sequencing in Sugar Beet Under Alkaline Stress. *Sugar Tech* **2021**, *23*, 57–64. [CrossRef]
72. Ding, Y.; Shi, Y.; Yang, S. Molecular Regulation of Plant Responses to Environmental Temperatures. *Mol. Plant* **2020**, *13*, 544–564. [CrossRef] [PubMed]
73. Hatfield, J.L.; Boote, K.J.; Kimball, B.A.; Ziska, L.H.; Izaurralde, R.C.; Ort, D.; Thomson, A.M.; Wolfe, D. Climate Impacts on Agriculture: Implications for Crop Production. *Agron. J.* **2011**, *103*, 351–370. [CrossRef]
74. Hatfield, J.L.; Prueger, J.H. Temperature extremes: Effect on plant growth and development. *Weather Clim. Extrem.* **2015**, *10*, 4–10. [CrossRef]
75. Moliterni, V.M.; Paris, R.; Onofri, C.; Orrù, L.; Cattivelli, L.; Pacifico, D.; Avanzato, C.; Ferrarini, A.; Delledonne, M.; Mandolino, G. Early transcriptional changes in *Beta vulgaris* in response to low temperature. *Planta* **2015**, *242*, 187–201. [CrossRef]
76. Hoffmann, C.; Kluge-Severin, S. Growth analysis of autumn and spring sown sugar beet. *Eur. J. Agron.* **2011**, *34*, 1–9. [CrossRef]
77. Jalilian, M.; Dehdari, M.; Fahliani, R.A.; Dehnovi, M.M. Study of cold tolerance of different sugar beet (*Beta vulgaris* L.) cultivars at seedling growth stage. *Environ. Stresses Crop Sci.* **2017**, *10*, Pe475–Pe490. [CrossRef]
78. Biancardi, E. *Genetics and Breeding of Sugar Beet*; CRC Press: Boca Raton, FL, USA, 2005; pp. 45–57. [CrossRef]
79. Stevanato, P. Resistance to abiotic stresses. In *Genetics and Breeding of Sugar Beet*; Biancardi, E., Campbell, L.G., Skaracis, G.N., De Biaggi, M., Eds.; Science Publisher Inc.: Enfield, NH, USA, 2005; pp. 116–119.
80. Rodrigues, C.M.; Müdsam, C.; Keller, I.; Zierer, W.; Czarnecki, O.; Corral, J.M.; Reinhardt, F.; Nieberl, P.; Fiedler-Wiechers, K.; Sommer, F.; et al. Vernalization Alters Sink and Source Identities and Reverses Phloem Translocation from Taproots to Shoots in Sugar Beet. *Plant Cell* **2020**, *32*, 3206–3223. [CrossRef]
81. Kirchhoff, M.; Svirshchevskaya, A.; Hoffman, C.; Schechert, A.; Jung, C.; Kopisch-Obuch, F. High degree of genetic variation of winter hardiness in a panel of *Beta vulgaris* L. *Crop Sci.* **2012**, *52*, 179–188. [CrossRef]
82. Hoffmann, C. Root Quality of Sugarbeet. *Sugar Tech* **2011**, *12*, 276–287. [CrossRef]

83. Barbier, H.; Nalin, F.; Guern, J. Freezing injury in sugar beet root cells: Sucrose leakage and modifications of tonoplast properties. *Plant Sci. Lett.* **1982**, *26*, 75–81. [CrossRef]
84. ElSayed, A.; Rafudeen, M.; Golldack, D. Physiological aspects of raffinose family oligosaccharides in plants: Protection against abiotic stress. *Plant Biol.* **2014**, *16*, 1–8. [CrossRef]
85. Paul, M.J.; Primavesi, L.F.; Jhurreea, D.; Zhang, Y. Trehalose metabolism and signaling. *Annu. Rev. Plant Biol.* **2008**, *59*, 417–441. [CrossRef] [PubMed]
86. Poschet, G.; Hannich, B.; Raab, S.; Jungkunz, I.; Klemens, P.; Krueger, S.; Wic, S.; Neuhaus, E.; Büttner, M. A Novel Arabidopsis Vacuolar Glucose Exporter Is Involved in Cellular Sugar Homeostasis and Affects the Composition of Seed Storage Compounds. *Plant Physiol.* **2011**, *157*, 1664–1676. [CrossRef]
87. Kumar, A.; Yang, F.; Goddard, L.; Schubert, S. Differing Trends in the Tropical Surface Temperatures and Precipitation over Land and Oceans. *J. Clim.* **2004**, *17*, 653–664. [CrossRef]
88. TeKrony, D.M.; Egli, D.B. Relationship of Seed Vigor to Crop Yield: A Review. *Crop Sci.* **1991**, *31*, 816–822. [CrossRef]
89. Fahad, S.; Bajwa, A.A.; Nazir, U.; Anjum, S.A.; Farooq, A.; Zohaib, A.; Sadia, S.; Nasim, W.; Adkins, S.; Saud, S.; et al. Crop Production under Drought and Heat Stress: Plant Responses and Management Options. *Front. Plant Sci.* **2017**, *8*, 1147. [CrossRef]
90. Havaux, M. Rapid photosynthetic adaptation to heat stress triggered in potato leaves by moderately elevated temperatures. *Plant Cell Environ.* **1993**, *16*, 461–467. [CrossRef]
91. Murakami, Y.; Tsuyama, M.; Kobayashi, Y.; Kodama, H.; Iba, K. Trienoic fatty acids and plant tolerance of high temperature. *Science* **2000**, *287*, 476–479. [CrossRef]
92. Malmir, M.; Mohammadian, R.; Sorooshzadeh, A.; Mokhtassi-Bidgoli, A.; Ehsanfar, S. The response of the sugar beet (*Beta vulgaris* L. ssp. vulgaris var. altissima Döll) genotypes to heat stress in initial growth stage. *Acta Agric. Slov.* **2020**, *115*, 39–52. [CrossRef]
93. Zandalinas, S.I.; Mittler, R.; Balfagon, D.; Arbona, V.; Gomez-Cadenas, A. Plant adaptations to the combination of drought and high temperatures. *Physiol. Plant* **2018**, *162*, 2–12. [CrossRef]
94. Albayrak, S.; Çamas, N. Effects of temperature and light intensity on growth of fodder beet (*Beta vulgaris* var. crassa Mansf.). *Bangladesh J. Bot.* **2007**, *36*, 1–12. [CrossRef]
95. Demmers-Derks, H.; Mitchell, R.A.C.; Mitchell, V.J.; Lawlor, D.W. Response of sugar beet (*Beta vulgaris* L.) yield and biochemical composition to elevated CO_2 and temperature at two nitrogen applications. *Plant Cell Environ.* **1998**, *21*, 829–836. [CrossRef]
96. Brown, R.H.; Byrd, G.T. Relationships between specific leaf weight and mineral concentration among genotypes. *Field Crops Res.* **1996**, *54*, 19–28. [CrossRef]
97. Hawkes, J.S. Heavy Metals. *J. Chem. Educ.* **1997**, *74*, 1374. [CrossRef]
98. Herawati, N.; Suzuki, S.; Hayashi, K.; Rivai, I.F.; Koyoma, H. Cadmium, copper and zinc levels in rice and soil of Japan, Indonesia and China by soil type. *Bull. Environ. Contam. Toxicol.* **2000**, *64*, 33–39. [CrossRef] [PubMed]
99. Khan, S.; Cao, Q.; Zheng, Y.M.; Huang, Y.Z.; Zhu, Y.G. Health risks of heavy metals in contaminated soils and food crops irrigated with wastewater in Beijing, China. *Environ. Pollut.* **2008**, *152*, 686–692. [CrossRef] [PubMed]
100. Zhang, M.K.; Liu, Z.Y.; Wang, H. Use of single extraction methods to predict bioavailability of heavy metals in polluted soils to rice. *Commun. Soil Sci. Plant Anal.* **2010**, *41*, 820–831. [CrossRef]
101. Carrillo-Chavez, A.; Salas-Megchun, E.; Levresse, G.; Munoz-Torres, C.; Perez-Arvizu, O.; Gerke, T. Geochemistry and mineralogy of mine-waste material from a "skarn-type" deposit in central Mexico: Modeling geochemical controls of metals in the surface environment. *J. Geochem. Explor.* **2014**, *144*, 28–36. [CrossRef]
102. Turgut, C.; Pepe, M.K.; Cutright, T.J. The effect of EDTA on Helianthus annuus uptake, selectivity, and translocation of heavy metals when grown in Ohio, New Mexico and Colombia soils. *Chemosphere* **2005**, *58*, 1087–1095. [CrossRef] [PubMed]
103. Dubey, R.S. Metal toxicity, oxidative stress and antioxidative defense system in plants. In *Reactive Oxygen Species and Antioxidants in Higher Plants*; Gupta, S.D., Ed.; CRC Press: Boca Raton FL, USA, 2011; pp. 177–203.
104. Gamalero, E.; Lingua, G.; Berta, G.; Glick, B.R. Beneficial role of plant growth promoting bacteria and arbuscular mycorrhizal fungi on plant responses to heavy metal stress. *Can. J. Microbiol.* **2009**, *55*, 501–514. [CrossRef] [PubMed]
105. Villiers, F.; Ducruix, C.; Hugouvieux, V. Investigating the plant response to cadmium exposure by proteomic and metabolomic approaches. *Proteomics* **2011**, *11*, 1650–1663. [CrossRef] [PubMed]
106. DalCorso, G.; Farinati, S.; Furini, A. Regulatory networks of cadmium stress in plants. *Plant Signal Behav.* **2010**, *5*, 663–667. [CrossRef] [PubMed]
107. Carrier, P.; Baryla, A.; Havaux, M. Cadmium distribution and microlocalization in oilseed rape (Brassica napus) after long-term growth on cadmium-contaminated soil. *Planta* **2003**, *216*, 939–950. [CrossRef] [PubMed]
108. Sharma, P.; Dubey, R.S. Involvement of oxidative stress and role of antioxidative defense system in growing rice seedlings exposed to toxic concentrations of aluminum. *Plant Cell Rep.* **2007**, *26*, 2027–2038. [CrossRef] [PubMed]
109. Reeves, R.D.; Baker, A.J.M. Metal accumulating plants. In *Phytoremediation of Toxic Metals: Using Plants to Clean Up the Environment*; Raskin, I., Ensley, B.D., Eds.; John Wiley and Sons: New York, NY, USA, 2000; pp. 193–229.
110. Greger, M.; Ögren, E. Direct and indirect effects of Cd^{2+} on photosynthesis in sugar beet (*Beta vulgaris*). *Physiol. Plant.* **1991**, *83*, 129–135. [CrossRef]
111. Kevrešan, S.; Petrović, N.; Popović, M.; Kandrač, J. Effect of heavy metals on nitrate and protein metabolism in sugar beet. *Biol. Plant* **1998**, *41*, 235–240. [CrossRef]

112. Trela, Z.; Burdach, Z.; Przestalski, S.; Karcz, W. Effect of trimethyllead chloride on slowly activating (SV) channels in red beet (*Beta vulgaris* L.) taproots. *Comptes Rendus Biologies* **2012**, *335*, 722–730. [CrossRef] [PubMed]
113. Sharma, P.; Dubey, R.S. Lead toxicity in plants. *Braz. J. Plant Physiol.* **2005**, *17*, 35–52. [CrossRef]
114. Larbi, A.; Morales, F.; Abadía, A.; Gogorcena, Y.; Lucena, J.; Abadía, J. Effects of Cd and Pb in sugar beet plants grown in nutrient solution: Induced Fe deficiency and growth inhibition. *Funct. Plant Biol.* **2002**, *29*, 1453–1464. [CrossRef] [PubMed]
115. Sagardoy, R.; Morales, F.; López-Millán, A.F.; Abadía, A.; Abadía, J. Effects of zinc toxicity on sugar beet (*Beta vulgaris* L.) plants grown in hydroponics. *Plant Biol.* **2009**, *11*, 339–350. [CrossRef] [PubMed]
116. Saleh, A.; El-Meleigy, S.; Ebad, F.; Helmy, M.; Jentschke, G.; Godbold, D. Base cations ameliorate Zn toxicity but not Cu toxicity in sugar beet (*Beta vulgaris*). *J. Plant Nutr. Soil Sci.* **1999**, *162*, 275–279. [CrossRef]
117. Gutierrez-Carbonell, E.; Lattanzio, G.; Sagardoy, R.; Rodríguez-Celma, J.; Ríos Ruiz, J.J.; Matros, A.; Abadía, A.; Abadía, J.; López-Millán, A.F. Changes induced by zinc toxicity in the 2-DE protein profile of sugar beet roots. *J. Proteom.* **2013**, *94*, 149–161. [CrossRef]
118. Greger, M.; Johansson, M.; Stihl, A.; Hamza, K. Foliar uptake of Cd by pea (*Pisum sativum*) and sugar beet (*Beta vulgaris*). *Physiol. Plant* **1993**, *88*, 563–570. [CrossRef]
119. Greger, M.; Johansson, M. Cadmium effects on leaf transpiration of sugar beet (*Beta vulgaris*). *Physiol. Plant* **1992**, *86*, 465–473. [CrossRef]
120. Greger, M.; Bertell, G. Effect of Ca^{2+} and Cd^{2+} on the Carbohydrate Metabolism in Sugar Beet (*Beta vulgaris*). *J. Exp. Bot.* **1992**, *43*, 167–173. [CrossRef]
121. Greger, M.; Lindberg, S. Effects of Cd^{2+} and EDTA on young sugar beets (*Beta vulgaris*). I. Cd^{2+} uptake and sugar accumulation. *Physiol. Plant* **1986**, *66*, 69–74. [CrossRef]
122. Lindberg, S.; Wingstrand, G. Mechanism of Cd^{2+} inhibition of (K^+ + Mg^{2+}) ATPase activity and K^+ ($^{86}Rb^+$) uptake in young roots of sugar beet (*Beta vulgaris*). *Physiol. Plant.* **1985**, *63*, 181–186. [CrossRef]
123. Chang, Y.-C.; Zouari, M.; Gogorcena, Y.; Lucena, J.J.; Abadía, J. Effects of cadmium and lead on ferric chelate reductase activities in sugar beet roots. *Plant Physiol. Bioch.* **2003**, *41*, 999–1005. [CrossRef]
124. Tran, T.A.; Popova, L.P. Functions and toxicity of cadmium in plants: Recent advances and future prospects. *Turk. J. Bot.* **2013**, *37*, 1–13.
125. Abbas, T.; Rizwan, M.; Ali, S.; Adrees, M.; Zia-ur-Rehman, M.; Qayyum, M.F.; Ok, Y.S.; Murtaza, G. Effect of biochar on alleviation of cadmium toxicity in wheat (Triticum aestivum L.) grown on Cd-contaminated saline soil. *Environ. Sci. Pollut. Res.* **2018**, *25*, 25668–25680. [CrossRef]
126. Pál, M.; Horváth, E.; Janda, T.; Páldi, E.; Szalai, G. Physiological changes and defense mechanisms induced by cadmium stress in maize. *J. Plant Nutr. Soil. Sci.* **2006**, *169*, 239–246. [CrossRef]
127. Hall, J.L. Cellular mechanisms for heavy metal detoxification and tolerance. *J. Exp. Bot.* **2002**, *53*, 1–11. [CrossRef]
128. Yazici, M.A.; Asif, M.; Tutus, Y.; Ortas, I.; Ozturk, L.; Lambers, H.; Cakmak, I. Reduced root mycorrhizal colonization as affected by phosphorus fertilization is responsible for high cadmium accumulation in wheat. *Plant Soil* **2021**, *468*, 19–35. [CrossRef]
129. Harada, E.; Kim, J.A.; Meyer, A.J.; Hell, R.; Clemens, S.; Choi, Y.E. Expression profiling of tobacco leaf trichomes identifies genes for biotic and abiotic stresses. *Plant Cell Physiol.* **2010**, *51*, 1627–1637. [CrossRef] [PubMed]
130. Poniedziałek, M.; Sękara, A.; Jędrszczyk, E.; Ciura, J. Phytoremediation efficiency of crop plants in removing cadmium, lead and zinc from soil. *Folia Hortic. Ann.* **2010**, *22*, 25–31. [CrossRef]
131. Liu, D.; An, Z.; Mao, Z.; Ma, L.; Lu, Z. Enhanced Heavy Metal Tolerance and Accumulation by Transgenic Sugar Beets Expressing Streptococcus thermophilus StGCS-GS in the Presence of Cd, Zn and Cu Alone or in Combination. *PLoS ONE* **2015**, *10*, e0128824. [CrossRef] [PubMed]
132. Yadav, R.K.; Minhas, P.S.; Lal, K.; Chaturvedi, R.K.; Yadav, G.; Verma, T.P. Accumulation of metals in soils, groundwater and edible parts of crops grown under long-term irrigation with sewage mixed industrial effluents. *Bull. Environ. Contam Toxicol* **2015**, *95*, 200–206. [CrossRef] [PubMed]
133. Farrell, M.; Jones, D.L. Use of composts in the remediation of heavy metal contaminated soil. *J. Hazard. Mater.* **2010**, *175*, 575–582. [CrossRef] [PubMed]
134. Dronnet, V.M.; Renard, C.M.G.C.; Axelos, M.A.V.; Thibault, J.-F. Binding of divalent metal cations by sugar-beet pulp. *Carbohydr. Polym.* **1997**, *34*, 13–82. [CrossRef]
135. Vasiljeva, S.; Smirnova, G.; Basova, N.; Babarykin, D. Cadmium-induced oxidative damage and protective action of fractioned red beet (*Beta vulgaris*) root juice in chickens. *Agron. Res.* **2018**, *16*, 1517–1526. [CrossRef]
136. Colovic, M.B.; Vasic, V.M.; Djuric, D.M.; Krstic, D.Z. Sulphur-containing amino acids: Protective role against free radicals and heavy metals. *Curr. Med. Chem.* **2018**, *25*, 324–335. [CrossRef] [PubMed]
137. Levall, M.W.; Bornman, J.F. Differential response of a sensitive and tolerant sugarbeet line to Cercospora beticola infection and UV-B radiation. *Physiol. Plant.* **2001**, *109*, 21–27. [CrossRef]
138. Young, A.R.; Claveau, J.; Rossi, A.B. Ultraviolet radiation and the skin: Photobiology and sunscreen photoprotection. *J. Am. Acad. Derm.* **2017**, *76*, S100–S109. [CrossRef]
139. Rahimzadeh, P.; Hosseini Sarghein, S.; Dilmaghani, K. Effects of UV-A and UV-C radiation on some morphological and physiological parameters in savory (*Satureja hortensis* L.). *Ann. Biol. Res.* **2011**, *2*, 164–171.

140. Panagopoulos, I.; Bornman, J.F.; Bjorn, L.O. Effects of ultraviolet radiation and visible light on growth, fluorescence induction, ultra-weak luminescence and peroxidase activity in sugar beet plants. *J. Photochem. Photobiol. B Biol.* **1990**, *8*, 73–87. [CrossRef]
141. Panagopoulos, I.; Bornman, J.F.; Björn, L. The effect of UV-B and UV-C radiation on Hibiscus leaves determined by ultraweak luminescence and fluorescence induction. *Physiol. Plant.* **1989**, *76*, 461–465. [CrossRef]
142. Rahimzadeh Karvansara, P.; Razavi, S.M. Physiological and biochemical responses of sugar beet (*Beta vulgaris* L) to ultraviolet-B radiation. *PeerJ* **2019**, *7*, e6790. [CrossRef] [PubMed]
143. Levall, M.W.; Bornman, J.F. Selection in vitro for UV-tolerant sugar beet (*Beta vulgaris*) somaclones. *Physiol. Plant.* **1993**, *88*, 37–43. [CrossRef]
144. Bornman, J.F.; Bornman, C.H.; Björn, L.O. Effects of Ultraviolet Radiation on Viability of Isolated *Beta vulgaris* and Hordeum vulgare Protoplasts. *Z. Für Pflanzenphysiol.* **1982**, *105*, 297–306. [CrossRef]
145. Bornman, J.F.; Evert, R.F.; Mierzwa, R.J. The effect of UV-B and UV-C radiation on sugar beet leaves. *Protoplasma* **1983**, *117*, 7–16. [CrossRef]

MDPI
St. Alban-Anlage 66
4052 Basel
Switzerland
Tel. +41 61 683 77 34
Fax +41 61 302 89 18
www.mdpi.com

Plants Editorial Office
E-mail: plants@mdpi.com
www.mdpi.com/journal/plants

www.ingramcontent.com/pod-product-compliance
Lightning Source LLC
Chambersburg PA
CBHW061759210926
51850CB00006B/594